Integrating Ecology and Poverty Reduction

Jane Carter Ingram · Fabrice DeClerck
Cristina Rumbaitis del Rio
Editors

Integrating Ecology and Poverty Reduction

Ecological Dimensions

Springer

Editors
Jane Carter Ingram
Wildlife Conservation Society
Bronx, New York, NY, USA
cingram@wcs.org

Cristina Rumbaitis del Rio
The Rockefeller Foundation
New York, NY, USA
crumbaitisdelrio@rockfound.org

Fabrice DeClerck
Centro Agronómico Tropical de
Investigación y Enseñanza (CATIE)
Turrialba, Costa Rica
fdeclerck@catie.ac.cr

ISBN 978-1-4419-0632-8 e-ISBN 978-1-4419-0633-5
DOI 10.1007/978-1-4419-0633-5
Springer New York Dordrecht Heidelberg London

Library of Congress Control Number: 2011943619

Printed on acid-free paper

Springer is part of Springer Science+Business Media (www.springer.com)

Foreword

Humanity has entered the Anthropocene. If ever there was a time when we could take nature's beneficence for granted, it has passed. With seven billion people on the planet, and the eight-billionth arrival expected by 2025, human pressures on every ecosystem have multiplied, in some cases to the breaking point. The famine in the Horn of Africa reminds us that productive and resilient ecosystems are important not only for human well-being but also for human survival, especially in the dire circumstances of impoverished populations.

The urgent need to sustain ecosystems in the face of climate change, growing human populations, and rising demands for a multitude of primary commodities and agricultural outputs is giving rise to a burgeoning new discipline of sustainable development. More than ever, we need to understand how society depends on a range of complex and subtle ecosystem functions, and conversely, how ecosystem functions are impacted by human activities. The intellectual challenge is enormous. Both ecosystems and human systems are immensely complex. Their interactions add further dimensions of complexity. And understanding natural and human systems requires a range of analytical tools that surpass traditional academics' disciplinary boundaries.

The present volumes, Integrating Ecology and Poverty Reduction, are a powerful and innovative addition to this vital field of research. These volumes are also a personal thrill for me, since their genesis is the multidisciplinary setting of the Earth Institute at Columbia University. I am most grateful to our former Earth Institute postdocs who conceived and carried out these studies. They and the contributors to these volumes have earned our admiration and gratitude.

Every chapter in these volumes shows that the emerging scientific discipline of sustainable development is both vital and difficult. This is especially the case when it is viewed as an applied science that aims to find practical solutions in specific human-ecological contexts. It is one thing to recognize that ecosystem functions are vital to a society's health and economic productivity (as explored in the first volume, *Integrating Ecology and Poverty Reduction: Ecological Dimensions*), and quite another to devise institutions and policies that protect ecosystems in the face of climate change, growing populations, and rising economic pressures (as explored in the second volume, *Integrating Ecology and Poverty Reduction: The Application of Ecology in Development Solutions*). The case studies in these volumes describe as many failures as successes in the policy sphere and illuminate the subtle and multidimensional approaches to both science and policy that are necessary for success in managing complex and interacting systems.

Despite the range of geographies, ecologies, and development challenges covered in these volumes, there is a unified and highly successful intellectual approach. This is development seen through the ecologist's eyes and with the ecologist's tools. The overriding theme is how the science of ecology – with its focus on complex systems, interacting components and networks, threshold effects, and strong nonlinearities – can and should inform development thinking and design.

As one would expect, the detailed ecological context of development looms large. The details of ecological stress, resource ownership, community organization, gender relations, migration patterns, biodiversity, land use patterns, transport conditions, and vulnerability to environmental hazards and climate change, all condition the interactions of society and ecosystems, and all shape the ways to find sustainable approaches to economic development. It is a vast challenge to understand these complex relations. It is an even greater challenge to ensure that the impacted communities themselves can appreciate the ecological and social context in which they operate, so that they can devise effective means to solve pressing problems.

The chapters put a great deal of emphasis on how ecological knowledge is shared and diffused within a community. There is need for formal training and scientific knowledge, of species, climate, and ecological changes. There is need for a deep understanding of the key actors in the communities. There is an especially vital need for gender awareness and women's empowerment. Women are often disempowered in local communities, and yet play the vital role in managing croplands, water resources, fuelwood, and other ecosystem services. Without women's empowerment, sustainable solutions are impossible to identify, much less to achieve.

Population dynamics, including the challenges of the demographic transition to low fertility rates and the management of migration, loom large in the challenges. Both the issues of natural population increase caused by continued high fertility rates in low-income settings and the challenges of massive migration, from rural to urban areas and across national boundaries, are among the most vexing problems of sustainable development. Population growth is highest in the poorest and most fragile ecosystems, such as the drylands of the Horn of Africa. Migration from such regions can also trigger social conflicts and violence. Migration is leading to a dramatic surge of urbanization, beyond the planning and management capacity of many sprawling urban areas. The second volume has excellent discussions of these dimensions of demographic-ecological interactions.

Many of the chapters in the second half of the second volume deal with various strategies for monetizing the social value of ecosystem services. The basic idea is straightforward: since ecosystem services provide great value to society, there ought to be a way to create economic incentives to sustain those services, and more generally to benefit poor communities that manage the services. Yet the wonderful case studies and analyses make clear that this strategy is much easier said than done. There is no off-the-shelf strategy for creating appropriate incentives. Each situation, type of ecosystem service, and pattern of local culture and politics calls for a tailored design.

The cases are fascinating. We gain insight into community-based management of forests, fisheries, non-forest products, biodiversity conservation, ecotourism, and much more. We learn about a fascinating project to "pay for ecosystem services" (PES) in a wildlife reserve in Tanzania. Even though the community receives very modest compensation for its conservation activities, and for forgoing other economic activities around the site, the project has proved to be very popular with the community and has successfully combined conservation with development initiatives; in short, PES proved to be "a highly cost-effective model for community-based conservation" (Chap. 12, Vol.2). In other cases, however, with different ecological and social dynamics, PES proved to be less robust and less effective.

What is most exciting about these volumes is the consistently high quality of ecological analysis combined with an equally high quality of keen social observation. This collection of chapters is, in short, sustainable development analysis at its best, drawing strength by acknowledging the complexity of biological and social systems, avoiding oversimplification, and always giving due attention to the interactions of nature, culture, and economy. Readers will savor these chapters as bold and cutting-edge approaches to a budding scientific discipline of enormous practical importance. The field of sustainable development is enormously enriched by this pioneering effort.

Professor Jeffrey D. Sachs, Director of the Earth Institute

Preface

The two volumes comprising the series *Integrating Ecology and Poverty Reduction* address the ecological dimensions of some of the major challenges in reducing poverty in developing countries (Vol. 1) and potential solutions and opportunities for more effectively leveraging ecological science and tools to address some of those challenges (Vol. 2). Collectively, we hope these volumes serve to foster a deeper, more nuanced understanding of the ecological dimensions of various aspects of poverty, particularly in rural areas of developing countries where some of the world's poorest people live, and a heightened appreciation for the role that ecological science and tools can play in poverty reduction efforts. We acknowledge that no development challenge is uniquely ecological in its provenance or its resolution, but posit that ecological science and tools are critical components of effective solutions to some of the world's most vexing international problems.

This first volume, *Integrating Ecology and Poverty Reduction: Ecological Dimensions*, identifies some of the most pressing challenges related to rural poverty in developing countries and critically reviews the ecological dimensions of those problems. Specifically, we address key ecological processes and principles associated with food security, water provisioning, human health, energy security, climate change, and disaster reduction. These topics are by no means an exhaustive list of the many challenges facing poor, rural communities nor are the ecological factors discussed the only ones to consider when evaluating a problem. Rather, these chapters address some of the major obstacles to human well-being and economic development and demonstrate how ecological thinking can be applied to understand and address these problems. The second volume, *Integrating Ecology and Poverty Reduction: The Application of Ecology in Development Solutions,* addresses ways in which ecology can be applied in the context of education, gender relations, demography, markets and governance to address many of the challenges addressed in the first volume.

Bronx, New York, NY	Jane Carter Ingram
Turrialba, Costa Rica	Fabrice DeClerck
New York, NY	Cristina Rumbaitis del Rio

Acknowledgments

The editors would like to thank CATIE, the Rockefeller Foundation and the Wildlife Conservation Society and our colleagues at these institutions for supporting us during the preparation of these volumes. We would also like to thank the Earth Institute for providing us with the inspiration, community, and resources that catalyzed this project. We are also grateful to the Ecological Society of America and its members for supporting discussion groups and symposia we have held on these topics, which greatly helped develop the ideas and concepts presented in these volumes. Finally, we would like to express tremendous gratitude to the many authors who contributed their time, resources, experiences and ideas to this effort.

Contents

Contributors

Colin Apse The Nature Conservancy, Brunswick, Maine, USA

Rob Bailis Yale School of Forestry and Environmental Studies, New Haven, Connecticut, USA

Jill Blockhus The Nature Conservancy, Arlington, Virginia, USA

Casey Brown Department of Civil and Environmental Engineering, University of Massachusetts at Amherst, Amherst, Massachusetts, USA

Louise E. Buck Department of Natural Resources, Cornell University, Ithaca, New York, USA

Jeff L. Chatellier Yale School of Forestry and Environmental Studies, Yale University, New Haven, Connecticut, USA

Gretchen Daily Center for Conservation Biology, Department of Biology, Stanford University, Stanford, California, USA

Fabrice DeClerck Centro Agronómico Tropical de Investigación y Enseñanza (CATIE), Turrialba, Cartago, Costa Rica

Susan C. Doll Department of Technology and Environmental Design, Appalachian State University, Boone, North Carolina, USA

Jessica Fanzo UN-Reach, World Food Programme, Rome, Italy

Thomas FitzHugh FitzHugh Consulting, Sacramento, California, USA

David J. Ganz Lowering Emissions in Asia's Forests (LEAF), Winrock International, Bangkok, Thailand

Adrian Ghilardi Center for Ecosystems Research at Universidad Nacional Autónoma de México, Morelia, Michoacán, Mexico

Ashley G. Hawes Department of Natural Resources Conservation, University of Massachusetts Amherst, Amherst, Massachusetts, USA

Jeremy Hess Department of Emergency Medicine, Department of Environmental Health, Rollins School of Public Health, Emory University, Atlanta, Georgia, USA

Jane Carter Ingram Wildlife Conservation Society, Bronx, New York, USA

Louise E. Jackson Department of Land, Air, and Water Resources, University of California, Davis, California, USA

Felicia Keesing Department of Biology, Bard College, Annandale, New York, USA

Bijan Khazai Karlsruhe Institute of Technology, Center for Disaster Management and Risk Reduction, Karlsruhe, Germany

Alexandra-Maria Klein Department of Environmental Sciences, Policy, and Management, University of California, Berkeley, California, USA

Craig Leisher The Nature Conservancy, Arlington, Virginia, USA

Karen Levy Department of Environmental Health, Rollins School of Public Health, Emory University, Atlanta, Georgia, USA

Julie A. March Office of U.S. Foreign Disaster Assistance, United States Agency for International Development, Washington, District of Columbia, USA

Jeffrey C. Milder Department of Natural Resources, Cornell University, Ithaca, New York, USA

Samuel S. Myers Department of Environmental Health, Harvard School of Public Health, Harvard Medical School, Boston, Massachusetts, USA

Richard S. Ostfeld Cary Institute of Ecosystem Studies, Millbrook, New York, USA

Cheryl A. Palm Tropical Agriculture Program of the Earth Institute, Columbia University, New York, USA

Stacy M. Philpott Department of Environmental Sciences, University of Toledo, Toledo, Ohio, USA

Timothy O. Randhir Department of Natural Resources Conservation, University of Massachusetts Amherst, Amherst, Massachusetts, USA

Roseline Remans Tropical Agriculture and Rural Environment Program, The Earth Institute, Columbia University, New York, USA

Cristina Rumbaitis del Rio The Rockefeller Foundation, New York, NY USA

David S. Saah Spatial Informatics Group, LLC and University of San Francisco, San Francisco, California, USA

John Sanderson The Nature Conservancy, Fort Collins, Colorado, USA

Sara J. Scherr Ecoagriculture Partners, Washington, District of Columbia, USA

Ridge Schuyler The Nature Conservancy, Charlottesville, Virginia, USA

Anton Seimon Wildlife Conservation Society, Bronx, New York, USA

Sean M. Smukler Tropical Agriculture Program, The Earth Institute, Columbia University, New York, USA

Leigh Winowiecki International Center for Tropical Agriculture, Tropical Soil Biology and Fertility (TSBF) Research Area, Nairobi, Kenya

Chapter 1
Introduction to Integrating Ecology and Poverty Reduction

Fabrice DeClerck, Jane Carter Ingram, and Cristina Rumbaitis del Rio

Background

At the writing of this book, the world is at a critical crossroads. The year 2010 was the United Nations (U.N.) year of biodiversity and the year when the targets of the Convention of Biological Diversity (CBD), which was signed in 2002, were supposed to have been met. The CBD aimed to achieve by 2010 a "significant reduction of the current rate of biodiversity loss at the global, regional and national levels as a contribution to poverty reduction and to the benefit of all life on Earth." However, progress remains elusive – species extinction rates continue to be 1,000 times greater than background rates in the geological record (Secretariat of the CBD 2006; Walpole et al. 2009, 2010; Butchart et al. 2010).

We are also at a critical stock-taking point on progress towards meeting the Millennium Development Goals (MDGs), a set of time-bound goals for achieving measurable improvements in the lives of the world's poorest people by the year 2015 (www.un.org/millenniumgoals/). The MDGs were agreed upon by every member nation of the United Nations in 2000 as a global commitment to reducing extreme poverty. Progress towards the goals was recently reviewed in an MDG summit convened during the 2010 annual United Nations General Assembly meeting. The eight goals can be summarized as follows: (1) eradicate extreme economic poverty and hunger; (2) achieve universal primary education; (3) promote gender equality and empower women; (4) reduce child mortality; (5) improve maternal health; (6) combat HIV/AIDS, malaria, and other diseases; (7) ensure environmental sustainability; and (8) develop a global partnership for development.

F. DeClerck (✉)
Centro Agronómico Tropical de Investigación y Enseñanza (CATIE),
Turrialba, Cartago, Costa Rica
e-mail: fdeclerck@catie.ac.cr

J.C. Ingram et al. (eds.), *Integrating Ecology and Poverty Reduction: Ecological Dimensions*, DOI 10.1007/978-1-4419-0633-5_1,

Despite the historical separation between biodiversity conservation and poverty reduction efforts (Adams et al. 2004; Sanderson and Redford 2003, 2004; Redford et al. 2008), there is increasing consensus that the maintenance of biodiversity is an integral part of reducing extreme poverty reduction. Biodiversity conservation is a core focus of the MDGs, in particular, MDG 7 that focuses on environmental sustainability and includes the CBD goal of achieving a significant reduction in the rate of biodiversity loss. Progress towards MDG 7 is measured in terms of the proportion of land area covered by forest, a reduction of carbon dioxide (CO2) emissions and of ozone-depleting substances, the proportion of fish stocks within safe biological limits, a reduction in the proportion of the total water resources used, an increase in the proportion of terrestrial and marine areas protected, and a reduction in the proportion of species threatened with extinction (Secretariat of the CBD 2006).

Despite widespread international commitment to all of these goals, including MDG 7, integrating environmental sustainability, and biodiversity conservation specifically, into development projects and national development strategies remains a challenge. In 2004, Adams et al. wrote that biodiversity conservation scientists face a dilemma as a result of the increasing global concern that international conservation efforts are in conflict with efforts to reduce poverty and that lasting positive outcomes of conservation-with-development projects are elusive. Indeed, many perceive biodiversity conservation and poverty reduction to be two completely disparate goals. Adams et al. (2004) addressed these perceived conflicts and proposed a typology for clarifying the different relationships between conservation and poverty reduction: (1) poverty and conservation are separate policy realms, (2) poverty is a critical constraint on conservation, (3) conservation should not compromise poverty reduction, and (4) poverty reduction depends on living resource conservation. Much of this discussion, however, has focused on the impact that protected areas and reserves have on poverty reduction – which in many cases will be minimal. For example, Redford et al. (2008) demonstrate that only about 0.25% of the world's poorest people are found in areas that are somewhat or extremely wild.

These two volumes focus predominately on the fourth typology proposed by Adams et al. (2004), that poverty reduction depends on living resource conservation. However, there are several important clarifications to be made. First, the chapters included in these volumes push beyond the notion that poverty reduction is disproportionately dependent on living species simply for production services obtained from nature, but that integrating ecological concepts into development strategies can be a useful approach for achieving multiple MDGs and improving livelihoods (Rumbaitis del Rio et al. 2005; DeClerck et al. 2006). Second, we distinguish between integrating ecological tools into development practice, and the conservation of critically endangered biodiversity. That is, many of the interventions and tools highlighted in these volumes address conserving ecological integrity in human-dominated landscapes with the specific aim of sustaining and restoring ecosystem services that contribute to human well-being. Multiple studies have demonstrated that practices that target biodiversity conservation in human-dominated landscapes can make significant contributions to biodiversity conservation (see Gardner et al. 2010), and to ecosystem services (Naeem et al. 2009), but that these interventions often fail to protect sensitive species (Milder et al. 2010). Thus, ecological science will continue to be important for informing conservation planning aimed at protecting threatened biodiversity, but will also be critical for successfully achieving the MDGs in human-dominated landscapes that may not be high priorities for biodiversity conservation, but where poverty is high and persistent (Kareiva and Mavier 2007). Finally, we acknowledge that poverty is a multidimensional condition resulting from a lack of access to material and non-material needs (Anand and Sen 2000; Alkire and Santos 2010). However, in these volumes, we have focused and expanded upon the multiple components of poverty represented by the MDG framework (Sachs 2005; UN 2010), while recognizing that some aspects of poverty may not be addressed explicitly in these volumes. Rather, these volumes can be viewed as a starting point for illustrating how ecology underpins certain components of poverty (Volume 1) and considering how several types of mediating social forces can be leveraged to increase the benefits that ecosystems provide to the poor (Volume 2).

Certainly, conservation and poverty reduction "win-win" situations are by no means commonplace nor easy to achieve, as they may require compromise with respect to one or both goals. For example, in a global meta-analysis using 11 case studies from Latin America, Africa, and Asia, Tekelenburg et al. (2009) investigated how biodiversity and poverty are related to each other by exploring the ways in which indicators of conservation and development changed over a 10-year period. In all but one example, gains in biodiversity were uncorrelated with poverty reduction. The single example of gains in both was found within the Chorotega Biological Corridor in the Guanacaste peninsula of Costa Rica. The Chorotega Biological Corridor is part of the greater Mesoamerican Biological Corridor (MBC), which aims to facilitate the movement of biodiversity from southern Mexico to northern Colombia. Although at its conception, the MBC consisted entirely of conservation goals (biological connectivity), recent analysis of the most functional corridors indicate that these goals have been supplemented with more development-focused goals such as ensuring water quantity and quality (Estrada and DeClerck 2010; see also Chap. 14 in Vol. 2). Although many factors have led to positive results for conservation and livelihoods in the Chorotega Biological Corridor, part of the success can be attributed to the integration of local needs (water) with conservation goals.

Achieving conservation and poverty reduction goals, as exhibited by the Chorotega example, will require cross-disciplinary approaches, which have been growing (NAS 2005; Ostrom et al. 2007; Ostrom 2009). Thus, it is now timely to ask what is and should be the role of ecology in efforts to alleviate poverty? Why should ecological understanding of the way in which biological communities work be relevant to solving complex development problems? How can ecological knowledge be integrated into cross-disciplinary approaches to support development planning? These questions are the central starting points for these volumes. While the importance of ecosystem services for human well-being is now widely accepted, the challenge remains as to how we can practically maintain biodiversity and ecosystem function alongside poverty reduction initiatives? This is the key challenge these volumes seek to explore across a range of development goals and through the lens of several potential solutions that may provide a way to achieve both conservation and poverty reduction. Specifically, these volumes explore what the role of ecologists and the science of ecology is in addressing these challenges and contributing to potential solutions.

The Science of Ecology

Ecology is the science of studying the interactions of organisms and their environment. During the relatively short history of ecology as a field of study, this has focused on understanding how populations of species are shaped and influenced by the environment (e.g., temperature, humidity, latitude, and elevation) and by interactions with other species (e.g., predation, competition for resources, and cooperation). Much of the early work of ecologists has specifically and intentionally focused on areas characterized by low human impact – relatively intact wilderness or protected areas, or laboratory microcosms – with the explicit goal of understanding how ecological communities are formed and operate in the absence of human influences. Much of this early ecological research documented the effects of human perturbations on ecosystems as an external forcing, but has not looked at humans as an important component in the system. In large part, traditional ecology has sought to minimize human influence and even to exclude the human footprint in our understanding of how the biosphere works, rather than disentangling the complex relationships between humans, other species, and the physical environment. For example, Real and Brown's (1991) edited volume "Foundations of Ecology" includes 40 classic ecological papers that form the theoretical foundation for most students of ecology. However, not a single one of these papers includes humans as a critical ecological player. In fact, much of the research about the interactions between humans and the ecosystems in which they live, also referred to as social-ecological systems, has occurred within disciplines such as geography and the burgeoning field of sustainability science (Kauffman 2009) and has been promoted within programs such as the International Human Dimensions Program (IHDP, www.ihdp.org). Only recently have ecologists shifted their focus to consider not only how humans impact the environment, but also how functional ecosystems contribute to human well-being (Daily 1997; Rumbaitis et al. 2005; DeClerck et al. 2006; Kareiva and Marvier 2007; Naeem et al. 2009).

An important first question is what are the contributions of ecology and its subdisciplines, beyond conservation implications? As previously stated, ecology is the science of studying organisms in their environment and of understanding the relationships between communities of organisms. This includes a multitude of branches such as population ecology that specializes in how organisms of the *same* species interact with one another to acquire resources and reproduce. In contrast, community ecology studies the interactions *among* species, which includes multiple classes of interactions such as predation and competition, but also facilitation and cooperation. Landscape ecology, one of the youngest branches of ecology, considers how spatial context or position in a landscape affects ecological interactions.

Early ecologists focused primarily on the impacts of the environment on the distribution of organisms and ecological communities through observations of how these communities changed from the poles to the tropics, or at smaller scales, from valley bottoms to mountaintops, rather than how organisms and communities influence the functioning of the environment in which they exist. Today, ecologists increasingly recognize that species are not just passive recipients of the environment, but that they play a very active role in shaping and driving ecological processes (Naeem 2002). This functional view of biodiversity is a significant paradigm shift that pushes the work of ecologists into the cross-disciplinary realm where biodiversity and ecosystems are understood to be essential contributors to human well-being through the provisioning of essential goods and services (Naeem et al. 2009). The chapters comprising these volumes reflect on that role with a particular focus on how ecological knowledge, tools, and understanding can contribute to improving the living conditions of the world's poorest people.

A Functional Role for Ecology in Poverty Reduction

The important distinction between the MDGs and other development initiatives is the renewed focus on cross-disciplinary (Fig. 1.1; Eigenbrode et al. 2007), and multi-scalar approaches. Past development interventions have been criticized for

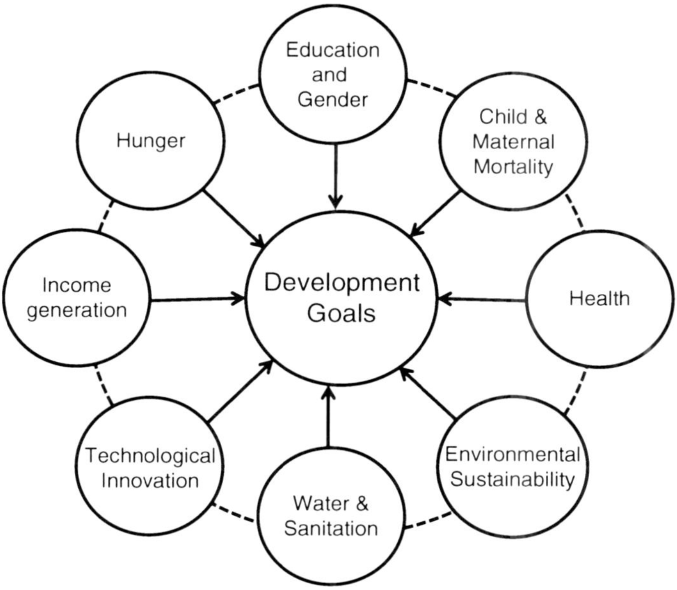

Fig. 1.1 The Millennium Development Goals were agreed upon by all member countries of the UN in 2000 and aim to significantly reduce poverty by 2015. This schematic illustrates development challenges the Millennium Development Goals aim to address

their shotgun approach. In many cases, there has been little to no interaction among different disciplines, or when there was, there have been negative impacts, where the advances made by one discipline negated the efforts made by another. Multiple development projects have resulted in unintended consequences, where well-meaning interventions have not considered the indirect, systemic effects of actions (Ranganathan et al. 2008). One example is agriculture's Green Revolution, which undoubtedly saved millions of lives by increasing the agricultural productivity of the world's most important grain crops, but with a large associated environmental cost (Tilman 1998). The question remains, whether the negative environmental impacts of the Green Revolution might have been reduced had ecosystem science been more developed as a discipline at the time, and had there been greater dialogue between ecologists and agronomists on the imperative to sustainably meet global food production needs without compromising the ecosystem services important for meeting other basic needs?

This is much more difficult than it might appear and although development goals, including the MDGs, may be multidisciplinary and combine several usually separate branches of learning; they are far from being truly interdisciplinary by fostering increased interaction and integration of contributing disciplines. The primary difference between the two, according to Eigenbrode et al. (2005), is that multidisciplinary research is conducted by scientists from different disciplines, but is designed to address a question pertaining to a single system. In contrast, interdisciplinary research requires a greater degree of coordination among disciplines from the start with research questions that often span several temporal and spatial scales and fields of study. When considering the MDGs as presented in Fig. 1.1, it is easy for a professional in a specific discipline to focus on the goal most relevant to his or her work. This approach, however, limits the opportunity for finding novel solutions and avoiding conflicts (NAS 2005).

We propose, however, that rather than identifying with individual goals, professionals consider each goal through the lens of their respective discipline (Fig. 1.2). For example, ecologists could consider their role not only in ensuring environmental sustainability, but also in reducing hunger, improving maternal health, or achieving universal primary education. Certainly, ecological expertise, knowledge, and methods, which we term the "ecological toolbox" (Rumbaitis et al. 2005), will have limited application in achieving some development goals, and greater application in others, but we may be surprised by the solutions that arise simply by looking at a problem in a new light. Such an exercise serves not only to identify how ecologists can contribute to areas outside of their typical remit, and to highlight the interaction between the fields, but also serves to highlight areas of potential conflict between fields where cross-disciplinary discussion and considerable negotiations will be needed to identify tradeoffs and/or negative impacts before they occur.

Of course, we do not suggest that ecology or any single approach is a panacea capable of solving all of the world's most pressing problems, or even a single problem alone (Ostrom et al. 2007; Ostrom 2009). However, we do strongly believe that ecology can make significant contributions to most of the MDGs, and

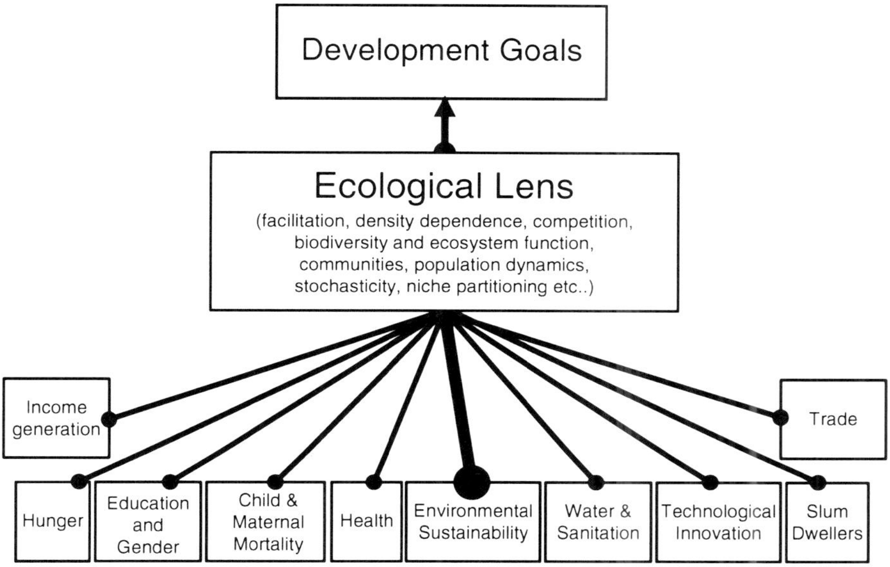

Fig. 1.2 The role of ecology in achieving poverty reduction should not be restricted to development goals that are explicitly environmental. Rather, ecology offers useful concepts and tools for achieving progress towards other development goals, as discussed throughout these volumes and illustrated in this figure. For some development goals, the role of ecology will be more direct and significant than for others. Nevertheless, considering a problem through the lens of multiple disciplines, as encouraged throughout these volumes and as demonstrated herein with the field of ecology, may lead to new, innovative solutions for addressing poverty

that the integration of the ecological perspective with that of other disciplines will present solutions that are novel, sustainable, and may result in fewer trade-offs in the long-term than quick-fix solutions that deliver immediate returns on a single development goal. Examples of this integrative thinking are becoming more popular. For instance, the increasing collaboration of ecologists with agronomists in the field of agroecology focuses largely on how ecological interactions can be used to reduce the need for agrochemicals while maintaining competitive yields (Smukler et al. Chap. 3, Vol. 1). Many ecologists also work directly alongside engineers and farmers to design riparian (riverside) forests whose functional role is to improve water quality before it enters rivers and streams, reducing the cost of water treatment for downstream communities. Interaction of ecologists with nutritionists and medical professionals has shed new light on how species composition, interactions, and distributions can be manipulated to decrease malnutrition (Chap. 4, Vol. 1) and the risk of infectious diseases (Chaps. 13 and 14, Vol. 1). The purpose of these volumes is to focus specifically on these issues in relation to major development challenges and how knowledge of interactions and trade-offs can be integrated into solutions.

Organization of These Volumes

To prepare the two volumes comprising the series *Integrating Ecology and Poverty Reduction*, we have asked authors to address a major development challenge or solution and to assess if/how an ecological approach is relevant within that context and the advantages and/or limitations of using the ecological toolbox. This task was more straightforward for some development goals and solutions than others. Nevertheless, all of the chapters have highlighted the utility of ecological science for addressing development problems and solutions through the direct application of ecological theory and tools, as well as the more indirect application of ecological thinking, which emphasizes the importance of spatial and temporal scales, feedbacks, and trade-offs. We recognize that entire books can be written on each of the topics presented herein and, thus, we do not attempt to cover all possible applications of ecology with respect to development challenges, a task that is beyond the scope of this project. Rather, these two volumes seek to highlight how major development challenges can be viewed through an ecological lens and addressed through the use and applications of the ecological toolbox. We do not propose that ecology alone will be able to answer many of these critical questions; rather, we suggest that ecological science combined with the tools of other disciplines can make a greater contribution to developing a sustainable future and reducing the tremendous poverty that persists in our world.

The series is divided into two volumes. The first volume, *Integrating Ecology and Poverty Reduction: Ecological Dimensions*, focuses on the ecological dimensions to global development challenges. The chapters in this volume deal with the biophysical aspects of ecology and demonstrate two primary points. First, that understanding the ecological foundations of human-dominated landscapes can provide a better understanding of how we are impacted by ecological processes. The American conservationist Aldo Leopold once famously stated that "to keep every cog and wheel is the first precaution of intelligent tinkering." We would add to that by stating that applying the right tool for the job should be the second rule of intelligent tinkering. In the chapters included in this volume, we explore the direct application of ecological tools to achieving distinct development goals of reducing hunger, improving human health and nutrition, decreasing vulnerability to extreme events, and increasing access to clean water and energy. These chapters present specific examples of the application of ecological principles in poverty reduction – or examples of how ecological tools fit and function in a development toolbox.

The second volume, *Integrating Ecology and Poverty Reduction: The Application of Ecology in Development Solutions*, focuses on mediating forces and solutions for poverty reduction and addresses the relevance and role of ecology in relation to these. We recognize that the mediating forces and the solutions that we have addressed – education, gender, demography, innovative financing, and ecosystem governance – represent far from an exhaustive list of topics that we could have covered in this volume. Nevertheless, these chapters collectively address many ways in which humans interact with each other and the ecosystems in which they live and how these interactions inform how ecological science and tools can be applied to positively influence forces shaping human societies and the creation of solutions that conserve biodiversity and ecological processes alongside poverty reduction. For example, demographic trends in population growth, urbanization, and migration influence the nature of human interactions with the environment (see the chapters on population, Vol. 2). Similarly, gender dynamics influence how the sexes perceive and interact with the environment and how natural resource access and management decisions are made (Gutierrez et al. Chap. 4, Vol. 2). An understanding of the dynamics underlying these forces must be factored into developing successful poverty reduction measures in communities that rely directly on natural resources for their livelihoods. In the section on Innovative Financing, the authors of various chapters demonstrate that implementation of mechanisms, such as Payments for Ecosystem Services, requires an application of sound ecological science and tools, in addition to an understanding of the social, economic, and governance constraints and opportunities where such programs may be developed, if they are to be effective. In the section on Ecosystem Governance, the authors emphasize the importance of strong ecological science, tools, and targets for governing and managing a land or seascape for multiple, often conflicting purposes. These chapters demonstrate that reducing poverty will require understanding the interplay of ecological, social, economic, and political systems and illustrate that ecologically sound solutions will require major shifts in conventional thinking, which society may or may not be willing to make. In a final concluding piece, Naeem critically addresses the overarching role of ecology in sustainable development, and states what this role is currently and proposes what it should be, if we are to have truly, sustainable development.

Conclusions

Traditionally, the science of ecology has not been an integral component of many aspects of international development for a variety of reasons. Increasingly, however, there has been a renewed interest in finding more sustainable means of development, grounded in ecological knowledge. Yet, a range of concepts and approaches that are becoming more widely used across a range of sectors, such as ecosystem services, resilience, and social-ecological systems thinking, are signs of a paradigm shift. Our goal with these volumes is to build upon this recent momentum at this important moment in time to increase the dialogue between ecologists and development practitioners. We have produced these volumes for both audiences in the hope that ecologists who read them will see the contribution that the field can make to poverty reduction and that development practitioners will gain an understanding of the contribution that ecology as a discipline can make to sustainable development. Our ultimate intention is that these volumes will facilitate increased dialogue among multiple disciplines, including ecology, and that this dialogue will result in a more effective use of ecological science and tools to improve the livelihoods of the world's poorest people alongside the conservation of functioning ecosystems.

References

Adams, W. M., R. Aveling, D. Brockington, B. Dickson, J. Elliott, J. Hutton, D. Roe, B. Vira, and W. Wolmer. 2004. Biodiversity conservation and the eradication of poverty. *Science* 306 (5699):1146–1149.

Alkire, S. and Santos, M.E. 2010. Acute Multidimensional Poverty: A New Index for Developing Countries. OPHI Working Paper 38.

Anand, S. and Sen, A. 2000. Human development and economic sustainability. World Development 28(12):2029–2049.

Butchart, S. H. M., M. Walpole, B. Collen, A. van Strien, J. P. W. Scharlemann, R. E. A. Almond, J. E. M. Baillie, B. Bomhard, C. Brown, J. Bruno, K. E. Carpenter, G. M. Carr, J. Chanson, A. M. Chenery, J. Csirke, N. C. Davidson, F. Dentener, M. Foster, A. Galli, J. N. Galloway, P. Genovesi, R. D. Gregory, M. Hockings, V. Kapos, J. F. Lamarque, F. Leverington, J. Loh, M. A. McGeoch, L. McRae, A. Minasyan, M. H. Morcillo, T. E. E. Oldfield, D. Pauly, S. Quader, C. Revenga, J. R. Sauer, B. Skolnik, D. Spear, D. Stanwell-Smith, S. N. Stuart, A. Symes, M. Tierney, T. D. Tyrrell, J. C. Vie, and R. Watson. 2010. Global Biodiversity: Indicators of Recent Declines. *Science* 328 (5982):1164–1168.

Daily, G.C. 1997. *Nature's Services: societal dependence on natural ecosystems*. Washington, DC: Island Press.

DeClerck, F., J. C. Ingram, and C. M. R. del Rio. 2006. The role of ecological theory and practice in poverty alleviation and environmental conservation. *Frontiers in Ecology and the Environment* 10 (4):533–540.

DeClerck, F.A.J., R Chazdon, K.D. Holl, J.C. Milder, B. Finegan, A. Martinez-Salinas, P. Imbach, L. Canet, and Z. Ramos. (2010). Biodiversity conservation in human-modified landscapes of Mesoamerica: past, present and future. Biological Conservation 143(2010): 2301–2313.

Eigenbrode, S. D., M. O'Rourke, J. D. Wulfhorst, D. M. Althoff, C. S. Goldberg, K. Merrill, W. Morse, M. Nielsen-Pincus, J. Stephens, L. Winowiecki, and N. A. Bosque-Perez. 2007. Employing philosophical dialogue in collaborative science. *Bioscience* 57 (1):55–64.

Gardner, T.A., Barlow, J., Sodhi, N.S. and C.A. Peres. 2010. A multi-region assessment of tropical forest biodiversity in a human-modified world. Biological Conservation 143:2293–2300.

Kareiva, P, and M. A. Marvier. 2007. Conservation for the people. Scientific American, October, 9.

Kauffman, J, 2009. Advancing sustainability science: report on the International Conference on Sustainability Science (ICSS). 2009. Sustainability Science 4:233–242.

Millennium Ecosystem Assessment. 2005. Ecosystems and Human Well-being: Biodiversity Synthesis.

Milder, J.C., F. DeClerck, A. Sanfiorenzo, D. Sanchez, D. Tobar, and B. Zuckerberg. (2010). Effects of tree cover, land use, and landscape context on biodiversity conservation in an agricultural landscape in western Honduras. Ecosphere 1(1):1–22.

Naeem, S. 2002. Ecosystem consequences of biodiversity loss: The evolution of a paradigm. *Ecology* 83 (6):1537–1552.

Naeem, S., D. E. Bunker, A. Hector, M. Loreau, and C. Perrings. 2009. *Biodiversity, ecosystem functioning, and human wellbeing: an ecological and economic perspective*. Oxford: Oxford Biology.

National Academy of Sciences, Committee on Facilitating Interdisciplinary Research and Committee on Science Engineering and Public Policy. 2005. Facilitating Interdisciplinary Research. Washington (DC): National Academies Press.

Ostrom, E. 2009. A General Framework for Analyzing Sustainability of Social-Ecological Systems. *Science* 325 (5939):419–422.

Ostrom, E., M. A. Janssen, and J. M. Anderies. 2007. Going beyond panaceas. *Proceedings of the National Academy of Sciences of the United States of America* 104 (39):15176–15178.

Ranganathan, J., K. Bennett, C. Raudsepp-Hearne, N. Lucas, F. Irwin, M. Zurek, N. Ash, and P. West, 2008. Ecosystem services: a guide for decision makers. world resources institute. Washington, DC.

Real, L.A., and J.H. Brown. 1991. *Foundations of Ecology: Classic Papers with Commentaries*. Chicago: The University of Chicago Press.

Redford, K. H., M. A. Levy, E. W. Sanderson, and A. de Sherbinin. 2008. What is the role for conservation organizations in poverty alleviation in the world's wild places? *Oryx* 42 (4):516–528.

Rumbaitis-Del Rio, C., J. C. Ingram, and F. Declerck. 2005. Leveraging ecological knowledge to end global poverty. *Frontiers in Ecology and the Environment* 3 (9):463–463.

Sachs, J.S. 2005. The End of Poverty: Economic Possibilities of Our Time. The Penguin Press, New York, New York.

Sanderson, S., and K. Redford. 2004. The defence of conservation is not an attack on the poor. *Oryx* 38 (2):146–147.

Sanderson, S. E., and K. H. Redford. 2003. Contested relationships between biodiversity conservation and poverty alleviation. *Oryx* 37 (4):389–390.

Secretariat of the Convention on Biological Diversity. 2006. Global Biodiversity Outlook 2. Montreal, 81 + vii pages.

Tekelenburg, A., B.J.E. ten Brink, and M.C.H. Witmer. 2009. How do biodiversity and poverty relate? – An explorative study. Bilthoven: Netherlands Environmental Assessment Agency.

Tilman, D. 1998. The greening of the green revolution. *Nature* 396 (6708):211–212.

United Nations (UN). 2010. Gateway to the UN System's work on the MDGs, http://www.un.org/millenniumgoals/maternal.shtml, accessed September 2010.

Walpole, M., R.E.A. Almond, C. Besancon, S.H.M. Butchart, D. Campbell-Lendrum, G.M. Carr, B. Collen, L. Collette, N.C. Davidson, E. Dulloo, A.M. Fazel, J.N. Galloway, M. Gill, T. Goverse, M. Hockings, D.J. Leaman, D.H. W. Morgan, C. Revenga, C.J. Rickwood, F. Schutyser, S. Simons, A.J. Stattersfield, T.D. Tyrrell, J.C. Vie and M. Zimsky. 2009. Tracking Progress Toward the 2010. Biodiversity Target and Beyond. *Science* 325: 1503–1504.

Chapter 2
Introduction to Ecological Dimensions of Hunger

Fabrice DeClerck

Introduction

Global hunger is a complex multi-faceted problem that typically has been the domain of agronomists, economists, and rural socialists. Less frequently, however, have we asked what ecology's, and ecologists' contribution might be to alleviating extreme and hidden hunger. Ecology has many contributions to make in finding, and developing ecologically based, sustainable solutions to this development goal. The hunger section of this book, in addition to several other chapters, provides compelling evidence of ecosystem services that are critical to sustaining agricultural production including pollination, pest control, and increasing the stability of agroecosystems (Chap. 3). Milder et al. (Chap. 5) demonstrate how taking a landscape scale approach permits the co-existence of production, conservation, and livelihood improvement goals in human dominated landscapes, and gives detailed information on relevant landscape measures appropriate to evaluating the conservation, production, and livelihood status of such landscapes. Remans et al. (Chap. 4) demonstrate that human nutrition is a critical ecosystem function with direct ties to farm-scale agrobiodiversity. All three chapters demonstrate, and define how integrated approaches, including a focus on ecosystem functions in agricultural landscapes, can reduce both acute and hidden hunger.

The UN Hunger Task Force Report, "Halving Hunger: It can be done," published in 2005 by Sanchez and colleagues, reminded us that too many people (852 million in 2005) are still chronically or acutely malnourished. They made a series of general recommendations including calling for the restoration and conservation of the natural

F. DeClerck (✉)
Centro Agronómico Tropical de Investigación y Enseñanza (CATIE), Turrialba, Cartago, Costa Rica
e-mail: fdeclerck@catie.ac.cr

J.C. Ingram et al. (eds.), *Integrating Ecology and Poverty Reduction: Ecological Dimensions*, DOI 10.1007/978-1-4419-0633-5_2,

resources essential for food security, securing local ownership, access, and management rights to forests, fisheries, and rangelands, developing natural resource-based "green enterprises" and paying poor rural communities for environmental services. Agricultural food production is, at its heart, an ecological enterprise that relies on the management of species interactions. Food production is also probably one of the few ecosystem services that has been fully exploited by humans and is the only service to be increasing rather than decreasing as highlighted by the Millennium Ecosystem Assessment. The ecological tradition of elucidating complex systems and relationships between species and the environment while working across multiple scales and disciplines equips ecologists with the tools necessary to tackle similarly and multifaceted problems associated with alleviating hunger.

Agricultural production and biological conservation have often been presented as diametrically opposed. However, agricultural food production is important not only in terms of being the source of our sustenance, it is also the human activity that occupies the greatest extension of the global terrestrial surface area. The "natural capital" of clean water, soils, fish, wildlife, and other resources encompassed in these landscapes provide about two-thirds of household income for the rural poor. As such, agriculture is not only critical to producing the food that sustains us, and to employing the rural poor, it is also essential in maintaining ecosystem services (Chap. 3).

The ecosystem services paradigm highlights synergies between conservation and production, and provides a framework for simultaneously meeting both essential development goals. Farming communities can put wild and associated biodiversity to use for a multitude of ecosystem services including pollination services, pest control services, and multiple soil and water related services (highlighted in Chap. 3). International attention on payment for ecosystem services (PES) has largely focused on payments for carbon, biodiversity conservation, and hydrological services. However, many ecosystem services operate on a much smaller spatial scale, such as the benefits of integrating biodiversity conservation into farm management. Fortunately, interventions that improve the provisioning of ecosystem services at the farm scale frequently are the same as those that improve ecosystem services at larger scales (see Chaps. 3 and 18).

Though there is increasing scientific knowledge regarding ecological integration in agroecosystems, several hurdles remain. For one, the dialog between scientific knowledge and local knowledge must be improved (Chap. 3, Vol. 2) to fill the gap between these distinct worldviews. Improving the role of ecology in education will be essential in bridging this gap, particularly if the future global economy is to be based on a sound understanding of the impact of humans on ecosystems and the environment. In a similar vein, March (Chap. 23) in this volume discusses a new model in seed distribution systems for rural farmers that stresses the importance of understanding farmers' seed needs and preferences, which helps build resilience into social and agroecological systems.

Second, the planning and management of ecosystem services for hunger alleviation must take a landscape scale perspective in order to be truly integrative as is highlighted by Milder et al. in Chap. 5. The landscape scale is the level at which many ecosystem processes operate and at which synergies and tradeoffs between and among environment and development objectives are often mediated (O'Neill et al. 1997). Bringing multiple stakeholders to the table to understand individual perspectives is the first step in building a shared vision that includes ecological integration leading to actions and policies that increase synergies while decreasing tradeoffs. Milder et al. in their chapter, demonstrate that by revealing landscape dynamics across multiple spatial and temporal scales, these processes can also help identify the "bottlenecks" to sustainable rural development, many of which may be non-obvious. The landscape highlighted by Milder et al. includes the explicit recognition that conservation and production goals are critical components of landscape management. Estrada and DeClerck (Chap. 14, Vol. 2) provide an example as to how landscape ecology can be used to locate farms that are critical to providing ecosystem services within a larger landscape.

Remans et al. (Chap. 4) adopt the concept of econutrition, highlighting the relationships between agricultural production, ecology, and human nutrition. The authors discuss how interdisciplinary approaches that combine the knowledge bases of ecologists, nutritionists, and agronomists to develop strategies to alleviate hidden hunger. Increasing crop functional diversity, for example, can alleviate anemia, particularly in communities that are strongly dependent on subsistence agriculture. Second, the integration of ecology with nutrition fosters environmental interventions that simultaneously have direct and indirect impacts on human health and nutritional well-being.

It is important to recognize that development interventions should not occur in isolation, and that finding opportunities for synergies between development goals are essential. Synergies are found when providing corridors for wild biodiversity simultaneously increases crop production by increasing pollinator services, creates barriers for crop pests (Chap. 3), improves water quality (Chap. 6–9), or improves human health (Chap. 10–14). Such examples for bundling ecosystem services present opportunities for reducing the cost of ecosystem services. Effective management of multiple services will require a strong ecological understanding of the relationship between land use and the provisioning of these services.

The world's poor do not solely depend on terrestrial landscapes as a primary source of food. Millions of medium and small-scale fishers and fish farmers, often very poor, depend on fishing and aquaculture, with over 97% of fishers living in developing countries. For these people, fish provide their primary source of protein. In contrast to terrestrial food production, where wild biodiversity makes a relatively small contribution to the human diet, marine and fisheries are populated by predominately wild species and managed through a complex framework of national and international agreements. More so than in terrestrial landscapes, management of fisheries must and has begun to take an ecosystems approach. McClennen details the use of ecosystem approaches in Chap. 16, Vol. 2 highlighting the ecological principles for sustainable fisheries.

In addition, agricultural landscapes and their management have impacts that reach far beyond alleviating hunger. Myers demonstrates the important consequences of land use change on human health in Chap. 11, and Keesing highlights the role of landscape and community ecology in understanding how fragmentation and altering of ecological communities affects the spread of infectious diseases. Agricultural lands are also one of the biggest users of water for irrigation. Food grows where water flows, as the saying goes. Access to water is often the biggest limitation to increasing production as well as maintaining year-round production in arid systems.

Increased agricultural productivity through "Green Revolution" technologies, including inorganic fertilizers, improved seed, and small-scale irrigation projects are likely to contribute significantly towards meeting the United Nation's goal of halving global poverty by 2015, but sustained poverty alleviation will require more. Ecology and ecologists alone will not be able to solve these critical development challenges; what is clear, however, is that the tools and skills that ecologists manage provide essential components to lasting, sustainable solutions. Ecologists must begin to consider how their field can contribute not only to the conservation of biodiversity, but to the relationship between conserved biodiversity and provisioning of nontraditional ecosystem services such as human health. Likewise, nutritionists, agronomists, and other development practitioners must recognize that many of the solutions to increasing human well-being and health can best be achieved by focusing on a healthy environment and the conservation of ecosystem services.

Chapter 3
Ecosystem Services in Agricultural Landscapes

Sean M. Smukler, Stacy M. Philpott, Louise E. Jackson, Alexandra-Maria Klein, Fabrice DeClerck, Leigh Winowiecki, and Cheryl A. Palm

Introduction

There is a tenuous relationship between the world's rural poor, their agriculture, and their surrounding environment. People reliant on farming for their livelihood can no longer focus on current food production without considering the ecosystem processes that ensure long-term production and provide other essential resources required for their well-being. Farmers are now expected to not only produce food, but also steward the landscape to ensure the provisioning of drinking water, wood products for construction and cooking, the availability of animal fodder, the capacity for flood attenuation, the continuity of pollination, and much more. Farmer stewardship of the landscape helps ensure ecological functions that, when beneficial to human well-being, are referred to as ecosystem services. Human activities strongly affect ecosystem services and there is often a resulting trade-off among their availability, which frequently results in the loss of many at the expense of few, most notably when producing food (Foley et al. 2005).

Ecosystem services have been categorized into four basic groups which include supporting services such as soil formation, photosynthesis, and nutrient cycling that are essential for providing provisioning services such as fodder, food, water, timber, and fiber production; regulating services that affect agricultural pests, climate, floods, disease, wastes, and water quality; and cultural services that provide recreational, aesthetic, and spiritual benefits (Fig. 3.1) (MA 2005a; Daily and Matson 2008). While provisioning services provided by agriculture have been and promise to remain a primary gateway for overcoming poverty, the dynamics of how to achieve poverty alleviation are changing rapidly with the declining availability of ecosystem services (MA 2005b; Witcover et al. 2006; Sachs 2008).

S.M. Smukler (✉)
Tropical Agriculture Program, The Earth Institute, Columbia University, 61 Route 9W, Lamont Hall, Room 2H, Palisades, NY 10964-8000, USA
e-mail: sas2242@columbia.edu

J.C. Ingram et al. (eds.), *Integrating Ecology and Poverty Reduction: Ecological Dimensions*, DOI 10.1007/978-1-4419-0633-5_3,

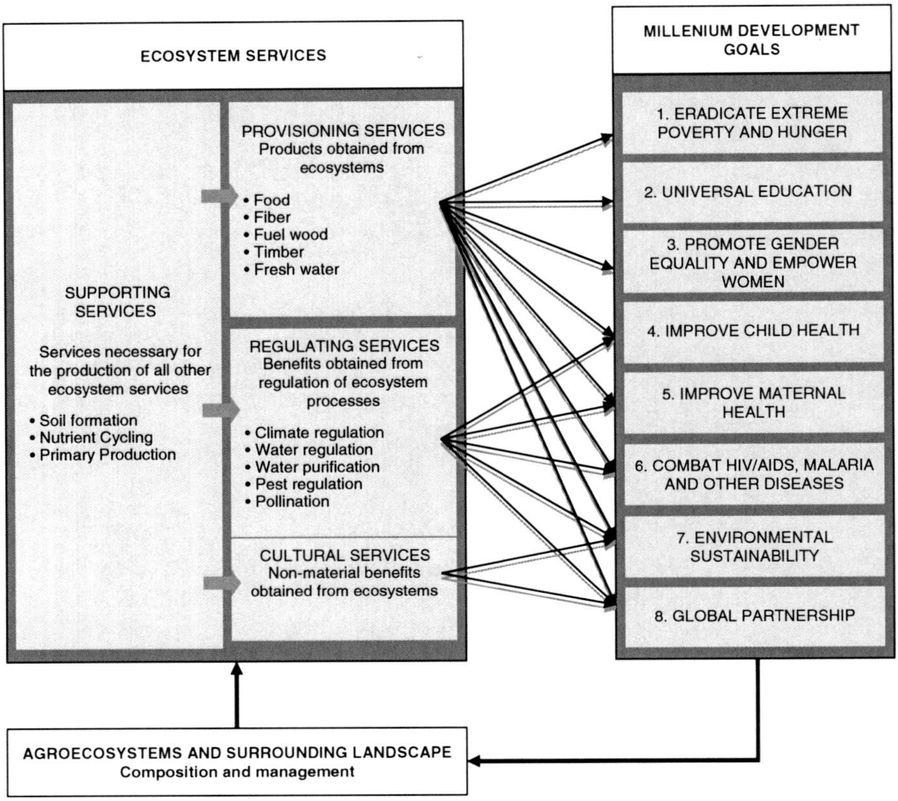

Fig. 3.1 Ecosystem services provided by agricultural landscapes and their connection to the Millennium Development Goals (Modified from the MA 2005a, b)

The strategy of using agriculture as the primary means for economic development has been successful primarily because of the availability and consumption of ecosystem services, many of which were supplied by natural ecosystems (Tilman et al. 2002; Evenson and Gollin 2003; Semwal et al. 2004; Robertson and Swinton 2005). For example, a farm's current yield may be largely dependent on nutrients supplied by thousands of years of plant decomposition and cycling (supporting services) prior to agricultural production and irrigation using water originating in natural watersheds (provisioning services). The availability of ecosystem services has both direct and indirect links to economic development and can be considered a form of capital. The processes by which this "natural capital" can be sustainably used to support the livelihoods of impoverished people requires an understanding of ecological and socioeconomic processes, particularly its links to other forms of capital, such as: (1) human capital: skills, knowledge, health, ability to labor and to pursue livelihood strategies; (2) financial capital: savings, livestock, supplies of credit, remittances; (3) physical capital: basic infrastructure (e.g., transport, energy, market access, communication, irrigation); and (4) social capital: membership of groups, networks, access to wider institutions of society (Bebbington 1999). These forms of capital are intrinsically interconnected in how they contribute to management of agricultural landscapes, but here we focus mainly on natural capital and more specifically on the ecosystem services that define this capital.

It is clear that around the world natural capital is diminishing and the availability of ecosystem services provided by natural ecosystems is becoming more uncertain as populations and urban development expand, and farmers seek new or better land for agriculture to meet increasing food demands (Bennett and Balvanera 2007; Srinivasan et al. 2008). Already farmers in many regions of the world are no longer able to expand into forests, wetlands, and savannahs to produce or gather food and fiber, and many can no longer expect to rely on the resources provided by the remaining intact natural ecosystems (DeFries et al. 2007). To exacerbate the situation, climate change is adding to the variability in availability of ecosystem services and will dramatically impact natural capital over the coming century (Schmitz et al. 2003; Rosenzweig et al. 2004). While those with financial means are able to purchase necessities such as cooking fuel, timber or even clean water produced in distant locations, the most impoverished do not have this option. They are restricted to utilizing what they can, where they are. With recent loss of natural ecosystems, looming threats of climate change, and rapid expansion and intensification of agricultural production using non-renewable resources, ecosystem services must be better managed within farmed landscapes to ensure the well-being of the poor.

Land management decisions are often framed as a dilemma that involves a trade-off between agricultural productivity and the preservation of natural ecosystems (Balmford et al. 2005; Green et al. 2005). However, the apparent trade-off between agricultural productivity and other ecosystem services is an over-simplification of management choices available; farmers may be able to manage for both (Foley et al. 2005; Perfecto and Vandermeer 2008). Some have questioned the validity of achieving win–win situations for both conservation and agricultural benefits and argue that the best way to preserve nature is to intensify or maximize agricultural productivity (Balmford et al. 2005; Green et al. 2005). Intensification may spare land from agricultural conversion but it by no means ensures the preservation of ecosystem services, especially when it is based on inputs of synthetic fertilizers and pesticides that have adverse effects on biota and resources in adjoining natural areas (Perrings et al. 2006). It is therefore critical that any effort towards agricultural intensification include goals beyond crop and livestock yields.

The objective of this chapter is to give agricultural, environmental, and development practitioners a basic understanding of the many ecosystem services provided by agricultural landscapes, their relationship to poverty alleviation, and a brief introduction to some of the ways to manage for optimum availability using a variety of strategies at multiple scales. In addition, the long-term consequences of environmental degradation for the rural poor will be highlighted (Dasgupta 2007). We have selected a series of examples that demonstrate the relationship between ecosystem services and poverty alleviation in agricultural landscapes. We also recognize that there is much yet to learn about this relationship and try to illustrate the areas in need of more research.

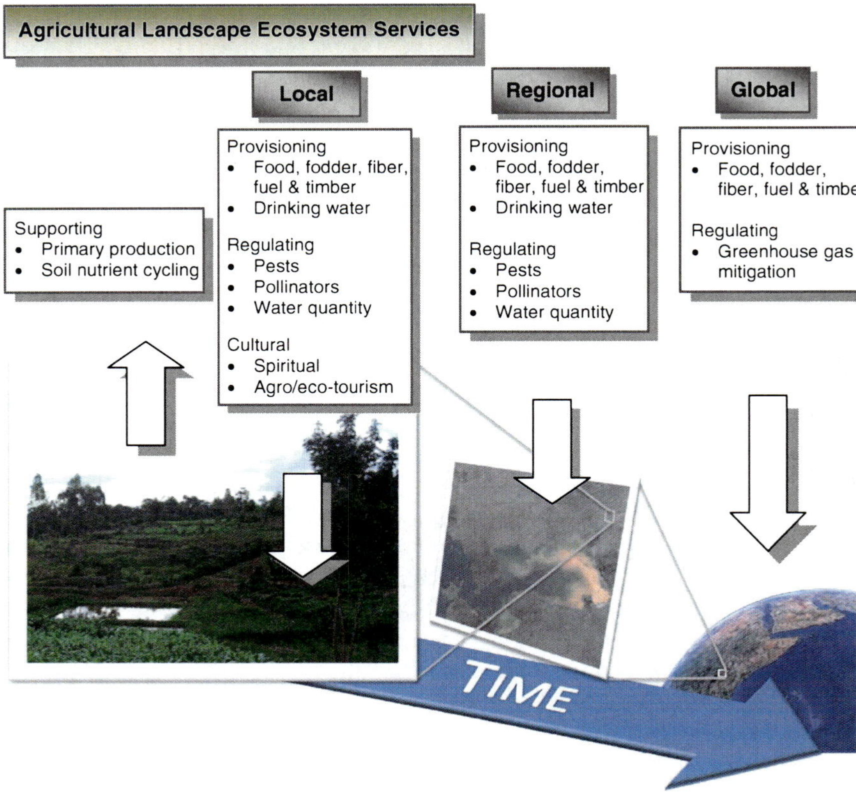

Fig. 3.2 Farm management when aggregated across a landscape can impact not only the ecosystem services available at a local scale but also the services available at regional and global scales. Here runoff from a farm in a watershed of Lake Victoria in western Kenya can negatively affect water quality and fisheries (image provide by Keith Shepherd). Practices such as agroforestry, if adopted widely, could reduce erosion, improve water quality, and mitigate greenhouse gas emissions but at time scales that may be beyond the farmer's season to season financial horizon

As agricultural landscapes are managed by numerous farmers, the overall supply of ecosystem services provided is dependent on aggregated practices, thus we discuss management decision made at scales ranging from the farm field, to the farmscape (the area of land managed by a farmer including the non-production areas), and the landscape. The aggregated results of farmers' management decisions impact the availability of ecosystem services not only for themselves but for people in distant places or in the distant future (Fig. 3.2). Our intent is to illustrate the origins and main beneficiaries of ecosystem services to show the various potential links of management decisions to poverty alleviation. In the first section of the chapter we discuss how agroecological management strategies at the field and farmscape scales can directly improve human well-being through increased agricultural production and other ecosystem services such as soil health, nutrient cycling, pest regulation and crop pollination. In the second part we describe how many of the same management practices designed to increase local benefits when scaled up to the landscape can benefit regional water quality and global greenhouse gas (GHG) mitigation. Because the adoption of services with regional or global impacts may not have immediate economic benefits we introduce the role of payments for ecosystem services (PES) for poverty alleviation. Finally, we discuss the need for conserving biodiversity and assessing the potential trade-offs and synergies of ecosystem services across scales.

Modes of Agricultural Intensification

There is growing recognition that agricultural intensification must entail ecologically sustainable management. This goes beyond past efforts to improve the living standard of the world's rural poor, e.g., by increasing agricultural productivity using improved crop varieties and inputs such as inorganic fertilizers, biocides and irrigation (Matson et al. 1997; Tilman et al. 2002). While agricultural intensification strategies reliant on high inputs increased food production in Latin America and Asia (Evenson and Gollin 2003), this "Green Revolution" strategy was often at a cost to other ecosystem services and their societal impacts continue to be debated (Kiers et al. 2008). Moreover, these strategies were not universally successful and in many regions, such as sub-Saharan Africa, they were largely not adopted (Smaling 1993; Smaling et al. 1993; Stoorvogel et al. 1993; Sanchez et al. 2009).

Agricultural intensification that has focused on provisioning of agricultural products has often resulted in severe impacts to other ecosystem services because of inappropriate use of irrigation, tillage, and fossil fuel-based industrial farm inputs such as pesticides and fertilizers. Such intensification has contributed to degradation of 1.9 billion ha impacting some and 2.6 billion people, the withdrawal of 70% of global freshwater (2.45% of rainfall) for irrigation (IAASTD 2008), biocide effects on health of humans and other organisms (Maumbe and Swinton 2003; Atreya 2008), the massive coastal anoxic zones (Prepas and Charette 2007), and biodiversity loss that has exceeded historical rates (Tilman et al. 2001; Cassman et al. 2003; MA 2005b; Robertson and Swinton 2005). These impacts, while prominent in the developed world, are often exacerbated in the developing world where institutional recognition of environmental impacts and agricultural extension services are limited, safety equipment is prohibitively expensive and regulation is often minimal (Dasgupta et al. 2002; Atreya 2008). Furthermore, industrial inputs, most of which are reliant on fossil fuels for production and delivery, are susceptible to increasing fluctuations in availability and price, which could negatively impact cost–benefit ratios and thus farmers' livelihoods as seen in recent food crises (Gomiero et al. 2008).

There is a growing recognition that increasing agricultural productivity at the expense of other ecosystem services is not a sustainable means of alleviating poverty.

M.S. Swaminathan, one of the founders of the "Green Revolution," has called for a new "Ever-Green Revolution" that encompasses the principles of ecology, economics, and social and gender equality (Swaminathan 2005). The United Nations has taken up this charge and on the eve of the millennium, in their commitment to halve poverty by 2015, included environmental sustainability as one of the eight goals designed to be road markers for meeting their objective. The eight Millennium Development Goals (MDGs) are an interdisciplinary task list; each task carefully selected to help reduce poverty. While there is much debate as to the best strategies to achieve the MDGs, it is likely the inextricable link between each MDG and the availability of ecosystem services (Fig. 3.1) has been underestimated as their role is ignored in many of the current poverty alleviation approaches (MA 2005b). Lack of conclusive evidence linking ecosystem services to human well-being is often cited as a reason for this neglect particularly in regions where the poor are most reliant on them. Meanwhile 60% of the earth's ecosystem services continue to be degraded or used unsustainably (MA 2005b).

Increasing Ecosystem Services for On-Farm and Local Benefits

Many of the one billion people who earn less than a dollar a day are reliant on agriculture for survival. The needs of these farmers are largely dependant on the success of very specific farm practices. Farmers often grow food for personal consumption on small pieces of land, and are typically challenged by uncertain land tenure, little access to capital for investments, labor constraints, no available agricultural extension, and limited access to markets (Sachs 2005). In this context it may be difficult for farmers to think about managing resources beyond their farm. We will examine how utilizing ecological principles for management at multiple scales to intensify on-farm production can improve the availability of ecosystem services for farmers adopting such practices.

Agroecological Management from the Field to the Farmscape

The aim of agroecological management strategies on and around the farm is to focus beyond immediate crop yields to increase the production of ecosystem services through efficient use of resources to create a farming system that is profitable over time without compromising future well-being (Altieri 2002; Ananda and Herath 2003; Witcover et al. 2006). Ecological intensification of farm management can include incorporating a diversity of species to fill available openings or "niches" and can help ensure critical supporting and regulating services such as soil formation, pest control, and nutrient cycling required for sustaining provisioning services. Essentially this corresponds to the idea of increasing the agricultural use and conservation of biodiversity (Jackson et al. 2007).

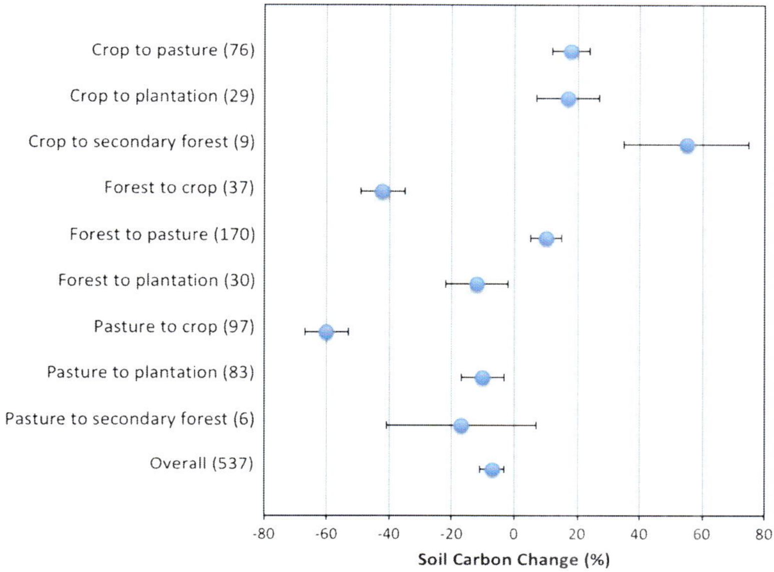

Fig. 3.3 Changes in soil organic carbon (SOC) an indicator of SOM resulting from changes in land management. This is from a study that reviewed 537 cases of land use changes. The bars indicate the 95% confidence intervals and the number of observations is shown in parentheses (Modified from Guo & Gifford 2002)

Supporting Services Example: Building Healthy Soil

Many of the ecosystem services provided by agricultural landscapes are contingent upon the status and management of soil, e.g., tillage negatively impacts soil aggregate stability affects and thus water infiltration rates (Cannell and Hawes 1994). Thus, collective management of on-farm soil properties can mediate the landscape's potential for supporting services underpinning numerous other ecosystem services (Fig. 3.1). Low soil fertility is a basic and immediate constraint limiting farmers' ability to meet consumption needs at the household level (Lal 2006; Sanchez et al. 2009). Identifying and improving the soil's ability to provide a favorable growing environment for crops is an important step in restoring ecological function and increasing food security, production, and income for smallholder farmers.

Converting natural ecosystems to simplified agroecosystems and maintaining a constant state of disturbance through annual cultivation and management causes drastic changes in soil conditions, including the decline of soil organic matter (SOM) (Fig. 3.3). Maintaining or increasing SOM may be the single most important management strategy designed to improve supporting services.

The addition of even small quantities of SOM can have substantial benefits for the biological, chemical, and physical properties of the soil (Stevenson and Cole

1999). Soil organic matter consists mainly of soil organic carbon (SOC), and provides the substrate required by bacteria to mineralize nitrogen and other nutrients stored in complex molecules (e.g., carbohydrates, amino acids, lignin) that would otherwise be unavailable to plants. SOM is a critical component of sustained nutrient cycling and has numerous other benefits such as increasing soil water holding capacity, infiltration rates, nutrient storage capacity, aggregation and reducing bulk density (Lal 2006). Continual cropping without sufficient organic matter inputs leads to depletion and unproductive soils resulting in reduced net primary productivity (NPP) or the rate an ecosystem accumulates energy or biomass. Eventually continued depletion causes a negative feedback loop where the landscape can no longer support the plants needed to provide the organic matter required for their growth (McDonagh et al. 2001).

Once soils have been degraded, improving their functions requires substantial active management and a clear understanding of the driving factors that led to degradation (Whisenant 1999). Using nitrogen fixing plants in agricultural fields can begin to rehabilitate soils. Alternatively inorganic fertilizers can be used to increase crop productivity and the amount of residue that can be returned to the soil (Palm et al. 2007). Incorporating residues can increase SOM resulting in a positive feedback loop, where yields continue to increase, and soil becomes responsive to further inputs. For example every 1 Mg ha^{-1} increase in the SOC pool in the rooting zone has been shown to increase wheat yields by 20–70 kg ha^{-1}, rice yields by 10–50 kg ha^{-1} and maize yields by 30–300 kg ha^{-1} over un-amended soils (Lal 2006).

Although inorganic fertilizer use may be an important tool for jump starting substantially degraded systems and increasing crop productivity, fertilizer use does not come without substantial trade-offs. The most important trade-off may be financial, as the use of inorganic fertilizer does not necessarily translate into increased livelihoods. Inorganic fertilizers are expensive, particularly for smallholder farmers, and their price may be increasingly volatile (Sogbedji et al. 2006; Gomiero et al. 2008; Huang et al. 2009). In 2008, for example, worldwide inorganic fertilizer prices rose as much as 200%. Furthermore, smallholder farmers living in impoverished regions may have limited access to inorganic fertilizers due to lack of infrastructure, capital or financing (Sachs 2005). The excessive or inappropriate use of inorganic fertilizers can also lead to substantial reductions in other ecosystem services. Therefore, developing locally derived sources of nutrients from plants and animals used instead of, or in combination with inorganic fertilizers may help farmers achieve greater economic returns over time and promote various ecosystem services.

Filling every available niche with carefully selected plants and animals can help balance nutrient losses from agricultural systems. Selecting appropriate plants for different niches can increase the availability of organic matter and retains nutrients otherwise lost through erosion and leaching (Fig. 3.4). To increase organic matter inputs, farmers may improve water management to increase NPP, incorporate woody biomass (trees or shrubs) within the cropping system (i.e., agroforestry), use cover crops to minimize erosion, green manures (nitrogen fixing legumes), mulch, or increase animal manure inputs. Once organic matter production is increased, combining inorganic fertilizers and organic materials can become even more effective in restoring fertility to nutrient-poor soils than either alone. Research in sub-Saharan

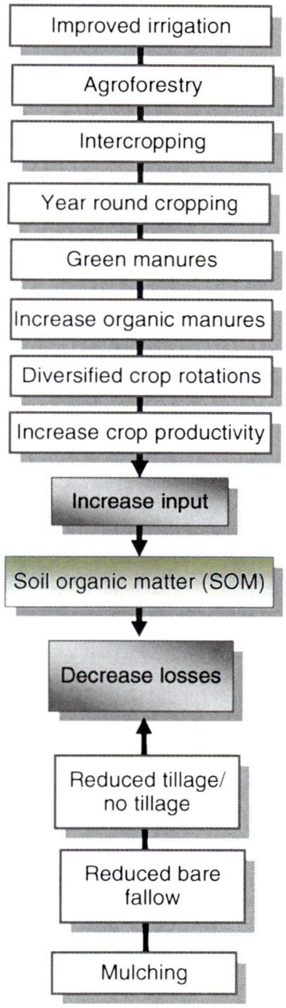

Fig. 3.4 Management practices that can either increase organic matter inputs across the agricultural landscape or reduce nutrient losses (Modified from Neider and Benbi 2008)

Africa shows that the application of inorganic fertilizers on soil amended with manure increased grain yields compared to plots to which just inorganic fertilizers or just manure was added (Kimani and Lekasi 2004). In another long-term study, maize yields in Kenya were compared along a climatic and amendment gradient; demonstrating the benefits of incorporating a combination of high-quality manure, crop residue, and inorganic fertilizers (Okalebo et al. 2004). Research on combining inorganic and organic soil amendments in smallholder systems is not new, but its current focus on the timing of nutrient availability, interactions between inorganic and organic sources, and the residual benefits to the system over the long-term is critical (Giller 2002; Vanlauwe et al. 2002; Drinkwater and Snapp 2007).

Utilizing plants that fix their own nitrogen (legumes) through symbiotic relationships with micro-organisms associated with their root system can also increase overall nitrogen capital of the site. Legumes can be planted as annuals or perennial either in the field as a cover, relay crop, or intercrop or in the field margins as a hedgerow. Estimates for N fixation from legumes range from 58–120 kg N ha^{-1} $year^{-1}$ for annuals to 228 kg N ha^{-1} $year^{-1}$ for perennial legumes such as alfalfa (Nieder and Benbi 2008). A meta-analysis of over 94 studies of maize yields in sub-Saharan Africa reports that while yields increased most significantly using inorganic fertilizer applications (2.3 t ha^{-1}), yields also increased by 1.6 t ha^{-1} using coppiced woody legumes, 1.3 t ha^{-1} using non-coppiced woody legumes, and 0.8 t ha^{-1} using herbaceous green manure legumes (Sileshi et al. 2008). These authors concluded that the combination of inorganic fertilizer and organic matter inputs increased efficiency of fertilizer use, as there were no significant differences in crop yield between organic inputs combined with 50% vs. 100% of the recommended fertilization rate in 48 studies with paired treatments (Sileshi et al. 2008). This study suggests that the use of organic inputs could increase the efficiency of inorganic fertilizer use by at least 50%.

In addition to increasing organic matter inputs, management practices that reduce erosion can also have substantial benefits for soil nutrients. Clearing land for production has resulted in erosion rates 1–2 orders of magnitude greater than rates of erosion in natural ecosystems systems (Montgomery 2007). Generally, management that reduces the amount of soil disturbance or the impacts of disturbance, are likely to improve nutrient retention on the site (Lal 2004a). To decrease nutrient losses through erosion and leaching, farmers can eliminate or limit soil disturbance by reducing the amount of tillage. They could also fill the niche opened by tillage during land preparation or harvesting by mulching or planting non-production crops such as improved fallows, cover crops, or green manures. Mulching with crop or agroforestry residues can protect otherwise bare soil from dislodging due to rainfall impact or wind (Roose and Ndayizigiye 1997). Cover crops also protect soil and have been shown to reduce nitrate losses as the roots scavenge the soil for any residual nutrients which can then be incorporated and made available for subsequent production crops (Jackson et al. 1993; Wyland et al. 1996; Sileshi et al. 2008).

Many of the practices just described are done so in the context of maintaining supporting services, mainly nutrient cycling required for primary productivity. Unfortunately, many of these practices generally introduce complexity and increased labor to the production system without necessarily increasing immediate economic returns. This creates an obvious draw towards the use of simpler methods relying solely on inorganic fertilizer inputs that promise immediate income even though the alternative integrated soil fertility management approach may have greater long-term financial benefits. In addition, it may not be apparent to those on such short time horizons that the soil provides the support services critical for the provisioning of many other ecosystem services, including ensuring a buffering capacity for environmental perturbations caused by climate change (e.g., resilience to drought through increased soil moisture capacity) (Pimentel et al. 2005). Therefore education, through demonstration and outreach or even short-term economic incentives, will help promote the adoption of soil management practices key to ensuring long-term agricultural productivity and well-being (Table 3.1).

Table 3.1 The importance, challenges, and strategies for securing provisioning ecosystem services key to rural well-being such as food, fiber, fuel, and timber production

Provisioning service	Importance	Challenges	Strategies
Food and fiber	Food security Independence Textiles Income	Biophysical constraints: water and nutrients Farm size Land tenure Labor	Identify most profitable agroecological crop management techniques Increase access to: markets, seeds, and soil amendments Diversify production Develop cooperatives
Fuel	Means to cook food Provide heat	Inefficiency of stoves Distance to available wood Labor	Improve stove efficiency Increase fuel resources Develop alternative fuel sources
Timber	Construction materials Carbon sequestration in biomass Litterfall inputs	Incorporation into farming system without impacting crops Land tenure Farm size Labor	Identify optimal design and management of agroforestry system

Provisioning Services Examples: Providing Food, Fiber, Fuel, and Timber

Small-holder farmers trying to escape poverty must find a way move beyond subsistence to produce goods and services that are profitable. For the rural poor their ability to profit is often constrained by limited availability of capital for investment in equipment, inputs, and land, limited access to markets both for sales and purchasing of inputs, and lack of infrastructure for irrigation or extension. While the focus of "Green Revolution" agricultural development on increasing the productivity of a few high yielding crops has had great success increasing the caloric output of the farm, it has not ensured profit, nutrition, or other provisioning services that the landscape may provide (Evenson and Gollin 2003; Kiers et al. 2008). These services can include timber, fodder, fuel wood, medicine, and many other products that are of direct use to farming communities or marketed for income generation. One approach for maximizing these services would be to fill all available niches with useful plant or animal species, capitalizing on synergistic interactions whenever possible.

Filling niches by managing farmland for multiple species or agroforestry has been promoted as a means of maximizing production potential for small landholders. Agroforestry systems improve soil conditions (Young 1997; Sanchez 1999), provide habitat for species (Somarriba et al. 2003; Schroth et al. 2004; Harvey and Villalobos 2007), mitigate against global climate change (Montagnini and Nair 2004;

Verchot et al. 2005), and provide alternative sources of income (Sanchez 1999; Alavalapati et al. 2004). Agroforestry systems is defined as multiple cropping of at least two plant species that interact biologically where one of the plant species is a woody perennial and at least one of the plant species is managed for forage, annual or perennial crop production and thus, can provide food, fuel, fiber, and timber resources (Somarriba 1992). Ecological interactions in agroforestry systems can have both positive (soil fertility improvement, improved nutrient cycling, carbon sequestration) and negative affects (shading, allelopathy) (Kohli et al. 2008); therefore, careful, ecologically based management that maximizes synergies is essential to ensure the greatest benefits at multiple scales.

There are numerous examples of the success of agroforestry systems to maintain ecosystem services from around the world, particularly in regions where there has been a strong infrastructure for research and extension (Graves et al. 2004). Remnant trees in otherwise treeless pastures in the highlands of Costa Rica provision food for birds and animals (90% of the tree species), timber (37%), firewood (36%), and fence posts (20%) (Harvey and Haber 1999). A spatial evaluation of agroforestry systems in Talamanca, Costa Rica demonstrated that cacao agroforestry systems can mimic the structural diversity of native tropical forests (Suatunce et al. 2003). Farmer interviews in Talamanca demonstrated that species incorporated into their agricultural landscapes had numerous ecosystem functions, for example fiber, timber, and fruit (Córdova et al. 2003; Somarriba and Harvey 2003). If farmers use their land as sources of fuel, fiber, and timber, the encroachment into native forests is minimized.

Agroecological Management from the Farmscape to the Landscape

Farmers can benefit directly from ecosystem services other than supporting and provisioning services by managing beyond the production field. There are a number of regulating services that farmers can manage that will directly enhance their well-being by maintaining or increasing crop productivity or the availability of other resources. These services can include the regulation of climate, water for drinking and irrigation, pollination and pest reduction. Most of these services require farmers to manage the farmscape and the surrounding landscape. Filling available niches around and between fields with a diversity of plants selected for multiple uses or maintaining already existing patches of natural vegetation can create windbreaks that generate favorable micro-climates for crops (Seck et al. 2005), maintain hydrologic processes that ensure water availability (e.g., percolation into ground water) (Rockström et al. 2004), reduce the prevalence of weeds (Bokenstrand et al. 2004), and provide habitat for pollinators and other beneficial organisms (Tscharntke et al. 2008). The management of areas of the agricultural landscape past the edge of the production field can result in a complex landscape mosaic of diverse vegetation types.

Regulating Services Example: Maintaining Pollinators

Recently, the importance of pollinators as a key element of agricultural diversity supporting human livelihoods has been increasingly recognized. The great majority of plant species benefit from pollination provided by bees, flies, birds, bats, and other taxa. For example, 90% of flowering species in tropical rainforests rely on animal pollination (Bawa 1990) and 75% of the most important crop species benefit from pollination services (Klein et al. 2007) accounting for €153 billion annually (Gallai et al. 2009). As much of the diversity of marketable horticultural products originates from developing countries (Aizen et al. 2008), crop diversification and pollinator-crop interactions could play a key role in poverty alleviation.

Agricultural intensification in many parts of the world has necessitated managed pollination services, as wild pollinators are not available in sufficient numbers to ensure crop yield (Potts et al. in press). At the global scale, populations of managed honey bees have not increased at the same rate as the proportion of agricultural crops that depend on pollinators (Aizen and Harder 2009). Hence, the demand for pollination services is likely to outstrip current honeybee hive numbers in many areas of the world (Aizen and Harder 2009). Because wild bees may be able to provide insurance against ongoing honeybee losses (Winfree et al. 2007), the demand for wild pollination is increasing. Concerns about the delivery of future pollination services are high for tropical countries as habitat isolation affects pollinators in tropical landscapes (Ricketts et al. 2008).

Managing pollination services solely reliant on the European honeybee is a risky strategy because it relies on single species management. Almond pollination in California relies on the European honey bee and oil palm pollinations in South-East Asia depend on a single imported African beetle (Kremen et al. 2002). Recent studies highlight the need to promote biodiversity to improve resilience in the provisioning of pollination services by buffering pollination against asynchronous annual fluctuations in bee abundance (Kremen et al. 2002; Winfree et al. 2007; Klein 2009). Different pollinating species also occupy different spatial, temporal, and conditional niches in which only a diversity of pollinator groups will lead to high quality and quantity services (Hoehn et al. 2008; Klein et al. 2008, 2009; Blüthgen and Klein 2011). These facts suggest pollinator diversity must be protected or restored across agricultural landscapes to ensure pollination services under various conditions and across space and time.

Local (farmscape) alterations that impact pollination services include changes in the abundance, diversity, distribution, and temporal continuity of floral resources (Klein et al. 2002; Greenleaf and Kremen 2006; Potts et al. 2006; Holzschuh et al. 2007; Williams and Kremen 2007) and availability of nesting sites and materials (Shuler et al. 2005). Therefore, local management practices must include planting and preservation of year-round foraging resources (e.g., understory blooming herb, flower strips), nesting resources (e.g., open ground, dead wood, and whole twigs), and should avoid agro-chemical use during pollinators' times of activity (Brittain et al. 2010).

Management recommendations should include possible risks and trade-offs or dis-services that might be promoted through the establishment of pollinator-friendly management practices. These include the promotion of pest species or the repelling of pollinators from crop species by planting flower resources (pollinators attracted to the flowering resources planted rather than to the crop species of interest). It is especially important to collect information and train local people to shift human labor from providing hand pollination services (e.g., passion fruit pollination in Brazil) to the management of pollinator-friendly agricultural practices.

The conservation of natural or semi-natural habitats that provide additional pollinator resources is essential to promote pollination (Kremen et al. 2002; Ricketts et al. 2008). These natural or semi-natural habitats are often large in area, which seems to be important for many rare pollinating species. Therefore, pollination management practices should be promoted not only at the farmscape but also at landscape scales. These practices include natural habitat protection and creating stepping stones or corridors to connect agriculture and bee (semi- and natural) habitats of high quality. The complexity of managing habitats across a landscape suggests the coordination by local farmers and organizations may be critical for maintaining wild pollination services and sustaining the continued health pollinators.

Regulating Services Example: Managing Agricultural Pests

Many of the strategies used to promote soil and pollinator functions can also be employed to increase pest regulation services. The importance of local and landscape factors in pest regulation services has long been a topic of ecological studies. In theory, strategies that limit density and diversity of herbivores and increase density and diversity of predators should enhance pest regulation services, but relationships between farm management, pests, and predators are complex (Tscharntke et al. 2005; Clough et al. 2010). In fact, crop diversity, landscape complexity, and predator diversity all have important implications for provisioning of pest regulation services within agroecosystems.

Diversity of crop and non-crop plants within agroecosystems affects pest regulation. Theory on the relationship between crop diversity and pest regulation hypothesizes that natural enemies are more diverse in polycultures because alternative prey or other food resources are available ("enemies hypothesis") and that prey populations are lower in polycultures where locating host plants is difficult ("resource concentration hypothesis") (Root 1973). Subsequently Andow (1991) reviewed >200 studies on pests, predators, and crop diversity finding that the majority of herbivore species (52%) had larger population sizes in monocultures than polycultures; only a few herbivore species (15.3%) were more abundant in polycultures. A number of different strategies such as habitat diversification (e.g., intercropping, use of cover crops and trap crops, allowing weed growth, crop rotation, inclusion of perennial plants), agroforestry with a diverse, dense mix of trees as or in addition to crop plants, and organic management (e.g., elimination of pesticides, and organic matter addition) all contribute to maintenance of diversity and density of natural enemies within agroecosystems leading in some cases to enhanced pest regulation (e.g., Altieri 1999; Rao et al. 2000; Östman et al. 2001; Eilers and Klein 2009).

Increased vegetation complexity in agroforests, for example, can harbor greater abundance and diversity of insectivorous birds enhancing pest control services (Perfecto et al. 2004; Van Bael et al. 2008). However, plant diversity may not always benefit pest regulation if, for example, trees serve as alternate hosts for pests, compete with crops for water or nutrients making crops more susceptible to pest problems, or limit the movement of olfactory cues that attract natural enemies to pest species (Rao et al. 2000).

Landscape context is extremely important in determining the strength and persistence of pest regulation within agricultural systems. Placement of non-crop habitats such as hedgerows, windbreaks and early succession fallow areas at edges of crop fields can enhance predation by providing overwintering sites for natural enemies, facilitating dispersal, and providing habitat for alternative hosts of parasitoids of crop pests and necessary food resources like pollen, and nectar (Altieri 1999; Rao et al. 2000; Tscharntke et al. 2005; Bianchi et al. 2006; Zhang et al. 2007). Non-crop habitats act as sources for intensively managed agricultural systems to maintain biodiversity of natural enemies (Tscharntke et al. 2005; Eilers and Klein 2009). At larger spatial scales, maintaining a balance between the proportion of crop and non-crop vegetation may be critical as diversity and abundance of some natural enemies declines far away from natural habitats (Perfecto and Vandermeer 2002). Similarly, a greater degree of landscape complexity can enhance natural enemy density and specifically increase predator activity, fecundity, oviposition rate, predation area, and condition of natural enemies, and can reduce pest pressure (Bianchi et al. 2006). In some cases, complex habitats with smaller crop fields relate to slow establishment of pests (Östman et al. 2001); however, complex habitats and landscapes may also benefit populations of pests that subsequently invade crops (Bianchi et al. 2006). Finally, both vertical intensification (large numbers of trees in same area) and horizontal intensification (great expanses planted with the same species) may increase pest outbreaks (Rao et al. 2000). Thus, diversified landscapes can enhance pest control services when they result in higher natural enemy diversity and density of enemies that colonize crop fields, reduce pest densities, reduce damage levels, and increase crop yields keeping in mind possible trade-offs in maintaining complex landscapes (Bianchi et al. 2006).

Finally, natural enemy diversity may actually enhance or hinder pest control, depending on the exact context examined. The combined effects of predators depend on two main groups of factors: (1) degree of complementarity in the assemblage of predators and parasitoids; and (2) the occurrence of predation or competition among predators and parasitoids of pests. In assemblages with a high degree of complementarity, there may be behavioral or diet differences among predator species that mean that multiple predator species are more effective than single predator species for controlling pests (Cardinale et al. 2003; Tscharntke et al. 2005; Snyder et al. 2006). For example, Cardinale et al. (2003) found synergistic effects of two ladybird beetles and one parasitoid species on aphid pests of alfalfa, and cascading effects on alfalfa yield, such that their effects were greater with all three natural enemies present than expectations based on their individual effects. Likewise, others have found additive effects of multiple predator species controlling cabbage aphids (Snyder et al. 2006). In other situations, however, multiple predators performed worse then each predator alone due to the prevalence of intraguild predation (one predator consuming another instead of feeding on herbivore species among the predator assemblage) (Finke and Denno 2005). Thus, the exact relationship between natural enemy diversity and pest suppression and crop benefit may depend on predator and parsitoid species composition, and the type of agroecosystem examined.

Managing Agricultural Landscapes for Regional and Global Ecosystem Services

Many practices that maximize ecosystem services that directly benefit farmers have impacts far beyond farm borders. Soil management, irrigation, use of pesticides, inorganic fertilizers, fuel, and non-production areas all can affect ecosystem services at the regional or even global scale (Tilman et al. 2001; Prepas and Charette 2007; Smith et al. 2008). From the perspective of those living in distant locations the impact of individual farms may seem small and irrelevant but collectively can be dramatic. Although the impacts may not necessarily directly affect the farmers that manage them the impacts are indirectly connected to their livelihoods and the alleviation of poverty. We focus on two of the most obvious ecosystem services resulting from management of agricultural landscapes that have impacts at multiple scales: the regulation of water and GHG emissions. One example shows how agricultural practices regulate quality and quantity of water from upland to downstream users to the coastal communities (Fig. 3.1, Table 3.2) (Prepas and Charette 2007). Another example shows how management of nutrients across the agricultural landscape collectively results in GHG reductions on a scale that influences global atmospheric conditions. In both of these examples the impacts most likely will not be directly or immediately felt. Water scarcity and climate change will be chronic, slowly developing problems that will invariably disproportionally affect the poor (Parry et al. 2004; Srinivasan et al. 2008). It is important to recognize that the agroecological management practices that benefit regional and global population may be challenging for poor farmers to adopt for a number of reasons (van Noordwijk et al. 2002). It is, therefore, in the best interest of the recipients, particularly those in rich countries, to see themselves as stakeholders in the management of these potentially distant agricultural landscapes and support policy and actions that promote practices that maximize ecosystem services and alleviate poverty.

Regional Regulating Service Example: Protecting Water Quality and Availability

The amount of rainfall intercepted by agricultural landscapes and the immense volume of water applied as irrigation makes agricultural production one of the most important modifiers of water-related ecosystem services. More than 15% of the

Table 3.2 Summary of agroecological management practices and their associated ecosystem service

Management practices		Ecosystem services											
		Local benefits									Regional benefits		Global benefits
		Soil sustainability	Food and fiber	Fuelwood and timber	Drinking water quality	Genetic resources	Pollination	Pest regulation	Irrigation availability	Cultural	Water regulation	Water purification	Climate regulation
Crop management	Cover crops	X	X		X	X	X	X	X		X	X	X
	Relay crops	X	X		X	X	X		X		X	X	X
	Green manure	X	X		X	X	X		X		X	X	X
	Crop rotations	X	X		X	X	X	X	X		X	X	X
	Reduced tillage	X	X		X			X	X		X	X	X
	Perennial cropping	X	X	X	X	X	X		X		X	X	X
	Mulching	X	X		X			X	X		X	X	X
	Composting	X	X		X			X	X		X	X	X
Agroforestry	Hedgerows	X	X	X	X	X	X	X		X	X	X	X
	Intercropping	X	X	X	X	X	X	X		X	X	X	X
	Wood lots	X	X	X	X	X	X	X		X	X	X	X

(continued)

Table 3.2 (continued)

Management practices		Ecosystem services											
		Local benefits									Regional benefits		Global benefits
		Soil sustainability	Food and fiber	Fuelwood and timber	Drinking water quality	Genetic resources	Pollination	Pest regulation	Irrigation availability	Cultural	Water regulation	Water purification	Climate regulation
Livestock	Improved grazing management	X	X		X	X				X	X	X	X
	Restoration of grasslands	X			X	X				X	X	X	X
	Fire management	X	X		X			X			X	X	X
Infrastructure	Terracing	X	X	X	X				X		X	X	X
	Ponds	X	X						X		X	X	
	Biogas digesters	X		X									X
	Improve cook stoves	X		X	X					X			X

water that runs across the global terrestrial surface area flows in or through cultivated landscapes and agriculture accounts for 70% of global water withdrawals (MA 2005a, b). This interaction with agriculture affects a number of water-related ecosystem functions, including water availability, water quality and flood attenuation, all of which impact local human well-being and regional populations. Ensuring an adequate supply of clean water for both human consumption and the maintenance of natural habitat will be among the most significant global challenges of this century (MA 2005a, b). More than 1.1 billion people do not have access to clean water (WHO/UNICEF 2004), yet already 5% to possibly 25% of water use exceeds the long-term accessible supply of global freshwater (MA 2005a, b; see the chapters on water in this volume). As demand increases, water scarcity will inevitably result in problems for food production, human health, economic development, and biodiversity (Postel et al. 1996; Rockström et al. 2004). In addition, the degradation of water quality from agricultural runoff further threatens the viability of freshwater and coastal ecosystems and people dependant on coastal fisheries for their livelihood (see chapter on Fisheries, Vol. 2).

One aspect of human well-being that can be directly measured is human health, which is tightly linked in a number of ways to water availability and quality (see the chapters on Health, this volume). Nitrate losses from fields into waterways and aquifers may adversely impact human health. For example, high nitrate concentrations in drinking water have been attributed to methemoglobinemia (blue baby syndrome) and impaired immune response (Fewtrell 2004). Due to a lack of reliable data and difficulties of proper analysis, the extent of high nitrate levels in drinking water is unknown but suspected to surpass the 10 ppm nitrate limits set by the World Health Organization (WRI 1989) in many regions or the world. High nitrate levels can also cause toxic algal blooms (toxic cyanobacteria) associated with chronic disease such as liver cancer (WCD 2000). Irrigation ditches, canals, rice fields, and reservoirs harbor and spread vector-borne diseases, particularly in tropical regions where Rift Valley Fever and Japanese encephalitis occur (WCD 2000). For people with little access to medical care, poor water management can threaten their very survival.

Impaired water quality in aquatic habitats also indirectly affects human well-being. Movement of soil, nutrients, and pesticides from agricultural fields to adjacent waterways has already caused massive reduction of ecosystem services. The physical impact of rainfall hitting bare soil dislodges soil particles and can result in sealing and runoff of soil particles, nutrients, and pesticide residues. Likewise, irrigation can dislodge soil, nutrients, and pesticides affecting water quality and flow. Runoff of nitrogen from terrestrial to aquatic ecosystems has doubled from 111 million tons per year in pre-industrial times to between 223 and 268 million tons per year (Galloway et al. 2004). Inputs of nitrogen (N) and phosphorus (P) into aquatic ecosystems stimulate primary production including algal growth, thereby consuming much of the water's dissolved oxygen. This depletion of oxygen causes eutrophication, kills fish in streams, rivers, lakes, and coastal regions, and leads to substantial economic impacts to communities reliant on fisheries

(Prepas and Charette 2007). For example, in Lake Victoria, eutrophication threatens fisheries and the livelihoods of rural poor living around the lake (Prepas and Charette 2007).The impacts of pesticide runoff are not well understood as their release into the environment often creates synergistic or unanticipated effects. Pesticide runoff negatively affects humans (Rola et al 1993; Polidoro and Bosque-Perez 2007; Luo and Zhang 2009) and non-humans alike (Hayes et al. 2002; Jarrard et al. 2004).

Strategies to improve water-related ecosystem services in agricultural landscapes might be as simple as changing crop management practices (e.g., growing cover crops) or extremely involved (e.g., constructing detention ponds with irrigation return capabilities). In general, water-related ecosystem services can be improved by efficient use of inputs (i.e., nutrients, pesticides and irrigation), limiting bare soil, reducing the speed or volume of irrigation water or runoff, or by capturing and retaining runoff. Agroecological management strategies for increasing local ecosystem services also serve to increase water-related services that ensure the availability of clean water on a larger scale.

Beyond the agricultural fields or pastures, management of water-related ecosystem services may require more investment in time, labor, and money since strategies may not show obvious benefits for farm production, or may require longer time horizons or complex construction. However, basic practices could include protecting waterways by filling niches and covering the soil surrounding the farm fields with organisms that can utilize excess nutrients or pesticides or slow the movement of water. Planting vegetation along pathways that lead to water bodies is effective at capturing sediments and phosphorus (Uusi-Kamppa et al. 2000). Grassed waterways, once mature, can reduce the flow rate of runoff and cause sediment to drop out of suspension before reaching aquatic systems (Maass et al. 1988). One study found 92% of sediment, and 71% of nutrients contained in irrigation runoff was reduced in the first 4 m of a grass filter strip (Blanco-Canqui et al. 2004). Another showed that an 8 m buffer of grasses decreased nitrate loads by 28% and a 16 m buffer decreased loads by up to 42% (Bedard-Haughn et al. 2004). More complex buffer zones can be created by planting woody vegetation either along waterways or even adjacent to fields as hedgerows. The deep roots of hedgerow plantings of trees and shrubs reach niches in the soil where nutrients otherwise untapped by shallow rooted annual crops would leach from and enter water bodies via sub-surface flow (Wigington et al. 2003). Riparian buffers that included switch grass and woody plants removed 97% of the sediments, 94% of the total nitrogen, 85% of the nitrate, 91% of the total phosphorus, and 80% of the phosphate in runoff (Lee et al. 2003).

Alternatively, existing wetlands could be used to filter agricultural runoff and can remove as much as 59% of the total phosphorus, 38% of the total nitrogen, and 41% of the total organic carbon (Jordan et al. 2003). Practices such as planting and maintaining woody vegetation or infrastructure development may not be adopted without financial capital, as there is little incentive for farmers lacking financial capital to make the large investment required to ensure their success, and this will be discussed in more general terms below.

Global Regulating Service Example: Mitigating Climate Change

Global climate change resulting from GHG emissions is another major upcoming challenge that will inevitably and disproportionately impact the poor (Srinivasan et al. 2008). Impoverished people reliant on agricultural production will be highly susceptible to fluctuating temperatures and rainfall, movement of disease and pests, and many other anticipated results of climate change (MA 2005b). Some of the agroecological management strategies described here have potential to help farmers adapt to climate impacts and to mitigate climate change. Agriculture directly contributed 13.5% of total global GHG emissions in 2004, and when the additional 17.4% of total emissions from deforestation (much of which is for farming and grazing) are considered, managing to reduce these emissions represent a major pathway for mitigation (Smith et al. 2008).

Agriculture-related emissions result from deforestation caused by extensification, use of pesticides and inorganic fertilizers, livestock production, fuel use for mechanized production including irrigation pumping, and transportation of goods. While these practices may be a sizeable proportion of the global emissions they also represent ways that emissions could be reduced if alternative practices are adopted. Although policy discussions are largely focused on carbon dioxide, other more powerful GHGs can also be mitigated through changes in agricultural practices. Agriculture accounts for about 60% of the nitrous oxide and 50% of the methane emitted globally by anthropogenic sources in 2005 (Smith et al. 2007). Much of the emissions from agriculture landscapes, like other nutrient losses, indicate inefficiency in management practices.

Emissions from agricultural landscapes can be categorized into primary, secondary, or tertiary emissions (Lal 2004b). Primary emissions are those due to mobile operations (e.g., tillage, harvesting) that are not largely used in small-holder farming reliant on manual labor for most operations. Secondary emissions result from stationary sources (e.g., crop drying, irrigation pumping) that may be more important for small-holder operations. Tertiary emissions are a result of the equipment manufacture, construction of farm buildings and acquisition of raw materials (Lal 2004b). The adoption of agroecological management strategies may help limit emissions at each of these stages (Fig. 3.5).

Agroforestry practices, i.e., woody perennial shrubs and trees in hedgerows significantly increase carbon storage across the landscape (Smith et al. 2008). The potential of agroforests to sequester carbon varies widely depending on soil type, climate, and rainfall. Values for biomass sequestration rates in agroforestry systems can range from 0.29 to 15.21 Mg C ha^{-1} $year^{-1}$ (Nair et al. 2009). Reforesting sites that are marginal for crop production could sequester between 0.8 and 18.8 Mg C ha^{-1} $year^{-1}$ depending on site conditions and management. (Richards and Stokes 2004) While both agroforestry and reforestation may substantially increase carbon storage, this may come at a cost to ecosystem services directly useful to the farmers (e.g., fuel wood, timber or food production). Alternatively, managing cropland soils for carbon may increase the mitigation potential and enhance crop production.

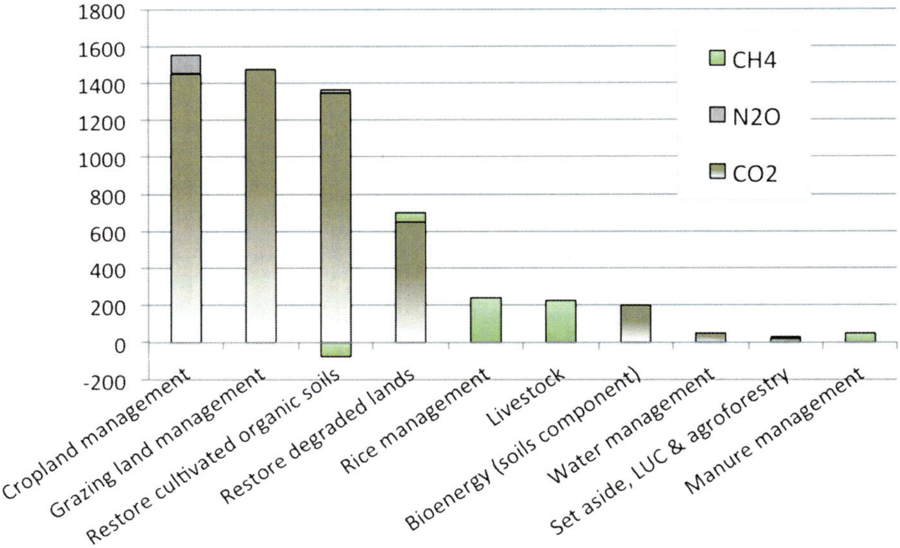

Fig. 3.5 The technical global mitigation potential of various agricultural management practices for nitrous oxide (N_2O), methane (CH_4), and carbon dioxide (CO_2) (Source Smith et al. 2007)

Reducing tillage and increasing residue inputs could sequester on average 0.2–0.7 Mg C ha^{-1} year^{-1} or improved management of grazing lands could sequester 0.1–0.8 Mg C ha^{-1} year^{-1} depending on climate zone (Smith et al. 2007). Managing soils for carbon sequestration, i.e., increasing SOM, may seem the most promising means of mitigating GHG emission in agricultural landscapes, but if additional benefits to farmers can materialize, then efforts to provide incentives should weigh the trade-offs in terms of labor and yield costs in the short-term.

Conservation of Biodiversity to Provide Ecosystem Services

Agrobiodiversity refers to the biological resources (genes, species, and habitats) that contribute to food, agriculture, and human well-being in agricultural landscapes (Qualset et al. 1995). It serves as a source of adaptation for crops and livestock, and in a larger context, for transformation to new production systems under unknown future environmental conditions. It also encompasses the biodiversity in natural ecosystem fragments and the organisms that move across the mosaic of ecosystems within agricultural landscapes (Jackson et al. 2007). For the poor, agrobiodiversity is one of the greatest sources of natural capital, especially in the form of traditional varieties of crops, medicinal plants, and useful wild species, e.g., that may be moved into home gardens (Jarvis et al. 2008). Evaluating the actual value of agrobiodiversity is difficult to assess, given the lack of knowledge on ecosystem functions of most of the biodiversity in agricultural landscapes, and the discrepancy between private vs. social benefits derived from it (Pascual and Perrings 2007).

The conservation of biodiversity can be managed for at multiple scales across the landscape and is thought to be critical for many other services that can benefit poor farmers, besides the provisioning services generated by crop genetic resources. In Daily's (1997) pioneering work on ecosystem services, she defines these services as "the conditions and processes through which natural ecosystems and the species that make them up, sustain and fulfill human life." There is increasing evidence that ecosystem services are regulated by ecological communities yet global trends indicate that biodiversity losses are resulting in losses in ecosystem functioning (Flynn et al. 2009; Jackson et al. 2009). There is a growing recognition of a strong relationship between the biodiversity of an agricultural system or landscape and the provisioning of services (Chan et al. 2006), and the willingness of the some consumers (especially in the developed world) to pay for biodiversity friendly products (Wendland et al. in press). A small, though increasing number of studies demonstrate that conserving both natural and semi-natural elements in agricultural landscapes can make significant contributions to biodiversity conservation. Harvey et al. (2006) found that the abundance of bird, beetle, tree, and butterfly diversity could be conserved by increasing tree cover and that even minimal number of trees on pastures can enhance biodiversity (Daily and Ehrlich 1995; Harvey and Haber 1999; Ricketts 2004; Schroth et al. 2004). Likewise, maintaining forest cover in agricultural landscapes, and minimizing the distance between forest fragments can make significant contributions to conserving wild biodiversity and their related services (Daily and Ehrlich 1995; Ricketts 2004). Management of these patches can also contribute to the provisioning of ecosystem services, particularly pollination and pest control (Schroth et al. 2004), and carbon storage (Smukler et al. 2010).

Petit and Petit (2003) demonstrated that although many species are conserved in managed landscapes, few of these are species of conservation concern. Flynn et al. (2009) used a metric of functional diversity, where species were classified not by their taxonomy or evolutionary relationships, but rather were classified by their contributions to ecosystem functions (insectivores, frugiviores, omnivores, canopy species, leaf gleaners, etc.) and found that functional diversity was lost more rapidly than species richness. This work suggests that the loss of biodiversity may have important consequences for the provisioning of ecosystem services.

There are several overarching ecologically based management strategies that can be used to guide development practitioners and land managers. These are outlined by McNeely and Scherr (2003) who provide six general recommendations: (1) create biodiversity reserves that also benefit local farming communities; (2) develop habitat connectivity in non-farmed areas; (3) reduce or reverse the conversion of natural ecosystems to agriculture by increasing farm productivity; (4) minimize agricultural pollution; (5) modify management of soil, water, and vegetation resources; and (6) modify farming systems to mimic natural ecosystems.

Tradeoffs and Synergies

The current situation of diminishing natural areas and the loss of biodiversity with its associated ecosystem services is coupled with increased demands for food production. This forces farmers to make difficult management choices that can have serious long-term consequences for economic prosperity of their farming operations, as well as broader impacts on ecological and human well-being (Jordan et al. 2007; Scherr and McNeely 2008). The decisions farmers make will inevitably incur trade-offs in the availability of ecosystem services. At the same time some decisions could actually have synergistic effects on multiple ecosystem services (Robertson and Swinton 2005; Swallow et al. 2009). Converting forests to agriculture is obviously a trade-off between forest goods and services for food (Fig. 3.6). In a landscape where trees and food are produced together, such as in multistrata agroforestry systems, important synergies can be achieved. Planting trees in agricultural landscapes may enhance water quality and quantity, provide habitat for pollinators and beneficial insects, store carbon, and improve long-term agricultural productivity.

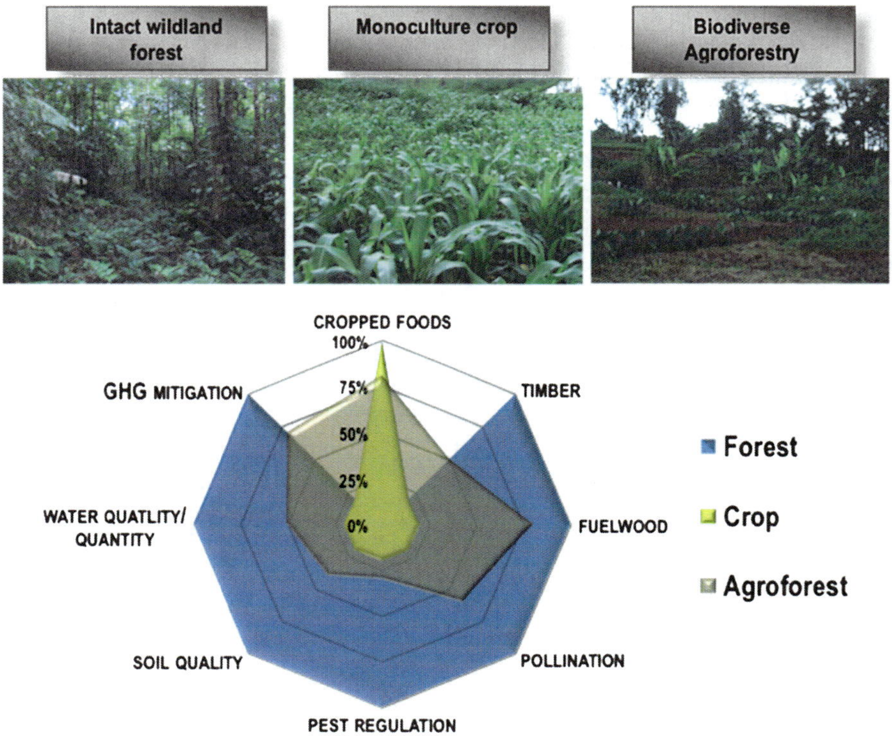

Fig. 3.6 A hypothesized comparison of the trade-offs in ecosystem services from a biodiverse intact wildland forest landscape, a monoculture farmed landscape, and a biodiverse farmed landscape (Modified from Foley 2005). The relative availability of ecosystem services could be compared as percentages of a reference landscape as indicated here

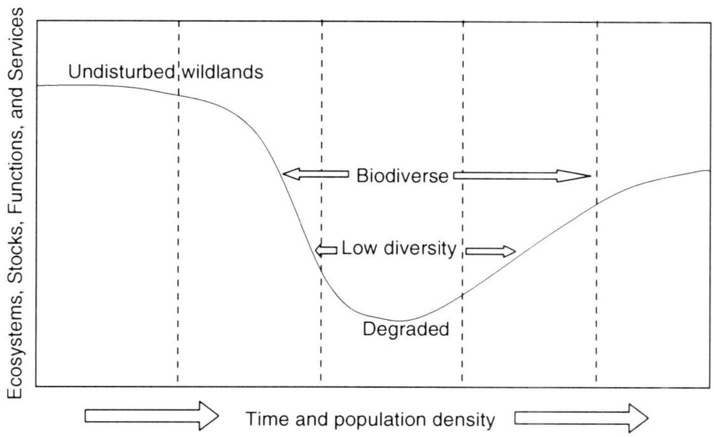

Fig. 3.7 A theoretical degradation and rehabilitation U-shaped trajectory curve. The possible relative degradation status of three categories of agricultural production, biodiverse, low diversity and degraded agricultural landscapes. Over time as agricultural production becomes more simplified (e.g. less biodiverse) due to intensification, ecological theory suggests that ecosystem stocks, functions and services are likely to decline. As biodiversity is restored to these systems, the stocks, functions and services may increase but never quite to the original state

Similarly planting trees may incur significant trade-offs in terms of a large short-term cash and labor investment to purchase seeds or seedlings and to ensure survival, or in some cases depending on the species, may actually reduce water availability (Farley et al. 2005).

Tradeoffs and synergies also occur among the communities that actually utilize the services. For example, people living at the top of a watershed may receive the benefits of food provisioning services from agricultural production while the people at the bottom of the watershed suffer from the loss of clean water that would have been provided by an intact forested watershed (Falkenmark and Folke 2002). This type of geographical disconnect between management practices and the availability of ecosystem services has lead to large-scale reduction in these services. Unless these two groups of stakeholders can interact to negotiate management strategies that have mutually beneficial outcomes, one group will continue to benefit from certain services and the trade-off will be loss of services to the other (Lant et al. 2008) Fig. 3.7.

These examples illustrate how evaluating the trade-offs and synergies for a particular management decision is extremely difficult given that many benefits may be felt in geographically or temporally distant places. Maximizing the benefits of ecosystem services requires an effective ecological assessment of the potential trade-offs and synergies of various agricultural land management options at a variety of scales that incorporate all those who could be potentially impacted. The need for an effective ecological assessment is clear but the methodologies for doing so are still in the development stage. To accurately assess the potential trade-offs and synergies requires simultaneous quantification of multiple ecosystem services over long periods of time. Given the difficulties of such research, most studies of ecosystem services are confined to 2 or 3 years and one or two services. To date there are only few examples of projects that have assessed multiple services required to observe trade-offs and synergies let alone generate accurate quantification at a meaningful scale (Gamfeldt et al. 2008; Tallis et al. 2008, Smukler et al. 2010). Currently analysis must rely on complex biophysical process models that still need development and validation for agroecosystems (Tallis et al. 2008). A number of other models and planning tools are being employed to provide financial incentives directly to land managers for the provisioning of ecosystem services in developed countries (Eigenraam et al. 2006), but models directly relevant to farmers in developing countries seem a long way off. Furthermore, in order for assessments to actually influence management decisions, rural communities must be involved through a participatory or economic approach, evaluating and prioritizing their ecosystem service needs. The infrastructure required for such an approach in impoverished regions may be prohibitively expensive and require institutional intervention not currently in place, but the financial promise of PES may help initiate some form of such analysis in these regions.

Conclusions

Increased agricultural productivity through "Green Revolution" technologies, including inorganic fertilizers, improved seed and small-scale irrigation projects will contribute towards meeting the United Nation's goal of halving global poverty by 2015 (Sachs 2005; Denning et al. 2009; Sanchez et al. 2009) but sustained poverty alleviation will require a diversity of interventions. We must manage agricultural landscapes for multiple ecosystem services, or else our efforts to alleviate rural poverty may in fact just transpose dire economic conditions to other regions (e.g., downstream), or even exacerbate poverty in coming generations.

Examining solutions to poverty through the lens of ecology means seeing the situation holistically both spatially and temporally. Ecological research illustrates a number of principles that can be applied to management of agricultural landscapes to ensure the availability of multiple ecosystem services. Management practices that increase soil organic matter, or maximize biodiversity by filling niches with a variety of plants that cycle nutrients and provide habitat for beneficial organisms, are examples of ways to increase natural capital across the landscape that have been effectively demonstrated. Given that many of these practices do not directly increase agricultural productivity or income immediately, novel approaches will be required to convince farmers to adopt them.

The plight of the rural poor cannot be isolated from either regional or global markets, and poverty has to be viewed as inextricably tied to regional and global ecosystem processes. Human well-being is reliant on a complex interconnection between the management of individual farms and the sustained availability of ecosystem services on a much broader scale. Those that recognize this relationship have a responsibility to help forge policy, markets, education, and community outreach required to plan for trade-offs and synergies that will empower impoverished rural farmers to maximize the availability of ecosystem services that will have long-term benefits for society at large.

References

Aizen, M. A., L. A. Garibaldi, S. A. Cunningham, and A. M. Klein. 2008. Long-Term global trends in crop yield and production reveal no current pollination shortage but increasing pollinator dependency. Current Biology **18**:1572–1575.

Aizen, M. A. and L. D. Harder. 2009. The global stock of domesticated honey bees is growing slower than agricultural demand for pollination. Current Biology **19**:915–918.

Alavalapati, J. R. R., R. K. Shrestha, G. A. Stainback, and J. R. Matta. 2004. Agroforestry development: An environmental economic perspective. Agroforestry Systems **61**:299–310.

Altieri, M. A. 1999. The ecological role of biodiversity in agroecosystems. Agriculture Ecosystems & Environment **74**:19–31.

Altieri, M. A. 2002. Agroecology: the science of natural resource management for poor farmers in marginal environments. Agriculture Ecosystems & Environment **93**:1–24.

Ananda, J. and G. Herath. 2003. Soil erosion in developing countries: a socio-economic appraisal. Journal of Environmental Management **68**:343–353.

Andow, D. A. 1991. Vegetational diversity and arthropod population response. Annual Review of Entomology **36**:561–586.

Atreya, K. 2008. Health costs from short-term exposure to pesticides in Nepal. Social Science & Medicine **67**:511–519.

Balmford, A., R. E. Green, and J. P. W. Scharlemann. 2005. Sparing land for nature: exploring the potential impact of changes in agricultural yield on the area needed for crop production. Global Change Biology **11**:1594–1605.

Bawa, K. S. 1990. Plant-pollinator interactions in tropical rain-forests. Annual Review of Ecology and Systematics **21**:399–422.

Bebbington, A. 1999. Capitals and Capabilities: A Framework for Analyzing Peasant Viability, Rural Livelihoods and Poverty. World Development **27**:2021–2044.

Bedard-Haughn, A., K. W. Tate, and C. van Kessel. 2004. Using nitrogen-15 to quantify vegetative buffer effectiveness for sequestering nitrogen in runoff. Journal of Environmental Quality **33**:2252–2262.

Bennett, E. M. and P. Balvanera. 2007. The future of production systems in a globalized world. Frontiers in Ecology and the Environment **5**:191–198.

Bianchi, F. J. J. A., C. J. H. Booij, and T. Tscharntke. 2006. Sustainable pest regulation in agricultural landscapes: a review on landscape composition, biodiversity and natural pest control. Proceedings of the Royal Society B **273**:1715–1727.

Blanco-Canqui, H., C. J. Gantzer, S. H. Anderson, E. E. Alberts, and A. L. Thompson. 2004. Grass barrier and vegetative filter strip effectiveness in reducing runoff, sediment, nitrogen, and phosphorus loss. Soil Sci Soc Am J **68**:1670–1678.

Blüthgen, N. and A. M. Klein. 2011. Functional complementarity and specialization: the role of biodiversity in plant-pollinator interactions. Basic and Applied Ecology **12**(4):282–291.

Bokenstrand, A., J. Lagerlof, and P. R. Torstensson. 2004. Establishment of vegetation in broadened field boundaries in agricultural landscapes. Agriculture, Ecosystems & Environment **101**:21–29.

Brittain, C. A., M. Vighi, R. Bommarco, J. Settele, and S. G. Potts. 2010. Impacts of a pesticide on pollinator species richness at different spatial scales. Basic and Applied Ecology **11**:106–115.

Cannell, R. Q. and J. D. Hawes. 1994. Trends in tillage practices in relation to sustainable crop production with special reference to termperate climates Soil & Tillage Research **30**:245–282.

Cardinale, B., C. Harvey, K. Gross, and A. Ives. 2003. Biodiversity and biocontrol: emergent impacts of a multi-enemy assemblage on pest suppression and crop yield in an agroecosystem. Ecology Letters **6**:857–865.
Cassman, K. G., A. Dobermann, D. T. Walters, and H. Yang. 2003. Meeting cereal demand while protecting natural resources and improving environmental quality. Annual Review of Environment and Resources **28**:315–358.
Chan, K. M. A., M. R. Shaw, D. R. Cameron, E. C. Underwood, and G. C. Daily. 2006. Conservation planning for ecosystem services. PLoS Biology **4**:2138–2152.
Clough, Y., S. Abrahamczyk, A. Adams M-O, A , N. Ariyanti, L. Betz, D. Buchori, D. Cicuzza, K. Darras, D. Putra, B. Fiala, S. Gradstein, M. Kessler, A.-M. Klein, R. Pitopang, B. Sahari, C. Scherber, C. Schulze, Shahabuddin, S. Sporn, K. Stenchly, S. Tjitrosoedirdjo, T. Wanger, M. Weist, A. Wielgoss, and T. Tscharntke. 2010. Biodiversity patterns and trophic interactions in human-dominated tropical landscapes in Sulawesi (Indonesia): plants, arthropods and vertebrates. Pages 15–71 *in* T. Tscharntke, C. Leuschner, E. Veldkamp, H. Faust, E. Guhardja, and A. Bidin, editors. Tropical Rainforests and Agroforests under Global Change. Ecological and Socio-economic Valuations. Springer Verlag, Berlin.
Córdova, L. T., E. Somarriba, and C. A. Harvey. 2003. Plantas útiles en las fincas cacaoteras de indígenas Bribri y Cabécar de Talamanca, Costa Rica. Agroforesteria en las Américas **10**:36–41.
Daily, G. C. 1997. Nature's services: societal dependence on natural ecosystems. Island Pr.
Daily, G. C. and P. R. Ehrlich. 1995. Preservation of biodiversity in small rain-forest patches rapid evaluations using butterfly trapping. Biodiversity and Conservation **4**:35–55.
Daily, G. C. and P. A. Matson. 2008. Ecosystem services: From theory to implementation. Proceedings of The National Academy of Sciences of The United States of America **105**:9455–9456.
Dasgupta, P. 2007. Nature and the economy. Journal of Applied Ecology:475–487.
Dasgupta, S., C. Meisner, D. Wheeler, and Y. H. Jin. 2002. Agricultural trade, development and toxic risk. World Development **30**:1401–1412.
DeFries, R., A. Hansen, B. L. Turner, R. Reid, and J. G. Liu. 2007. Land use change around protected areas: Management to balance human needs and ecological function. Ecological Applications **17**:1031–1038.
Denning, G., P. Kabambe, P. Sanchez, A. Malik, R. Flor, R. Harawa, P. Nkhoma, C. Zamba, C. Banda, C. Magombo, M. Keating, J. Wangila, and J. Sachs. 2009. Input Subsidies to improve smallholder maize productivity in Malawi: Toward an African Green Revolution. PLoS Biology **7**:2–10.
Drinkwater, L. E. and S. S. Snapp. 2007. Nutrients in agroecosystems: Rethinking the management paradigm. Advances in Agronomy **92**:163–186.
Eigenraam, M., L. Strappazzon, N. Lansdell, A. Ha, C. Beverly, and J. Todd. 2006. EcoTender: Auction for multiple environmental outcomes, Project Final report for National Action Plan for Salinity and Water Quality National Market Based Instruments Pilot Program. Department of Primary Industries, Victoria, Australia.
Eilers, E. J. and A. M. Klein. 2009. Landscape context and management effects on an important insect pest and its natural enemies in almond. Biological Control **51**:388–394.
Evenson, R. E. and D. Gollin. 2003. Assessing the impact of the Green Revolution, 1960 to 2000. Science **300**:758–762.
Falkenmark, M. and C. Folke. 2002. The ethics of socio-ecohydrological catchment management: towards hydrosolidarity. Hydrology and Earth System Sciences **6**:1–9.
Farley, K. A., E. G. Jobbagy, and R. B. Jackson. 2005. Effects of afforestation on water yield: a global synthesis with implications for policy. Global Change Biology **11**:1565–1576.
Fewtrell, L. 2004. Drinking-water nitrate, methemoglobinemia, and global burden of disease: A discussion. Environmental Health Perspectives **112**:1371–1374.
Finke, D. L. and R. F. Denno. 2005. Predator diversity and the functioning of ecosystems: the role of intraguild predation in dampening trophic cascades. Ecology Letters **8**:1299–1306.
Flynn, D. F. B., M. Gogol-Prokurat, T. Nogeire, N. Molinari, B. T. Richers, B. B. Lin, N. Simpson, M. M. Mayfield, and F. DeClerck. 2009. Loss of functional diversity under land use intensification across multiple taxa. Ecology Letters **12**:22–33.

Foley, J. A., R. DeFries, G. P. Asner, C. Barford, G. Bonan, S. R. Carpenter, F. S. Chapin, M. T. Coe, G. C. Daily, H. K. Gibbs, J. H. Helkowski, T. Holloway, E. A. Howard, C. J. Kucharik, C. Monfreda, J. A. Patz, I. C. Prentice, N. Ramankutty, and P. K. Snyder. 2005. Global consequences of land use. Science **309**:570–574.

Gallai, N., J.-M. Salles, J. Settele, and B. E. VaissiËre. 2009. Economic valuation of the vulnerability of world agriculture confronted with pollinator decline. Ecological Economics **68**:810–821.

Galloway, J. N., F. J. Dentener, D. G. Capone, E. W. Boyer, R. W. Howarth, S. P. Seitzinger, G. P. Asner, C. C. Cleveland, P. A. Green, E. A. Holland, D. M. Karl, A. F. Michaels, J. H. Porter, A. R. Townsend, and C. J. Vorosmarty. 2004. Nitrogen cycles: past, present, and future. Biogeochemistry **70**:153–226.

Gamfeldt, L., H. Hillebrand, and P. R. Jonsson. 2008. Multiple functions increase the importance of biodiversity for overall ecosystem functioning. Ecology **89**:1223–1231.

Giller, K. E. 2002. Targeting management of organic resources and mineral fertilizers: Can we match scientists' fantasies with farmers' realities? Pages 155–171 *in* B. Vanlauwe, J. Diels, N. Sanginga, and R. Merckx, editors. Integrated plant nutrient management in sub-Saharan Africa: From concept to practice. CABI, Wallingford.

Gomiero, T., M. G. Paoletti, and D. Pimentel. 2008. Energy and environmental issues in organic and conventional agriculture. Critical Reviews in Plant Sciences **27**:239–254.

Graves, A., R. Matthews, and K. Waldie. 2004. Low external input technologies for livelihood improvement in subsistence agriculture. Pages 473–555 Advances in Agronomy, Vol 82.

Green, R. E., S. J. Cornell, J. P. W. Scharlemann, and A. Balmford. 2005. Farming and the fate of wild nature. Science **307**:550–555.

Greenleaf, S. S. and C. Kremen. 2006. Wild bees enhance honey bees' pollination of hybrid sunflower. Proceedings of The National Academy of Sciences of The United States of America **103**:13890–13895.

Guo and Gifford. 2002. Soil carbon stocks and land use change: a meta analysis. Global Change Biology **8**(4):345–360.

Harvey, C. A. and W. A. Haber. 1999. Remnant trees and the conservation of biodiversity in Costa Rican pastures. Agroforestry Systems **44**:37–68.

Harvey, C. A., A. Medina, D. M. Sanchez, S. Vilchez, B. Hernandez, J. C. Saenz, J. M. Maes, F. Casanoves, and F. L. Sinclair. 2006. Patterns of animal diversity in different forms of tree cover in agricultural landscapes. Ecological Applications **16**:1986–1999.

Harvey, C. A. and J. A. G. Villalobos. 2007. Agroforestry systems conserve species-rich but modified assemblages of tropical birds and bats. Biodiversity and Conservation **16**:2257–2292.

Hayes, T. B., A. Collins, M. Lee, M. Mendoza, N. Noriega, A. A. Stuart, and A. Vonk. 2002. Hermaphroditic, demasculinized frogs after exposure to the herbicide atrazine at low ecologically relevant doses. Proceedings of the National Academy of Sciences of the United States of America **99**:5476–5480.

Hoehn, P., T. Tscharntke, J. M. Tylianakis, and I. Steffan-Dewenter. 2008. Functional group diversity of bee pollinators increases crop yield. Proceedings of the Royal Society B-Biological Sciences **275**:2283–2291.

Holzschuh, A., I. Steffan-Dewenter, D. Kleijn, and T. Tscharntke. 2007. Diversity of flower-visiting bees in cereal fields: effects of farming system, landscape composition and regional context. Journal of Applied Ecology **44**:41–49.

Huang, W. Y., W. McBride, and U. Vasavada. 2009. Recent volatility in U.S. fertilizer prices. Amber Waves : The Economics of Food, Farming, Natural Resources, and Rural America **March**.

International Assessment of Agricultural Science and Technology for Development (IAASTD). 2008. Island Press, Washington, DC.

Jackson, L., T. Rosenstock, M. Thomas, J. Wright, and A. Symstad. 2009. Managed ecosystems: biodiversity and ecosystem functions in landscapes modified by human use.*in* S. Naeem, D. E. Bunker, and A. Hector, editors. Biodiversity, Ecosystem Functioning, and Human Wellbeing: An Ecological and Economic Perspective. Oxford Univ Pr.

Jackson, L. E., U. Pascual, and T. Hodgkin. 2007. Utilizing and conserving agrobiodiversity in agricultural landscapes. Agriculture Ecosystems & Environment **121**:196–210.

Jackson, L. E., L. J. Wyland, and L. J. Stivers. 1993. Winter cover crops to minimize nitrate losses in intensive lettuce production. Journal of Agricultural Science **121**:55–62.

Jarrard, H. E., K. R. Delaney, and C. J. Kennedy. 2004. Impacts of carbamate pesticides on olfactory neurophysiology and cholinesterase activity in coho salmon (Oncorhynchus kisutch). Aquatic Toxicology **69**:133–148.

Jarvis, D. I., A. H. D. Brown, P. H. Cuong, L. Collado-Panduro, L. Latournerie-Moreno, S. Gyawali, T. Tanto, M. Sawadogo, I. Mar, M. Sadiki, N. T. N. Hue, L. Arias-Reyes, D. Balma, J. Bajracharya, F. Castillo, D. Rijal, L. Belqadi, R. Ranag, S. Saidi, J. Ouedraogo, R. Zangre, K. Rhrib, J. L. Chavez, D. Schoen, B. Sthapit, P. De Santis, C. Fadda, and T. Hodgkin. 2008. A global perspective of the richness and evenness of traditional crop-variety diversity maintained by farming communities. Proceedings of The National Academy of Sciences of The United States of America **105**:5326–5331.

Jordan, N., G. Boody, W. Broussard, J. D. Glover, D. Keeney, B. H. McCown, G. McIsaac, M. Muller, H. Murray, J. Neal, C. Pansing, R. E. Turner, K. Warner, and D. Wyse. 2007. Environment - Sustainable development of the agricultural bio-economy. Science **316**:1570–1571.

Jordan, T. E., D. F. Whigham, K. H. Hofmockel, and M. A. Pittek. 2003. Nutrient and sediment removal by a restored wetland receiving agricultural runoff. Journal of Environmental Quality **32**:1534–1547.

Kiers, E. T., R. R. B. Leakey, A. M. Izac, J. A. Heinemann, E. Rosenthal, D. Nathan, and J. Jiggins. 2008. Ecology - Agriculture at a crossroads. Science **320**:320–321.

Kimani, S. K. and J. K. Lekasi. 2004. Managing manures throughout their production cycle enhances their usefulness as fertilisers: A review. Pages 187–197 *in* A. Bationo, editor. Managing nutrient cyles to sustain soil fertility in sub-Saharan Africa. AfNet-CIAT, Nairobi.

Klein, A. M. 2009. Nearby rainforest promotes coffee pollination by increasing spatio-temporal stability in bee species richness. Forest Ecology and Management **258**:1838–1845.

Klein, A. M., S. A. Cunningham, M. Bos, and I. Steffan-Dewenter. 2008. Advances in pollination ecology from tropical plantation crops. Ecology **89**:935–943.

Klein, A. M., C. M. Mueller, P. Hoehn, and C. O. U. P. Kremen, pp. 195–208. 2009. Understanding the role of species richness for pollination services.*in* S. Naeem, D. E. Bunker, A. Hector, M. Loreau, and C. Perrings, editors. The consequences of changing biodiversity – solutions and scenarios.

Klein, A. M., I. Steffan-Dewenter, D. Buchori, and T. Tscharntke. 2002. Effects of land-use intensity in tropical agroforestry systems on coffee flower-visiting and trap-nesting bees and wasps. Conservation Biology **16**:1003–1014.

Klein, A. M., B. E. Vaissiere, J. H. Cane, I. Steffan-Dewenter, S. A. Cunningham, C. Kremen, and T. Tscharntke. 2007. Importance of pollinators in changing landscapes for world crops. Proceedings of the Royal Society B-Biological Sciences **274**:303–313.

Kohli, R. K., H. P. Singh, D. R. Batish, S. Jose. 2008. Ecological Interactions in Agroforestry: An Overview. Page 382 *in* D. R. Batish, R. K. Kohli, S. Jose, H. P. Singh. Ecological Basis of Agroforestry. CRC Press, Boca Raton.

Kremen, C., N. M. Williams, and R. W. Thorp. 2002. Crop pollination from native bees at risk from agricultural intensification. Proceedings of The National Academy of Sciences of The United States of America **99**:16812–16816.

Lal, R. 2004a. Agricultural activities and the global carbon cycle. Nutrient Cycling in Agroecosystems **70**:103–116.

Lal, R. 2004b. Carbon emission from farm operations. Environment International **30**:981–990.

Lal, R. 2006. Enhancing crop yields in the developing countries through restoration of the soil organic carbon pool in agricultural lands. Pages 197–209.

Lant, C. L., J. B. Ruhl, and S. E. Kraft. 2008. The tragedy of ecosystem services. Bioscience **58**:969–974.

Lee, K. H., T. M. Isenhart, and R. C. Schultz. 2003. Sediment and nutrient removal in an established multi-species riparian buffer. Journal of Soil and Water Conservation **58**:1–8.

Luo and Zhang. 2009. Multimedia transport and risk assessment of organophosphate pesticides and a case study in the northern San Joaquin Valley of California. Chemosphere **75**(7):969–978.

Maass, J. M., C. F. Jordan, and J. Sarukhan. 1988. Soil-Erosion and nutrient losses in seasonal tropical agroecosystems under various management-techniques. Journal of Applied Ecology **25**:595–607.

Matson, P. A., W. J. Parton, A. G. Power, and M. J. Swift. 1997. Agricultural Intensification and Ecosystem Properties. Science **277**:504–509.

Maumbe, B. M. and S. M. Swinton. 2003. Hidden health costs of pesticide use in Zimbabwe's smallholder cotton growers. Social Science & Medicine **57**:1559–1571.

McDonagh, J. F., T. B. Thomsen, and J. Magid. 2001. Soil organic matter decline and compositional change associated with cereal cropping in southern Tanzania. Land Degradation & Development **12**:13–26.

McNeely, J. A. and S. J. Scherr. 2003. Ecoagriculture: strategies for feeding the world and conserving wild biodiversity. Island Press, Washington, DC.

Millennium Ecosystem Assessment. 2005a. Ecosystems and Human Well-Being: Global Assessment Reports Island Press, Washington, DC.

Millennium Ecosystem Assessment. 2005b. Ecosystems and Human Well-being: Synthesis. Island Press, Washington, DC.

Montagnini, F. and P. K. R. Nair. 2004. Carbon sequestration: An underexploited environmental benefit of agroforestry systems. Agroforestry Systems **61**:281–295.

Montgomery, D. R. 2007. Soil erosion and agricultural sustainability. Proceedings of The National Academy of Sciences of The United States of America **104**:13268–13272.

Nair, P. K. R., B. M. Kumar, and V. D. Nair. 2009. Agroforestry as a strategy for carbon sequestration. Journal of Plant Nutrition and Soil Science-Zeitschrift Fur Pflanzenernahrung Und Bodenkunde **172**:10–23.

Nieder, R. and D. K. Benbi. 2008. Carbon and nitrogen in the terrestial environment. Springer.

Okalebo, J. R., C. Palm, J. K. Lekasi, C. O. Othieno, M. Waigwa, and K. W. Ndungu. 2004. Use of organic and inorganic resources to increase maize yields in some Kenyan infertile soils: A five-year experience. Pages 359–372 *in* A. Bationo, editor. Managing nutrient cycles to sustain soil fertility in sub-Saharan Africa. AfNet-CIAT, Nairobi.

Östman, Ö., B. Ekborn, and Bengtsson. 2001. Landscape heterogeneity and farming practice influence biological control. Basic and Applied Ecology **2**:365–371.

Palm, C., P. Sanchez, S. Ahamed, and A. Awiti. 2007. Soils: A contemporary perspective. Annual Review of Environmental Resources **32**:99–129.

Parry, M. L., C. Rosenzweig, A. Iglesias, M. Livermore, and G. Fischer. 2004. Effects of climate change on global food production under SRES emissions and socio-economic scenarios. Global Environmental Change-Human and Policy Dimensions **14**:53–67.

Pascual, U. and C. Perrings. 2007. Developing incentives and economic mechanisms for in situ biodiversity conservation in agricultural landscapes. Agriculture Ecosystems & Environment **121**:256–268.

Perfecto, I. and J. Vandermeer. 2002. Quality of agroecological matrix in a tropical montane landscape: Ants in coffee plantations in southern Mexico. Conservation Biology **16**:174–182.

Perfecto, I. and J. Vandermeer. 2008. Biodiversity conservation in tropical agroecosystems - A new conservation paradigm. Pages 173–200 Year in Ecology and Conservation Biology 2008.

Perfecto, I., J. H. Vandermeer, G. L. Bautista, G. I. Nunez, R. Greenberg, P. Bichier, and S. Langridge. 2004. Greater predation in shaded coffee farms: The role of resident neotropical birds. Ecology **85**:2677–2681.

Perrings, C., L. Jackson, K. Bawa, L. Brussaard, S. Brush, T. Gavin, R. Papa, U. Pascual, and P. De Ruiter. 2006. Biodiversity in agricultural landscapes: Saving natural capital without losing interest. Conservation Biology **20**:263–264.

Petit, L. J. and D. R. Petit. 2003. Evaluating the importance of human-modified lands for neotropical bird conservation. Conservation Biology **17**:687–694.

Pimentel, D., P. Hepperly, J. Hanson, D. Douds, and R. Seidel. 2005. Environmental, energetic, and economic comparisons of organic and conventional farming systems. Bioscience **55**:573–582.

Polidoro and Bosque-Perez. 2007. Incorporating livelihoods in biodiversity conservation: a case study of cacao agroforestry systems in Talamanca, Costa Rica. Biodiversity and Conservation **16**:2311–2333.

Postel, S. L., G. C. Daily, and P. R. Ehrlich. 1996. Human appropriation of renewable fresh water. Science **271**:785–788.

Potts, S. G., J. C. Biesmeijer, C. Kremen, P. Neumann, O. Schweiger, and W. E. Kunin. Global pollinator declines: trends, impacts and drivers. Trends in Ecology & Evolution **In Press, Corrected Proof**.

Potts, S. G., T. Petanidou, S. Roberts, C. O'Toole, A. Hulbert, and P. Willmer. 2006. Plant-pollinator biodiversity and pollination services in a complex Mediterranean landscape. Biological Conservation **129**:519–529.

Prepas, E. E. and T. Charette. 2007. Chapter 9.08 Worldwide Eutrophication of Water Bodies: Causes, Concerns, Controls. Pages 311–331 *in* B. S. Lollar, D. H. Heinrich, and K. T. Karl, editors. Treatise on Geochemistry. Elsevier, Oxford.

Qualset, C. O., P. E. McGuire, and M. L. Warburton. 1995. Agrobiodiversity key to agricultural productivity. California Agriculture:45–49.

Rao, M. R., M. P. Singh, and R. Day. 2000. Insect pest problems in tropical agroforestry systems: Contributory factors and strategies for management. Agroforestry Systems **50**:243–277.

Richards, K. R. and C. Stokes. 2004. A review of forest carbon sequestration cost studies: A dozen years of research. Climatic Change **63**:1–48.

Ricketts, T. H. 2004. Tropical forest fragments enhance pollinator activity in nearby coffee crops. Conservation Biology **18**:1262–1271.

Ricketts, T. H., J. Regetz, I. Steffan-Dewenter, S. A. Cunningham, C. Kremen, A. Bogdanski, B. Gemmill-Herren, S. S. Greenleaf, A. M. Klein, M. M. Mayfield, L. A. Morandin, A. Ochieng, and B. F. Viana. 2008. Landscape effects on crop pollination services: are there general patterns? Ecology Letters **11**:499–515.

Robertson, G. P. and S. M. Swinton. 2005. Reconciling agricultural productivity and environmental integrity: a grand challenge for agriculture. Frontiers in Ecology and the Environment **3**:38–46.

Rockström, J., C. Folke, L. Gordon, N. Hatibu, G. Jewitt, F. Penning de Vries, F. Rwehumbiza, H. Sally, H. Savenije, and R. Schulze. 2004. A watershed approach to upgrade rainfed agriculture in water scarce regions through Water System Innovations: an integrated research initiative on water for food and rural livelihoods in balance with ecosystem functions. Physics and Chemistry of the Earth, Parts A/B/C **29**:1109–1118.

Rola et al. 1993. Pesticide use, rice productivity, and human health impacts in the Philippines. Agricultural Policy and Sustainability: Cases Studies from India, Chile, Philippines, and the United States pp. 47–62.

Roose, E. and F. Ndayizigiye. 1997. Agroforestry, water and soil fertility management to fight erosion in tropical mountains of Rwanda. Soil Technology **11**:109–119.

Root, R. B. 1973. Organization of a plant-arthropod association in simple and diverse habitats: The Fauna of collards (*Brassica Oleracea*). Ecology **43**:95–124.

Rosenzweig, C., K. M. Strzepek, D. C. Major, A. Iglesias, D. N. Yates, A. McCluskey, and D. Hillel. 2004. Water resources for agriculture in a changing climate: international case studies. Global Environmental Change-Human and Policy Dimensions **14**:345–360.

Sachs, J. 2005. The End of Poverty. Penguin, London.

Sachs, J. 2008. Common Wealth Economics for a Crowded Planet. Penguin Group, New York, NY.

Sanchez, P. 1999. Delivering on the promise of agroforestry. Environment,Development and Sustainability **1**:275–284.

Sanchez, P., G. Denning, and G. Nziguheba. 2009. The African Green Revolution moves forward. Food Security.

Scherr, S. J. and J. A. McNeely. 2008. Biodiversity conservation and agricultural sustainability: towards a new paradigm of 'ecoagriculture' landscapes. Philosophical Transactions of the Royal Society B-Biological Sciences **363**:477–494.

Schmitz, O. J., E. Post, C. E. Burns, and K. M. Johnston. 2003. Ecosystem responses to global climate change: Moving beyond color mapping. Bioscience **53**:1199–1205.

Schroth, G., G. A. B. da Fonseca, C. Harvey, C. Gascon, H. Vasconcelos, and A. N. Izac. 2004. Introduction: The role of agroforestry in biodiversity conservation in tropical landscape.*in* G. Schroth, G. A. B. da Fonseca, C. Harvey, C. Gascon, H. Vasconcelos, and A. N. Izac, editors. Agroforestry and Biodiversity Conservation in Tropical Landscapes. Island Press, Washington.

Seck, M., M. N. Abou Mamouda, and S. Wade. 2005. Case study 4: Senegal - Adaptation and mitigation through "produced environments": The case for agriculture intensification in Senegal. Ids Bulletin-Institute of Development Studies **36**:71-+.

Semwal, R. L., S. Nautiyal, K. K. Sen, U. Rana, R. K. Maikhuri, K. S. Rao, and K. G. Saxena. 2004. Patterns and ecological implications of agricultural land-use changes: a case study from central Himalaya, India. Agriculture, Ecosystems & Environment **102**:81–92.

Shuler, R. E., T. H. Roulston, and G. E. Farris. 2005. Farming practices influence wild pollinator populations on squash and pumpkin. Journal of Economic Entomology **98**:790–795.

Sileshi, G., F. K. Akinnifesi, O. C. Ajayi, and F. Place. 2008. Meta-analysis of maize yield response to woody and herbaceous legumes in sub-Saharan Africa. Plant and Soil **307**:1–19.

Smaling, E. M. A. 1993. Soil nutrient depletion in sub-Saharan Africa. Pages 53–67 *in* van Reuler and W. Prins, editors. The Role of Plant Nutrients for Sustainable Food Crop Production in sub-Saharan Africa. Dutch Association of Fertilizer Producers, Leidschendam, Netherlands.

Smaling, E. M. A., J. J. Stoorvogel, and P. N. Windmeijer. 1993. Calculating Soil Nutrient Balances in Africa at Different Scales .2. District Scale. Fertilizer Research **35**:237–250.

Smith, P., D. Martino, Z. Cai, D. Gwary, H. Janzen, P. Kumar, B. McCarl, S. Ogle, F. O'Mara, C. Rice, B. Scholes, O. Sirotenko, M. Howden, T. McAllister, G. Pan, V. Romanenkov, U. Schneider, S. Towprayoon, M. Wattenbach, and J. Smith. 2008. Greenhouse gas mitigation in agriculture. Philosophical Transactions of the Royal Society B-Biological Sciences **363**:789–813.

Smith, P., D. Martino, Z. Cai, D. Gwary, H. Janzen, P. Kumar, B. McCarl, S. Ogle, F. O'Mara, C. Rice, B. Scholes, and O. Sirotenko, editors. 2007. Agriculture. In Climate Change 2007: Mitigation. Cambridge University Press Cambridge, United Kingdom and New York, NY, USA.

Smukler, S., L. Jackson, S. Sánchez Moreno, S. Fonte, H. Ferris, K. Klonsky, A. O'Geen, K. Scow, and K. Steenwerth. 2010. Biodiversity and multiple ecosystem functions in an organic farmscape. Agriculture Ecosystems and Environment.

Snyder, W. E., G. B. Snyder, D. L. Finke, and C. S. Straub. 2006. Predator biodiversity strengthens herbivore suppression. Ecology Letters **9**:789–796.

Sogbedji, J. M., H. M. van Es, and A. K.L. 2006. Cover cropping and nutrient management strategies for maize production in western Africa. Agronomy Journal **98**:883–889.

Somarriba, E. 1992. Revisting the past: an essay on agroforestry definition. Agroforestry Systems **19**:233–240.

Somarriba, E. and C. Harvey. 2003. ¿Cómo integrar producción sostenible y conservación de biodiversidad en cacaotales orgánicos indígenas? Agroforesteria en las Américas **10**:12–17.

Somarriba, E., M. Trivelato, M. Villalobos, A. Súarez, P. Benavides, K. Moran, L. Orozco, and A. López. 2003. Diagnóstico agroforestal de pequeñas fincas cacaoteras orgánicas de indígenas Bribri y Cabécar de Talamanca, Costa Rica. Agroforesteria en las Américas **10**:24–30.

Srinivasan, U. T., S. P. Carey, E. Hallstein, P. A. T. Higgins, A. C. Kerr, L. E. Koteen, A. B. Smith, R. Watson, J. Harte, and R. B. Norgaard. 2008. The debt of nations and the distribution of ecological impacts from human activities. Proceedings of The National Academy of Sciences of The United States of America **105**:1768–1773.

Stevenson, F. J. and M. A. Cole. 1999. Cycles of soil: carbon, nitrogen, phosphorus, sulfur, micronutrients. John Wiley & Sons Inc.

Stoorvogel, J. J., E. M. A. Smaling, and B. H. Janssen. 1993. Calculating soil nutrient balances in Africa at different scales 1. Supra-National scale. Fertilizer Research **35**:227–235.

Suatunce, P., E. Somarriba, C. Harvey, and B. Finegan. 2003. Composición florística y estructura de bosques and cacaotales en los Territorios Indígenas de Talamanca, Costa Rica. Agroforesteria en las Américas **10**:31–35.

Swallow, B. M., J. K. Sang, M. Nyabenge, D. K. Bundotich, A. K. Duraiappah, and T. B. Yatich. 2009. Tradeoffs, synergies and traps among ecosystem services in the Lake Victoria basin of East Africa. Environmental Science & Policy **12**:504–519.

Swaminathan, M. 2005. Lessons from the Past and Policies for the Future: An Ever-green Revolution. An International Dialogue on Agricultural and Rural Development in the 21st Century, Beijing.

Tallis, H., P. Kareiva, M. Marvier, and A. Chang. 2008. An ecosystem services framework to support both practical conservation and economic development. Proceedings of The National Academy of Sciences of The United States of America **105**:9457–9464.

Tanner, C. C., M. L. Nguyen, and J. P. S. Sukias. 2005. Nutrient removal by a constructed wetland treating subsurface drainage from grazed dairy pasture. Agriculture, Ecosystems & Environment **105**:145–162.

Tilman, D., K. G. Cassman, P. A. Matson, R. Naylor, and S. Polasky. 2002. Agricultural sustainability and intensive production practices. Nature **418**.

Tilman, D., J. Fargione, B. Wolff, C. D'Antonio, A. Dobson, R. Howarth, D. Schindler, W. H. Schlesinger, D. Simberloff, and D. Swackhamer. 2001. Forecasting Agriculturally Driven Global Environmental Change. Science **292**:281–284.

Tscharntke, T., A. M. Klein, A. Kruess, I. Steffan-Dewenter, and C. Thies. 2005. Landscape perspectives on agricultural intensification and biodiversity - ecosystem service management. Ecology Letters **8**:857–874.

Tscharntke, T., C. H. Sekercioglu, T. V. Dietsch, N. S. Sodhi, P. Hoehn, and J. M. Tylianakis. 2008. Landscape constraints on functional diversity of birds and insects in tropical agroecosystems. Ecology **89**:944–951.

Uusi-Kamppa, J., B. Braskerud, H. Jansson, N. Syversen, and R. Uusitalo. 2000. Buffer zones and constructed wetlands as filters for agricultural phosphorus. Journal of Environmental Quality **29**:151–158.

Van Bael, S. A., S. M. Philpott, R. Greenberg, P. Bichier, N. A. Barber, K. A. Mooney, and D. S. Gruner. 2008. Birds as predators in tropical agroforestry systems. Ecology **89**:928–934.

van Noordwijk, M., T. P. Tomich, and B. Verbist. 2002. Negotiation support models for integrated natural resource management in tropical forest margins. Conservation Ecology **5**.

Vanlauwe, B., J. Diels, K. Aihou, E. N. O. Iwuafor, O. Lyasse, N. Sanginga, and R. Merckx. 2002. Direct interactions between N fertilizer and organic matter: Evidence for trials with ^{15}N-labelled fertilizer. Pages 173–183 *in* B. Vanlauwe, J. Diels, N. Sanginga, and R. Merckx, editors. Integrated plant nutrient management in sub-Saharan Africa: From concept to practice. CABI, Wallingford.

Verchot, L. V., J. Mackensen, S. Kandji, M. Van Noordwijk, T. Tomich, C. Ong, A. Albrecht, C. Bantilan, K. V. Anupama, and C. Palm. 2005. Opportunities for linking adaptation and mitigation in agroforestry systems.*in* C. Robledo, M. Kanninen, and L. Pedroni, editors. Tropical forests and adaptation to climate change: In search of synergies. Center for International Forestry Research (CIFOR), Bogor.

WCD (World Commission on Dams). 2000. Dams and Development: A New Framework for Decision-Making. Water and Environment Journal : The Journal. World Commission on Dams, Earthscan, London, UK.

Wendland, K. J., M. Honzák, R. Portela, B. Vitale, S. Rubinoff, and J. Randrianarisoa. Targeting and implementing payments for ecosystem services: Opportunities for bundling biodiversity conservation with carbon and water services in Madagascar. Ecological Economics **In Press, Corrected Proof**.

Whisenant, S. G. 1999. Repairing Damaged Wildlands. Cambridge University Press, Cambridge.

WHO/UNICEF. 2004. Meeting the MDG Drinking Water and Sanitation Target: A Mid-term Assessment of Progress. Page 33. WHO, Geneva, Switzerland/ UNICEF, New York, NY.

Wigington, P. J., S. M. Griffith, J. A. Field, J. E. Baham, W. R. Horwath, J. Owen, J. H. Davis, S. C. Rain, and J. J. Steiner. 2003. Nitrate removal effectiveness of a riparian buffer along a small agricultural stream in Western Oregon. Journal of Environmental Quality **32**:162–170.

Williams, N. M. and C. Kremen. 2007. Resource distributions among habitats determine solitary bee offspring production in a mosaic landscape. Ecological Applications **17**:910–921.

Winfree, R., N. M. Williams, J. Dushoff, and C. Kremen. 2007. Native bees provide insurance against ongoing honey bee losses. Ecology Letters **10**:1105–1113.

Witcover, J., S. A. Vosti, C. L. Carpentier, and T. C. D. A. Gomes. 2006. Impacts of soil quality differences on deforestation, use of cleared land, and farm income. Environment and Development Economics **11**:343–370.

WRI (World Resources Institute). 1989. World Resources 1998–99: Environmental change and human health. WRI, UNEP, UNDP, and The World Bank, New York.

Wyland, L. J., L. E. Jackson, W. E. Chaney, K. Klonsky, S. T. Koike, and B. Kimple. 1996. Winter cover crops in a vegetable cropping system: Impacts on nitrate leaching, soil water, crop yield, pests and management costs. Agriculture Ecosystems & Environment **59**:1–17.

Young, A. 1997. Agroforestry for soil management. 2nd edition. CAB International, Wallingford.

Zhang, W., T. H. Ricketts, C. Kremen, K. Carney, and S. M. Swinton. 2007. Ecosystem services and dis-services to agriculture. Ecological Economics **64**:253–260.

Chapter 4
Ecology and Human Nutrition

Roseline Remans, Jessica Fanzo, Cheryl A. Palm, and Fabrice DeClerck

One of the very nicest things about life is that we must regularly stop whatever it is we are doing and devote our attention to food (Pavarotti)

Introduction

Adequate nutrition lies at the heart of the fight against hunger and poverty (Sanchez et al. 2005). Great strides in reducing hunger through increases in agricultural productivity have been made worldwide; however, more than 900 million of people remain chronically underfed, i.e. do not have access to continuously meet dietary requirements (FAO 2008). It has long been known that malnutrition undermines economic growth and perpetuates poverty (World Bank 2006). Healthy individuals contribute to higher individual and country productivity, lower health care costs, and greater economic output by improving physical work capacity, cognitive development, school performance, and health (Grosse and Roy 2008). Unrelenting malnutrition is contributing not only to widespread failure to halve poverty and hunger, the first of the eight Millennium Development Goals (MDG), but if not appropriately eradicated, many of the other MDGs such as reducing maternal and child health, HIV/AIDS, universal education, and gender equity will be difficult to achieve (World Bank 2006). Yet the international community and most governments in developing countries continue to struggle in tackling malnutrition in all its complexity.

Malnutrition has many dimensions, including not only insufficient amount of food and calories, but also lack of essential nutrients, poor absorption, and excessive loss of nutrients. It is increasingly recognized that the current global crisis in malnutrition finds roots in dysfunctional agricultural and food systems that do not

R. Remans (✉)
Tropical Agriculture and Rural Environment Program, The Earth Institute, Columbia University, 2E Lamont Hall, 61 Route 9W, PO Box 1000, Palisades, NY 10964-8000, USA
e-mail: rremans@ei.columbia.edu

J.C. Ingram et al. (eds.), *Integrating Ecology and Poverty Reduction: Ecological Dimensions*, DOI 10.1007/978-1-4419-0633-5_4,

deliver enough essential nutrients to meet the dietary requirements of all (Graham et al. 2007; Pollan 2008). Agricultural practices are almost always directed at maximizing production while minimizing costs. Recently, preserving the environment has become one of the more prominent goals of agriculture worldwide (see Chap. 3), but maximizing nutrient output of farming systems has never been a primary objective in modern agriculture, human health, or public policy.

Increased crop production during the Asian Green Revolution prevented mass starvation in many nations. The focus, however, was primarily on cereal crops (rice, wheat, and maize), which are mainly sources of carbohydrates and contain only modest amounts of protein and a few other nutrients essential to meet human nutritional requirements. The change in agricultural production from diversified cropping systems towards ecologically more simple cereal-based systems may have contributed to poor diet diversity, significant micronutrient deficiencies, and resulting malnutrition (Graham et al. 2007).

One of the areas often not associated with malnutrition is ecology, the study of the interactions between organisms and their environment. Yet the relationship between organisms, in this case humans, and resource acquisition (nutrients) is fundamentally an ecological question. The environment is a critical determinant of which species occur in an area, and the interactions among species results in a local assemblage of species or communities. As humans modify their environment, they select and protect some species and exclude and eradicate others. Many have argued that the relationship between humans and their crops were critical to the development of civilization as we know it (Diamond 1997; Quinn 1999). In natural communities, the structure and function of the population revolves around food as an energy source (Elton 1927). This situation is also true for human societies. Early societies used foods as the medium of exchange long before currency was used. They traded surpluses of crops and in this way not only improved their own diets by obtaining more diverse foods but also had the opportunity to interact with other groups.

Increasingly, ecologists have focused on the impact of communities and their interactions on ecological processes, functions and ecosystem services. These studies, known as biodiversity and ecosystem function studies, explore the relationship between the number and kinds of organisms in a community and the ecosystem services that are derived from them. Though many ecologists have focused on the relationship between biodiversity and ecosystem functioning, there has been little focus on the role that ecosystems play in providing the essential elements of human diets. How does the combination of environment, communities, and species, and human modification of these assemblages impact human nutrition? How can ecological knowledge of species–environment interactions be used as a means of improving human nutritional well-being? What is gained through increased interactions between ecologists, agronomists and nutritionists? These are questions that this chapter tackles.

We distinguish between agricultural production as an ecosystem service, which is covered in Chap. 3, and focus on the capacity of ecosystems to provide the diversity of elements required for complete human diets resulting in healthy and productive lives. Figure 4.1 presents the cyclical approach schematically of ecosystem services, human nutrition, and agroecosystems using the framework of the Millennium Ecosystem Assessment (2005).

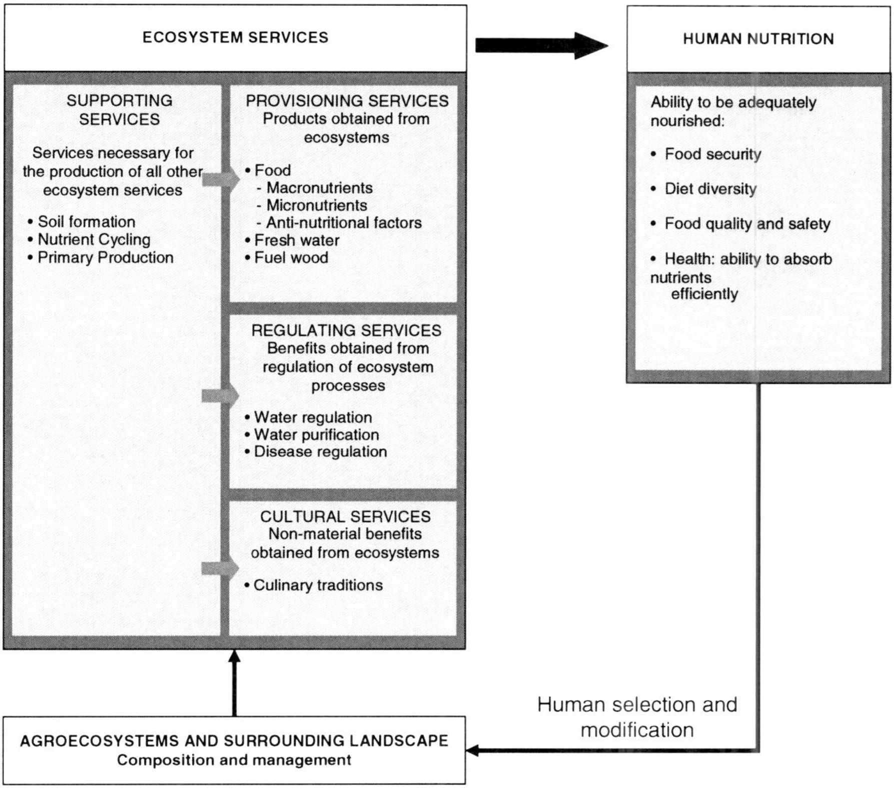

Fig. 4.1 A schematic diagram of the cyclical approach to human nutrition as an ecosystem service of agroecosystems using the Millennium Ecosystem Assessment framework (Millennium Ecosystem Assessment 2005)

The notion that nutrition, human and agricultural productivity, and environmental sustainability are interrelated was discussed by Deckelbaum et al. (2006) and has been described as "econutrition." Deckelbaum et al. (2006) argued that to tackle malnutrition much can be gained by linking agriculture and ecology to human nutrition and health. Biodiversity hotspots and hunger hotspots almost directly overlap, and although the intellectual paths of agronomists, ecologists, and nutritionists rarely cross, their geographic extensions are the same. The areas where there is hunger, loss of biodiversity, and a need for improved agricultural systems are largely identical. Cyclical feedbacks between soil fertility decline, biodiversity loss, decreased food production, and malnutrition can be identified and need to be turned around through an interdisciplinary approach (Deckelbaum et al. 2006; Sanchez et al. 2005). Indeed, these disciplines share a common concern, notably the rapid loss of biodiversity that typically accompanies agricultural intensification. While ecologists tend to focus on nondomesticated species, agriculturalists on improving crop yields of a few crops, and nutritionists on the availability and utilization of food crops and specific nutrients, there is no reason to think that lessons learned by ecologists on the functional consequences of species losses (Flynn et al. 2009) should not apply within a nutritional framework. Several subdisciplines of ecology pertain to the linkages and use of ecology to improve human nutrition. These areas include community ecology, biogeochemistry, and human ecology.

In this chapter, we explore how ecological knowledge, models, and tools can be used in tackling questions of human nutrition, more specifically how to address human malnutrition that is linked to extreme poverty.

Malnutrition and Hidden Hunger

Massive numbers of people are dying each year because of a lack of sufficient nutrients to thrive. Symptoms of micronutrients deficiency including vitamin A, zinc, iodine, iron, and B vitamins, are not always visible which has led to the term hidden hunger in contrast to acute hunger often associated with lack of calories and starvation. The magnitude of this crisis in human health is enormous with over 50% of annual deaths worldwide associated with malnutrition (WHO 2003). In 2003, approximately 30 million deaths, mostly among the resource-poor people in developing countries, were associated or directly caused by malnutrition (Muller and Krawinkel 2005; WHO 2003).

In the developing world, malnutrition is often defined as insufficient food intake that includes being underweight for one's age, too short for one's age (stunted), dangerously thin (wasted), or deficient in vitamins and minerals (micronutrient malnutrition). In sub-Saharan Africa (SSA), 28% of children are underweight, and in Eastern/Southern Africa and West/Central Africa, 40% and 36% of children are stunted, respectively (UNICEF 2008). Malnutrition has devastating impacts on the health of individuals, particularly women and children. During the first 2 years of life, malnutrition significantly affects child survival, growth, and development (World Bank 2006). Malnourished children are more likely to die from common yet preventable childhood sicknesses such as diarrhea and airway respiratory infections, have lowered resistance to infection, and have irreversible harm to their cognitive development.

To live healthy and productive lives, humans require a diversity of at least 51 known nutrients in adequate amounts consistently (Table 4.1, Graham et al. 2007).

In order to consider how ecological tools can be used to redress malnutrition, it is important to understand which nutrient deficiencies are drivers of malnutrition. We now briefly outline critical nutrients that are frequently lacking in the diets of the poor.

Probably one of the most devastating nutritional problems is protein-energy malnutrition (PEM) which affects about 800 million people globally (WHO 2002). Failure to grow is the first and most important manifestation of protein-energy deficiency. The current view is that most PEM is the result of inadequate intake or poor utilization

Table 4.1 The known 51 essential nutrients for sustaining human life (Adapted from Graham et al. 2007)

The known 51 essential nutrients for sustaining human life[a]					
Air, water and energy	Protein (amino acids)	Lipids-fat (fatty acids)	Macrominerals	Trace minerals	Vitamins
Oxygen	Histidine	Linoleic acid	Na	Fe	A
Water	Isoleucine	Linolenic acid	K	Zn	D
Carbohydrates	Leucine		Ca	Cu	E
	Lysine		Mg	Mn	K
	Methionine		S	I	C (absorbic acid)
	Phenylalanine		P	Fe	B_1 (thiamine)
	Threonine		Cl	Se	B_2 (riboflavin)
	Tryptophan			Si	B_3 (niacin)
	Valine			Mo	B_5 (pantothenic acid)
				Co (in B_{12})	B_6 (pyroxidine)
				B[b]	B_7/H (biotin)
				Ni[b]	B_9 (folic acid, folacin)
				Cr[b]	B_{12} (cobalamin)
				V[b]	
				As[b]	
				Li[b]	
				Sn[b]	

[a]Numerous other beneficial substances in food are also known to contribute to good health

[b]Not generally recognized as essential but some supporting evidence published

of food and energy, not a deficiency of one nutrient and usually not simply a lack of dietary protein (FAO 1997). Over three billion people are known to be afflicted with one or more micronutrient deficiencies (Mason and Garcia 1993). An estimated 100–140 million children are afflicted with vitamin A deficiency (VAD), and an additional 43 million children in SSA are at risk of VAD, a deficiency which is a major cause of blindness, lowered immunity, and increased risk of maternal mortality (Aguayo and Baker 2005). Although difficult to quantify, approximately two billion people are thought to be zinc-deficient (Hotz and Brown 2004), which leads to stunted growth and immunodeficiency syndromes.

Anemia is one of the most common and intractable nutritional problems (WHO 2008). The World Health Organization estimates that some two billion people are anemic defined as hemoglobin concentrations that are below recommended thresholds. Anemia occurs at all stages of the life cycle, but is more prevalent in pregnant women and young children. It has negative consequences on cognitive and physical development of children, and on physical performance, particularly work productivity in adults and severe anemia can lead to death.

The primary cause for anemia is iron deficiency, but it is seldom present in isolation. More frequently it coexists with a number of other causes, such as malaria, parasitic infection, hemoglobinopathies, and other nutritional deficiencies such as Vitamin A deficiency (WHO 2008; McLean et al. 2009). Low dietary intake of iron as well as poor absorption from diets high in phytate and phenolic compounds, increase the risk of iron deficiency anemia. Phytate and phenolic compounds are, among others, considered as anti-nutritional factors. They interact with essential nutrients to form biochemical complexes that cannot be absorbed by humans. Heme food sources, predominantly red meats, on the other hand, contain highly absorbable iron and promote the absorption of iron from other less bioavailable food sources (Stolzfus and Dreyfuss 1998).

Although malnutrition is most commonly associated with gross deficiencies, overnutrition is increasingly becoming a global problem. Overnutrition and subsequent obesity, or what has been coined the 'nutrition transition' occurs as people in low-income countries with more disposable income change their eating habits and move from locally produced foods to more processed foods or a "western diet." As poor countries become more affluent, they acquire some of the benefits along with some of the problems of industrialized nations which include obesity due to changes in diet, physical activity, health, and nutrition (Popkin 2001; Popkin and Ng 2007).

Urban areas in developing countries are much further along in this nutritional transition, but rural areas are susceptible as well. Increased mechanization of farm activity leads to reduced physical activity at the same time that more food, but not necessarily a better variety of foods, becomes available (Popkin 2004). Many rural farmers have changed from farming multiple crops that provide a more balanced diet in favor of a single, high-yielding cash crop. The transition is often linked to a growing trend in increased noncommunicable disease risk, and this can occur quite rapidly as demonstrated in Brazil and China (Monteiro et al. 2004).

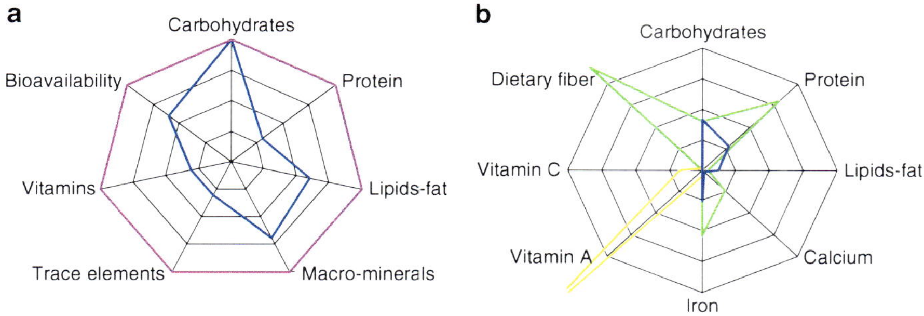

Fig. 4.2 Ecological spider's web presenting nutrient diversity requirements in a human diet. (**a**): Nutrient composition of an ideal diet that meets all nutritional needs is shown in pink. An example of nutrient composition of a diet that meets carbohydrate demand but lacks protein and micronutrients or trace elements is shown in blue. (**b**): Nutrient composition data of three food crops are shown as % of daily requirement (100%). The blue line represents one cup of white corn (166 g), the green line one cup of black beans (194 g) and the orange line one cup of pumpkin (116 g) (nutrition facts from www.nutritiondata.com). The spider diagram shows the complementarity between the three food crops for carbohydrates, proteins, dietary fiber and vitamin A

Nutritionists have long recognized that malnutrition is a complex phenomenon, involving different nutrient deficiencies. It is clear that current global nutrition gaps can not be corrected by increased supply of only one or a couple of nutrients. The role of micronutrients in human health and the synergies in their physiologic functions are being increasingly recognized and support the notion that nutrient deficiencies rarely occur in isolation (Frison et al. 2006). The challenge is to provide the diversity and adequate amount of nutrients required for a complete human diet. This urges a multidimensional approach. Ecologists also work in multidimensional systems, composed of organisms, energy, and the physical environment interacting at various spatial and temporal scales, which can be described in terms of composition, structure, functions, fluxes, resilience, or other dynamics (Pickett and Cadenasso 2002; DeClerck et al. 2006). Optimizing for nutrient diversity can be presented schematically as maximizing the various arms of an ecological spider diagram as illustrated in Fig. 4.2.

While supplementation and food fortification programs have been quite successful in improving nutrition conditions, these programs are costly and strongly dependent on external funding including government programs, overseas development assistance, or access to clinics and markets. In contrast, improving local crop and livestock production combined with adequate processing and storage methods, can provide a more sustainable and long-term solution to nutrition security. The strategic linkages between agriculture and nutrition to combat malnutrition have been described as "food systems," "field fortification," and "food based" strategies (Kataki 2002; Kataki and Babu 2002). It is in these food systems that ecological concepts can be of particular use to study and optimize the sustainability and nutritional value of the underlying food-systems.

Food Systems and System Diversity

In the past, food-based interventions in developing countries have been mostly single-nutrient oriented (Frison et al. 2006). This approach may in part be attributed to a lack of knowledge in earlier years of the interactions among nutrients in human physiology and metabolism. From various recommendations for high-protein diets (Brock et al. 1955) and later for high-carbohydrate diets (McLaren 1966, 1974), to more recent efforts directed at the elimination of micronutrient deficiencies (UN committee on nutrition 2000; Ruel and Levin 2002), the attention was generally concentrated on single nutrient approaches. The introduction of crops focusing on single nutrients serves as an important means to address specific nutrients (macro or micronutrients), but caution must be exercised as any single crop, including fruit or vegetable crops, does not address the complex nutritional needs that humans require (Graham et al. 2007). The importance of nutrient diversity for human well-being, as discussed above, calls for dietary diversification.

Diet diversity is often defined as the number of certain food groups consumed by and individual or family Table 4.2. Many studies done on several different age groups show that an increase in individual dietary diversity is related to increased nutrient adequacy of the diet. Diet diversity has been positively correlated with micronutrient density of diets of non-breastfed children, adolescents, and adults (Hatloy et al. 2000; Ruel et al. 2004; Steyn et al. 2006; Kennedy et al. 2007; Mirmiran et al. 2004; Foote et al. 2004). More research is ongoing to improve the understanding of the association between dietary diversity and micronutrient uptake (FAO, IFPRI, Bioversity International and others).

One means of assuring adequate dietary diversity for all would be to manage agroecosystems in ways that will result in a plentiful and diversified nutrient output of farming systems. In our opinion, achieving such dietary diversity in agroecosystems is best achieved through interdisciplinary collaborations between nutritionists, agronomists, ecologists, and local communities as we demonstrate below.

Agricultural biodiversity and diet diversity illustrate the nexus of nutrition and ecology. Ecologists have studied the effects of species removals or additions in ecological communities. For instance, several large-scale grassland studies in the U.S. and in Europe have demonstrated that as the number of species in a grassland area increases, so does the net primary productivity. In addition, increasing species richness has increased the stability of the community; as indicated during drought years, species-rich communities exhibited less reduction in biomass produced than the species-poor communities (Rees et al. 2001).

The mechanisms that drive these relationships between species richness and enhance ecological performance are still heavily debated, but are largely due to two processes. The first is known as the sampling effect, and simply argues that as you increase the number of species in a plot, the probability of including a highly productive species is greater. From a nutritional point of view, this is analogous to considering that as you increase the number of crops produced on a farm, or in a

Table 4.2 Fourteen food groups as described by FAO/FANTA (2007). The diet diversity score is calculated as the number of food groups consumed

	Food group	Rich in	Examples
1	Cereals	Carbohydrates	Maize, millet, sorghum, rice
2	Vitamin A rich vegetables and tubers	Vitamin A	Carrots, orange flesh sweet potatoes
3	White tubers and roots	Carbohydrates	White potatoes, white yams, cassava
4	Dark green leafy vegetables	Trace elements (iron), vitamins	Kale, amaranth, black nightshade
5	Other vegetables	Vitamins, trace elements	Onions, cabbage, mushrooms
6	Vitamin A rich fruits	Vitamins (vitamin A)	Mangoes, papaya, guava, passion fruit, pumpkin
7	Other fruits	Vitamins	Oranges, pineapple, apples, plums, grapes
8	Organ meat	Protein, trace elements (iron), vitamin B, fatty acids	Liver, kidney, heart or other organ meats or blood-based foods
9	Flesh meat	Protein, minerals, trace elements, vitamin B, fatty acids	Beef, pork, lamb, goat, rabbit, chicken
10	Eggs	Protein	
11	Fish	Protein, minerals, trace elements, vitamin B, fatty acids	Fresh or dried fish or shellfish
12	Legumes, nuts and seeds	Protein, minerals, trace elements	Beans, peas, lentils, groundnuts
13	Milk and milk products	Minerals (calcium)	Milk, cheese, yogurt
14	Oils and fats	Fatty acids	Butter, sunflower oil, palm oil

region, the probability that one of those crops be high in a particular nutrient, for example Vitamin A, also increases. Thus, simply by chance, if we increase the number of crops available to local communities, we increase the probability that they will obtain the nutrients needed for healthy, productive lives.

The second mechanism is known as the complementary effect, where interactions between species result in a yield or function greater than expected by chance, also called over-yielding. There are numerous possible interactions that can lead to complementarity; these interactions range from resource partitioning where different organisms use resources differently thus reducing competition, to symbiotic and mutual interactions where species facilitate the presence or success of another.

Probably one of the best known examples of such ecological complementarity that also results in net nutritional benefit comes from the Mesoamerican "three sisters." The combination of corn (a grass), beans (a nitrogen-fixing legume) and squash (a low-lying creeper) maximizes trait differences for growth and resource use efficiency between species (Risch and Hansen 1982), resulting in higher yields compared to those obtained through three monocultures of these crops. The corn is a grass species particularly efficient in maximizing photosynthesis in warm environments. In structure, the corn grows straight and tall adding a vertical dimension to the system. The vine-like bean takes advantage of the growth form of the corn for structural support that also enables it to reach more sunlight. The beans are also unique in their capacity to bring atmospheric nitrogen in the system by symbiotic nitrogen fixation; this nitrogen becomes available to the corn in subsequent cropping seasons. The interaction between the corn and beans is an example of complementarity where the over-yielding is due to a positive interaction between the species. The third member of this assemblage, squash, does not perform as well as corn in direct sunlight, and thus occupies the remaining space near the ground where light is somewhat reduced, and humidity is increased reducing photorespiration (Gliessman 2006). The addition of squash can decrease the amount of soil lost to erosion by its low lying nature and broad leaves ensuring greater soil coverage. The added productivity from squash does not so much come from positive interactions with the beans and maize, but rather comes from the capacity of the squash to use resources (namely light) that are not captured by the corn and beans, an example of resource partitioning.

It is not only that these crops are ecologically complementary that is notable, but also that they are nutritionally complementary. The corn is an important source of carbohydrates and some amino acids. By adding the beans, the set of essential amino acids for a human diet becomes complete and important contributions in carbohydrates, dietary fiber, vitamin B_2 and B_6, zinc, iron, manganese, iodine, potassium, magnesium and phosphorus are made. Squash in contrast can be an important source of vitamin A depending on the variety. It is important to note that each of these crops can make an important contribution to human diet; however, none of these crops in isolation provides total nutrition.

Ecological and Nutritional Functions of Compounds in the Plant World

Why is there an association between crop diversity and human nutrition? Or the question can be rephrased as why is there such a great diversity of nutritional compounds within the plant world? The evolution of nutritional traits is purely a function of either rewarding us or other animals for dispersing their seeds in the case of almost every piece of fruit we consume, a defense against plant pests in the case of chili peppers and mint for example, or ensuring that their seeds are best prepared for the ultra-competitive world of seedlings as in the case of beans. The point is that ecological interactions are at the heart of the nutritional content of most species we consume.

Members of the genus *Capsicum*, more commonly known as chili peppers, are frequently consumed in the tropics and enjoyed by many in either their sweet or spicy form. Why are these chili peppers so pungent? Birds consume the fruit and facilitate the dispersal of chili seeds, apparently unaffected as mammals are by the spiciness. However, recent research (Tewksbury et al. 2008) has shown that plants with greater rates of insect piercing on the fruit had higher levels of the phytochemical capsaicin and that the plant uses this chemical primarily as a defense against fungi which enter the fruit on the backs of insects to consume the seed. From a nutritional point of view, *Capsicum* has amongst the highest levels of vitamins A, C, and beta-carotene of crops commonly consumed in poverty hotspots. Capsaicin has been shown to have an antibacterial function (Billing and Sherman 1998; Molina-Torres et al. 1999; Xing et al. 2006) and some researchers propose that the prevalence of spicy foods in tropical regions is no coincidence but rather a means of preserving food or killing off bacteria in food (Billing and Sherman 1998; Sherman and Billing 1999).

Another example of ecological application in human nutrition is the use of nitrogen-fixing plants in agricultural systems. Nutritionists, development specialists, and most farmers recognize that legumes, such as common beans, groundnuts, and soybeans are important sources of protein. This comes as no surprise to agronomists or ecologists, who recognize that all three of these food items come from a unique and third largest plant family, the legumes or *Fabaceae*. This plant family is also a major player in the nitrogen cycle in terrestrial ecosystems and is recognized as a driver of several ecosystem functions including primary productivity in natural systems. From a nutritional point of view, legumes contain five times the amount of high quality protein of maize, and 18 times the protein content of potatoes and are also superior to cereals as a source of micronutrients (Broughton et al. 2003).

It is worth exploring the ecological foundation for these high protein levels. Manufacturing protein has a high nitrogen demand, and although 80% of our atmosphere is composed of nitrogen (N_2), none of this is available to plants. To further exacerbate the problem, most soils are nitrogen-limited. Many species in the legume family have developed the unique symbiotic association with *Rhizobium*, a type of soil bacteria found in the roots of most legumes that allows the plant to convert atmospheric dinitrogen gas into ammonium which the plant then uses to form amino acids, the building blocks of proteins. The plant in return provides the bacteria with photosynthetic sugars. This relationship is energetically expensive to legumes; this cost, however, provides unique access to one of the nutrients most limiting to primary production in terrestrial ecosystems. This access to nitrogen allows legumes to colonize soils that are inhospitable to many other plant families or to outcompete other plants in nitrogen-poor environments. The high protein content of legume seeds provides the plants progeny with a competitive advantage for growth in systems low in nitrogen (Andersen et al. 2004; Andersen 2005). Humans have learned to take advantage of this high nitrogen content both in terms of our own nutritional well-being, as well as a natural source of organic nitrogen fertilizer (see Chap. 3, this volume).

Legumes are often advocated in diets because of their beneficial effects and because they are a low cost source of protein (Borade et al. 1984). However, compared to other food crops, legumes also show high content of secondary metabolites with antinutritional effects, such as amylase inhibitors, lectins, and trypsin inhibitors, which can cause adverse physiological responses or diminish the availability of certain nutrients (Wink 2004; Muzquiz 2004). This raises the question as to why do legumes combine such attractive nutritional characteristics like high protein and mineral content with relatively high contents of antinutritional factors. Secondary metabolites including antinutrients have shown to provide natural mechanisms of defense for plants against microbes, insects, and herbivores (Wink 1997, 2004). In many agricultural crops which have been optimized for yield, their original lines of defense have often been selected out because the underlying metabolites were unpalatable or toxic for humans or livestock. But in legumes, the numerous nitrogen-containing metabolites with antinutritional properties appear to function both as chemical defenses and as nitrogen storage compounds that facilitate germination in low N systems. Legume genotypes selected for low amounts of to no antinutrients show reduced germination power and thus a general selection advantage (Savelkoul et al. 1992; Wink 2004). During germination, however, these antinutrients degrade to a lower level by the action of several enzymes resulting in improved digestibility of bean sprouts for humans as compared to dry beans. This example illustrates how enhanced knowledge of underlying ecological functions can benefit human nutrition.

Though we tend to consider humans outside of natural systems, the examples above demonstrate that the interactions between species are literally the spice of human life. Long-term interaction between plants and animals and the active selection of plants from various families by humans have resulted in a large diversity of nutritional traits. It is proposed that the long-term approach toward diversification of nutrient rich crops will address the significant deficits in micronutrients amongst the diets and the particular nutrition needs of communities. A promising example is the tree species *Moringa oleifera*. *Moringa oleifera* can grow well in the humid tropics or hot dry lands, can survive destitute soils, and is little affected by drought (Morton 1991). It tolerates a wide range of rainfall (Palada and Changl 2003) and is native of the western and sub-Himalayan tracts, India, Pakistan, Asia Minor, Africa, and Arabia (Somali et al. 1984; Mughal et al. 1999). In addition, the leaves, fruit, flowers, and immature pods of this tree are highly nutritive. *Moringa* leaves have been reported to be a rich source of b-carotene, protein, Vitamin C, calcium, and potassium and act as a good source of natural antioxidants (Siddhuraju and Becker 2003; Sabale et al. 2008). However, more research on its ecological function and its impact on human health and nutrition is required.

Nutritional Diversity

Community ecology has demonstrated that increases in biodiversity can lead to increases in plant community productivity when species complement each other, or use resources differently as discussed above. Though increasingly ecologists have focused on the relationship between biodiversity and ecosystem functioning, there has been very little focus on the capacity, or role that ecosystems play in providing the essential elements and nutrients of the human diet. Many studies of biodiversity and ecosystem function have demonstrated that there is much variance that cannot be explained by species richness. This leads to the question of whether the relationship between the taxonomic identity of a species and its functional identity are not completely related. For example, does it matter that an ecosystem has five species, or would it be more important that a system has five different functional groups? Is a field with maize, rice, wheat, sorghum, and millet the equivalent of a field with maize, beans, squash, sweet potato, and guava? Both have five species, but the latter contains five functional distinct species from a nutritional point of view in contrast to the former where all of the species are from the grass family, high in carbohydrates, but poor in essential nutrients.

To illustrate, a field survey on 30 farms in western Kenya identified over 146 plant species including 39 edible species important to the local diet. Edible plant diversity was relatively high in farmer fields with an average of 14 edible plants ranging between 5 and 22 species per field. Rather than simply look at the relationship between crop diversity and nutrition, we classified the edible species according to their nutritional content for seven nutrients of importance: protein, carbohydrates, vitamin A, vitamin C, iron, zinc, and folate. With this classification, species high in protein (beans, peanuts and amaranthus) form a distinct cluster in the dendrogram; species high in vitamin A (sweet potato and chili) form a second important cluster and species with high carbohydrate content (sugar, sorghum, and maize) also form a unique cluster (Fig. 4.3).

Using this same dendrogram, where functional diversity is measured as branch length (see Petchey and Gaston 2002 for details), we regressed functional diversity (FD) against species richness of each of the 30 farms. Several patterns became apparent through this regression. The first is that there is a relatively strong relationship between functional richness and species richness. That is, as the number of edible species increases, the functional richness of that farm also increases.

Fig. 4.3 Agrobiodiversity of 30 farms of a typically western Kenya village. Edible plant diversity was relatively high in farmer fields with an average of 14 edible plants ranging between 5 and 22 species per field. Here we classify the edible species according to their nutritional content for seven nutrients of importance: protein, carbohydrates, vitamin A, vitamin C, iron, zinc, and folate identifying distinct cluster of species high in protein (beans, peanuts and amaranthus), high in vitamin A (sweet potato and chili) and species with high carbohydrate content (sugar, sorghum and maize)

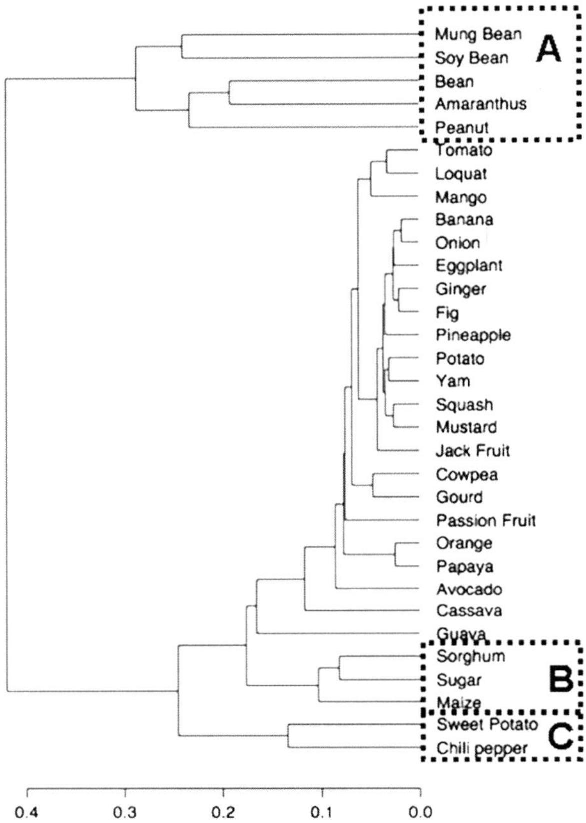

This confirms the notion that increasing farm agrobiodiversity increases the capacity of the farm to provide a multitude of nutritional functions to its owner. The second notable pattern is that although species richness and functional richness are correlated, it is possible for a farmer to have a field with many species but low nutritional diversity, or for a farm to have fewer species but greater nutritional diversity in addition to the general trend of increasing nutritional functional diversity with species richness. A look at Fig. 4.4 shows that an important cluster made up of amaranth, soy bean, and mung bean is entirely missing from Farm "A," and that the absence of this cluster suggests that an important nutritional function may also be absent from this farm.

Why is there a relationship between species richness and functional richness and human nutrition? It would be reasonable to expect that nutrients would be normally distributed between the crops grown in this example with a few species that have low content for any particular nutrient, many species with moderate nutrient levels, and a few species that have high nutritional values. However, this is not the case, and as with many ecological variables, nutritional content for any particular element is log-normally distributed with most species containing low levels of any particular

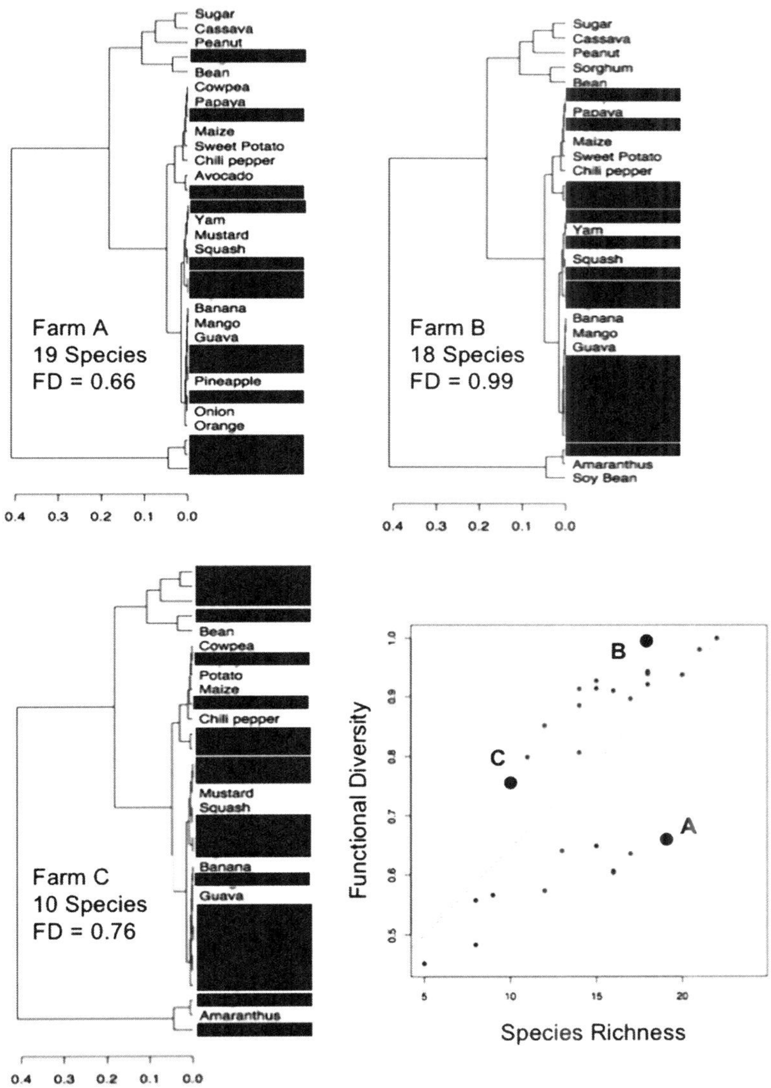

Fig. 4.4 Comparisons of the species diversity and the functional nutritional diversity of 30 farms in Kenya demonstrating the positive relationship between species richness and nutritional diversity. Note however that when nutritional functions of the crops are taken into consideration it is possible to have a field with high nutritional diversity but low species richness as in farm "C." Random assemblages of plants can lead to high species richness but low nutritional diversity as in farm "C"

nutrient, and few containing high levels. The other distinct pattern in nutrient distribution amongst these species is that there is little redundancy of plants that contain the same nutritional content and there is no single species that is capable of providing all the nutrients needed for a human diet. While a single species may be good for providing a single nutrient, there are no crop species capable of providing all essential nutrients. The importance of a diversified diet increases with the number of nutritional functions we expect agricultural systems to provide. In the example of this Kenyan village, community members that had low agricultural functional agrobiodiversity demonstrated statistically significant levels of anemia; as functional agrobiodiversity increased, anemia became absent.

Focusing on those nutritional elements most severely missing in many developing countries – energy, protein, iron, vitamin A, vitamin C, folate, and zinc-based on the findings from this village survey, corn is the greatest provider of carbohydrates, beans, and amaranthus are by far the greatest providers of iron, protein, and folate, orange-fleshed sweet potato for vitamin A, guava for vitamin C, and amaranthus, by far the greatest source of zinc. In order to have a diet sufficient in these essential nutrients, a family would have to consume at least five different plant species, but more importantly, not just any five species, but those five species that belong to the different nutritional functional groups that together make a nutritious diet.

As in agricultural systems, past interventions have focused on the most pressing deficiencies (nitrogen in agriculture, carbohydrates in human hunger). The focus must now shift to increasing functional agricultural diversity to tackle these micronutrient deficiencies, or hidden hunger. A clear understanding of which crops play specific nutritional as well as ecological functions shows tremendous promise for selecting plants and managing agricultural systems that provide numerous functions by identifying and combining species assemblages that maximize functions.

Food-Systems and Biogeochemistry

Since plants obtain their nutrients from the soil, soils play a critical role in fueling the entire food chain (Daily 1997). Soils are a key source of nutrients and are crucial for nutrient cycling. The provisioning of food from crops and livestock has increased by 170% in the past four decades. The large increases in production, however, have come with trade-offs that include degradation of soils and many of the regulatory and supporting ecosystem services they provide, such as the regulation of hydrological and nutrient cycles. These trade-offs between provisioning and regulatory services can ultimately undermine the ability of the ecosystems to provide the essential nutrients for human diets (Palm et al. 2007).

Agriculture practiced in many poor regions without replenishing nutrients soon results in soils that for crop production are deficient in nitrogen (N), available phosphorus (P) and to a lesser extent, potassium (K) and sulfur (S). In addition, Sillanpaa (1990) estimated that of the important agricultural soils of the world, 49% are deficient in zinc, 31% deficient in boron, 15% deficient in molybdenum, 14% deficient in copper, 10% deficient in manganese ,and 3% deficient in iron for crop production. These figures may be compared with corresponding figures for the human population that depends on the same soils. Many of those countries, where human micronutrient deficiencies are a problem, are also the countries that have large areas of micronutrient-poor/deficient soils (White and Zasoski 1999; Graham 2008; Welch 2008).

Plants are able to supply all the known essential minerals for human diets even though they may not necessarily require all of them for their own growth. In particular, plants contain Se, I, and Co in concentrations high enough to fully satisfy human requirements if the soils on which they grow are not too poor in these same elements. However, probably half of all soils are deficient in one of these three ultra-micronutrients (daily requirements about 100 times less than those of Fe and Zn) and although plant production is not restricted by this deficiency, human diets based on the crops grown on these soils can be deficient. 'Linking unhealthy people and unhealthy soils' was emphasized by Sanchez and Swaminathan (2005) in reference to integrated approaches to tackling hunger in Africa. The critical point is that crop diversity alone may not be sufficient to meeting nutritional needs. The health of the soils upon which these crops are grown can play an important role in ensuring human health.

Food Systems and Soil Ecology

Soils are ecosystems unto themselves with numerous ecological interactions of important consequences to the capacity of crop plants to be both productive and nutritious. Management interventions that alter the soil environment have an impact on these ecological interactions and can change the nutritional value of crops. The use of farmyard manure and other forms of organic matter can increase plant-available micronutrients by changing both the physical and biological characteristics of the soil (Allaway 1975, 1986). These changes improve soil physical structure and water-holding capacity resulting in more extensive root development and enhanced soil microfloral and faunal activity, all of which can increase available micronutrient levels in soils that impact plants and then humans (Stevenson 1991, 1994). Very few controlled experiments have been conducted to determine which types of organic matter inputs and practices significantly enhance or depress trace elements levels, or micronutrients in the edible portions of major food crops. More research should be carried out to understand the impact of various types of organic matter on crop nutritional quality with respect to trace elements (Welch 2008), and its impact on human nutrition.

Another interesting relation between soil practices and the nutrient value of crops is found in the anti-nutritional activity of oxalate in plants. Oxalate is considered as an anti-nutrient because it forms insoluble precipitates with many minerals, a prominent example being calcium, and thereby reduces the availability of these minerals for human absorption. In plants, such as the iron-rich leafy vegetable Amaranth, two dominant fractions of oxalate are found, a boiling water soluble fraction that is predominantly in the form of potassium and magnesium oxalate, and an associated insoluble residue which is predominantly the calcium oxalate form (Vityakon and Standal 1989). The insoluble oxalate cannot further inactivate Ca in the diet, whereas the soluble forms can combine with Ca from other foods and reduce its availability. Interestingly, oxalate also interacts with minerals in the soil and the amount of oxalate and the soluble/insoluble ratio is dependent on soil conditions. Soil practices that manipulate the soil nutrient environment, such as liming of acid soils to increase soil calcium, have been suggested to reduce soluble oxalates (Vityakon and Standal 1989; Bakr and Gawish 1997) and thereby increase availability of Ca in diets. Fertilization with urea, on the other hand, has shown to result in higher oxalate levels in spinach and lettuce as compared to plants that did not receive urea supply reducing the nutritional quality of the crop (Bakr and Gawish 1997).

Though the term biodiversity conjures images of toucans, jaguars, giant sequoias, and panda bears, a significant portion of biodiversity consists of largely invisible micro-organisms inhabiting the below-ground realm. A pinch of soil is said to hold more than ten billion bacteria (Wilson 1999) contributing tremendously to numerous ecosystem services. Plants live in intimate association with this tremendous diversity of soil microorganisms that can result in deleterious, beneficial, or no effects for the plant. Plant species composition and diversity play a significant role in shaping soil microbial communities and in determining the biological outcome of such associations. Beneficial plant-associated microorganisms can contribute to plant growth and/or nutritional value by suppressing disease, promoting root development, fixing atmospheric nitrogen, and enhancing or facilitating nutrient uptake from the soil.

An example of microorganisms affecting the nutritional outcome of the plant is illustrated through iron acquisition by plants. Under nonsterile soil system, plants show no iron-deficiency symptoms and have fairly high iron level in roots in contrast to plants grown in sterile system (Masalha et al. 2000). *Pseudomonas spp.* are bacteria that produce siderophores (Crowley et al. 1992; Gamalero et al. 2003; Carrillo-Castaneda et al. 2005), which are low molecular weight chelators with high binding affinity and specificity for iron (Fe^{3+}). Siderophores are released by the bacteria under iron-limited conditions and creates iron starvation conditions for phytopathogens but also establishes a crucial competition for iron in the rhizosphere (Glick 1995). This competition ultimately determines the population structure of organisms living around the roots of the plant and some of the microbial siderophores can be utilized by plant systems resulting in increased Fe content in the plant (Ma and Nomoto 1996; Masalha et al. 2000).

These examples illustrate some of the effects that soil ecosystems can have on the nutritional value of crops and ultimately on humans diets. In present-day farming systems, soil is often treated as if it were mainly a medium for physically supporting the plant. When soil is managed for sustainable production and emphasis is placed on the role of soil ecosystems, however, the role of soil is greatly expanded.

Additional Ecosystem Services Affecting Human Nutrition

In addition to the provision of the diversity of nutrients required for complete human diets, ecosystems can provide services that help increasing the bioavailability and absorption of nutrients by the consumer. Above, we addressed the issue of antinutritional factors present in food crops and the role they may play in ecosystems. However, seemingly unrelated topics such as access to firewood or other sources of energy and water (see chapters on energy and water, this volume) are also directly related to nutritional well-being.

Access to water and fuel allows cooking of food which reduces the amount of antinutritive factors and facilitates digestion and nutrient absorption of many food items. The simple process of cooking thereby enlarges the spectrum of edible foods enormously. For example, dry beans contain relatively high amounts of antinutrients and require 1–2 h of prolonged cooking to enable digestion and nutrient absorption by humans (Elsheikh et al. 2000). In communities dependent on firewood for cooking, forest and woodland degradation occur impacting many ecosystem services. In Ruhiira, in Southwestern Uganda, the landscape is dominated by banana plantations and few trees remain. Women in Ruhiira comment that regularly they lack firewood to prepare harvested beans (Remans, unpublished data). Their diet diversity is therefore directly impacted by firewood shortage. Human nutrition could benefit in this area from reforestation and woodlot that provide sustainable wood supply or from alternative energy sources.

Access to clean drinking water is also critical to human nutrition aiding in digestion and nutrient uptake. Diarrhea, often caused by pathogens present in contaminated water sources, disturbs the human gut ecosystem and drastically reduces nutrient uptake efficiency (Petri et al. 2008). Without adequate access to clean, potable drinking water children fail to thrive, regardless of diet diversity. The maintenance or rehabilitation of ecosystem services that filter and clean water and provide drinking water quality are a critical component of an integrated approach to ensuring improved human nutrition.

Conclusion

The global health crisis of malnutrition afflicts massive numbers of people and urges changes in global food systems to provide adequate nutrition for all.

In this chapter we argue that ecological knowledge, tools, and models have an important role to play in efforts to direct food systems at improved human nutrition. Undernutrition has many dimensions and the complex nature of human nutrition calls for dietary diversification. If agricultural practices are directed at improving the nutritional quality and diversity of their output, they must encompass a holistic system perspective to assure that the intervention will be sustainable. It is here where ecology through studying interactions between species and their environment can identify synergies and tradeoffs between agriculture and nutrition and has an important role to play in guiding agriculture interventions for improved human nutrition. Agricultural biodiversity and diet diversity illustrate the nexus of nutrition and ecology. Examples in community ecology, biogeochemistry, and soil ecology described in this chapter pertain to the linkages between ecology, nutrition and agriculture and are only a start of using ecology to improve food systems for human nutrition.

A clear understanding of which species play specific nutritional as well as ecological functions shows tremendous promise for managing agricultural systems that provide numerous functions by identifying and combining species assemblages that maximize functions. If you say you want a truly new green revolution in agriculture, you'd better invite ecologists at the table.

The excitement of vitamins, nutrition and metabolism permeated the environment.

(Paul D. Boyer)

References

Aguayo VM and Baker SK (2005) Vitamin A deficiency and child survival in sub-Saharan Africa: A reappraisal of challenges and opportunities. Food Nutr Bull 26(4): 348–355

Allaway WH (1975) The effects of soils and fertilizers on human and animal nutrition. USDA ARS Agriculture Information Bulletin No. 378. Washington, DC, US Government Printing Office. Ref Typ: Pamphlet

Allaway WH (1986) Soil-plant-animal and human interrelationships in trace element nutrition. In Mertz W (ed.) Trace elements in Human and Animal Nutrition (5th edition) Academic Orlando, San Diego, CA, New York, Austin, London, Montreal, Sydney, Tokyo, Toronto. pp. 465–488

Andersen MK, Hauggaard-Nielsen H, Ambus P and Jensen ES (2004) Biomass production, symbiotic nitrogen fixation and inorganic N use in dual and tri-component annual intercrops. Plant Soil 266: 273–287

Andersen MK (2005) Competition and complementarity between intercropped barley, rape and fieldpea in ecological cropping systems – the role of plant available nitrogen and sulphur as well as cropping design. PhD Dissertation

Bakra AA and Gawishb RA (1997) Trials to Reduce Nitrate and Oxalate Content in Some Leafy Vegetables. 2. Interactive Effects of the Manipulating of the Soil Nutrient Supply, Different Blanching Media and Preservation Methods Followed by Cooking Process J Sci Food Agric 73: 169–178

Billing J and Sherman PW (1998) Antimicrobial functions of spices: Why some like it hot. Quart Rev Biol 73:3–49

Borade VP, Kadam SS and Salunkhe DK (1984) Changes in phytate phosphorus and minerals during germination and cooking of horse gram and moth bean. Qualitas Plantarum- *Plant Foods Hum Nutr* 34(2): 151–157

Brock JF, Hansen JDL, Howe EE, Pretorius PD, Daval JGA, Hendricks RG (1955) Kwashiorkor and protein malnutrition. A dietary therapeutic trial. Lancet 1955;2: 355–60

Broughton WJ, Hernández G, Blair MW, Beebe S, Gepts P, Vanderleyden J (2003) Beans (*Phaseolus* spp.) – model food legumes. Plant Soil 252: 55–128

Carrillo-Castaneda G, Juarez Munoz J, Jose R Peralta-Videa JR, Gomez E and Gardea-Torresdey JL (2005) Modulation of Uptake and Translocation of Iron and Copper from Root to Shoot in Common Bean by Siderophore-Producing Microorganisms. J Plant Nut 28: 1853–1865

Crowley DE, Romheld V, Marschner H and Szanizlo PJ (1992) Root microbial effects on plant iron uptake from siderophores and phytosiderophores. Plant Soil 142: 1–7

Daily GC (1997) Nature's Services: Societal Dependence on Natural Ecosystems. Island Press, Washington. 392pp

Deckelbaum RJ, Palm C, Mutuo P and DeClerck F (2006) Econutrition: Implementation models from the Millennium Villages Project in Africa. Food Nutr Bull 27 (4): 335–342

DeClerck F, Ingram JC, and Rumbaitis del Rio CM (2006) The role of ecological theory and practice in poverty alleviation and environmental conservation. Front Ecol Environ 4(10): 533–540

Diamond J (1997) Guns, Germs, and Steel: The Fates of Human Societies. WW Norton publisher. 480pp

Elsheikh EAE, Fadul IA and Tinay AHEI (2000) Effect of cooking on anti-nutritional factors and in vitro protein digestibility (IVPD) of faba bean grown with different nutritional regimes. Food Chemistry 68(2): 211–212

Elton CS (1927) Animal ecology. Sidgwick & Jackson, London.

FAO (1997) Human Nutrition in the Developing world. Food and Agriculture Organization of the United Nations, Rome, 508pp

FAO/FANTA (2007) Guidelines for Measuring Household and Individual Dietary Diversity. FAO, Rome

FAO (2008) The State of Food Insecurity in the World. Food and Agriculture Organization of the United Nations, Rome, 57pp

Flynn DFB , Gogol-Prokurat M, Nogeire T ,Molinari N ,Trautman Richers B, Lin BB , Simpson N, Mayfield MM and DeClerck F (2009) Loss of functional diversity under land use intensification across multiple taxa. Ecology Letters 12(1): 22–33

Foote J, Murphy S, Wilkens L, Basiotis P and Carlson A (2004) Dietary variety increases the probability of nutrient adequacy among adults. Journal of Nutrition 134: 1779–1785

Frison E, Smith IF, Johns T, Cherfas J and Eyzaguirre PB (2006) Agricultural biodiversity, nutrition, and health: Making a difference to hunger and nutrition in the developing world Food Nutr Bull 25: 143–155

Gamaleroa E, Fracchiaa L, Cavalettoa M, Garbayeb J, Frey-Klettb P, Varese GC M. G. Martinotti MG (2003) Characterization of functional traits of two fluorescent pseudomonads isolated from basidiomes of ectomycorrhizal fungi. Soil Biol Biochem 35: 55–65

Glick BR (1995) The enhancement of plant growth by free-living bacteria. Candian Journal of Microbiology pp. 109–117

Gliessman SR (2006) Agroecology: The ecology of Sustainable Food systems. CRC Press, 384pp

Graham RD, Welch RM, Saunders DA, IOrtiz-Monasterio I, Bouis HE, Bonierbale M, de Haan S, Burgos G, Thiele G, Liria R, Meisner CA, Beebe SE,Potts MJ, Kadian M, Hobbs PR, Gupta RK and Twomlow S (2007) Nutritious subsistence food systems. Advances Agronomy 92: 1–72

Graham RD (2008) Micronutrient Deficiencies in Crops and Their Global Significance. In Alloway BJ (eds) Micronutrient Deficiencies in Global Crop Production. Springer, pp. 41–62

Grosse SD and Roy K (2008) Long-term economic effect of early childhood nutrition. The Lancet 371(9610): 365–366

Hatloy, A., Hallund, J., Diarra, MM. and Oshaug A (2000) Food variety, socioeconomic status and nutritional status in urban and rural areas in Koutiala (Mali). Public Health Nutrition 3:57–65

Hotz C and Brown K (2004) Assessment of the Risk of Zinc Deficiency in Populations and Options for Its Control. International Zinc Nutrition Consultative Group (IZiNCG) Technical Document 1, 114pp

Kataki PK (2002) Food systems for improved human nutrition: Linking Agriculture, Nutrition, and Productivity. In Kataki PK and Babu SC Eds. Food systems for improved human nutrition: Linking Agriculture, Nutrition, and Productivity. pp. 7–29

Kataki PK and Babu SC (2002) Eds. Food systems for improved human nutrition: Linking Agriculture, Nutrition, and Productivity. The Haworth Press, Inc, Binghamton, 394pp

Kennedy G, Pedro MR, Seghieri C, Nantel G and Brouwer I (2007) Dietary diversity score is a useful indicator of micronutrient intake in non breast-feeding Filipino children. Journal of Nutrition 137: 1–6

Ma JF and Nomoto K (1996) Effective regulation of iron acquisition in graminaceous plants. The role of mugineic acids as phytosiderophores. Physiologia Plantarum 97 (3): 609–617

Masalha J, Kosegarten H, Elmaci O and Mengel K (2000) The central role of microbial activity for iron acquisition in maize and sunflower. Biol Fertil Soil 30: 433–439

Mason JB, and Garcia M (1993) Micronutrient deficiency—the global situation. SCN News 9, 11–16

McLaren DS (1966) A fresh look at protein calorie malnutrition. Lancet 1966;2:485–488

McLaren DS (1974) The great protein fiasco. Lancet 1974;2:93–96

McLean E, Cogswella M, Eglia I, Wojdylaa D and de Benoista B (2009) Worldwide prevalence of anaemia, WHO Vitamin and Mineral Nutrition Information System, 1993–2005. Public Health Nutrition doi:10.1017/S1368980008002401

Millennium Ecosystem Assessment (2005) Ecosystems and human well-being. Summary for decision makers. Island Press, Washington DC, 109pp

Mirmiran P, Azadbakht L, Esmaillzadeh A and Azizi F (2004) Dietary diversity score in adolescents- a good indicator of the nutritional adequacy of diets: Tehran lipid and glucose study. Asia Pacific Journal of Clinical Nutrition 13(1): 56–60

Molina-Torres J, García-Chávez A, Ramírez-Chávez E (1999) Antimicrobial properties of alkamides present in flavouring plants traditionally used in Mesoamerica: Affinin and capsaicin. J Ethnopharmacology 64: 241–248

Monteiro C, Conde WL, Lu B, Popkin BM (2004) Obesity and inequities in health in the developing world. Int J Obesity 28 (9): 1181–1186

Morton JF (1991) The horseradish tree, Moringa pterygosperma (Moringaceae)—A boon to Arid Lands? Econ Botany 45(3): 318–333

Mughal MH, Ali G, Srivastava PS, Iqbal M (1999) Improvement of drumstick (Moringa pterygosperma Gaertn.) - a unique source of food and medicine through tissue culture. Hamdard Med. 42: 37–42

Müller O and Krawinkel M (2005) Malnutrition and health in developing countries. Can Med Assoc J 173(3): 279–286

Muzquiz M (2004) Recent Advances of Research in Antinutritional Factors in Legume Seeds and Oilseeds. CHIPS books 370pp

Palada MC and Changl LC (2003) Suggested Cultural Practices for Moringa. International Cooperators' Guide AVRDC. AVRDC pub # 03-545, pp. 1–5

Palm CA, Sanchez PA, Ahamed S, Awiti A (2007) Soils: A Contemporary Perspective. Ann Rev Environ Res 32: 99–121

Petchey OL and Gaston KJ (2002) Functional diversity (FD), species richness, and community composition. Ecology Letters, 5, 402–411

Petri WA, Miller M, Binder HJ, Levine MM, Dillingham R and Guerrant RL (2008) Enteric infections, diarrhea, and their impact on function and development. J Clinic Invest 118(4): 1277–1290

Pickett STA and Cadenasso ML (2002) The ecosystem as a multidimensional

Pollan M (2008) Farmer in chief. New York Times Magazine. Ocober 12, 2008

Popkin BM (2001) The Nutrition Transition and Obesity in the Developing World. J Nutrition 131: 871S–873S

Popkin BM (2004) The nutrition transition: An overview of world patterns of change. Nutrition reviews 62 (7) S140–S143

Popkin B and Ng SW (2007) The nutrition transition in high- and low-income countries: what are the policy lessons? Agricultural Economics 37: S199–211

Quinn, D (1999) Beyond Civilization: Humanity's next great adventure. Harmony Books, 208pp.

Rees M, Condit R, Crawley M, Pacala S, Tilman D (2001) Long-Term Studies of Vegetation Dynamics. Science 27 293 (5530): 650–655

Risch SJ and Hansen MK (1982) Plant growth, flowering phenologies, and yields of corn, beans and squash grown in pure stands and mixtures in Costa Rica. J Appl Ecol 19: 901–916

Ruel MT and Levin CE (2002) Food based approaches for alleviating micronutrient malnutrition—an overview. J Crop Prod 2002; 6: 31–35

Ruel M, Graham J, Murphy S and Allen L (2004) Validating simple indicators of dietary diversity and animal source food intake that accurately reflect nutrient adequacy in developing countries. Report submitted to GL-CRSP

Sanchez P, Swaminathan MS, Dobie P, Yuksel N (2005) UN Millennium Project Task Force on hunger. Halving hunger: it can be done. Millennium Project, 245pp

Sanchez P and Swaminathan M (2005) Hunger in Africa: the link between unhealthy people and unhealthy soils. Lancet 365: 442–444

Savelkoul FH, van der Poel AF and Tamminga S (1992) The presence and inactivation of trypsin inhibitors, tannins, lectins and amylase inhibitors in legume seeds during germination. A review. Plant Foods Hum Nutr 42(1):71–85

Sabale V, Patel V, Paranjape A, Arya C, Sakarkar SN and Sabale PM (2008) Moringa Oleifera (Drumstick): An Overview. Phcog Rev 2(4): 7–13

Sherman PW, Billing J (1999) Darwinian gastronomy: Why we use spices. Bioscience 49: 453–463

Siddhuraju P and Becker K (2003) Antioxidant properties of various solvent extracts of total phenolic constituents from three different agro-climatic origins of drumstick tree Moringa oleifera Lam.). J. Agric. Food Chem. 15: 2144–2155

Sillanpää M (1990) Micronutrient assessment at the country level: An international study, FAO Soils Bulletin 63. FAO/Finnish International Development Agency, Rome, Italy

Somali MA, Bajnedi MA and Al-Fhaimani SS (1984) Chemical composition and characteristics of Moringa peregrina seeds and seeds oil. J. Am. Oil Chem. Soc. 61, pp. 85–86

Stevenson FJ (1991) Organic matter-micronutrient reactions in soil. In Micronutrients in Agriculture. 2nd edition. Soil Sci. Soc. Amer., Inc., Madison, WI.

Stevenson FJ (1994) Humus Chemistry 2nd edition. John Wiley & Sons, Inc, New York

Steyn NP, Nel JH, Nantel G, Kennedy G and Labadarios D (2006) Food variety and dietary diversity scores in children: are they good indicators of dietary adequacy? Public Health Nutrition 9(5): 644–650

Stoltzfus RJ and Dreyfuss ML (1998) Guidelines for the Use of Iron Supplements to Prevent and Treat Iron Deficiency Anemia.

Tewksbury J, Reagan K, Machnicki N, Carlo T, Haak D, Peñaloza AC, Levey D (2008) Evolutionary ecology of pungency in wild chilies. Proc Natl Acad Sci USA 105: 11808–11811

UNICEF (2008) The State of the World's Children 2009. United Nations Children's Fund, New York 47pp

United Nations Administrative Committee on Coordination/ Subcommittee on Nutrition (ACC/SCN) (2000) Fourth report on the world nutrition situation: ACC/ SCN in collaboration with the International Food Policy Research Institute (IFPRI)

Vityakon P, Standal BR (1989) Oxalate in vegetable amaranth (*Amaranthus gangeticus* L). Forms, contents and their possible implications for human health. J Sci Food Agric 48 469È474.

Welch RM (2008) Linkages Between Trace Elements in Food Crops and Human Health. In Alloway BJ (eds) Micronutrient Deficiencies in Global Crop Production. Springer, pp. 287–310

White JG and Zasoskib RJ (1999) Mapping soil micronutrients. Field Crops Research 60: 11–26

Wilson EO (1999) The Diversity of Life (2nd edition) WW Norton & Company, 424pp

Wink M (1997) Resistance of grain legumes towards insects and other pests: Role of plant secondary metabolites. Grain Legumes 15: 16–18

Wink M (2004) Evolution of toxins and anti-nutritional factors in plants with special reference to Leguminosae. In T. Acamovic, C.S. Stewart, T.W. Pennycott (eds). Poisonous plants and related toxins. CABI Publishing,Wallingford, GB, pp. 1–15

World Bank (2006) Repositioning Nutrition as central to development. The World Bank, Washington DC, 23pp

World Health Organization (2002) The World Health Report 2002. Reducing Risks, Promoting Healthy Life (B. Campanini, Ed.) (2002). World Health Organization, Geneva, Switzerland, 168pp

World Health Organization (WHO) (2003) Diet, Nutrition and the Prevention of chronic diseases. Report of a joint WHO/FAO Expert Consulation. World Health Organization, Geneva, Switzerland, 160pp

World Health Organization (WHO) (2008) Worldwide prevalence of amaemia 1993–2005. WHO Global Database of Anaemia. World Health Organization, Geneva, Switzerland, 51pp

Xing FB, Cheng GX, Yi KK (2006) Study on the antimicrobial activities of the capsaicin microcapsules. J Appl Polym Sci 102: 1318–1321

Chapter 5
Landscape Approaches to Achieving Food Production, Natural Resource Conservation, and the Millennium Development Goals

Jeffrey C. Milder, Louise E. Buck, Fabrice DeClerck, and Sara J. Scherr

The Río Copán watershed in western Honduras is not unlike many agricultural landscapes throughout the developing world. A journey through this 800 km^2 watershed reveals a mixture of small and mid-sized farms producing cattle, coffee, and subsistence crops. Residents here face many challenges: recent population growth has led to deforestation and water pollution, while agricultural productivity is generally low and poverty levels remain high, especially among the indigenous Mayan population.

Environmental degradation is both a cause and a consequence of these problems. Poverty has driven many local people to cut wood in the vanishing native pine-oak forests or to cultivate or graze hillsides that are too steep for these purposes. Such practices, in turn, contribute to silted rivers unsuitable for human or livestock consumption and to landslides that routinely close roads and isolate villages from needed goods and services for weeks or months at a time. To meet the Millennium Development Goals (MDGs) in the Río Copán watershed will require not just new schools, new health centers, and new crop varieties; it will require a suite of coordinated activities, many of them focused on environmental restoration and natural resource management.

Fortunately, unlike many rural communities that address poverty issues piecemeal at the household or village level, Copán's communities have recognized that these challenges grow from—and, in turn, influence—key dynamics and ecosystem processes operating at the scale of the entire watershed, and sometimes beyond. For local leaders, the wake-up call that spurred this landscape-level thinking arrived suddenly, drenching them, quite literally, like a bucket of cold water from above. In 1998, Hurricane Mitch tore through the region, wreaking havoc not just on devegetated hillsides but on the farms, villages, waterways, and infrastructure below.

J.C. Milder (✉)
Department of Natural Resources, Cornell University, Ithaca, NY 14853, USA
e-mail: jcm85@cornell.edu

J.C. Ingram et al. (eds.), *Integrating Ecology and Poverty Reduction: Ecological Dimensions*, DOI 10.1007/978-1-4419-0633-5_5,

After taking stock of the extensive damage, the four municipalities in the watershed decided to band together to form a regional coalition aimed at preventing such devastation in the future, and at finding solutions to shared problems such as erosion, water pollution, and poor human health. They created a vision and plan for the watershed's future and, for the past several years, have been using this plan to target and guide externally-funded rural development activities. The problems and challenges in the watershed are not solved, but their root causes and interactions are now better understood. This knowledge encourages leaders to find solutions that do not trade off one landowner's wellbeing for another's, or one development objective for another, but that seek to maintain and restore the landscape's natural and human capital for the benefit of all.

Integrating Rural Development and Natural Resource Management

Leaders in the Río Copán watershed have learned through experience what thousands of scientists have documented over the past two decades: ecosystem services are critical to human wellbeing, especially in rural landscapes in developing countries. The Earth's natural capital of clean water, soils, fish, wildlife, and other resources provides about two-thirds of household income for the rural poor (MA 2005) and 26% of all wealth in low-income countries (World Bank 2006). Environmental causes are responsible for nearly one-fourth of the global disease burden, and more than four million children die each year from illnesses such as diarrhea, malaria, and respiratory infections that could be significantly mitigated by improved environmental management (Prüss-Üstün and Corvalán 2006). In light of the fundamental role of natural capital in supporting human wellbeing, it is especially worrisome that 15 of the 24 key ecosystem services upon which humans depend are being degraded or used unsustainably (MA 2005).

World leaders and major development funding agencies have acknowledged that environmental factors are either at the root of, or closely linked to, MDGs 1–6—those relating to food security, human health, education, and gender equality (Sachs and Reid 2006; DFID et al. 2002; DFID 2006). This connection means that much of the recent progress toward meeting the MDGs (UN 2008) is likely to be fleeting if the natural capital that underlies these improvements continues to decline (WRI 2005). Yet, despite these well-documented linkages, the treatment of the environment in the MDGs "...harkens back to old, outmoded ways of thinking" (WRI 2005:154). Rather than being framed as a cross-cutting theme that underlies the long-term achievement of other poverty alleviation goals, the environment is addressed only in MDG 7. And although the revised MDG targets and indicators issued in 2008 provide more specific measures of success for MDG7, these measures still fail to address many of the aspects of environmental management that are most relevant for sustaining the ecosystem services that are critical for poverty alleviation (WRI 2005; DFID 2006).

Unfortunately, this inattention to natural capital as a foundation of human well-being has been reflected in global funding priorities and implementation frameworks for poverty alleviation. For example, Poverty Reduction Strategy Papers (PRSPs)—the vehicle by which national governments formulate objectives for meeting the MDGs and establish their priorities for international aid—have often paid insufficient attention to the environment (Bojö et al. 2004; WRI 2005). This undervaluing of environmental factors is likely a result both of the stated priorities of the aid agencies themselves (World Bank and IMF 2005) and of the apparent tendency of some governments preparing PRSPs to favor more fundable infrastructure projects over environment and agriculture projects identified as priorities by local communities and district-level agencies (Swallow 2005). The general result, at the field level, has been an overly sectoral approach to rural development that neither integrates environmental and livelihood objectives nor adequately addresses the environmental drivers underlying development goals (Sanderson 2005).

In light of these shortcomings, many have argued that the rural development agenda must be reformulated to integrate environmental sustainability at all scales, from international funding priorities to on-the-ground projects. This chapter suggests that such integration needs to include a strong focus on the landscape scale—the level at which many ecosystem processes operate and at which interactions among environment and development objectives are often mediated (O'Neill et al. 1997) (see Box 5.1). For example, in many landscapes, conservationists and rural development advocates have both targeted the same land or water resources for advancing their respective objectives—often with little communication or recognition of the conflicts between these aspirations (Wood et al. 2000; McNeely and Scherr 2003). In such situations, landscape-scale assessment, negotiation, planning, and monitoring can help identify actions and policies that increase synergies while decreasing tradeoffs (Palm et al. 2005). On the other hand, if tradeoffs are not explicitly acknowledged and addressed through negotiated solutions, sectoral programs and investments will move forward in isolation, leading to composite outcomes that are likely to be far sub-optimal, especially for less powerful stakeholders.

The purpose of this chapter is to explore the theory and practice of landscape approaches to sustainable rural development and to illustrate the ways in which this paradigm can be applied to address the MDGs. The chapter begins by introducing and reviewing existing landscape approaches. Next, we present the Landscape Measures framework, a landscape approach that we developed specifically for use in 'ecoagriculture' landscapes where food production is a key objective. We then introduce some tools for implementing the Landscape Measure approach, focusing on those that apply ecological knowledge and methods. We illustrate the use of such tools by elaborating on the Copán case study introduced above as well as a recent project in Kenya. Finally, we conclude by identifying important actions for mainstreaming landscape approaches to help achieve the MDGs.

Box 5.1 Why Use a Landscape Perspective to Address Food Security and Rural Poverty?

The reasons for working at a landscape scale stem not only from the biophysical realities of how natural resource-dependent systems function, but also from the growing interdependence and interconnectedness of rural regions. Motivations include:

1. *Scale of key ecological functions and processes*. Recent scientific research has demonstrated that flows of water, nutrients, sediment, plants, animals, and disease organisms in agricultural regions often operate beyond the farm or village level to encompass the entire landscape (Forman 1995). Many of these flows are critical to human wellbeing, providing ecosystem services such as clean water for human consumption, irrigation water, and natural pest control. Major threats, such as insect-borne diseases, crop and livestock predation, and various natural disasters, are also mediated at the landscape scale.
2. *Scale of key institutional frameworks*. In many developing nations, government authority and social programs have been devolved to smaller units of government operating at the district level (Molnar et al. 2007). At the same time, villages, communities, and NGOs are increasingly forming partnerships, networks, and alliances to addressed shared objectives (Pretty and Ward 2001). Both trends create opportunities to analyze and address challenges at a landscape scale. Conversely, inaction or ineffective policies at the landscape or sub-regional levels can keep rural households mired in "poverty traps" even when effective action is taken at the farm or village scale (Barrett and Swallow 2006). Thus meso-scale institutional arrangements are especially important in determining whether rural communities can spring out of self-reinforcing poverty traps.
3. *Changing face of the rural agricultural economy*. Throughout the world, the role of subsistence farming is in decline, while market-linked agriculture becomes more widespread, even among small farmers. This trend is being reinforced by development and aid agencies, many of whom emphasize market access and rural enterprise development in their programs (WRI 2008). As rural communities become more tied to one other, more dependent on physical infrastructure and regional markets, and more influenced by global economic forces, it is necessary to widen the lens through which rural livelihoods are understood and advanced.
4. *New market opportunities*. Markets are beginning to place value on rural land uses that protect or enhance ecological values. Eco-certification allows producers to receive price premiums for ecologically friendly production practices, while payments for ecosystem services compensate land stewards for protecting carbon stocks, biodiversity, or watershed functions.

(continued)

Box 5.1 (continued)

These new market opportunities will shift incentives for rural land managers and motivate a greater focus on management at the landscape or watershed scale, where many ecosystem services are mediated.

5. *Climate change*. Resulting largely from anthropogenic forcing mechanisms, climate change is occurring faster and more dramatically than at any time in recent history. Without greater emphasis on resilience, adaptation, and regional cooperation to accommodate shifting patterns of agricultural suitability, water availability, and habitat quality, these rapid climate shifts could easily undermine local development or conservation successes (Fairhead 2004).
6. *Increased emphasis on resilience and adaptation*. The reality of climate change combined with ecologists' recognition of ecosystems as dynamic, non-equilibrium systems has led to an increased interest in resilience and adaptation as important objectives for rural landscapes (Sayer and Campbell 2004). As population growth and ecosystem degradation combine to create increasingly thin margins of error for human wellbeing in many landscapes, the ability to re-evaluate circumstances and adapt management solutions based on new information will be critical for human wellbeing (Diamond 2004). Doing so requires the continual development and use of knowledge at appropriate scales within an adaptive management framework (Röling and Wagemakers 1998; Plummer and Armitage 2007).

An Introduction to Landscape Approaches

Notwithstanding the limitations of current mainstream rural development priorities, many rural land stewards, non-governmental organizations (NGOs), researchers, and supporters have come to embrace the complexity of rural landscapes and have developed evidence-based management approaches that address the spatial, thematic, and human scope of the challenges themselves (Lal et al. 2001). We refer to these as landscape approaches and suggest that they have five defining characteristics: (1) a landscape-scale focus, (2) treatment of landscapes as complex systems, (3) management for multiple objectives, (4) adaptive management, and (5) management through participatory processes of social learning and multi-stakeholder negotiation. Each of these characteristics is discussed below.

First and most obviously, landscape approaches seek to address livelihood needs and environmental challenges at a landscape scale. There are many possible ways to define landscapes, but for management purposes it is helpful to define them functionally according to the objectives at hand and the physical extent of the features and processes that mediate these objectives (Buck et al. 2006). Precise boundaries are often ambiguous because the various biophysical gradients, socio-cultural attributes, and political jurisdictions found on the land operate at multiple scales and rarely coincide with one another. Thus, landscape approaches incorporate multi-scale linkages, helping to coordinate small-scale management efforts while considering relevant aspects of the landscape's regional and global context.

Second, landscapes are analyzed as complex systems—that is, assemblies of interconnected components that are expected to fulfill a specific set of purposes (Collins et al. 2007). Recent research on coupled human and natural systems has solidified the analytical foundations for understanding the reciprocal influences between humans and their environment at multiple scales (Liu et al. 2007). This field proposes increased emphasis on indirect linkages, feedbacks, and multi-temporal analysis when investigating or managing properties of interest such as the resilience and vulnerability of agroecosystems, which, by definition, encompass human goals, human behavior, and ecosystem dynamics. A range of methods for aiding in such analysis already exists, including system dynamics modeling, agent-based modeling, and various GIS-based tools. For example, Parker and colleagues (2003) illustrate how multi-agent system models of land use/land cover change can elucidate feedbacks between land stewards and the environment in the (very common) circumstance where landscape change is largely a composite outcome of numerous of household-level decisions. In practice, coupled systems thinking can help policy makers anticipate future trends, manage interactions among landscape components, and expose "blind spots" that can emerge from unanticipated feedbacks (Maarleveld and Dangbegnon 1999).

Third, landscape approaches manage for multiple objectives, among which there are likely to be both synergies and tradeoffs. Multi-objective management is essential when landscapes are expected to provide more than one type of product or service—as indeed most landscapes are—and when stakeholders disagree on the goals of management and their relative importance. Furthermore, indicators for the various management goals are likely to be non-commensurable ('apples and oranges') such that it is difficult to define any aggregate measure of landscape success even if the relative importance of each goal can be ascertained (Munda 2005; López-Ridaura et al. 2005). For this reason, multi-objective management is rarely amenable to the type of optimization algorithms that have transformed the management of single-objective initiatives such as maximizing corporate profitability or designing the most cost-effective system of nature reserves (Röling 2002). Instead, multi-objective initiatives are likely to be understood and reported using a combination of quantitative and qualitative metrics that track whether the landscape is progressing toward the sustainable provision of the desired environmental and socioeconomic outcomes (Buck et al. 2006).

Fourth, landscape approaches are predicated on adaptive management: "...a formal, systematic, and rigorous approach to learning from the outcomes of management actions, accommodating change and improving management" (Nyberg 1999). Adaptive management is essentially the scientific method applied to real-world challenges. Resource managers begin by hypothesizing models of cause and effect, then test these models through specific interventions and policies, monitor the outcomes of these interventions, and use the resulting information to refine the causal models and improve the interventions. Over time, managers become more knowledgeable about the system and better able to respond to changing conditions, thereby increasing the resilience of ecosystems and communities in the face of natural and anthropogenic dynamics (Folke et al. 2002). Adaptive management has its intellectual roots and early experience in ecosystem management (Holling 1978) and is now widely viewed as the preferred approach for addressing complex natural resource management challenges amid incomplete information (Lee 1993; Salafsky et al. 2001). More recent formulations of this paradigm recognize that resource management is not simply a technical puzzle to be solved through better information, analysis, and planning. It is a social dilemma in which the perceptions, priorities, capabilities, and negotiation capacity of land stewards and institutions determine sustainability at least as much as the management practices themselves (Ison et al. 2007; Röling 2002). These ideas underlie the practice of adaptive collaborative management, which positions 'experts' and their technical tools in the role of facilitators or technical advisors to assist a process that is guided by stakeholders themselves (Buck et al. 2001; Colfer 2005).

This leads to the fifth and final characteristic of landscape approaches: an ongoing, participatory process of 'social learning' through which stakeholders iteratively discover and generate relevant knowledge, negotiate goals and objectives, implement management plans, and evaluate outcomes (Leeuwis and Pyburn 2002; Steyaert et al. 2007). In the context of adaptive management, social learning encourages stakeholders to articulate and discuss their understanding of reality and mental models of cause and effect when formulating goals, objectives, and plans (van Noordwijk et al. 2001). These understandings are refined over time based on evidence from project monitoring as well as external sources. Because it provides a built-in mechanism for incorporating new information and responding to novel circumstances, social learning is essential for ensuring the sustainability and resilience of human and natural systems (Röling and Wagemakers 1998; Olsson et al. 2004).

Contemporary Uses of Landscape Approaches

We conducted a literature review to identify the ways in which landscape approaches have been used to address rural poverty and natural resource conservation challenges. This section provides a brief history of the development of landscape approaches and some leading examples of recent practice.

The roots of landscape approaches can be traced to the emergence of the sustainable development concept in the late 1980s (WCED 1987; Lele 1991). This framework ushered in a wave of Integrated Conservation and Development Projects (ICDPs) that included both rural development and environmental (particularly biodiversity protection) objectives. However, the outcomes of ICDPs proved generally to be disappointing. In many projects, the nexus between the development activities and conservation objectives was poorly conceived or fallacious: win-win solutions were assumed rather than acknowledging and addressing tradeoffs. Furthermore, local participation was often token, resulting in mis-directed efforts yielding transient benefits that evaporated when project funding ended (McShane and Wells 2004). Some observers blamed these failures on fundamental flaws in the integrated project model itself (Terbough 1999) while others argued that the basic ideas were sound but had not been fully embraced in most first-generation ICDPs (Brechin et al. 2003). In retrospect, we can say that these projects aspired to multi-objective rural land management but typically lacked most of the other attributes of landscape approaches, such as adaptive management in a social learning context. These omissions were often important causes of the projects' shortcomings.

The disappointing results of early ICDPs coincided with a growing awareness of ecosystem services and their role in sustaining society (Daily 1997; Costanza et al. 1998). This theme was echoed in the 1998 systemwide review of the Consultative Group on International Agricultural Research (CGIAR), which urged the 16 CGIAR centers to move beyond crop research to advance the field of natural resource management to support global food production (CGIAR 1998). Building on earlier formative work by the World Agroforestry Center and Center for International Forestry Research, the centers responded by adopting a program on Integrated Natural Resource Management (INRM), which they defined as a research and management approach that "...aims at improving livelihoods, agroecosystem resilience, agricultural productivity and environmental services [by] augment[ing] social, physical, human, natural and financial capital" (ICARDA 2005). While INRM is not specifically a landscape approach, it envisions management and analysis at multiple nested scales including that of the landscape (Campbell et al. 2001; Izac and Sanchez 2001). Recent INRM initiatives by several of the CGIAR centers have included a strong landscape emphasis and illustrate how 'action research' can facilitate stakeholder dialogue, planning, and management for conservation, food production, and livelihood objectives (Gottret and White 2001; Frost et al. 2006; Pfund et al. 2008).

Despite these promising initiatives, landscape-level planning and analysis does not yet play a significant role in mainstream agricultural investment, management, or policy. Nevertheless, there is some tradition of spatial thinking in agriculture, and this is gradually expanding to encompass larger scales and broader disciplinary foci. For instance, agricultural investment decisions are commonly made using spatially-sensitive methods such as agroecological suitability classification (based on factors such as altitude, rainfall, and soil type) and market analysis (based on transportation costs, access to inputs, value chain mapping, and distance to storage or processing facilities). Spatial zoning for agriculture is now becoming more nuanced, with certain agricultural uses contingent on the adoption of conservation management practices. Farmers are increasingly choosing to coordinate across sites to address challenges such as pest control, salinization, and limited availability of irrigation water. Such efforts are being supported by new scientific tools such as spatial modeling of nutrient flows, and by new policy instruments such as nutrient trading systems. The concept of foodsheds has encouraged more systematic spatial analysis of food supplies and value chains around major population centers (Kloppenburg et al. 1996). All of these approaches are beginning to increase the scale at which agricultural management is considered as well as the level of integration among production, conservation, and livelihood dimensions.

Concurrently, conservationists have begun to implement landscape approaches such as biological corridors, landscape-scale conservation planning, and green infrastructure planning to address the challenges of habitat fragmentation and ecosystem degradation in populous regions (Rosenberg et al. 1997; Benedict and McMahon 2006). Many such projects seek to address livelihood needs in concert with biodiversity conservation by engaging private and communal land stewards in transitioning to more conservation-friendly agriculture and livelihood strategies (e.g., Miller et al. 2001). A new generation of multi-objective landscape-scale projects by groups such as WWF and the Wildlife Conservation Society can be seen as a maturation of the ICDP concept to embrace genuine local participation and a broader set of spatial and temporal scales to address the poverty-biodiversity nexus (USFS 2006; Redford and Fearn 2007; COMACO 2009). For instance, the IUCN/WWF Forest Landscape Restoration initiative aims to restore ecosystem goods and services by increasing tree cover in degraded landscapes while engaging stakeholders to address institutional barriers at multiple scales (Barrow et al. 2002; Sayer and Buck 2008). A complementary process for landscape monitoring and adaptive management has also been developed, which uses the Capital Assets Framework (Carney 1998) to track multiple landscape variables and to use this information to aid in participatory decision-making (Sayer et al. 2007).

The preceding examples were of landscape approaches initiated by international NGOs and research centers. However, much of the impetus for landscape-level planning and management emerges from local and regional initiatives. For example, the practice of participatory watershed management arose as an alternative to ineffective top-down watershed planning. In this approach, priorities are negotiated at the watershed scale but implemented at the community level through micro-watershed plans focused on practices such as re-vegetation, soil management, and erosion control (Hinchcliffe et al. 1999; Kerr 2002). More generally, the concept of community-based natural resource management has been widely applied to forest, water, wildlife, rangeland and other common property or state-owned resources to secure tenure rights and support collective management and shared benefits (Borrini-Feyerabend et al. 2000; Leach et al. 1999). At a larger scale, the concept of territorial management has been used to assert local control over rural development processes, including land and resource use. This approach is best developed in Latin America, where it has been applied in the context of indigenous reserves as well as mainstream planning for rural areas (Sepúlveda et al. 2003).

Overall, our analysis revealed many instances of both community-led and externally driven initiatives that met three or four of the characteristics of landscape approaches described above, but relatively few that met all five. Of those cases that exhibited all five characteristics, most were being carried out in forested landscapes where the objective was to reconcile biodiversity conservation and poverty alleviation. To our knowledge, landscape approaches have rarely been applied to areas where cropland or rangeland is a major land use and where food production for a large local population is a central goal.

Ecoagriculture and the Landscape Measures Approach

The lack of methods and tools for landscape-scale management and monitoring of agroecosystems was a frequent theme at the first Ecoagriculture Conference and Practitioners' Fair in Nairobi in 2004. Many of the researchers, government and NGO representatives, community leaders, donors, and farmers at the meeting were involved in implementing or promoting ecoagriculture—that is, efforts to simultaneously achieve food production, conservation, and rural livelihood goals at a landscape level (McNeely and Scherr 2003; Scherr and McNeely 2008). Conference participants could point to many examples where ecoagriculture principles had been implemented successfully. Yet, their ability to sustain, document, and scale up these successes was limited by the dearth of existing frameworks or processes for planning and monitoring ecoagriculture landscapes. What was needed was a landscape approach that spoke to the particular issues and challenges of ecoagriculture contexts where food production (cropping, livestock, agroforestry, or fisheries) comprises a significant portion of the land base and the local economy.

The Landscape Measures approach (LM), which we describe and illustrate in the remainder of this chapter, addresses this need. Developed as part of Ecoagriculture Partners' Landscape Measures Initiative (LMI), the LM consists of a set of processes and tools for negotiating, planning, implementing, and evaluating ecoagriculture practices and innovations (Buck et al. 2006). Like other landscape approaches, the LM is predicated on stakeholder-driven adaptive management embedded in a social learning process (see Fig. 5.1). However, the LM is designed around the four major goals of ecoagriculture: (1) conserving biodiversity and ecosystem services, (2) producing food, (3) improving rural livelihoods, and (4) building effective institutions for cross-sector planning, analysis, and action. As such, the LM includes monitoring tools and methods specifically oriented toward these goals and toward measuring and negotiating the interactions among them.

The Landscape Measures Framework

One of the salient challenges of working at a landscape scale is to incorporate the important goals, processes, and dynamics into adaptive management without getting mired in excessive detail and layers of complexity (Lynam et al. 2007). To address this challenge, the LMI conducted a year-long consultative process that engaged scientists and practitioners from diverse disciplines and sectors in conversations about how to track change across multiple dimensions at landscape scale (Buck et al. 2006). One outcome of these conversations was a set of "20 Questions" about landscape performance that represented the key variables that are likely to be important in ecoagriculture landscapes worldwide (Buck et al. 2006; see Box 5.2). The 20 Questions offer tangible criteria for assessing progress toward the four broad goals of ecoagriculture. In turn, stakeholders can answer the questions by selecting

Fig. 5.1 Key roles of the Landscape Measures approach (*LM*) for guiding adaptive management for food production, conservation, and livelihoods in rural landscapes. The standard adaptive management cycle is depicted in *gray*, while key LM processes and tools for each phase of the cycle are shown as *black ovals*

Box 5.2 Twenty Questions for Assessing the Performance of Ecoagriculture Landscapes

Conservation goal: The landscape conserves, maintains, and restores wild biodiversity and ecosystem services.

Criterion C1: Does the landscape contain an adequate quantity and suitable configuration of natural and semi-natural habitat to protect native biodiversity?

Criterion C2: Do natural and semi-natural habitats in the landscape approximate the composition and structure of the habitats historically found in the landscape?

Criterion C3: Are important species within the landscape biologically viable?

Criterion C4: Does the landscape provide locally, regionally, and globally important ecosystem services?

Criterion C5: Are natural areas and aquatic resources degraded by productive areas and activities?

(continued)

Box 5.2 (continued)

Production goal: The landscape provides for the sustainable production of crops, livestock, fish, forests, and wild edible resources.

Criterion P1: Do production systems satisfy demand for agricultural products (crops, livestock, fish, wood) by consumers inside and outside the landscape?

Criterion P2: Are production systems financially viable and can they adapt to changes in input and output markets?

Criterion P3: Are production systems resilient to disturbances, both natural and human?

Criterion P4: Do production systems have a neutral or positive impact on wild biodiversity and ecosystem services in the landscape?

Criterion P5: Are species and varietal diversity of crops, livestock, fisheries and forests adequate and maintained?

Livelihoods goal: The landscape sustains or enhances the livelihoods and wellbeing of all social groups who reside there.

Criterion L1: Are households and communities able to meet their basic needs while sustaining natural resources?

Criterion L2: Is the value of household and community income and assets increasing?

Criterion L3: Do households and communities have sustainable and equitable access to critical natural resource stocks and flows?

Criterion L4: Are local economies and livelihoods resilient to change in human and non-human population dynamics?

Criterion L5: Are households and communities resilient to external shocks such as flooding, drought, changes in commodity prices, and disease epidemics?

Institutions goal: The landscape hosts institutions that support the planning, negotiation, implementation, resource mobilization, and capacity-building needed to integrate conservation, production and livelihood functions.

Criterion I1: Are mechanisms in place and functioning for cross-sectoral interaction at landscape scale?

Criterion I2: Do producers and other community members have adequate capacity to learn and innovate about practices that will lead to integrated landscapes?

Criterion I3: Does public policy support integrated landscapes?

Criterion I4: Are market incentives conducive to integrated landscapes?

Criterion I5: Do knowledge, norms, and values support integrated landscapes?

Source: Buck et al. (2006) and LMI (2009).

Table 5.1 Hierarchical framework of the Landscape Measures approach (LM) for identifying and tracking progress toward landscape objectives. Similar to other recent methods for landscape evaluation (e.g., CIFOR 1999; LAC-Net 2006), this hierarchical approach helps ensure that all major system components are considered while leaving room to interpret these components in relation to the landscape's specific biophysical and socio-cultural context

Hierarchical level	Selection process	Description
Goals	Universal; part of the LM framework	Comprises the four broad goals of ecoagriculture: sustainable food production, viable rural livelihoods, conservation of biodiversity and ecosystem services, and effective supporting institutions
Criteria	Universal; part of the LM framework	The 20 Questions, which enumerate five specific sub-goals for each of the four ecoagriculture goals
Indicators	Place-specific; selected by stakeholders	Tangible factors or characteristics in the landscape that are measured to reveal how well each criterion is being fulfilled. Stakeholders select indicators that are relevant to the landscape context and to their specific objectives
Means of measure	Place-specific; selected by stakeholders	Methods or techniques for evaluating indicators, such as land cover analysis or household interviews. Stakeholders select means of measure that are appropriate to the desired level of precision and availability of monitoring resources

and evaluating context-appropriate indicators and means of measure (see Table 5.1). Because many of the 20 Questions focus explicitly on the interactions among conservation, food production, rural livelihoods, and supporting institutions, they can help spur cross-sector dialogue and encourage stakeholders to negotiate tradeoffs among competing interests rather than avoiding such important conversations.

The 20 Questions provide a useful complement to the MDG goals, targets, and indicators for monitoring the performance and sustainability of rural landscapes. Whereas the targets for MDGs 1–6 are focused on specific human wellbeing outcomes, the 20 Questions help elucidate some of the ecological drivers that undergird long-term human wellbeing in rural landscapes. In addition, the 20 Questions offer a more detailed framework for monitoring MDG 7 (environmental sustainability) by focusing on local and landscape-scale ecosystem structure and function. The LM thus helps to address recent calls for improved monitoring of ecosystem services in assessing progress toward the MDGs—for example, by tracking soil fertility, hydrological function, and the maintenance of biodiversity, as well as the ways in which local people value, utilize, and sustain such ecosystem services (WRI 2005).

The LM is designed to be used in all phases of the adaptive management cycle, including goal setting, planning, and monitoring (see Fig. 5.1):

Goal setting and stakeholder negotiation. The framework and 20 Questions provide a 'roadmap' to landscape multi-functionality, identifying those functions that local and external stakeholders typically expect a landscape to fulfill. In our experience, nearly all of these 20 factors have proven relevant in landscapes across a diverse range of contexts. By providing a broad view of what would constitute successful landscape management, the framework can also help ensure that goals are not skewed too far toward or away from any single interest group. Under-represented stakeholders are given greater legitimacy in negotiations while all participants are encouraged to consider landscape processes or objectives that may be outside their ordinary purview.

Landscape planning. In rural landscapes in developing countries, there is a significant history of spatial planning for single objectives or projects (plantation forestry, large-scale agriculture, conservation networks, and so forth), but much less experience with multi-functional landscape planning (Selman 2002). Such planning can identify and promote synergies among disparate landscape objectives to a much greater degree than sectoral plans that optimize for a single outcome. Essentially, multi-functional landscape planning for ecoagriculture is the process of making the 20 Questions spatially explicit by establishing land and resource use parameters that can be implemented locally. The resulting spatial plans will often have a high proportion of multi-use zones (such as agroforestry or rotational grazing), substantial integration of activities on the landscape, and a relatively fine spatial resolution, reflecting the knowledge-intensive, ecosystem-based management that is proposed (Scherr et al. 2009). Integrated planning can also help ensure that sectoral plans are consistent with broader goals and will register positively against multiple criteria in the LM framework. Although landscape planning requires technical expertise, the process need not be controlled by outside experts; indeed, facilitated multi-objective planning processes can be an effective vehicle for engaging diverse stakeholders to influence management and policy outcomes (Wollenberg et al. 2000).

Landscape monitoring. One constraint to the use of ecosystem-based approaches to poverty alleviation is the inadequacy of environmental monitoring systems in many parts of the developing world (WRI 2005:161). Tracking landscape change requires going beyond project-based evaluation monitoring that focuses on a small set of landscape variables that the project expects to influence. Instead, monitoring should track all key system components such that it can reveal unexpected results of interventions as well as complex interactions of policy or management changes with other landscape dynamics. The LM helps define the scope of landscape monitoring by identifying a series of objectives for which stakeholders can select context-appropriate indicators for measuring progress over time. Data on these indicators then feeds back into the social learning process, expanding the base of information upon which future plans and decisions are made (Sayer and Campbell 2004).

Implementation Process

As with other landscape approaches, the LM is implemented through a process of social learning and negotiation among landscape stakeholders to adaptively manage land, natural resources, capital assets, and market and policy structures.

Consistent with the multi-scaled nature of landscapes, adaptive management must engage participants at many levels. Local participation and leadership are essential, but external stakeholders and higher-level agencies must also be represented to the extent that they have a legitimate interest in the landscape. Processes that fail to engage external actors who have the will and power to exert significant influence (such as agri-business companies or international NGOs) are naïve and unlikely to be successful. Instead, conflict and trade-offs between local and external interests must be acknowledged and clarified so that negotiation can occur.

Implementation of the LM usually requires a 'landscape facilitator'—individual(s) or organization(s) who work on a systematic and sustained basis to convene stakeholders, guide negotiation, manage information, and promote collective action (Laumonier et al. 2008; Buck and Scherr 2009). Steyaert and Jiggins (2007) define facilitation as "...a combination of skills, activities and tools used to support and guide learning processes among multiple interdependent stakeholders [to] bring about systemic change in complex situations...." Ideally, the landscape facilitator should be a neutral party that is dedicated only to the social learning process itself, as guided by the 20 Questions—not to any specific outcomes. Truly disinterested parties are rarely available as they have little incentive to participate; instead, facilitators are often drawn from the ranks of NGOs and research organizations, which often have a disciplinary or normative bias, if not a deliberate agenda. In these cases, facilitators must be scrupulous in acknowledging their biases and working to subordinate them to the larger process.

One key role of the landscape facilitator is to integrate stakeholders' disparate knowledge systems, data needs, and ways of communicating and using information. Past experience indicates that for scientific information to support sustainable development, greater efforts are needed to bridge the realms of knowledge generation and decision-making by ensuring that information is credible, salient, and legitimate to decision makers (Cash et al. 2003; Dietz et al. 2003). Yet, farmers, government agencies, and international donors each have very different conceptions of credibility, salience, and legitimacy. Furthermore, knowledge of rural landscapes can be rooted in many different epistemologies. Landscape level innovation systems integrate experiential or 'tacit' knowledge—gained by people who live in the landscape and are intimately familiar with aspects of its workings over time—with evidence of phenomena that are revealed through scientific inquiry and likely to be less visible to local people. Combining these approaches can provide a richer understanding of the landscape, and one that is credible to local and external stakeholders alike (Bell and Morse 2001).

Although the LM is predicated on significant coordination among sectors and scales in rural landscapes, the goal is not to establish a centralized landscape 'secretariat' but rather a web of activity nodes that are knit together by shared purpose, shared information, and dedication to evidence-based decision making. These nodes come together from time to time to negotiate and establish broad-level goals, formulate plans, identify needed collaborations, and share monitoring results to understand the interactive effects of different projects and programs on the landscape. Actual management and policy interventions are carried out at a range of scales—from the household to the region or beyond—but these interventions occur within the context of the landscape planning and monitoring process (see Fig. 5.2).

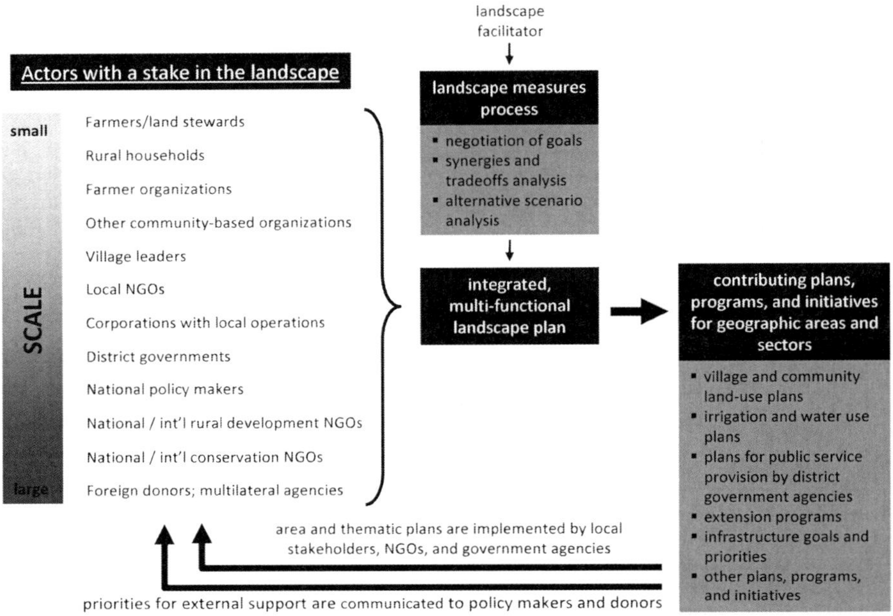

Fig. 5.2 Idealized representation of the interactions among stakeholder groups in the Landscape Measures approach (*LM*). Moving from *left to right* in the diagram: (*1*) A wide range of actors—operating at multiple scales—have a stake in rural landscapes. Many of these groups are already linked to each other through social networks, joint projects, and so forth, and the LM can strengthen or augment such linkages. (*2*) These diverse actors come together to participate in the LM under the auspices of a landscape facilitator. Negotiation and social learning supported by technical analysis lead to the formulation of an integrated, multi-functional landscape plan. (*3*) Landscape actors then incorporate information, insights, and agreed-upon goals and objectives from the broader LM process into their geographically- and sectorally-focused activities, programs, and plans. These activities are implemented on the ground and communicated to stakeholders operating at other scales (especially donors and policy makers). Over time, the relationships depicted here are sustained and strengthened in an iterative process, while the resulting plans and activities are frequently revisited in light of new circumstances, new priorities, and new landscape monitoring data

Ecologically-Based Tools for Implementation

To aid in the implementation of the LM, we assembled, developed, and tested a set of tools for landscape planning and monitoring. These are described in an online portal for practitioners known as the Landscape Measures Resource Center (LMRC) (LMI 2009). Some of the more promising tools draw on recent thinking in the field of ecology to quantify the performance and resilience of rural land-use systems to advance conservation, food production, and livelihood goals. Given this book's focus on the contribution of ecology to rural development, here we highlight some of the LMRC monitoring tools in which ecological science offers an especially valuable perspective.

As discussed above, a key challenge of multi-stakeholder adaptive management is to bridge different types and uses of knowledge by different landscape actors.

One way to do so is through landscape monitoring programs that incorporate both scientific and community knowledge (Place and Were 2005). For example, several of the methods in the LMRC combine social learning with scientifically rigorous sampling and analysis methods to add external credibility to community-generated datasets while bringing local relevance to monitoring data demanded by outside donors and program evaluators. A second challenge is to generate sufficient knowledge about multi-faceted landscape systems with limited funding and personnel resources. We therefore advocate approaches that derive additional value from existing monitoring efforts, employ participatory monitoring, and take advantage of new low-cost data collection and analysis tools.

One such method—repeat ground-based photo-monitoring—can be a cost-effective way to track changes in vegetation and land use when aerial imagery is unavailable or unaffordable (Lassoie et al. 2006). In this method, scientists use stratified sampling to establish points throughout the landscape from which digital photographs are taken in all directions. The photos are analyzed according to a standard protocol that yields quantitative descriptors, which are entered into a database. As the photo points are re-visited over the course of months and years, the data begin to reveal trends in land use, agricultural management, vegetation condition, and other factors. The digital photographs themselves can be taken by local people, providing a credible and easily interpretable data source for household- and village-level adaptive management while generating 'research quality' data through systematic aggregation across the network of photo points. More generally, participatory monitoring and evaluation can often yield data that are widely credible if it follows a scientifically designed protocol (Bonney et al. 2009).

A second method in the LMRC toolkit achieves the opposite type of knowledge transfer, taking data that are collected for external evaluators and making them relevant to local land stewards to use in adaptive management. On eco-certified farms throughout the world, large amounts of data are collected annually to meet the auditing requirements of various certification systems. Yet much of this information is filed away, never to be used by land stewards in the service of improved management. For these data to be useful to landscape stakeholders, they must be entered into appropriate information systems, aggregated, analyzed, and communicated effectively. For example, monitoring data on agrochemical usage, cover cropping, or soil erosion potential could be spatially plotted in a geographic information system (GIS) to visualize trends across space and time. This information could then be combined with downstream water quality monitoring data to track the relationship between on-farm practices and watershed-level ecosystem services. Again, approaches from the field of ecology can be used to help establish appropriate sampling protocols, aggregation methods, and analysis techniques.

Central to the LM is the use of integrative indicators that provide answers to several of the 20 Questions at once. An important integrative indicator in almost every landscape is the composition and configuration of land use and land cover. Basic land cover maps can be created by interpreting aerial imagery or by compiling data from field surveys or repeat ground-based photography. Maps can then be analyzed quantitatively to derive key measures of composition (e.g., area under native forest) and structure (e.g., degree of interspersion of complementary or conflicting land uses). Often, these measures can be further extrapolated to estimate outcomes related to food production, species viability, hydrological functions, and other key landscape parameters.

Given the great interpretive power of such composition and structure measures, landscape design principles have been proposed as heuristics for maintaining ecological integrity in the context of endeavors such as regional planning (Forman 1995; Dramstad et al. 1996; Lindenmeyer et al. 2008) and agroecosystem management (Fischer et al. 2006; Harvey 2008). We believe that similar principles and proxies could be developed for other objectives of landscape multi-functionality, including increased agricultural production, decreased disease burden attributable to environmental factors, and other goals related to the long-term fulfillment of the MDGs. Recent work on ecosystem service mapping has begun to relate landscape composition, ecological integrity, livelihood potential, and economic value in a spatially explicit manner (e.g., Troy and Wilson 2006; Egoh et al. 2008). These efforts suggest how GIS-based analyses can be used to track many of the 20 Questions with relatively fine spatial and temporal resolution.

A final tool that we wish to highlight is the use of systems dynamics modeling—computer applications that allow a user to simulate complex systems by tracking numerous interacting variables over time (Sterman 2000). Although system dynamics modeling is based on a mechanistic view of systems, its great advantage is that it can account for much higher levels of complexity than is possible through human intuition and *ad hoc* methods, making it valuable for landscape approaches. Key applications include understanding causal relationships in the landscape, identifying high-leverage 'pressure points' for landscape change, determining thresholds at which dramatic changes may occur, exploring alternative scenarios through participatory modeling, and measuring the success of interventions by comparing actual landscape outcomes to simulated outcomes under alternative management programs (Campbell et al. 2001; Sandker et al. 2007).

Case Study 1: Applying the Landscape Measures Approach in Copán, Honduras

The Copán case study illustrates the application of the Landscape Measures framework to conduct a broad-reaching baseline evaluation of landscape conditions, elucidate and prioritize community needs, and track progress toward all four ecoagriculture goals. Honduras currently has the highest poverty rate in Central America (70%) and ranks 115 out of 170 countries globally in the index of human development (Programa Estado de la Nación 2008). The Copán region is somewhat insulated from the worst poverty due to the significant tourism revenue associated with local Mayan ruins. Ironically, however, the most impoverished landscape residents remain the Chorti Maya, whose ancestors built these temples. As such, the landscape contains a diverse mix of stakeholders, ranging from wealthier landowners concentrated around the colonial town of Copán Ruinas—whose income is principally drawn from ecological and cultural tourism—to coffee and cattle farmers and the *campesinos* they hire to work their lands, to the Chorti Maya, who are largely segregated from the Mestizo majority and work as farm laborers or depend on subsistence agriculture.

As discussed earlier, Copán already has some institutional capacity for carrying out landscape approaches to natural resource management and community development. A regional governing body known as the MANCOSARIC represents the watershed's four municipalities and works to improve basic human services while facilitating adaptive co-management with an emphasis on improving flows of ecosystem services and reducing risks from natural disasters such as flooding and landslides. The MANCOSARIC also helps empower local governments to take responsibility for natural resource stewardship through integrated watershed management.

In 2007, the MANCOSARIC and its partners decided to implement the Landscape Measures approach and the 20 Questions to provide a baseline evaluation of the watershed that would help them understand the current status of the landscape, identify priorities, and refine current landscape management plans. The landscape was particularly suitable for such evaluation because of the existence of the MANCOSARIC governing body, which was well positioned to utilize the information generated. The evaluation also promised to offer a wider perspective on the region and a staring point for initiating critical discussion on stakeholder priorities.

The baseline evaluation conducted by Bejarano (2009) was designed to synthesize useful information from pre-existing studies while generating strategic new data to answer some of the 20 Questions deemed most critical by local stakeholders. Consistent with the Landscape Measures approach, many landscape performance measures were derived or extrapolated from land use patterns and dynamics. In this regard, the MANCOSARIC was fortunate to have a 1-m resolution IKONOS satellite image of the landscape taken in 2007 that was classified into land uses at the plot scale (Sanfiorenzo 2008). This land use map provided a foundation for much of the landscape evaluation, allowing stakeholders to analyze information on production, conservation and livelihood indicators in a spatially explicit manner to understand where interventions and improvements were most needed.

One application, for example, was the interpretation of land use patterns to estimate the provision of ecosystem services throughout the watershed (see Fig. 5.3). While land use is not a precise proxy for such services, prior study has yielded enough information on the relationships between land use, biodiversity conservation, and carbon storage to help identify hotspots where ecosystem services have been eroded and where restoration efforts could address both conservation and livelihood goals. The spatially explicit nature of these maps facilitates negotiation by identifying specific property owners and municipalities that could benefit from interventions.

While landscape composition and structure metrics were an important part of the landscape evaluation, it was critical to supplement these measures with household interviews and plot-level field studies to answer many of the 20 Questions. For example, one of the surrogate measures for Conservation criteria 1 and 3 (Box 5.2) was to ask farmers when they had last seen a wild deer. Representative patches of each forest type in each community were also surveyed to evaluate vegetation

a

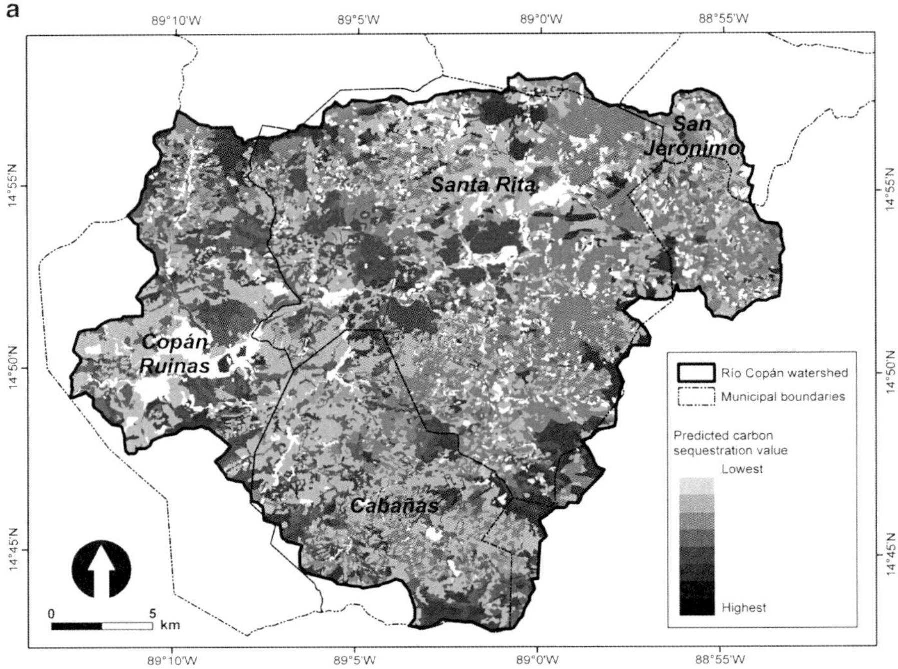

b

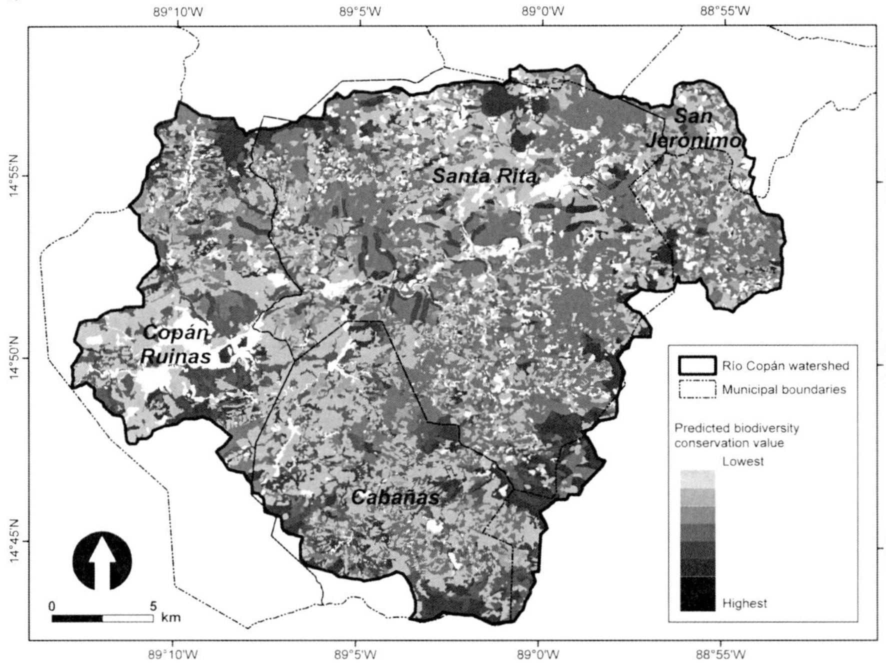

Fig. 5.3 Current status of the provisioning of carbon sequestration (**a**) and biodiversity conservation (**b**) in the Copán landscape based on estimates of the capacity of each land use to provide each of these ecosystem services. Indices of carbon sequestration and biodiversity conservation were adapted from Murgueitio and colleagues (2004)

structure and evidence of degradation from grazing, timber or fuelwood extraction, and other human interventions. This study indicated that forests are more degraded in Cabañas—where the economy is heavily based on natural resources—than in Copán Ruinas, a larger town with a more diversified economy.

The evaluation of livelihood indictors was primarily based on household interviews (45 per municipality), but these were spatially stratified and located with GPS coordinates to allow spatially explicit analysis of the relationships among multiple goals. Interviews revealed household members' education levels, production activities, agricultural yields, farm income, total income, and other factors. Results were integrated with those from earlier household surveys focusing on farm-level conservation practices and access to water and energy resources. Both sets of interviews also assessed the degree to which local social service and resource management entities were providing households with services, training, or sharing of ideas—or even the degree to which farmers were aware of relevant projects. These data helped define the effectiveness and sphere of influence of local institutions relative to their mission and objectives. The data also revealed spatial patterns of wealth and poverty—including both current income and capacity to improve and adapt household livelihood strategies. Again, the evaluation documented greater levels of poverty and need in the more resource-dependent communities outside of the tourism nexus (and MANCOSARIC headquarters) in Copán Ruinas.

The landscape evaluation reported answers to each of the 20 Questions individually but also amalgamated outcomes into the four basic 'axes' of ecoagriculture to help frame stakeholder discussion about landscape priorities (see Fig. 5.4). This type of synthesis is rife with challenges and value judgments (How do you weigh each indicator? Can landscape outcomes be traded off against each other, or must some or all objectives be met at a basic level?). But rather than forming an insurmountable barrier, such value questions can provide a starting point for dialogue about synergies and tradeoffs among disparate objectives.

In addition to providing a baseline assessment of landscape performance, the evaluation also explored various policy alternatives for improving outcomes to several of the 20 Questions. Framing policy analysis in terms of the 20 Questions is an alternative to sectoral analyses that predict the direct results of interventions while ignoring their indirect or feedback effects. For example, Sanfiorenzo (2008) conducted landscape modeling to evaluate the effects on biodiversity of proposed policies for reducing erosion, landslides, and water pollution in the landscape, which hinder progress toward several of the MDGs. A baseline analysis evaluated forest patch size, fragmentation, and functional connectivity of the existing landscape from the perspective of the genus *Trogon*—forest dependent birds that are also highly sought after by ecotourists. Existing forest cover in the landscape was both limited (comprising only 22% of the 680 km^2 landscape) and highly fragmented into 145 isolated patches. Sanfiorenzo (2008) then evaluated the effects of three potential policies: (1) enforcing the Honduran law to protect 10-meter forested buffers alongside all rivers and streams, (2) converting steep slopes (14–40%) to agroforestry systems such as shaded coffee or pasture with high tree density, and (3) re-vegetating all very steep slopes (>40%) to natural forest or timber plantations.

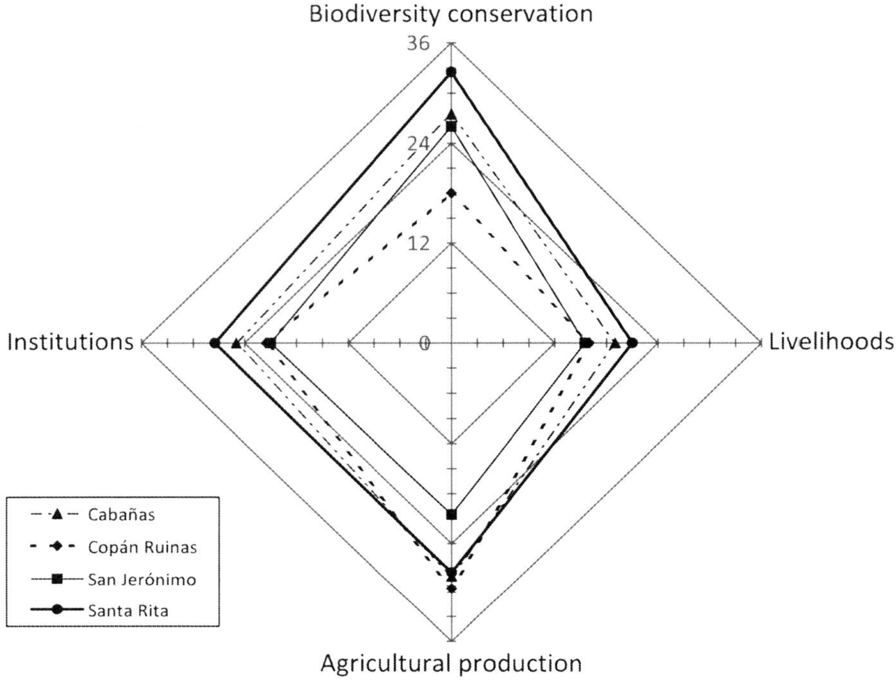

Fig. 5.4 Spider diagram indicating current conditions in each of the four municipalities in the Río Copán watershed with respect to each of the four principal axes of ecoagricultural development (food production, conservation, livelihoods, and institutional support). These indices are derived from mixed methods including household interviews, ecological field sampling, and land use analyses, as described in the text. The diagram provides a simplified performance evaluation to help assess progress toward community goals, set priorities for future projects, and evaluate progress over time

The models revealed that riparian buffers would decrease the number of isolated forest fragments from 145 to less than 40, while the three policies in combination would increase suitable trogon habitat from 22% to 38% of the landscape. The analysis not only sheds light on several of the 20 Questions (e.g., C1, C4, P4, L3, and L5); it also identifies the most promising target areas for restoration.

Reflecting on the LM evaluation in Copán, the approach at first glance seems similar to standard assessment methods—such as Rapid Rural Appraisal—that combine interviews and other forms of baseline data collection to identify needs and priorities. However, on closer examination, several key differences emerge. One is the use of an integrative framework to steer communities and field technicians to consider the possible importance or feedback effects of issues that have been neglected locally. Second is an emphasis on land use and landscape patterns as durable—though manageable—underlying drivers of many of the socioeconomic themes that are often the focus of rural appraisals. Third is a focus on quantitative indicators that can be readily and cost-effectively measured on a regular basis to track both the direct and indirect effects of landscape interventions, as well as the feedbacks between these interventions and exogenous policy and market forces. Based on the cost of the initial assessment, we estimate that repeating LM evaluations every 2–3 years as part of a landscape planning and adaptive management program would cost $50,000–$70,000. As the MANCOSARIC has learned, however, such up-front investment can pay for itself many times over by helping to attract and target foreign assistance to communities that have a clear vision for the future and understand which projects and interventions will help them achieve this vision.

Case Study 2: Applying the Landscape Measures Approach in Kijabe, Kenya

The second case study documents the use of the 20 Questions and two participatory evaluation tools within the Landscape Measures Resource Center as a basis for initiating dialogue about landscape dynamics and priorities. The case takes place in the Kijabe landscape on the eastern slopes of the Aberdare Mountains, just northwest of Nairobi, Kenya. Here lies the Kikuyu Escarpment Forest, a hotspot for plant and bird diversity that is also the watershed supplying water to more than a million of Nairobi's inhabitants. The landscape is a mosaic of ancient forests, tree plantations, and diverse agricultural plots, supporting a mixed agricultural economy and extensive tea production. However, recent population growth had led to increased pressure on the forest: cattle and sheep were killing seedlings, residents were cutting wood for charcoal production, and illegal loggers were exploiting the forest.

Recognizing the dependence of local livelihoods on the health of the forest, local leaders, with financial support from BirdLife International, established the Kijabe Environmental Volunteers (KENVO) to educate, train and support local residents in forest conservation and restoration efforts. KENVO began with a seedling initiative that organized landscape residents to plant and protect native trees to restore the ailing forest. By raising and selling the trees to KENVO, women and youth groups were able to earn income while supplying their farms with useful agroforestry trees. Meanwhile, a growing contingent of innovative farmers was building on KENVO's ideas by diversifying and intensifying their production systems to integrate small animals, bees, and fish farming and by utilizing organic wastes to enhance soil fertility. As these farmers increased their incomes and were able to realize prized education and health benefits for their families, others took notice and the ideas began to spread.

By 2007, KENVO had enjoyed significant success, ridding the area of illegal loggers and spawning numerous community-led forest restoration groups where none had existed before. KENVO's founder, David Kuria, remarked on residents' deep pride in these achievements but emphasized that "for conservation in this area to succeed, communities must continue to benefit."

Participatory Landscape Evaluation

In this context, KENVO was interested in using the Landscape Measures approach to re-assess its strategic direction and provide local stakeholders a forum in which to express their needs and priorities. For its part, Ecoagriculture Partners' Landscape Measures team sought to apply and evaluate the 'landscape performance scorecard' and 'institutional performance scorecard' tools, which it had recently designed for a Ugandan landscape with similar land use and livelihood dynamics. Both scorecards are based on the 20 Questions and offer a format for discussion and participatory evaluation of these questions to initiate dialogue on landscape dynamics.

KENVO convened a group of 22 stakeholders for a 5-h workshop at its strategically-located office and meeting space in the landscape. About two-thirds of the participants were farmers, while others represented public agencies of forestry and natural resources, agriculture and livestock, and social services as well as leaders of church groups and other local organizations. KENVO's multi-lingual professional staff issued the invitations, arranged for teas and lunch to be provided, and co-led the workshop with Landscape Measures Initiative (LMI) staff. The LMI team prepared color-coded copies of scorecards, data capture forms, and written instructions for the exercise.

The group began by translating each of the 20 Questions (see Box 5.2) into terms that made sense in the Kijabe landscape, a process that involved discussing various local examples that were meaningful to participants. Next, each participant filled out a copy of the landscape performance scorecard, which required evaluating each question on a five-point scale for the Kijabe landscape. The group then prepared for the institutional scoring exercise by brainstorming to identify all public, private, civic, or hybrid organizations that they considered to have an effect on the landscape's current status and future direction. Using a similar scorecard format, participants scored each institution based on its fulfillment of its mission and its contribution to the objectives articulated in the 20 Questions. The meeting facilitators entered all scorecard data into a Microsoft Excel data capture form, computed summary results, and generated illustrative spider diagrams of the results, all of which were projected for the group to view. Discussion ensued about the results and what they implied about the landscape's current balance among conservation, food production, and livelihood performance. Following the meeting, a group of Kenyan participants met with the LMI team to review the workshop process, assess the relevance and usability of the scoring tools, and determine whether the landscape perspective was helpful or viewed by participants as abstract and irrelevant.

Outcomes of the Landscape Evaluation

The landscape evaluation process exceeded the expectations of KENVO and the LMI team in three respects. First, the level of engagement and application of participants' knowledge to the tasks at hand were impressive and inspiring. Participants devoted much more time and effort to the institutional scoring than we had anticipated, producing an institutional map of the landscape that KENVO and its members have used subsequently in publications, presentations, and discussions with collaborators.

Second, the exercise stimulated creative thinking and discussion about strategic new directions for KENVO's activities. For example, the landscape scorecard made evident the fact that Kijabe was performing better with respect to conservation goals than livelihood goals. Reflecting on this result, participants realized that recent external investment in the landscape had been driven for some time by the agendas of conservation groups whose aims were to restore forest habitat for wildlife. While participants were proud of their conservation achievements, they articulated a need to pursue parallel improvements in food production and livelihood security. This discussion generated a list of concrete steps toward which the group agreed to organize, including improving farmers' access to markets for specialty products and securing credit for new enterprises. Results of the institutional scoring exercise stimulated participants to target private sector organizations—particularly companies dealing in agricultural products—for recruitment into KENVO's activities. They also used the newly-created institutional map to explore the potential of linking organizations to create agri-eco-tourism enterprises that would benefit entrepreneurs and the community by taking advantage of the landscape's strategic location and dramatic views into the rift valley.

A third outcome of the exercise was KENVO's decision to invest in the development of additional tools and analyses for assessing landscape performance and promoting 'landscape literacy' among residents and stakeholders. This decision stemmed partly from a growing realization—supported by the landscape scoring process—that important conservation benefits and other ecosystem services were being provided in the agricultural mosaic itself, not just in the Kikuyu forest. With encouragement and a modest seed grant from Ecoagriculture Partners, KENVO's leaders generated sufficient resources to commission the National Museums of Kenya to conduct a biodiversity inventory in the agricultural portions of the landscape to complement the previous inventory of the forest. KENVO also commissioned a socio-economic study of farming households to increase their understanding of local livelihood strategies and generate baseline information against which change could be measured over time. And KENVO worked with the Ecoagriculture Working Group at Cornell University to create a land use/land cover map that they could used to communicate with residents about land use dynamics and opportunities for forest restoration to provide conservation and livelihood benefits.

Conclusion

The post-workshop evaluation revealed that the landscape and institutional scoring tools—and the process by which they were implemented—were relevant and worthwhile. Participants were visibly engaged throughout the workshop and contributed impressive knowledge and insight from their individual perspectives. The discussion and use of the scorecards ran smoothly, with no apparent confusion, and the resulting baseline evaluations were judged to be credible by the people and organizations who participated. At the same time, however, the landscape evaluation did not merely reiterate what participants already knew. New information was brought forward through the multi-stakeholder forum and, more importantly, participants were able to organize and understand existing knowledge in new ways that made the trajectory, opportunities, and threats in the Kijabe landscape more apparent. This new understanding helped generate ideas about KENVO's future priorities for landscape level planning and management while solidifying KENVO's commitment to continuing to invest in strategic landscape information to support such planning and management. A further measure of impact, to be assessed later, would be KENVO's repeat use of the scorecard tools to evaluate changes in landscape performance attributable to its programs and to other factors.

Toward Mainstreaming of Landscape Approaches

The case studies from Honduras and Kenya illustrate the ways in which landscape-scale negotiation, planning, and monitoring will be crucial for meeting the MDGs on a sustained basis in rural landscapes. As documented in this chapter, landscape approaches have begun to be used in recent years, but further work is needed to continue to develop the science and practice of multi-stakeholder, multi-objective adaptive management at the landscape scale. Mainstreaming landscape approaches will also require the adoption of favorable policy, market, and institutional frameworks at the national and international levels. Many of these changes will entail substantial re-allocations of power, authority, and resources, and could take years or decades to achieve. Key actions needed to support landscape approaches include:

1. Shift power over land and resource management to landscape-level institutions that have (or can develop) the capacity to carry out such management. Continued devolution of government authority will be an important part of this process in many countries.
2. Legitimize and provide sustained support for multi-stakeholder processes in landscapes. Re-orient government line agencies toward a service role in which they provide technical resources and facilitation for these processes and subsequently incorporate landscape-level goals and plans into agency priorities and programs. Recognize roles for business, NGOs, farmers' organizations and citizen groups in implementing action and tracking progress based on these plans.
3. Expand opportunities for training and knowledge sharing around landscape-scale analysis, planning and monitoring, moving beyond fixed-curriculum extension to include demand-driven programs and peer-to-peer networks, with learning across sectors. Support action learning through partnerships between practitioners and researchers.

4. Clarify and adjust land and resource tenure arrangements so that households and communities are motivated and able to implement concepts or plans that emerge from landscape-level adaptive management processes.
5. Create more equitable approaches to the governance of natural resources so that corporate and government interests are required to participate in multi-stakeholder planning processes rather than shortcutting such negotiations through inside channels. This applies to both common-pool resources such as forests and oceans and privately-owned resources whose management affects public goods like water supply and biodiversity.
6. Eliminate market-distorting policies and subsidies that hinder evidence-based management of water, soil, crops, and land use. Establish markets for ecosystem services to internalize externalities associated with the management of rural landscapes, and encourage public and private procurement of agricultural products from farmers using ecoagriculture practices.
7. Re-align the priorities of government agencies, donors and NGOs to incorporate environmental sustainability and ecosystem management into agricultural and rural development programs, and to track human welfare in a way that accounts for the stocks and flows of natural capital that support rural livelihoods.

Historically, the link between environmental sustainability and the wealth of rural communities has been widely ignored or neglected, especially in the fertile, productive landscapes that supply much of the world's food. Technological innovations, inexpensive farm inputs, large subsidies from nature, and the relief valve of the agricultural frontier have all held crisis at bay in many rural landscapes. Going forward, however, we expect this picture to change. As population pressures mount, suitable vacant land diminishes, and productivity gains from technological innovation plateau in post-Green Revolution areas, healthy ecosystems will become increasingly fundamental to human wellbeing. As the margin of error for meeting livelihood needs in rural landscapes shrinks, the demand for effective landscape approaches will grow. Acting now to develop the science, the tools, and the institutional support mechanisms for landscape-scale adaptive management will ensure that such processes are fully functional at the time they are most needed.

References

Barrett, C.B. and B.M. Swallow. 2006. Fractal poverty traps. *World Development* 34(1):1–15.

Barrow, E., D. Timmer, S. White, and S. Maginnis. 2002. *Forest Landscape Restoration: Building Assets for People and Nature - Experience from East Africa*. IUCN, Gland, Switzerland and Cambridge, UK.

Bell, S. and S. Morse. 2001. Breaking through the glass ceiling: who cares about sustainability indicators? *Local Environment* 6:291–309.

Bejarano, L.F. 2009. Evaluación metodológica del enfoque de Ecoagricultura para medir el desempeño de un paisaje con matriz agropecuaria en la subcuenca del Río Copán, Honduras. Master's thesis. Centro Agronómico Tropical de Investigación y Enseñanza, Turrialba, Costa Rica.

Benedict, M.A. and E.T. McMahon. 2006. *Green Infrastructure: Linking Landscapes and Communities*. Island Press, Washington, DC.

Bojö, J., K. Green, S. Kishore, S. Pilapitiya, and R. Reddy. 2004. *Environment in Poverty Reduction Strategies and Poverty Reduction Support Credits*. World Bank Environment Department Paper No. 102. World Bank, Washington, DC.

Bonney, R., C.B. Cooper, J. Dickinson, S. Kelling, T. Phillips, K.V. Rosenberg, and J. Shirk. 2009. Citizen science: a developing tool for expanding science knowledge and scientific literacy. *Bioscience* 59(11):977–84.

Borrini-Feyerabend, G., M.T. Farvar, J.C. Nguinguiri, and V.A. Ndangang. 2000. *Co-management of Natural Resources: Organising, Negotiating and Learning-by-Doing*. GTZ and World Conservation Union, Kasparek Verlag, Heidelberg, Germany.

Brechin, S.R., P.R. Wilsusen, C.L. Fortwangler, and P.C. West, eds. 2003. *Contested Nature: Promoting International Biodiversity with Social Justice in the 21st Century*. SUNY Press, Albany, NY.

Buck L., C.C. Geisler, J. Schelhas, and E. Wollenberg, eds. 2001. *Biological Diversity: Balancing Interests through Adaptive Collaborative Management*. CRC Press, Boca Raton, FL.

Buck, L.E., J.C. Milder, T.A. Gavin, and I. Mukherjee. 2006. Understanding Ecoagriculture: A Framework for Measuring Landscape Performance. Ecoagriculture Discussion Paper #2, Ecoagriculture Partners, Washington, DC.

Buck, L.E. and S.J. Scherr. 2009. Building innovation systems for managing complex landscapes. In K.M. Moore, ed. *The Sciences and Art of Adaptive Management: Innovating for Sustainable Agriculture and Natural Resource Management*. Soil and Water Conservation Society, Ankeny, Iowa.

Campbell, B., J.A. Sayer, P. Frost, S. Vermeulen, M.R. Pérez, A. Cunningham, and R. Prabhu. 2001. Assessing the Performance of Natural Resource Systems. *Conservation Ecology* 5(2):22.

Carney, D. ed. 1998. *Sustainable Rural Livelihooods. What Contribution Can We Make?* Department for International Development, London.

Cash, D.W., W.C. Clark, F. Alcock, N.M. Dickson, N. Eckley, D.H. Guston, J. Jäger, and R. B. Mitchell. 2003. Knowledge systems for sustainable development. *Proceeding of the National Academy of Science* 100(14): 8086–8091.

CGIAR [Consultative Group on International Agricultural Research System Review Secretariat]. 1998. Third System Review of the Consultative Group on International Agricultural Research (CGIAR). CGIAR, Washington, DC.

CIFOR [International Center for Forestry Research]. 1999. *Guidelines for Developing, Testing and Selecting Criteria and Indicators for Sustainable Forest Management*. CIFOR Toolbox, Part 1. CIFOR, Bogor, Indonesia.

Colfer, C.J.P. 2005. *The Complex Forest: Communities, Uncertainty, and Adaptive Collaborative Management*. Resources for the Future and CIFOR, Washington, DC.

Collins, K., C. Blackmore, D. Morris, and D. Watson. 2007. A systematic approach to managing multiple perspectives and stakeholding in water catchments: some findings from three UK case studies. *Environmental Science & Policy* 10:564–74.

COMACO [Community Markets for Conservation]. 2009. COMACO website. Online: www.itswild.org, accessed February 19, 2009.

Costanza, R., R. d'Arge, R. de Groot, S. Farber, M. Grasso, B. Hannon, K. Limburg, S. Naeem, R.V. O'Neill, J. Paruelo, R.G. Raskin, P. Sutton and M. van den Belt. 1998. The value of the world's ecosystem services and natural capital. *Ecological Economics* 25(1): 3–15.

Daily, G.C. 1997. *Nature's Services*. Island Press, Washington, DC.

DFID [Department for International Development], European Commission, UN Development Programme, and World Bank. 2002. *Linking Poverty Reduction and Environmental Management*. World Bank, Washington, DC.

DFID [Department for International Development]. 2006. DFID's approach to the Environment. DFID, London.

Dietz, T., E. Ostrom, and P.C. Stern. 2003. The struggle to govern the commons. *Science* 302:1907–12.

Diamond, J. 2004. *Collapse: How Societies Choose to Fail or Succeed*. Viking, New York.

Dramstad, W.E., J.D. Olson, and R.T.T. Forman. 1996. *Landscape Ecology Principles in Landscape Architecture and Land-Use Planning*. Harvard University Graduate School of Design, Cambridge, MA.

Egoh, B., B. Reyers, M. Rouget, D.M. Richardson, D.C. Le Maitre, and A.S. van Jaarsveld. 2008. Mapping ecosystem services for planning and management. *Agriculture, Ecosystems and Environment* 127: 135–140.

Fairhead, J. 2004. Achieving sustainability in Africa. Pages 292–306 in R. Black and H. White, eds. *Targeting Development: Critical Perspectives on the Millennium Development Goals*. Routledge, London.

Fischer, J., D.B. Lindenmayer, and A.D. Manning. 2006. Biodiversity, ecosystem function, and resilience: ten guiding principles for commodity production landscapes. *Frontiers in Ecology and the Environment* 4(2): 80–86.

Folke, C., S. Carpenter, T. Elmqvist, L. Gunderson, C.S. Holling, and B. Walker. 2002. Resilience and sustainable development: building adaptive capacity in a world of transformations. *Ambio* 31(5): 437–40.

Forman, R. T. T. 1995. *Land Mosaics: The Ecology of Landscapes and Regions*. Cambridge University Press, Cambridge, UK.

Frost, P., B. Campbell, G. Medina, and L. Usongo. 2006. Landscape-scale approaches for integrated natural resource management in tropical forest landscapes. *Ecology and Society* 11(2):30.

Gottret, M.V. and D. White. 2001. Assessing the impact of integrated natural resource management: challenges and experiences. *Conservation Ecology* 5(2):17.

Harvey, C.A. 2008. Designing agricultural landscapes for biodiversity conservation. Pages 146–165 in S.J. Scherr and J.A. McNeely. *Farming with Nature*. Island Press, Washington, DC.

Hinchcliffe, F., J. Thompson, J.N. Pretty, I. Guijt and P. Shah. 1999. *Fertile Ground: The Impacts of Participatory Watershed Management*. Intermediate Technology Publications, London.

Holling, C.S. 1978. *Adaptive Environmental Assessment and Management*. Wiley, New York.

ICARDA [International Center for Agricultural Research in the Dry Areas]. 2005. Current definition of INRM. Online: http://www.icarda.cgiar.org/INRMsite/index.htm.

Ison, R., N. Röling, and D. Watson. 2007. Challenges to science and society in the sustainable management and use of water: investigating the role of social learning. *Environmental Science & Policy* 10:499–511.

Izac, A.N. and P.A. Sanchez. 2001. Towards a natural resource management paradigm for international agriculture: the example of agroforestry research. *Agricultural Systems* 69:5–25.

Kerr, J. with G. Pangare and V.L. Pangare. 2002. *Watershed Development Projects in India: An Evaluation*. IFPRI Research Report 127. International Food Policy Research Institute, Washington, DC.

Kloppenburg, J., J. Hendrickson, and G.W. Stevenson. 1996. Coming in to the foodshed. *Agriculture and Human Values* 13: 33–42.

LAC-Net [Regional Model Forest Network for Latin America and the Caribbean]. 2006. Estándar de Principios, Criterios e Indicadores (PC&I) de la Red Regional de Bosques Modelo para América Latina y el Caribe (LAC-Net). Online: www.bosquesmodelo.net/foro/documentos/Estandares_principios_criterios.doc.

Lal, P., H. Lim-Applegate, and M. Scoccimarro. 2001. The adaptive decision-making process as a tool for integrated natural resource management: focus, attitudes, and approach. *Conservation Ecology* 5(2): 11.

Lassoie, J.P., R.K. Moseley, and K. E. Goldman. 2006. Ground-based photomonitoring of ecoregional ecological changes in northwestern Yunnan, China. In: C. Aguirre-Bravo, P.J. Pellicane, D.P. Burns, and S. Draggan. (eds.) *Monitoring Science and Technology Symposium: Unifying Knowledge for Sustainability in the Western Hemisphere*. Proceedings RMRS-P-42CD. U.S. Department of Agriculture, Forest Service, Rocky Mountain Research Station, Fort Collins, CO. Online: http://www.fs.fed.us/rm/pubs/rmrs_p042.html.

Laumonier, Y., R. Bourgeois, and J. Pfund. 2008. Accounting for the ecological dimension in participatory research and development: lessons learned from Indonesia and Madagascar. *Ecology and Society* 13(1):15.

Leach, M., R. Mearns, and I. Scoones. 1999. Environmental entitlements: dynamics and institutions in community-based natural resource management. *World Development* 27: 225–47.

Lee, K.N. 1993. *Compass and Gyroscope: Integrating Science and Politics for the Environment*. Island Press, Washington, DC.

Lele, S.M. 1991. Sustainable development: a critical review. *World Development* 19:607–621.

Leeuwis, C. and R. Pyburn. 2002. Social learning in rural resource management. Pages 11–22 in C. Leeuwis and R. Pyburn, eds. *Wheelbarrows Full of Frogs: Social Learning in Rural Resource Management*. Uitgeverij Van Gorcum. Assen, The Netherlands.

Lindenmayer, D. et al. 2008. A checklist for ecological management of landscapes for conservation. *Ecology Letters* 11:78–91.

Liu, J., T. Dietz, S.R. Carpenter, C. Folke, M. Alberti, C.L. Redman, S.H. Schneider, E. Ostrom, A.N. Pell, J. Lubchenco, W.W. Taylor, Z. Ouyang, P. Deadman, T. Kratz, and W. Provencher. 2007. Coupled human and natural systems. *Ambio* 36: 639–49.

LMI [Landscape Measures Initiative]. 2009. The Landscape Measures Resource Center. Online: www.landscapemeasures.info. Ecoagriculture Partners, Washington DC, and Cornell University, Ithaca, NY.

López-Ridaura, S., H. van Keulen, M.K. van Ittersum, and P.A. Leffelaar. 2005. Multiscale methodological framework to derive criteria and indicators for sustainability evaluation of peasant natural resource management systems. *Environment, Development and Sustainability* 7:51–69.

Lynam, T., W. de Jong, D. Sheil, T. Kusumanto, and K. Evans. 2007. A review of tools for incorporating community knowledge, preferences, and values into decision making in natural resources management. *Conservation Ecology* 12(1):5.

MA [Millennium Ecosystem Assessment]. 2005. *Ecosystems and Human Well-being: Synthesis*. Island Press, Washington, DC.

Maarleveld, M. and C. Dangbegnon. 1999. Managing natural resources: a social learning perspective. *Agriculture and Human Values* 16:267–80.

McNeely, J.A. and S.J. Scherr. 2003. *Ecoagriculture*. Island Press, Washington DC.

McShane, T.O. and M.P. Wells, eds. 2004. *Getting biodiversity projects to work: towards better conservation and development*. Columbia University Press, New York.

Miller, K., E. Chang, and N. Johnson. 2001. *Defining Common Ground for the Mesoamerican Biological Corridor*. World Resources Institute, Washington, DC.

Molnar, A., S.J. Scherr and A. Khare, 2007. Community stewardship of biodiversity. In: S.J. Scherr and J. McNeely, eds. *Farming with Nature: The Science and Practice of Ecoagriculture*. Island Press, Washington, DC.

Munda, G. 2005. "Measuring sustainability": a multi-criterion framework. *Environment, Development, and Sustainability* 7:117–34.

Murgueitio, E., M. Ibrahim, E. Ramirez, A. Zapata, C.E. Mejia, and F. Casasola. 2004. Land use on cattle farms: guide for the payment of environmental services. Integrated Silvopastoral Approaches to Ecosystem Management project. CIPAV, Cali, Colombia.

Nyberg, B. 1999. An introductory guide to adaptive management for project leaders and participants. British Columbia Forest Service, Victoria, Canada.

O'Neill, R.V., C.T. Hunsaker, K.B. Jones, K.H. Riitters, J.D. Wickham, P.M. Schwartz, I.A. Goodman, B.L. Jackson, and W.S. Baillargeon. 1997. Monitoring environmental quality at the landscape scale. *BioScience* 47(8): 513–19.

Olsson, P., C. Folke, and F. Berkes. 2004. Adaptive comanagement for building resilience in social-ecological systems. *Environmental Management* 34(1):75–90.

Palm, C.A., S.A. Vosti, P.A. Sanchez, and P.J. Ericksen, eds. 2005. *Slash-and-burn Agriculture*. Columbia University Press, New York.

Parker, D.C., S.M. Manson, M.A. Janssen, M.J. Hoffmann, and P. Deadman. 2003. Multi-agent systems for the simulation of land-use and land-cover change: a review. *Annals of the Association of American Geographers* 93: 314–337.

Pfund, J.-L., P. Koponen, T. O'Connor, J. Boffa, M. van Noordwijk, and J. Sorg. 2008. Biodiversity conservation and sustainable livelihoods in tropical forest landscapes. Pages 297–322 in

R. Lafortezza, J. Chen, G. Sanesi, and T.R. Crow, eds. *Patterns and Processes in Forest Landscapes*. Springer, New York.

Place, F. and E. Were, eds. 2005. Proceedings of 5th Workshop of the Integrated Natural Resource Management (INRM) Stakeholder Group. October 20-21, 2003. World Agroforestry Centre, Nairobi.

Plummer, R. and D. Armitage. 2007. A resilience-based framework for evaluating adaptive co-management: linking ecology, economics and society in a complex world. *Ecological Economics* 61: 62–74.

Pretty, J. and H. Ward. 2001. Social capital and the environment. *World Development* 29(2): 209–229.

Programa Estado de la Nación. 2008. Estado de la Región en Desarrollo Humano Sostenible: Un Informe desde Centroamérica y para Centroamérica. Programa Estado de la Nación, San José, Costa Rica.

Prüss-Üstün, A. and C. Corvalán. 2006. Preventing disease through healthy environments. Towards an estimate of the environmental burden of disease. World Health Organization, Geneva.

Redford, K.H. and E. Fearn, eds. 2007. *Protected Areas and Human Livelihoods*. WCS Working Paper No. 32. Wildlife Conservation Society, New York.

Röling, N. 2002. Beyond the aggregation of individual preferences. Pages 25–48 in C. Leeuwis and R. Pyburn, eds. *Wheelbarrows Full of Frogs: Social Learning in Rural Resource Management*. Uitgeverij Van Gorcum. Assen, The Netherlands.

Röling, N.G and M.A.E. Wagemakers. 1998. A new practice: facilitating sustainable agriculture. Pages 3–22 in N.G. Röling and M.A.E. Wagemakers, eds. *Facilitating Sustainable Agriculture*. Cambridge University Press, Cambridge, UK.

Rosenberg, D.K, B.R. Noon, and E.C. Meslow. 1997. Biological corridors: form, function and efficacy. *BioScience* 47: 677–687.

Sachs, J.D. and W.V. Reid. 2006. Investments toward sustainable development. *Science* 312:1002.

Salafsky, N., R. Margoluis, and K. Redford. 2001. *Adaptive management: a tool for conservation practitioners*. Biodiversity Support Program, Washington DC.

Sanderson, S. 2005. Poverty and conservation: the new century's "peasant question?" *World Development* 33(2):323–332.

Sandker, M., A. Suwarno, and B.M. Campbell. 2007. Will forests remain in the face of oil palm expansion? Simulating change in Malinau, Indonesia. *Ecology and Society* 12(2):37.

Sanfiorenzo, A.R. 2008. Contribución de diferentes arreglos silvopastoriles a la conservación de la biodiversidad, mediante la provisión de hábitat y conectividad en el paisaje de la sub-cuenca del Río Copán, Honduras. Master's thesis. Centro Agronómico Tropical de Investigación y Enseñanza, Turrialba, Costa Rica.

Sayer, J. and B. Campbell. 2004. *The Science of Sustainable Development: Local Livelihoods and the Global Environment*. Cambridge University Press, Cambridge, UK.

Sayer, J., Campbell, B., Petheram, L., Aldrich, M., Ruiz Perez, M., Endamana, D., Dongmo, Z, Defo, L., Mariki, S., Doggart, N., Burgess, N. 2007. Assessing environment and development outcomes in conservation landscapes. *Biodiversity and Conservation* 16: 2677-2694.

Sayer, J. and L. Buck. 2008. Learning from Landscapes. Arbor*vitae* Special. IUCN Forest Conservation Program and Ecoagriculture Partners, Gland, Switzerland.

Scherr, S.J. and J. McNeely. 2008. Biodiversity conservation and agricultural sustainability: towards a new paradigm of 'ecoagriculture' landscapes. *Philosophical Transactions of the Royal Society B* 363: 477–494.

Scherr, S., J.A. McNeely, and S. Shames. 2009. Ecoagriculture: agriculture, environmental conservation, and poverty reduction at a landscape scale. Pages 64–86 in N. Gillis and P. Galizzi eds. *The Role of the Environment in Poverty Alleviation*. Fordham University Press, New York.

Selman, P.H. 2002. Multi-function landscape plans: a missing link in sustainability planning? *Local Environment* 7(3): 283–94.

Sepúlveda, S., A. Rodríguez, R. Echeverri, and M. Portilla. 2003. *El Enfoque Territorial del Desarrollo Rural*. Instituto Interamericano de Cooperación para la Agricultura, San José, Costa Rica.

Sterman, J.D. 2000. *Business Dynamics: Systems Thinking and Modeling for a Complex World.* McGraw-Hill, New York.

Steyaert, P., M. Barzman, J. Billaud, H. Brives, B. Hubert, G. Ollivier, and B. Roche. 2007. The role of knowledge and research in facilitating social learning among stakeholders in natural resources management in the French Atlantic coastal wetlands. *Environmental Science & Policy* 10: 537–550.

Steyaert, P. and J. Jiggins. 2007. Governance of complex environmental situations through social learning: a synthesis of SLIM's lessons for research, policy and practice. *Environmental Science & Policy* 10: 575–86.

Swallow, B. 2005. Potential for poverty reduction strategies to address community priorities: case study of Kenya. *World Development* 33(2):301–21.

Terborgh, J. 1999. *Requiem for Nature.* Island Press, Washington, DC.

Troy, A. and M.A. Wilson. 2006. Mapping ecosystem services: practical challenges and opportunities in linking GIS and value transfer. *Ecological Economics* 60: 435–49.

UN [United Nations]. 2008. The Millennium Development Goals Report. United Nations, New York.

USFS [United States Forest Service]. 2006. *US Forest Service Guide to Integrated Landscape Land Use Planning in Central Africa.* Unpublished paper.

van Noordwijk, M., T.P. Tomich, and B. Verbist. 2001. Negotiation support models for integrated natural resource management in tropical forest margins. *Conservation Ecology* 5(2):21.

WCED [World Commission on Environment and Development]. 1987. *Our Common Future* (Bruntland Report). Oxford University Press, Oxford. UK.

Wollenberg, E., D. Edmunds, and L. Buck. 2000. Using scenarios to make decisions about the future: anticipatory learning for the adaptive co-management of community forests. *Landscape and Urban Planning* 47:65–77.

Wood, S., K. Sebastian, and S.J. Scherr. 2000. *Pilot Analysis of Global Ecosystems: Agroecosystems.* Report prepared for the Millennium Assessment of the State of the World's Ecosystems. International Food Policy Research Institute and World Resources Institute, Washington, DC.

World Bank and IMF [International Monetary Fund]. 2005. Global Monitoring Report 2005: Millennium Development Goals: From Consensus to Momentum. World Bank, Washington, DC.

World Bank. 2006. *Where is the Wealth of Nations? Measuring Capital for the 21st Century.* World Bank, Washington, DC.

WRI [World Resources Institute in collaboration with United Nations Development Programme, United Nations Environment Programme, and World Bank]. 2005. *World Resources 2005: The Wealth of the Poor—Managing Ecosystems to Fight Poverty.* WRI, Washington, DC.

WRI [World Resources Institute in collaboration with United Nations Development Programme, United Nations Environment Programme, and World Bank]. 2008. *World Resources 2008: Roots of Resilience—Growing the Wealth of the Poor.* WRI, Washington, DC.

Chapter 6
Introduction to Water, Poverty, and Ecology: A Vision for Sustainability

Casey Brown

The chapters in this section address the issue of water and the challenge presented by its complex roles within ecology and poverty. Essential to all life on earth and seemingly connected to all facets of our lives, water defies succinct summary and easy solutions. Panacea have arrived and departed and our water systems exhibit some incremental changes, but typically only the slogans endure. Meanwhile, water grows scarcer around the world, the poor suffer limited access to water, aquatic ecosystems are among the most threatened by human alteration, societies remain vulnerable to water hazards such as floods and drought and most water use is inefficient in both physical and economic terms. Why is this so? The reasons are myriad, and in water issues, the details are critical and too often superseded by the force of strong convictions. Nonetheless, one often finds that a root cause of many complex water issues relates to the difficulty of managing the ever present intricacies of poverty and ecology. These are the issues that make water different from other, more easily commoditized, mediums.

People have long been challenged by the need for a practical valuation of water. Early economists, including Adam Smith, were perplexed by the water-diamond paradox. Why was water which is essential to life worth so little, while a diamond which has little practical use worth so much? The answer lay in the concept of marginal value, which means that a good is valued at its marginal use, that is, the use that would be the first to go if less water were available. Since water was so plentiful, the very lowest valued uses of water, such as unused river flows, were of no value to this early economic thinking. Since then we have learned much more about the value of water including the value of its instream flows. These include the importance of streamflow patterns for aquatic ecosystems that have evolved in concert with the variability that characterizes natural flows. It also includes the services that these

C. Brown (✉)
Department of Civil and Environmental Engineering,
University of Massachusetts Amherst, 37 Mather Drive, Amherst, MA 01003, USA
e-mail: cbrown@ecs.umass.edu

J.C. Ingram et al. (eds.), *Integrating Ecology and Poverty Reduction: Ecological Dimensions*, DOI 10.1007/978-1-4419-0633-5_6,

ecosystems provide for many poor and often indigenous communities such as sources of protein through fish harvesting and irrigation through flood recession agriculture. Now that we are reaching the limits of water availability, we face the prospect of choosing which of our current water uses will not receive water in the future.

As the authors in this section describe, there are no easy choices. Those who lose access to water when it is scarce suffer real damages regardless of our ability to quantify or moderate those damages. Yet these chapters also describe how careful thinking and rethinking of our objectives and the means of achieving those objectives can lead to better outcomes and, in many cases, avoid water conflicts. From this viewpoint, we have yet to fully rise to the challenge that water issues present and much remains that can be done.

The first chapter by Timothy Randhir and Ashley Hawes provides an overview of the watershed as the medium through which our decisions are translated to water quality and quantity with ecological and socioeconomic ramifications. This includes population growth and land use, the construction of dams, and the withdrawal of water for agriculture and other purposes. The concept of the connectedness of all things within a watershed is then used to explore its sustainability. The authors discuss the benefit of integrated approaches that incorporate social and ecological factors in addition to the usual economic ones. Finally, they discuss the management challenges that characterize most water discussions, issues that reappear in the following chapters.

The chapter by Thomas FitzHugh, Colin Apse, Ridge Schuyler, and John Sanderson focuses on the real issues that arise when the objectives of urban water supply appear to be in conflict with the water needs of the environment and/or the poor. As they describe, the robust trend in the population growth of urban centers and the accompanying increase in water demand implies that these challenges are not only pressing now, but will certainly increase in the foreseeable future. Given the typical approach to the management of urban water supply systems, a growth in these conflicts could easily lead to substantial damage to affected ecology and the people directly dependent on it. In response, the authors outline a process called "ecologically sustainable water management" that is designed to lessen negative impacts while maintaining the ability of the water supply systems to meet their objectives. In several cases, they describe how the process often leads to resolutions through consensus building and common sense approaches that multiple stakeholder groups can embrace. By focusing on the details they find small changes that can have large benefits in ecological terms.

The final chapter by Casey Brown focuses on the variability of water supplies and the challenges that it poses to the management of water resources and ultimately to economic growth in poor countries. With expectations that hydrologic variability will increase with climate change, there is likely to be a greater demand for water infrastructure with greater impact on the poor and ecosystems in the places where it is built. However, there are underutilized tools available, such as economic mechanisms for water allocation and hydrologic forecasting derived from remote sensing and global modeling data sources that have the potential to improve the management of water variability and decrease the negative impacts of infrastructure. If the potential of these tools are to be realized, a concerted research effort is needed that focuses on updating the science of water management to be consistent with our current priorities and to facilitate the integration of new technology.

In the end, the value of water will not be quantified easily. However, perhaps this impediment to the successful resolution of water issues can be a source of strength for finding enduring solutions. For water forces us to face the limits of the reductionism upon which our analytical house of cards is often built. As our authors demonstrate, practical solutions to water challenges can be found through discussions of diverse groups with varied and at times conflicting interests. Together they find commonalities and points of agreement that become a basis for a "shared vision" of the future. Models provide our best understanding of the system and are a practical means for organizing the data that is available. Yet the model is not expected to provide the solution, but rather to support the conversation. As a result, resolutions include a group of committed stakeholders with an interest in seeing a plan succeed. Through the research agenda and the experience of practical application described in these chapters, the messy realities of the water challenge can be met with a scientific rigor designed to solve it. A vision for sustainability in water may just emerge.

Chapter 7
Ecology and Poverty in Watershed Management

Timothy O. Randhir and Ashley G. Hawes

Introduction

As population growth expands throughout the globe, there is an increasing demand on watershed systems for goods and services that are vital to the survival of human population and ecosystems. These goods and services are innumerable and include food, timber, genetic resources, medicines, water purification, flood control, coastline stabilization, carbon sequestration, waste treatment, biodiversity conservation, soil generation, disease regulation, maintenance of air quality, and esthetic and cultural benefits (Ayensu et al. 1999). Due to impacts to watershed systems, many parts of the world are facing increasing economic and environmental problems. Poverty, unemployment, access to fresh water, poor sanitation, loss of biodiversity, and depletion of habitat are some of the difficulties societies are facing today.

There is an increasing need to improve or maintain the structure and function of watersheds to enhance their role in supporting human populations while simultaneously maintaining ecosystem needs. Palmer et al. (2004) observes that current trends show a growth in both population and resource consumption, indicating an increase in anthropogenic impacts on the natural environment. As development intensifies in response to population growth, rates of resource consumption will continue to increase. Sustaining these crucial resources will require an international effort towards a more integrated knowledge of development impacts on natural systems (Ayensu et al. 1999).

Land use, infrastructure development, and technology have substantially modified natural systems beyond their ability to recover. Required to sustain watershed

T.O. Randhir (✉)
Department of Natural Resources Conservation,
University of Massachusetts Amherst, Amherst, Massachusetts, USA
e-mail: randhir@eco.umass.edu; trandhir@gmail.com

J.C. Ingram et al. (eds.), *Integrating Ecology and Poverty Reduction: Ecological Dimensions*, DOI 10.1007/978-1-4419-0633-5_7,

services in the long term are integrated policies that address the relationships between human activity and the environment. Considering human activities as part of the watershed ecosystem encourages sustainable approaches to resource management and socioeconomic development. This chapter aims to provide a background on the ecological dimension of watershed management and the implications for sustainable development and alleviation of poverty.

Importance of Watershed Systems to Poverty and Development

A background on the state of population and poverty is useful for understanding the pressure on watershed systems and the need to integrate developmental issues into watershed management. World population continues to grow at an annual rate of 1.3%, adding 78 million people to the planet each year (United Nations 2000). At this rate, global population is expected to reach 7.2 billion in 2015 and 8.9 billion in 2050 (United Nations 1999, 2000). Eighty percent of the population lives in developing countries, and less developed regions are expected to absorb 98% of future population growth (United Nations 2000). A large proportion of the population in the developing world still lives without access to safe water and sanitation, limiting the health and productivity of those populations (United Nations 2000). Johnson et al. (2001) estimates that 2.3 billion people currently live in stressed river basins (where per capita water supply is below 1,700 m^3) and this is expected to increase to 3.5 billion by 2025.

Population densities are unevenly spread throughout the world, intensifying pressure on natural systems. The global average population density has been estimated at 48 persons per sq. km (square kilometer) in 2005 and is expected to increase to 67 persons per sq. km by 2050 under medium variant scenario (UN 2008). The population density of developed countries was 24 persons per sq. km in 2005; in less developed countries, the density was 64 persons per sq. km during the same period, and is expected to increase to 95 persons per sq. km by 2050 (UN 2008). UNEP (2007) estimates that land available to each person is shrinking from 7.91 ha in 1900 to 2.02 ha in 2005 and is expected to reduce to 1.63 ha in 2050. This illustrates the increased pressure watersheds within less developed countries are expected to face in the coming decades from increasing population growth and densification.

Many countries where the population is expected to increase are currently burdened by high levels of soil, water, and forest degradation. With increasing environmental degradation, these countries are more susceptible to declining incomes, increased unemployment rates, higher poverty rates, habitat loss, decreased biodiversity, reduced sanitation facilities, and reduced access to safe water.

Increasing population densities tend to aggravate watershed resource management problems. Population growth affects the extent of resource use, changes land cover, alters hydrology, and influences ecosystem services of a watershed. Higher population densities often result in higher per capita use of watershed resources, and result in increasing allocation of land to human uses. Expanding development alters the land cover conditions toward cropland, pasture, or urban cover and changes the characteristics of watershed ecosystems. Increased demand for water for agricultural and industrial purposes can reduce the availability of safe drinking water for all.

High human population densities intensify pressure on wildlife and tend to reduce native biodiversity in watersheds. Construction and other urbanization activities, while improving economic livelihoods, can increase soil loss in watershed systems. Soil loss reduces the capacity of water systems to provide abundant clean water by increasing sedimentation and contaminant loads in water bodies. Johnson et al. (2001) observes that population growth and associated human activity have impacted ecosystems by fragmenting river basins (altering flow paths), diverting water (altering water quantity), and decreasing water quality.

In addition to impacts derived from human densification, extensification of human land uses can also have negative impacts on watershed systems. For instance, forest loss for agricultural production depletes habitat necessary for native species. Forest loss also makes ecosystems vulnerable to the impact of storm events and hurricanes, which can result in large-scale land-form modifications, such as landslides, and associated economic losses. In intensive agricultural and logging land use areas, soil loss can reduce fertile top soil and increase sediment loading into water bodies.

Impacts from resource depletion have a direct effect on people dependent on natural resources for survival. For instance, poor water and sanitation facilities increase health costs and reduce productivity of the population. Waterborne diseases continue to be the leading cause of death of people throughout the world (WHO 2002). Every year 1.5 million people die from water related diseases, out of which 1.3 million are from developing countries (Prüss-Üstün et al. 2008). Similarly, depleted soils and poor water quality can also impact agricultural yield in watersheds throughout the world, thereby reducing safe and reliable food supplies for human consumption and decreasing food available to the poor.

Often, land use and technology have greatly altered ecosystems beyond their ability to recover. Johnson et al. (2001) indicates that ecosystem services, particularly watershed services, can be irreplaceable or cost prohibitive to replace, as is the case with production of fresh water through desalinization. Given that watersheds provide economic goods and ecosystem services that affect the livelihoods of people, a proactive approach is needed that incorporates information on tradeoffs in economic and ecological impacts (Randhir and Shriver 2009). As resource consumption shows no sign of slowing, compatible use is necessary to meet human needs (Palmer et al. 2004). Properly designed rural and ecosystem-based enterprises can create economic, social, and environmental resilience that can be first steps on the path out of poverty (World Resources Institute 2008). To design these interventions properly, though, it is important to adequately understand local watershed and poverty contexts.

System Dimensions of Watersheds and Poverty

Watershed systems are connected and differentiated internally and externally by several interrelated dimensions – lateral, longitudinal, vertical, and temporal (Ekness and Randhir 2007). Watershed dimensions, especially longitudinal and lateral, play a vital role in the distribution of amphibians, reptiles, birds, and mammals in a watershed (Ekness and Randhir 2007). In addition to ecological changes, the extent of poverty and potential for poverty alleviation can also change along these dimensions. The spatial variability in income and risk levels in watershed systems is important information to consider in watershed management for economic and environmental objectives (Randhir et al. 2000). Incorporating these dimensions into watershed analysis provides an integrated examination of watershed functions and efficient modeling of tradeoffs in support of developing sound poverty alleviation strategies. This, in turn, allows for more informed and balanced policy decisions to reduce poverty and the risk of unforeseen ecological costs.

The spatial boundaries of watershed ecosystems are generally determined by topographic and physical characteristics that delineate an area draining to a particular water body. This lateral dimension is not always static, but may be altered by natural or anthropogenic processes. An example of large-scale lateral boundary modification may be found in the periodic cycles of glaciation. Large erosive forces produced by a moving mass of ice hundreds of meters thick are capable of carving a landscape into valleys that preferentially direct surface water toward a receiving water body. Less imposing, but not insignificant, are the daily wind and water erosional processes that remove sediment from one area of a watershed and deposit it to another area, which may be outside the watershed entirely. Another lateral component of a watershed system is the riparian area, which is the interface between the aquatic and upland environments. The riparian area is connected hydrologically to the terrestrial environment by surface runoff, as well as groundwater, and is strongly influenced by the quantity and duration of water in the system.

Most human activities that depend on water tend to be located in close proximity to water bodies. Communities that are dependent on agriculture, fisheries, and natural resource industries are sited in proximity to water bodies for irrigation, transport, and waste servicing. Thus, riparian areas provide food and resources for communities and improve the livelihoods of those with access to riparian resources. The flood plains also have higher frequency of floods than other areas of the watersheds, thereby increasing the uncertainty in income. Other regions in the lateral dimension are headwaters, the areas of higher slope angles within first-order streams. These regions, if deforested, can be vulnerable to landslides and increased topsoil erosion. As was demonstrated by the impact of hurricane Mitch in Honduras, the lack of protection of headwaters can make these regions more prone to landslides following hurricanes and other natural disasters with catastrophic consequences for poor communities and lasting effects on water pollution.

The quantity of water in a system is altered by the introduction of dams, increased water usage for agriculture, and drought management programs. While dam construction can increase economic productivity through increased availability and stability of water supply, expansion in irrigated area, and flood control benefits, ecosystem impacts are often severe. The ecosystem impacts can affect livelihoods of the poor and natural resource based communities. Thus, there are clear tradeoffs between economic and ecological benefits that require evaluation for effective watershed management.

The duration of water in a system may be modified by activities that increase or decrease the rate at which water enters the stream channel, such as an increase in the area of an impervious surface. These variable hydrologic conditions within riparian systems create distinctive conditions, such as hydric soils, that encourage unique vegetation communities and provide habitat for many aquatic and terrestrial species during various life cycles stages. Riparian areas have also been recognized as performing such functions as removing contaminants from surface water, reducing stream temperature, and contributing coarse woody debris into aquatic communities. Loss of the riparian habitat can have severe impacts on income, employment, and subsistence resources of local people.

Human activity has the potential to greatly alter watershed boundaries by creating obstacles to surface and groundwater flow. These obstacles may be created by various developmental activities that include placement of roads, constructing impervious areas, and deforestation. These developmental activities can intercept and redirect precipitation outside of the catchment. Large scale earth-moving activities can also redirect hydrologic flows. Watersheds and riparian boundaries may be altered to increase the proportion of tillable or otherwise developable land. Development activities that ignore these critical ecosystem processes and dimensions may result in changes that are irreversible without direct restoration activity.

Rivers and streams longitudinally connect upstream areas of the watershed with the downstream areas. Hence, the impacts from each upstream section compound the effects on the sections below. The interplay between land use and poverty in watersheds is important to understand for effective management. Poverty can result in increased incentives for land uses that can rapidly exploit resources for quicker gains. For example, vegetation cover may be reduced by forest clearing for agriculture or urban development. Forests may also be thinned by fuel wood collection or other harvestable materials. Decreasing cropfield fallow periods can result in decreased crop yields that do not meet the nutritional needs of the human population. The effect of poverty on changing land use can be substantial where land use policies are lacking or not implemented and enforced. In turn, land use can affect poverty by reducing access to resources and income or employment potential. Furthermore, these land use changes can also create ecosystem impacts that alter natural hydrology and ecosystem health that can in turn reduce water quality, negatively impacting human health.

Changes in land use, particularly within the headwaters, have the potential to greatly impact the economic and ecological conditions of downstream locations. Increased impervious surface resulting from development activities in headwaters increases the quantity and rate at which runoff enters a stream during a storm event, ultimately increasing the rate of discharge at the watershed outlet. In addition to water, sediment, nutrients, and contaminants carried in the stormwater can be more rapidly transported from upstream reaches of the watershed with flash flood and other potentially catastrophic impacts. The potential for conflict arises when upstream management policies do not incorporate strategies to mitigate impacts to downstream communities which are often separated by political boundaries. As with the lateral dimension of watersheds, the longitudinal dimension may vary depending on climatic and hydrologic conditions, separating watershed communities that share perennial streams during times of drought and connecting communities linked by intermittent streams during periods of high flow.

The atmosphere and groundwater systems are connected vertically by processes such as precipitation, evapotranspiration, and infiltration. These processes influence the quantity, quality, and duration of water entering the watershed system. Although, largely controlled by climatic conditions, human activities can also alter these processes. Increasing development and impervious surfaces decrease infiltration rates, increase the quantity of water entering surface systems, and introduce anthropogenic contaminants. Development also has the potential to alter the micro-climate of the area, as is observed in urban communities. Policy decisions, such as incorporating rooftop gardens, community gardens or urban agriculture, and use of permeable building materials in community planning may decrease the effects of human activities on the watershed system.

The lateral, longitudinal, and vertical dimensions described above vary temporally as the system adjusts to varying physical, biological, and chemical conditions within the watershed. The significance of changes in these watershed conditions is dependent on the scale at which the watershed is managed. Single storm or erosion events may have significant impacts on a system managed for short term objectives, such as monthly water quality goals. These single events, however, could have less significance over the long term. Cumulative effects occur when the combination of individual activities and events within the watershed result in significant changes to watershed functioning. As with the temporal scale, the spatial scale may also modify the significance of events or activities. Cumulative effects may be observed when activities that are inconsequential on a small scale are of great consequence when combined over a larger scale, and, thus, there is a need to account for economic and ecological tradeoffs to achieve sustainable watershed management. This will allow management of watershed resources to alleviate poverty with minimal effects on the ecosystem that sustains local people and resources.

Ecological Components of Watershed Systems

The ecological health of a watershed influences the functions of the watershed as well as its value to the human population. The components that comprise watershed ecology include interrelated abiotic, biotic, and socioeconomic elements, many of which may be measured, modeled, and analyzed to evaluate current and potential management decisions. Abiotic elements of a watershed are nonliving physical factors, such as climate, hydrology, geochemistry, bedrock geology, and soils. These elements may be represented by precipitation statistics and trends in seasonal mean temperature, stream hydrographs and hydrologic response, water quality, and soil chemistry. Biotic elements of a watershed are comprised of the living organisms and their associated communities, including vegetation, microbes, wildlife, and people. Biotic elements may be represented by species number and distribution surveys, stress evaluations, and interaction analyses.

Abiotic and biotic elements interact to produce biogeochemical processes that greatly influence watershed ecology (Moldan and Cerny 1994). These physical and weathering processes may alter the physical boundaries of the watershed and vegetation communities throughout the catchment. Biota in riparian areas remove nutrients and contaminants from water originating in agricultural and developed regions of the watershed, improving the water quality in stream and downstream (Lowrance et al. 1984). The permeable structure of watershed ecosystems allows species mobility and hydrologic connections, permits the transfer of abiotic and biotic elements between ecosystems, and creates dynamic communities. This transfer of material is modified by socioeconomic factors that fragment watershed structure.

The socioeconomic component of the watershed affects the ecosystem through management decisions that reflect the values and priorities of the human population. Conversion of undeveloped areas to suburban and urban communities can result in direct habitat loss, fragmentation, and alteration. Increasing development often increases the cover of impermeable surfaces that increase runoff and reduce vegetation cover and groundwater infiltration. Natural features, such as streams, wetlands, and forests, may be altered through channelization, industry management for silviculture or aquaculture, and to increase the land available for agriculture. Increasing population densities can place uneven or unsustainable demand on local resources that may not support the population under times of natural environmental stress. The complex relationships among abiotic, biotic, and anthropogenic factors require careful assessment and planning to assess ecosystem impacts in the watershed system. A multi-attribute framework involves the combined assessment of multiple attributes of the watershed ecosystem (Randhir and Shriver 2009). This approach can be helpful to identify baseline conditions, prioritize policies, and adapt community planning to maintain ecosystems (Shriver and Randhir 2006).

Maintaining biodiversity requires habitat management techniques that include both the preservation of individual species, as well as, the processes necessary for overall ecologic health (Naiman et al. 1995). Human activities have the potential to cause ecosystem simplification by inhibiting the flow of species individuals between watershed boundaries (Doppelt et al. 1993). Fostering high water quality, maintaining natural dynamic biogeochemical processes, and reducing human impact on the ecosystem encourage biological integrity and promote overall biodiversity. Maintenance of ecosystem integrity is vital to balanced economic development and requires careful inventory of ecosystem resources. Tradeoffs in economic and ecological objectives are essential to the decision-making framework. For this, a participatory approach that informs and involves the people directly affected by a decision can enable sustainable solutions to watershed management. While there is a wide range of management ideals, from pure economic development and pure ecological conservation, a balance between these objectives is necessary to achieve poverty alleviation and sustainable development in watersheds. In the issue of poverty, the success of sustainable watershed management relies on the ability of stakeholders to consider the concerns of others and arrive at mutually supported policies.

Watershed Sustainability

The sustainability of watersheds becomes important as we consider economic and ecological services critical to support the increasing demands of human populations and wildlife. The World Commission on Environment and Development (Brundtland Report) in 1987 defines sustainability as "meeting the needs of the present without compromising the ability of future generations to meet their own needs." While the definition is straightforward, it is more complicated in implementation because of the primacy of current needs over those of future generations.

Watershed sustainability requires the use of watershed goods and services without impairing ecosystem integrity and without compromising the ability of future generation to use these services. This is a difficult and complex task requiring extensive information about ecosystem components and dynamics, stakeholder values, and resource constraints faced by watershed managers. The watershed context is defined by the state of ecosystem components, their interactions, depletion rates, recovery rates, the nature of stressors, and system resilience. The stakeholders' context is defined through population demography, community norms, values, institutions, organizations, community interaction, and public participation. Sustainability is influenced by such economic parameters as include the nature of markets, resource constraints, incentive structures, technology, and prices.

Achieving watershed sustainability requires balancing economic and ecological goods and services provided by a watershed system through time, while considering intergenerational ethics. This balance defines the rate of net depletion in watershed resources and services that can be maintained without compromising the future use of watershed services.

Watershed activities of the current generation can produce negative externalities that compromise intergenerational equity. Examples of negative externalities include accumulation of pollutants, deforestation, soil loss, excess water use, habitat loss, and reduction in biodiversity. These activities by the current generation can reduce watershed benefits for future generations. For instance, accumulated pollutants increase the cost of water treatment and public health costs of a future generation. However, current generations may not want to bear the cost of maintaining watershed services for future generations. Values and priorities change among generations, making it difficult to evaluate the cost-effectiveness of a particular activity over multiple generations.

Most ecosystem goods and services provided by watershed systems do not have market-based transactions. This makes it difficult to value ecosystem goods and services, since there are no revealed preferences for them through prices. Thus, ecosystem functions that are vital to future and current generations are difficult to value. This results in undervaluation and in inequitable distribution of damages from environmental degradation. However, new efforts have arisen to value watershed services through water markets. There are relatively few examples of functioning water markets, one kind of which is described further in Chap. 14, Vol. 2. Non-market valuation techniques are also being developed to monetize watershed services. These techniques use stated preference and other indirect methods to quantify the nominal value of watershed services. Non-market valuation of watershed commodities and services quantify ecosystem values that could be used to optimize management practices in watershed systems. Such integrated assessments based on both ecological and economic information are the basis of ecological economic approaches to developing sustainable management plans for watershed systems.

Integration Towards a Systems Framework

Integrating ecological and socioeconomic information into a systems framework based on watershed boundaries promotes management decisions that incorporate resource capabilities (Montgomery et al. 1995). Traditionally, resource and environmental policy has followed a single aspect or reactive approach, focusing on a particular resource or management factor without recognizing other interrelated components (Ayensu et al. 1999). This approach incorrectly isolates individual components of the system and creates an unrealistic system vacuum where external inputs and outputs are not adequately considered. Policy decisions based on this model create an unbalanced policy framework that is ultimately unsuccessful in achieving the initial management goal.

By incorporating ecological and socioeconomic factors, watershed management identifies environmental issues within and across various dimensions and facilitates an integrated policy response. Recognition of various interests and stakeholders encourages the development of adaptive solutions across political boundaries, clarifying jurisdiction and improving the success of management decisions. These solutions can be specifically structured to anticipate future environmental problems and thereby encourage a proactive management approach.

Challenges toward implementing an integrated system include incorrectly incorporating the multiple distinct elements, thereby creating a faulty framework (Omernick and Baily 1997). Holistic knowledge of the system, including ecology, resource potential, and management goals, requires research, model generation, and model validation across traditional boundaries. An integrated systems approach also necessitates coordination among experts across various scientific and socioeconomic fields and among regulatory agencies, which often have conflicting resource goals. The direct monetary cost of implementing the system presents challenges as the execution often produces non-market benefits that are difficult to recognize and measure. An integrated approach also needs to account for dynamic changes in the system attributes and people's values and norms, including factors, such as uncertainty in budgets, technological constraints, and the impact of natural disasters. Thus, an adaptive planning approach is needed to develop sound integrated solutions. In this frame, new information on the ecosystem, values, and constraints are constantly added to update management strategies.

Challenges

Several challenges in watershed management increase the complexity of decision making. Multiple boundaries, common pool resource issues, valuation of watershed services, the need to maintain watershed resilience, and the need for integrated modeling, and incorporating participatory approaches all add complexity to management plans. Each of these will be discussed in turn.

Handling Boundaries

One aspect of watershed management that becomes complex is dealing with a variety of boundaries. Most local, state or provincial, and national boundaries are administrative in purpose and do not typically coincide with watershed or other ecosystem-based boundaries. Watersheds can cross several administrative and ecosystem boundaries, increasing the complexity of watershed assessments and management, and requiring cooperation among different administrative units and agencies.

Three natural units in ecosystem management are watersheds, ecosystems, and biomes. Ecosystems are defined as a community of organisms interacting with one another and with the environment. Biomes are major regional communities of plants and animals with similar characteristics and environmental conditions. Given a variety of landscape units, a hierarchical approach can be used to accommodate multiple boundaries and units in watershed management. For example, biomes can be divided into river basins and further into watersheds. This requires an overlay of administrative lines to identify the nature of jurisdictions. The overlay can be hierarchically nested to smaller scales: sub-basins, reaches, and ecosystem patches. A hierarchical approach can be used to link small groups to larger groups as they relate to river reaches, sub-watersheds, and larger regions. This allows decisions at an appropriate scale, while also understanding scales below and above the current decision scale.

Valuing Watershed Services

Ecosystem services are often not transacted in commercial markets, which make them difficult to value. The rural poor rely on these services for supporting their livelihoods and meeting their basic needs. Costanza et al. (1997) observed that the economic value of ecosystem services and natural capital in the world ranges from $16–54 trillion each year. The economic valuation of watershed services and natural capital is useful in assessing changes in benefits and costs of restoration. This is useful in developing protection programs and in identifying levels of damage to watershed ecosystem. Compensation and other incentive programs can be designed for restoring impairment in watershed systems. Non-market valuation methods that include the Contingent Valuation Method, travel cost, and hedonic methods are common in arriving at stated values of ecosystem services. These values are vital to watershed management to prioritize and protect watershed goods and services that are vital to poverty alleviation.

Common Pool Resource Management

Most watershed resources can be classified as common pool resources, where users cannot be excluded and the resource use is rivalrous. Thus, watershed resources have a tendency to be overused when there are no management strategies in place. Understanding the nature of common resources and their successful governance is an important aspect of watershed management. Several factors influence successful management of the commons – size of the resource, boundary, number of users, monitoring, mobility of users, cooperation, mutual trust, and the nature of the resource itself. Optimal management of the common pool thus requires an institutional design that includes characteristics of both the resource and of the users, as well as a performance assessment based on efficiency, sustainability, and equity (Dietz et al. 2002).

Ecosystem Resilience

Resilience is the ability of a watershed system to recover from the impact of stressors by human, climatic, and other factors. Resilience of a watershed can be associated with the extent of wetlands, riparian ecosystems, forest cover, natural river flows, soil conditions, and geological conditions. While some of the factors are beyond human control, watershed resilience could be increased by restoring to historic conditions and minimizing characteristics that reduce resilience. For example, maintaining forest cover, wetlands, and riparian systems increases resilience of watersheds to damage from floods and hurricanes. River systems are dynamic and often resilient to large impacts on water quality. For example, rivers in the United States with high pollution levels in the 1960s recovered substantially to fishable and swimmable levels after the Clean Water Act.

Integrated Modeling

Palmer et al. (2004) observe that understanding how natural systems provide ecological services entails measuring services and assessing their dynamics at scales that match ecosystems properties. Modeling is an important tool in watershed management to address complexities and make effective decisions. Quantifying impacts and exploring the relationships among various ecosystem properties often requires statistical and simulation methods. To identify optimal practices, mathematical optimization can be used to identify efficient practices in watersheds. Integrated modeling that combines a variety of quantitative tools to address a problem is useful in watershed management. Ayensu et al. (1999) observe that integrated models allow policy makers to explore the consequences of management decisions and promote pro-active policy approaches. Decision support systems are being used to provide timely and appropriate information that is useful in arriving at particular watershed decisions.

Participatory Approaches

Participation is an important part of effective and sustainable solutions to watershed problems. This includes involvement and interaction of stakeholders, scientists, and managers in planning and management of watershed systems (Shriver and Randhir 2006). Good ecosystem management deals with conflicting goals and takes into account the linkages among environmental problems that require stakeholder participation to resolve (Ayensu et al. 1999). For example, management of watersheds for poverty alleviation need to include representatives of the poor, business community, planners, banks, and development agencies. This enables the processing of tradeoff information and development of outcomes that are consensus based. Incorporation of information and acceptance of a project by all stakeholders reduces conflicts, increases participation, and increases the effectiveness of the project. Public participation is often difficult to garner due to a lack of interest or disenfranchisement of community members from resource protection. Participation is also hindered by difficulties in organization, higher costs of participatory approaches, lack of awareness, and poor leadership. Some strategies that increase participation include public awareness campaigns, voluntary programs, compensation for time, and increased transparency in organizational procedures. Other methods that are useful in increasing public participation include working groups, public meetings, feedback sessions, watershed teams, and use of the Internet.

Conclusion

Watershed systems provide valuable goods and services to support human populations and ecosystems. There is an increasing need to improve and maintain the structure and function of watershed systems to achieve sustainable resource use.

Evaluating human activities as a part of the ecosystem is necessary to reach this goal.

Many countries with rapid population growth are already burdened with high rates of degradation in soil, water, and forest resources. With increased environmental degradation, further negative impacts are likely on income, unemployment, poverty, biodiversity, and water resources protection. Watershed management allows careful planning of strategies that mitigate these impacts and enable sustainable growth.

Watershed ecosystems are comprised of four major dimensions that are dynamic and interactive: the longitudinal, lateral, vertical, and temporal dimensions. These dimensions should be incorporated into watershed analysis for integrated assessment and efficient modeling of the system. Recognizing changes in biotic and abiotic processes along multiple dimensions is important for understanding the ecosystem transformations in a watershed. Similarly, incorporating the spatial and temporal changes is important in understanding cumulative impacts in a watershed system.

The sustainability of watershed systems becomes important as we consider increasing the extent and efficiency of economic and ecological services to satisfy the needs of human beings and wildlife. Information on the state of the ecosystem, interactions, rates of depletion, recovery, extent of stressors, and resilience are important characteristics of watershed systems. An integrated assessment of ecological and economic aspects is useful in developing sustainable watershed management plans.

Several challenges in ecosystem-based watershed management include multiple boundaries, the nature of common pool resources, ecosystem valuation, resilience, integrated modeling, and participatory approaches. Some potential solutions to these issues are also discussed. Understanding the ecology and economics behind watershed systems is critical for effective management. Complex information, dynamics in ecosystem processes, and human dimensions make watershed management a difficult task. Adaptive planning, integrated modeling, and participatory solutions provide hope for addressing complexity in watershed decision making.

References

Ayensu, E., D. Claasen, M. Collins, A. Dearing, L. Fresco, M. Gadgil, H. Gitay, G. Glaser, C. Juma, J. Krebs, R. Lenton, J. Lubchenco, J. McNeely, H. Mooney, P. Pinstrup-Anderson, M. Ramos, P. Raven, W. Reid, C. Samper, J. Sarukhan, P. Schei, J. Tundisi, R. Watson, X. Guanhua, and A. Zakri. 1999. Ecology: International Ecosystem Assessment. Science 286 (5440). 685–685.

Bruntland, G. (ed). 1987. Our Common Future: The World Commission on Environment and Development, Oxford: Oxford University Press.

R. Costanza, R. d'Arge, R. de Groot, S. Farber, M. Grasso, B. Hannon, S. Naeem, K. Limburg, J. Paruelo, R.V. O'Neill, R. Raskin, P. Sutton and M. van den Belt, 1997. The value of the world's ecosystem services and natural capital. Nature 387 (1997), pp. 253–260

Dietz, T., N. Dolsak, E. Ostrom, and P.C. Stern. 2002. The Drama of the Commons. In Ostrom, E., T. Dietz, N. Dolsak, P.C. Stern, S. Stovich, and E.U. Weberr, Eds. 2002. The Drama of the Commons, National Academy Press, Washington, DC.

Doppelt, B.M., M. Scurlock, C. Frissell, and J. Karr. 1993. Riverine Systems: The Stepchild of Environmental Efforts. In Doppelt, B., M. Scurlock, C. Frissell, and J. Karr. Entering the Watershed: A New Approach to Save America's River Ecosystems. Island Press, Washington, DC, pp. 7–30.

Ekness, P., and T.O. Randhir. 2007. Watershed-scale Influence of Spatial Dimensions and Landuse Disturbance on Habitat Potential: An Ecohydrologic Approach to Policy. Journal of American Water Resources Association (JAWRA), December, 43(6): 1468–1482.

Johnson, N., C. Revnga, and J. Echeverria. 2001. Managing Water for People and Nature. Science, May, 292(5519): 1071–1072.

Lowrance, R., R. Todd, J. Fail, Jr., O. Hendrickson, Jr., R. Leonard, and L. Asmussen. 1984. Riparian Forests as Nutrient Filters in Agricultural Watersheds. BioScience (34)6 pp. 374: 366.

Moldan, B. and J. Cerny. (1994) Small Catchment Research, in Biogeochemistry of Small Catchments: A Tool for Environmental Research. John Wiley and Sons, pp. 1–29.

Montgomery, D.R., G.E. Grand, and K. Sullivan. 1995. Watershed Analysis as a Framework for Implementing Ecosystem Management. Journal of the American Water Resources Association 31(3), 369–386 doi:10.1111/j1752-1688.1995.tb04026.x.

Naiman, R., J. Magnuson, D. McKnight, J. Stanford, and J. Karr. 1995. Freshwater Ecosystems and Their Management: A National Initiative. Science Vol. 270 pp. 584–585.

Omernick, J. and R. Baily. 1997. Distinguishing Between Watersheds and Ecoregions. Journal of the American Water Resources Association 33:5, 935–949.

Palmer, M., E. Bernhardt, E. Chornesky, S. Collins, A. Dobson, C. Duke, B. Gold, R. Jacobson, S. Kingsland, R. Kranz, M. Mappin, M. L. Martinez, F. Micheli, J. Morse, M. Pace, M. Pascual, S. Palumbi, O.J. Reichman, A. Simons, A. Townsend, and M. Turner. 2004. Ecology for a Crowded Planet. Science 304(5675): 1251–1252.

Randhir, T. O., and D. M. Shriver. 2009, Multiattribute optimization of restoration options: Designing incentives for watershed management, Water Resources Research, 45, W03405, doi:10.1029/2008WR007169.

Randhir, T. O., J. G. Lee, and B. Engel. 2000. Multiple Criteria Dynamic Spatial Optimization to Manage Water Quality at a Watershed Scale. Transactions of the American Society of Agricultural Engineers. 43(2): 291–299.

Shriver, D.M. and T. O. Randhir. 2006. Integrating stakeholder values with multiple attributes to quantify watershed performance. Water Resources Research 42, W08435, doi: 10.1029/2005 WR004413.

United Nations, 1999. World Population Prospects: The 1998 Revision, Volume 1. Comprehensive Table, No. E.99.XIII.9. United Nations, New York.

United Nations, 2000. Charting the progress of populations. Department of Economic and Social Affairs, United Nations, New York.

United Nations. 2008. World Population Prospects: The 2008 Revision, Population Division of the Department of Economic and Social Affairs of the United Nations Secretariat, URL: http://esa.un.org/unpp, Friday, May 08, 2009; 11:52:01 AM.

United Nations Environment Programme (UNEP). 2007. Global Environmental Outlook Environment for Development: Summary for Decision Makers. Progress Press Company, Valletta, Malta.

Prüss-Üstün A, Bos R, Gore F, Bartram J. 2008. Safer water, better health: costs, benefits and sustainability of interventions to protect and promote health. World Health Organization, Geneva, 2008.

World Health organization (WHO), 2002. The World Health Report 2002 : Reducing Risks, Promoting Healthy Life. Geneva, World Health Organization, 2002

World Resources Institute, 2008. World Resources – Roots of Resilience. Growing the wealth of the poor. World Resources Institute, Washington, DC.

Chapter 8
Balancing Human and Ecosystem Needs for Water in Urban Water Supply Planning

Thomas FitzHugh, Colin Apse, Ridge Schuyler, and John Sanderson

Introduction

Worldwide population and water use trends indicate that management of water supplies for urban areas will be a critical issue in the twenty-first century. Between 2005 and 2050, global population is expected to increase from 6.5 billion to over 9.2 billion, with population living in urban areas increasing from 3.2 billion to 6.4 billion (UN 2006; 2007). Such population growth will place increasing pressure on available water supplies for cities, as will water demands for industrial production associated with economic development. Current predictions of global non-agricultural water demands indicate a continuing uptrend in withdrawals, assuming continuation of existing consumption patterns (Rosegrant et al. 2002; Shen et al. 2008; 2030 Water Resources Group 2009). Given that 90% of the 3.2 billion person increase in urban populations by 2050 will be in developing countries (UN 2007), clearly there will be immense pressure to build additional infrastructure to supply water for cities over the next 40 years in the developing world.

These trends in urban water use are an important issue from an ecological perspective because of the negative impacts that urban water management and infrastructure typically have on natural freshwater ecosystems and biodiversity (FitzHugh and Richter 2004). The typical pattern of urban water development in the twentieth century has been expansion of water supply infrastructure without regard for ecological needs. Continuation of this pattern will have serious consequences for not only the health of freshwater ecosystems, but also the ecosystem services that those

T. FitzHugh (✉)
FitzHugh Consulting, 2430 O St. #6, Sacramento, CA 95816, USA
e-mail: tfitzhughil@yahoo.com

J.C. Ingram et al. (eds.), *Integrating Ecology and Poverty Reduction: Ecological Dimensions*, DOI 10.1007/978-1-4419-0633-5_8,

systems provide for people. Because populations in developing countries are generally more directly dependent on freshwater ecosystem services than most people in the developed world, managing any negative effects of infrastructure development will be particularly important, and even more so for poor people who often do not benefit from engineered urban water provision.

To manage negative ecological effects, it will be critical to implement a more informed process of water planning and management, so as to store and extract water for urban use in much less ecologically damaging ways than those used in the past. The goal of this chapter is to describe such a process, building on recommendations made previously in FitzHugh and Richter (2004). The level of investment that will be needed to supply water for growing urban populations in the developing world will clearly be immense, and this provides an opportunity to follow a different path that better takes into account ecosystem needs. After summarizing some typical ecological impacts of urban water systems, we will describe methods for incorporating ecosystem flow needs into water supply planning and management, both in general and through use of three case studies of environmental flow projects involving The Nature Conservancy (TNC). The case studies described here are projects where TNC has worked and is working with urban water suppliers to better incorporate ecosystem needs into system planning and operations. TNC is a global biodiversity conservation organization whose work includes the conservation and restoration of natural freshwater ecosystems.

The three case studies to be described are all projects in the United States, but we believe that the same process can be applied in the developing world. In fact, when construction of new infrastructure is necessary to extend water supplies to un-served populations, in one sense, the process can actually be used more fruitfully, since environmental needs can be incorporated into system planning from the beginning, rather than having to retrofit existing systems and restore already damaged ecosystems. In situations where data is lacking, which may be common in the developing world, there will have to be greater emphasis on field data collection to assess site conditions and investigate flow–ecology relationships, use of expert opinion to make up for the lack of quantitative information, and adaptive management to improve environmental flow recommendations and their implementation over time. But the same general framework can still be applied. The concluding sections will summarize the three case studies and further expand on the applicability of this framework in the developing world.

Ecological Impact of Urban Water Supply Systems

The focus of this chapter will be on the impact of water withdrawal and storage infrastructure on natural patterns of flow, which is one of the primary ecological impacts of urban water supply systems. The natural flow regime is defined as the characteristic pattern of water quantity, timing, and variability that occurs in freshwater ecosystems. It has been called a "master variable" in these systems, because of its direct influence on ecological integrity and also its relation to other important ecological processes, such as provision of water quality, energy sources, physical habitat, and biotic interactions (Poff et al. 1997). Consequently, alteration of natural flow regimes can have serious ecological impacts, and in fact damming of rivers and the accompanying impact on natural patterns of flow and sediment is recognized as a leading cause of the decline of freshwater biodiversity globally (Richter et al. 2006).

FitzHugh and Richter (2004) documented flow impacts of urban water development in five cities in the USA: Los Angeles, Phoenix, New York, San Antonio, and Atlanta. Ecological impacts in ecosystems affected by these cities range from extirpation and endangerment of native fish and other aquatic organisms, to elimination of riparian vegetation because of dewatering of rivers and streams, to reduction of habitat for migrating and nesting birds. Impact was most severe in situations where water was transferred out of a basin, which occurred in almost all of these cities due to the growth of the urban area beyond the water supply capacity of the local river basin or aquifer. The sources of alteration of natural flow regimes were water withdrawals from both river and aquifer systems, and operations of reservoirs to provide storage of water for use during drier parts of the year. These practices are increasingly common in developing world cities that are rapidly out-growing their local water supplies.

Weiskel et al. (2007) have introduced a useful typology for classifying the hydrologic impact of human activities. Affected systems can be classified according to changes in the quantity of flows, or alteration of flow patterns over time (without changes in quantity), as analyzed on an annual, monthly, or other time-scale. Generally, the impact of urban water supply systems involves either a decrease in the quantity of flow or a change in the pattern of flow (depending on the time scale being analyzed), because of the need to transfer and use water in the city. Less common are increases in annual flows, which can occur when water is transferred from one basin to another and a natural river course is used to move the water to another location.

Figures 8.1–8.3 show graphs of streamflow alteration from FitzHugh and Richter (2004), exemplifying some of the different types of flow alteration that can occur from water supply system operations. Figure 8.1 shows the reduction of flood flows below dams operated in the Salt River in Arizona, which is a primary water source for the city of Phoenix. Reduction or elimination of natural floods has created serious problems for native cottonwood forests in the Salt River Valley, and all of the native fish species have been extirpated from the Salt River. Figure 8.2 shows the projected impacts of future water use on dry season flows in the lower Apalachicola River in Florida, which is affected by water withdrawals for the city of Atlanta. Reduction of low flows in the river could have serious impacts on the productivity of the downstream Apalachicola Bay estuary. Figure 8.3 shows increases in streamflow in the upper Owens River in California, which was used to transfer water from an adjacent basin prior to entering an aqueduct to the city of Los Angeles. These increased inflows during all months of the year contributed to severe alterations of river habitat, including changes in important temperature and sediment regimes and accelerated rates of bank erosion.

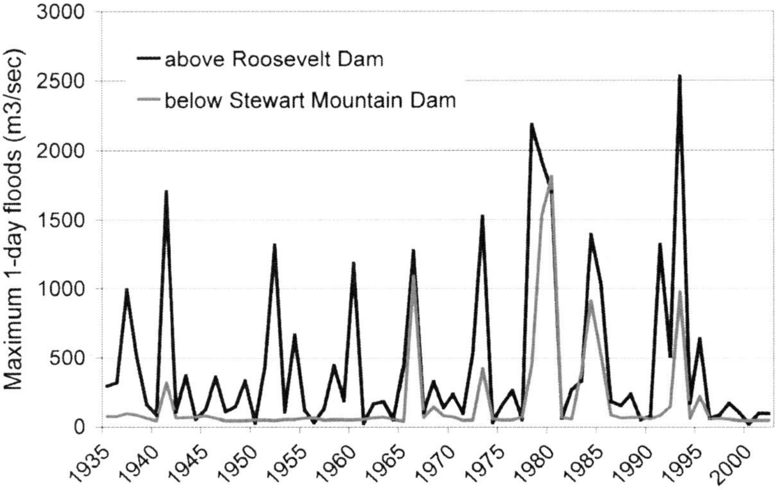

Fig. 8.1 Maximum 1-day floods on the Salt River, above the uppermost dam (Roosevelt Dam) and below the lowest storage reservoir (Stewart Mountain Dam) near Phoenix, Arizona. Data are from US Geological Survey stream gauging stations 09498500 and 09502000 (Reprinted from FitzHugh and Richter 2004)

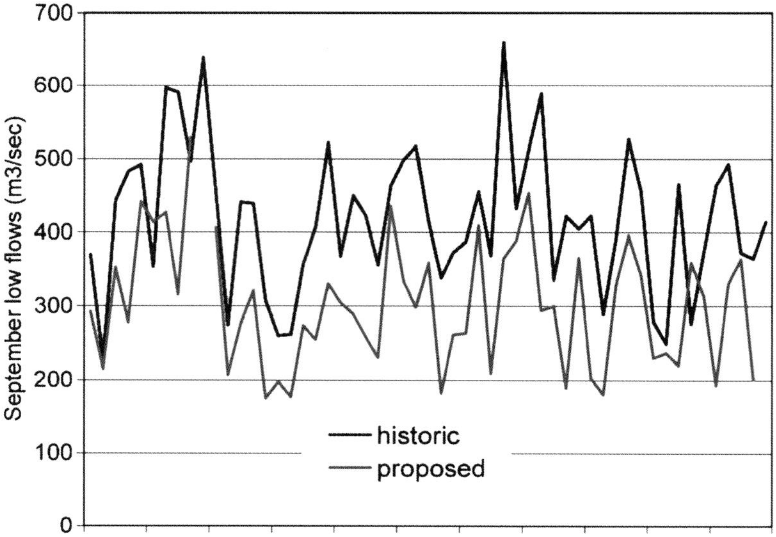

Fig. 8.2 Simulated impacts of future water use on flows in the lower Apalachicola River. The 1939–1993 measured September low flows are compared with model simulated low flows based on projected 2030 levels of water use in the basin, as submitted by the state of Florida during interstate water compact negotiations in February 2002 (Reprinted from FitzHugh and Richter 2004)

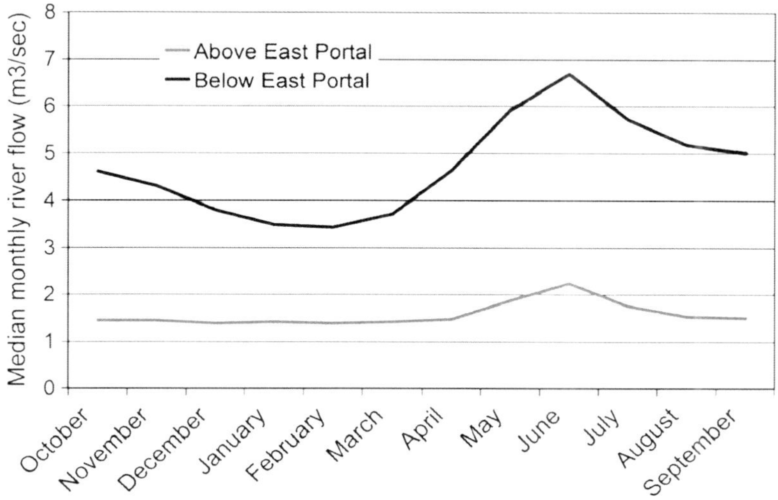

Fig. 8.3 Median monthly flows in the upper Owens River, California, above and below the East Portal, where water being transferred from the Mono Basin enters the Owens River, 1940–1989. Flows are drastically increased during all months of the year (Data are from CSWRCB (1994) Reprinted from FitzHugh and Richter 2004)

In addition to the impacts on ecosystem processes and native biota, development of water resources in these cities has affected or threatened important ecosystem services, including fisheries production, wildlife habitat, and mitigation of water and air pollution. A few examples can be cited here. Water development in Atlanta and San Antonio threatens productive estuarine fisheries along the Texas and Florida coast. One of the potential threats is to the productivity of the Apalachicola Bay estuary, which accounts for 90% of Florida's oyster production (ACOE 1998) and depends on freshwater inflows. Similarly, the Delaware River tributaries below New York City's dams harbor a sport fishery that provides significant economic benefits for the local economy (Maharaj et al. 1998). Removal of water for urban supply can also exacerbate pollution problems. Dry lake beds at Owens Lake and Mono Lake in California, dewatered because of transfer of water to Los Angeles, led to dust storms and serious air pollution (Wiens et al. 1993; Hundley 2001). The dams that New York City operates on the upper Delaware River tributaries must now be carefully managed to protect recreational fisheries, federally endangered species, and to control the salt front intrusion in the Delaware estuary (NYCDEP 1998), because of the potential impacts of increased salinity on downstream water supplies.

In sum, patterns of water development in these five cities have been typical of the twentieth century water development paradigm, which has focused on construction of physical infrastructure such as dams and aqueducts, relied on unsustainable use of groundwater, and included little attention to ecological values (Gleick 2000; Revenga et al. 2000). But trends in these cities, in recent decades, also point toward a solution, showing that improved planning and management that takes into account ecosystem needs, coupled with greater attention to water productivity and conservation, are both feasible and essential as a solution in the face of the continued pressures on ecosystems from urban water development. Litigation in both the Los Angeles and San Antonio cases has forced those cities to improve conditions in unique ecosystems such as Mono Lake and the Edwards Aquifer, and urban water supplies have not suffered as a result. Los Angeles, New York, and San Antonio have all significantly reduced per capita water use in recent decades through water conservation programs, making it easier to incorporate ecosystem needs into their water system operations.

Ecological Sustainable Water Management for Cities

Richter et al. (2003) presented a framework for ecologically sustainable water management (ESWM), which is defined as "protecting the ecological integrity of affected ecosystems while meeting intergenerational human needs for water and sustaining the full array of other products and services provided by natural freshwater ecosystems." In terms of the application of this framework to urban water management, an important point is that this is not a proposal to reduce water withdrawals and greatly increase release volumes everywhere to restore freshwater ecosystems to their natural state. Clearly this would be both undesirable from an economic development perspective and unrealistic. What is being proposed is that by using existing technologies and management tools, and by spurring further innovations in their use, urban water managers can do a better job of protecting freshwater ecosystems while meeting current and future human needs.

The six-step ESWM process to accomplish this goal consists of: "(1) [develop] initial numerical estimates of key aspects of river flow necessary to sustain native species and natural ecosystem functions; (2) [account] for human uses of water, both current and future, through development of a computerized hydrologic simulation model that facilitates examination of human-induced alterations to river flow regimes; (3) [assess] incompatibilities between human and ecosystem needs with particular attention to their spatial and temporal character; (4) collaboratively [search] for solutions to resolve incompatibilities; (5) [conduct] water management experiments to resolve critical uncertainties that frustrate efforts to integrate human and ecosystem needs; and (6) [design and implement] an adaptive management program to facilitate ecologically sustainable water management for the long term" (Richter et al. 2003). The process is designed to be continually adaptive, with steps (5) and (6) feeding back into improvements in the earlier steps. This chapter will primarily discuss steps (1) to (4) in the above process, as applied to management of urban water supplies.

There are a variety of methods for conducting step (1), specifying the environmental flow needs of the ecosystem (see Tharme 2003 for a review). Richter et al. (2006) describes the process used in many TNC environmental flow projects, which is a holistic process aimed at quantifying the flow needs of all parts of the ecosystem, using available literature and expert opinion on important flow–ecology relationships (see Fig. 8.4). This holistic approach can also incorporate habitat modeling

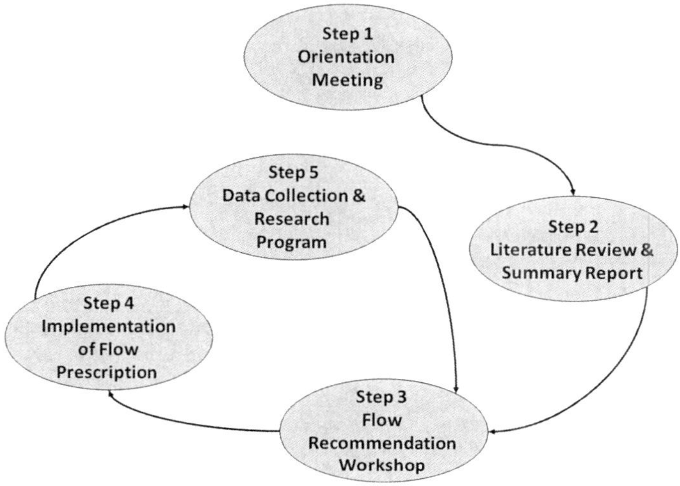

Fig. 8.4 The scientific process for developing environmental flow recommendations comprises five steps. Steps 3–5 are repeated indefinitely to enable iterative refinement of the flow recommendations over time (Reprinted from Richter et al. 2006)

for key species and communities of concern, as illustrated in the Upper Delaware basin case study later in this chapter. This approach reflects the growing understanding in recent decades that environmental flow standards need to move beyond just maintaining minimum flows, to include all components of the hydrograph (Poff et al. 1997; Richter et al. 1997). A particularly useful framework that has emerged in recent years is to describe natural flow regimes in terms of five different flow components: extreme low flows, low flows, high flows, small floods, and large floods (Mathews and Richter 2007). Recommendations will ideally be specified in these terms, describing the relevant characteristics – magnitude, frequency, duration, timing, rates of change – of these different events. Also, it is very important for flow recommendations to be expressed quantitatively, to facilitate implementation and analysis of trade-offs between human and ecosystem needs.

Step (2) in the process involves accounting for human uses of water. Many water supply systems in the United States are beginning to engage in quantitative water supply planning using demand forecasts and hydrologic modeling, which is sufficiently detailed for use in an analysis of the trade-offs between human and ecosystem needs. But methods for actually conducting the trade-off analysis (steps (3) and (4)) are much less developed. Vogel et al. (2007) found that the existing literature on methods for managing and quantifying trade-offs between water supply and environmental flow goals is fairly limited, especially if studies that used fixed minimum flow targets are excluded. A few of these studies that do exist and are most relevant to this chapter are as follows: Homa et al. (2005) introduced the concept of an "eco-deficit," a metric for quantifying the impact of human water withdrawals on the natural flow regimes, and showed that an optimization procedure can be used to maintain the reliability of water supply yield, while also improving the satisfaction of environmental flow requirements. Vogel et al. (2007) showed that the choice of reservoir operating rules has a significant impact on both water yield and environmental flows, and that incorporating a drought management policy can improve the overall trade-off between these two goals. Shiau and Wu (2006, 2007a, b) showed that environmental flow standards that incorporated monthly and inter-annual variability inflows can increase the ability to meet both water supply and ecological needs.

The subsequent sections will describe three projects that TNC has been involved in, working with municipal water supply agencies to specify environmental flow recommendations and to incorporate those recommendations into system operations. The rivers involved in these projects are the Upper Delaware basin, a major water source for the city of New York; the Rivanna River basin, the water source for Charlottesville and Albemarle County, Virginia; and the Fraser River watershed on the west slope of the Rocky Mountains, an important water source for the city of Denver, Colorado. The exact approaches used in these projects vary, due to the particular circumstances of each project and also the lack of a standardized approach for evaluating the trade-offs between water supply and environmental goals. The Delaware and Rivanna case studies cover the evolution of environmental flow policies in two watersheds that contain reservoirs, aqueducts, and other water supply infrastructure. The Fraser River case study, a project that is not as far along, details how environmental flow releases could be prioritized in a watershed with a number of diversions from smaller streams.

Moving Towards Environmental Flow Management Below New York City's Delaware System Reservoirs

New York City has the largest unfiltered surface water supply system in the world, delivering over 4.6 billion liters of high-quality water to a population of over nine million people (NYC DEP 2007). As an unfiltered system, it is analogous to many in the developing world. The system was continually expanded from the nineteenth to the middle twentieth centuries, culminating with the construction, between 1937 and 1965, of four large reservoirs and aqueducts to transfer water over 150 km from the three upper tributaries of the Delaware River. Currently, New York City obtains nearly half of its drinking water from the Delaware basin (USGS 2008) and its Delaware System is considered its highest quality source. Figure 8.5 shows the reservoirs in the Delaware system.

Controversy over early development of the Delaware System led to a US Supreme Court case in 1931, which allowed New York City to divert an average of up to 1670 million liters per day (mld) from the Delaware Basin. Over the objections of New Jersey and Pennsylvania, it was amended in 1954 to allow New York City to increase its diversions to 3,030 mld. The organization responsible for oversight and management of waters in the basin is the Delaware River Basin Commission (DRBC), which was created in 1961 by a unique federal-interstate compact.

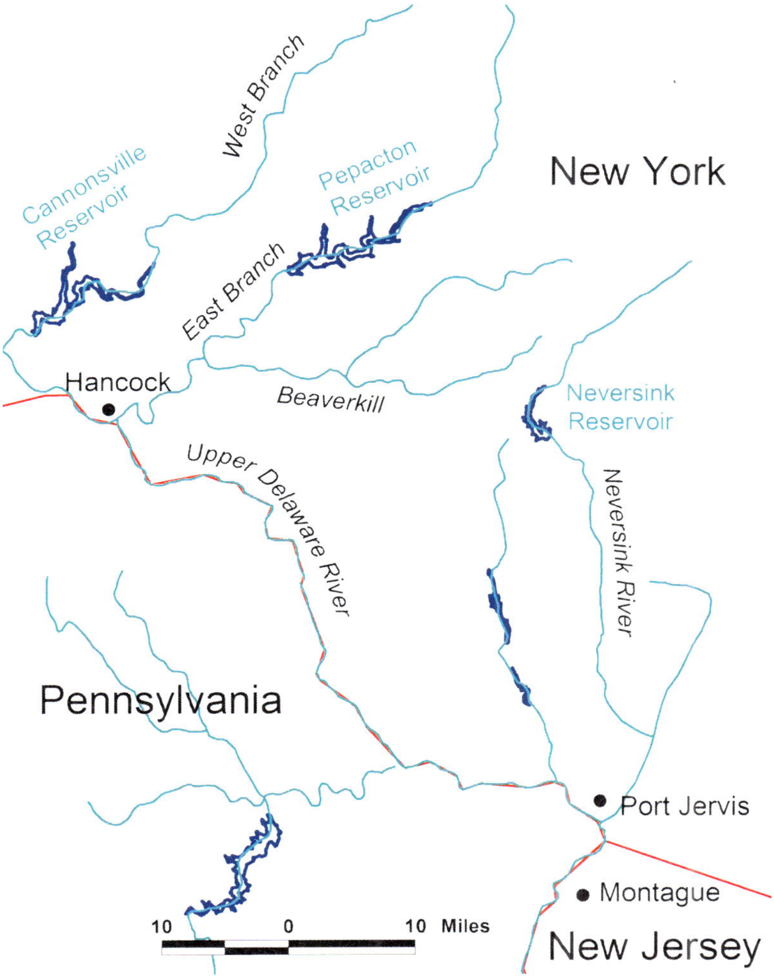

Fig. 8.5 Map of the Delaware basin

The history of environmental flows in the Delaware basin goes back to these initial court decisions, as do prescient concerns about the impacts of out-of-basin diversions on fisheries (including anadromous American shad), recreation, oyster resources in the lower basin, and water quality (Hogarty 1970). The 1954-amended decree established a "streamflow objective" of 49.6 cms, equivalent to 0.0055 cms per square kilometer for a 9,065 km^2 watershed, at Montague, New Jersey (just downstream of Port Jervis, see Fig. 8.5), which was to be maintained in all non-drought periods and administered by a River Master for the Delaware (USGS 2008). In the 1960s, non-regulatory standards for summertime minimum releases below the reservoirs were established, but these were as low as 0.00018 cms for every square kilometer of watershed area, over thirty times less than the Montague streamflow objective (Elliot 1998). In the 1970s, emerging science and a public push for increased releases to protect aquatic insects and fish (primarily trout) led New York State to enact regulations to increase minimum reservoir releases by 3–14 times the original values, with additional releases to meet maximum summer temperature targets, and these releases were made by New York City on an experimental basis starting in 1977 (USGS 2008). Over the subsequent 25 years, a number of other relatively minor experimental revisions were made to this release framework.

But despite these incremental changes to the release requirements over the years, reservoir operations and out-of-basin transfers of water have had severe impacts on the natural pattern of flows in the Delaware River tributaries for a half century (FitzHugh and Richter 2004). The ecosystems directly downstream have suffered habitat stresses from low water when the reservoirs were not spilling, high summer temperatures, and a "boom and bust cycle" of low minimum releases followed by high flows from spills or releases to meet the Montague objective (Elliot 1998). This boom and bust cycle was particularly pronounced on the West Branch Delaware below Cannonsville Reservoir, which was heavily relied on for releases to meet the Montague objective due to its relatively poor water quality. Alteration of natural flow conditions gained additional scrutiny with the discovery in 2000 of the federally endangered dwarf wedgemussel (*Alasmidonta heterodon*) in the Upper Delaware River, in an area heavily influenced by flow releases from New York City's Delaware System reservoirs (Lellis 2001).

The growing evidence of the impacts of flow alteration, combined with the increasing prominence of ecological and natural resource protection issues, occurred at the same time as New York City was becoming more efficient in its use of water, reducing its total annual water consumption and per capita water consumption by over 26% from 1980 to 2005 (NYC DEP 2008). All of these trends converged to create an opportunity for improved environmental flow management in the Delaware, which culminated in 2003 with the creation of the Delaware River Basin's Subcommittee on Ecological Flows (SEF). The SEF is made up of state, federal, non-profit, and academic representatives engaged in resource management and assessment in the Delaware Basin. The mission of this volunteer committee is to develop ecological flow requirements for the maintenance or restoration of healthy, self-sustaining, and managed aquatic ecosystems in the Delaware Basin. Recommendations from the SEF are forwarded to the basin states and New York City representatives that make up the Parties to the 1954 Supreme Court Decree. Creation of the Subcommittee on Ecological Flows allowed for a new, collaborative focus on flow management issues. The SEF was chaired by TNC for the first 5 years of its existence.

The SEF's work began with a strong vote of confidence from the federal government in the form of a $500,000 Congressional appropriation to the US Geological Survey (USGS), a federal science agency with a mission to provide "information that contributes to the wise management of the Nation's natural resources," to analyze issues of flow and temperature management in relation to the aquatic resources of the Upper Delaware. This funding was used to complete "A Decision Support Framework for Water Management in the Upper Delaware River" (Bovee et al. 2007). This USGS study broke new ground for the Delaware Basin by directly assessing the habitat needs of a wide range of species and natural communities in relation to flow and temperature patterns in the West Branch Delaware, East Branch Delaware, Neversink, and upper mainstem Delaware Rivers. The study resulted in a decision support system, available for use in 2006 on the Internet, which presented species/community habitat metrics and temperature metrics in a manner permitting comparison of different water management alternatives. Water management alternatives were modeled through basin-customized software called OASIS, a water management model allowing for virtual experiments with varying reservoir release regimes (Hydrologics, Inc. 2008).

Given this set of tools, the Parties to the Decree, the SEF and interested environmental groups such as Trout Unlimited set about re-examining the underlying structure of reservoir release management from the Delaware System. The new framework, eventually released to the public as the "Flexible Flow Management Program" (FFMP) in early 2007. The specific release provisions of FFMP included eight release seasons, or "bioperiods," that allowed for varying releases across each year from each reservoir, designed to protect ecological flow conditions without threatening water supply. More importantly, the new framework included seven release zones, with increasing release volumes directly tied to increasing levels of storage in each of the reservoirs (USGS 2008). Thus, during wetter years, more water would be automatically released from each of the reservoirs in a manner designed to provide high-quality habitat conditions for the ecosystem and particularly for species of interest. In drier years, the framework moves to a "share the pain" approach, in which some degree of stress is accepted on the river ecosystem in recognition that extended drought benefits neither human or ecological health. Releases from all reservoirs are, on average, increased from previous plans through a program that restores some of the inter- and intra-annual variability considered so important for ecosystem health (Power et al. 1995; Bunn and Arthington 2002; Postel and Richter 2003). The simplistic temperature targets that were a feature of the previous release regimes are removed in FFMP, eliminating unnatural fluctuations in releases for temperature management, and facilitating implementation. Finally, recommendations from the SEF on rate-of-change in reservoir releases have been used as guidance to modify practices from New York City's reservoirs.

The FFMP was approved by DRBC and the Parties to the Decree and was implemented beginning in September 2007. This is an accomplishment and a cause for optimism among water mangers and stakeholders in the basin. Yet much can be done to improve FFMP in the future to increase the ecological benefits. Potential improvements include further increasing spring releases to promote American shad spawning conditions and trout habitat, examining how flows can be managed to improve estuarine health, formalizing limits on rate-of-change in releases, and creating flexibility in the seven decade-old Montague streamflow objective. Also critical to the future is a commitment from relevant resource agencies to adaptive management, especially in light of climate change. As the Flexible Flow Management

Program is implemented, new investment in the collection and interpretation of basin-wide ecological monitoring information would enable refinements over time based on data, rather than modeling results and expert opinion.

Environmental flows in the Delaware Basin have moved from an afterthought, as with the provisions of 1954 Supreme Court Decree, into a central role in shaping the future of water management in the basin. Justice Oliver Wendell Holmes, Jr., in his opinion on the 1931 Supreme Court case dividing the waters of the Delaware, stated it well. "A river is more than an amenity, it is a treasure. It offers a necessity of life that must be rationed among those who have power over it." This "rationing" can now be informed by environmental flow science and sophisticated water management models. These tools, combined with institutional structure, stakeholder involvement, and continuing political will, make it possible to envision ecologically sustainable water management in a basin as disputed and treasured as the Delaware.

Integrating Water Supply and Environmental Flow Objectives in the Rivanna River Basin, Virginia

From its headwaters in the Blue Ridge Mountains, the Rivanna River winds its way through central Virginia's Piedmont plateau region to Columbia, where it joins the historic James River on its way eastward to the Atlantic Ocean. The river and its tributaries provide many benefits to the residents of this watershed, including water quality, fishing, wildlife viewing, and other forms of recreation. The basin is also the water source for the city of Charlottesville and Albemarle County, Virginia, providing water for more than 90,000 residents in a fast-developing part of the state. The current public water supply system was initiated in 1885 with construction of the Ragged Mountain Reservoir to the west of Charlottesville. This reservoir was later expanded in size and connected by pipeline to the headwaters of the Moormans River, where a second reservoir (Sugar Hollow) was also built, all to reduce the refill time of a system that originally drained only 4.66 km^2. When this proved insufficient due to the growth in water demand, in 1966, Charlottesville built a dam on the South Fork Rivanna River, which essentially doubled the storage capacity of the system.

In the early 2000s, after suffering through a severe drought, the community initiated an integrated water supply planning effort. Expanding the community's water storage capacity was essential for two reasons. First, burgeoning growth in the region is projected to increase demand from 41.6 mld in 2000 to 70.8 mld in 2055. Second, the South Fork Reservoir receives substantial sediment from an eroding landscape, and is projected to lose 75% of its capacity over the next 50 years. This effort to increase the community's water storage coincided with a parallel effort by TNC to protect and restore the Rivanna River ecosystem—with a focus on restoring its vital headwaters. The primary flow-related issue in the Rivanna basin is maintaining sufficient low flows during drier parts of the year, as the dams in the system are not large enough to control the higher flows. When the reservoirs are not spilling, the established streamflow standard was to release only minimum flows from the reservoirs, 0.017 cms on the Moormans River and 0.34 cms on the South Fork Rivanna River. These reservoir releases are equal to less than 20% of the flows in these rivers under normal summer conditions (Richter 2007).

Over the course of 2 years beginning in late 2004, TNC facilitated a comprehensive scientific assessment of the Rivanna River Basin, including evaluating the environmental flows needed to sustain a healthy river ecosystem. Following the process outlined in Richter et al. (2006), the assessment began with an orientation session where scientists throughout the region could learn about and shape the process for developing flow recommendations. Next, scientists from two regional universities were commissioned to produce a report summarizing the known information about the watershed and its ecological needs. Armed with this "summary report," the experts developed a set of flow principles during a two- and one-half day working session. More than 40 regional scientists were involved in this workshop. Despite the paucity of data in this watershed, the experts were able to use their best professional judgment to generate flow targets for low flows (baseflow) during each month, as well as minimum allowable extreme low flows that must be maintained during severe droughts. An overarching recommendation from the scientists was to manage reservoir releases to mimic natural, non-depleted flows to the greatest extent possible. After development of these recommendations, TNC worked with the local water utility, the Rivanna Water and Sewer Authority, to analyze out how to best integrate the recommendations into water supply planning for the basin.

One of the important decisions that had to be made was how to expand the capacity of the system, and to this end, the OASIS model (Hydrologics Inc. 2008) was used to simulate various water storage expansion options and determine which infrastructure improvements would supply the needs of people, restore variable flows to the rivers for wildlife, and cost the least for ratepayers. Four final infrastructure options were evaluated, including: (1) dredging the South Fork Rivanna Reservoir, (2) building a pipeline to another nearby river, the James River, (3) expanding the storage capacity of the South Fork Rivanna Reservoir, and (4) expanding the Ragged Mountain Reservoir. Ultimately, the community opted to expand the Ragged Mountain Reservoir and also link it through a new pipeline to the South Fork Rivanna Reservoir. From the standpoint of environmental flows, this option restored a natural flow regime to the Moormans River by ending the 81-year-old practice of diverting water from the headwaters of the river to the Ragged Mountain reservoir in another drainage 21 km away. And it also made sense from the perspective of replacing unsafe and aging infrastructure (the Ragged Mountain dam and pipeline from Sugar Hollow) and reducing costs, because by connecting the system's two main reservoirs, it allowed for sharing of treatment plant capacity and avoided costly treatment plant expansion in the future.

Some changes were made to the environmental flow recommendations from the workshop, to make it easier to integrate them into the water supply planning process. The primary modification was to express the desired releases in terms of a percentage of the amount of water flowing into the reservoirs. The original recommendations had set fixed standards for flow that would vary only depending on whether a year was classified climatically as wet, average, or dry. By basing releases on the percent of incoming daily flow, the system could more easily mimic the natural variability of flows, even when it cannot achieve the full volume. Review by scientists involved in the original flow recommendation indicated that standards expressed in such a way would satisfy the spirit of the original recommendations, if not the exact numbers.

Modeling was again conducted to assess the ability to meet various percent of flow standards, while also supplying sufficient water for the system. In the case of the Moormans River, where the dam is at the headwaters of a sensitive system, the flow standard set was that flow releases can never be less than 90% of inflow. Forecasting indicates that this standard can be met out to the year 2050, and that even with increased water demands, the river will have a natural flow 99% of the time, compared to 64% currently (Richter 2007). In the case of the South Fork Rivanna Reservoir, the imperative of balancing human demands and nature's needs is more acute, and the flow requirements reflect those competing interests. Forecasting indicates that for the South Fork Reservoir environmental flow releases will range from 70% to 100% of natural inflow at least 90% of the time, dropping to 30–50% of natural inflow only during extreme droughts (Richter and Thomas 2007). On the mainstem Rivanna River, compared to current releases, there will be essentially no change in flows 89% of the time based on 2020 demand projections, and 76% of the time based on 50-year demand projections.

The environmental flow releases being implemented in the Rivanna will substantially restore natural flow variability, as compared to the static environmental flow releases provided historically. Critical to the success of this effort was a coming together of the community in a stakeholder-driven process backed up by detailed information about the needs of the environment and needs for water supply. The stereotypical battle pitting so-called "developers" against "environmentalists" was avoided because the vast majority of citizens were able to come together around a plan that met the needs of both the human and aquatic communities. This was accomplished because the community itself set the standards it wanted from its expanded water supply system: the community wanted its drinking water system to meet its growing needs, it wanted its drinking water to come from within its watershed, and it wanted the water to be withdrawn in the most ecologically sustainable manner possible. And the plan that was adopted met all three objectives.

Evaluating Flow Management in Denver Water's Moffat Collection System

Colorado's oldest and largest water utility, Denver Water, currently serves 1.1 million people in the Denver, Colorado, metropolitan area, who use 351 million cubic meters of water annually. By mid-twenty-first century, population and water use are projected to grow an additional 73% and 63%, respectively (Denver Water 2002). Denver Water's Moffat Collection System (MCS) provides 20–30% of Denver

Water's total demand (ACOE 2003). This proportion may increase as additional storage facilities are built. The MCS operates by diverting water from eight small streams in the upper part of the Fraser River watershed, located 80 km west of Denver in the headwaters of the Colorado River. Water diverted from the streams is transported to the Moffat Tunnel, and thereby out of the Colorado River basin and into east-slope storage reservoirs prior to transfer to the city of Denver. Snowmelt dominates the hydrographs of these streams, with annual high flows occurring around June 1.

Unlike the two previous cases studies, the MCS has no storage capacity in the basin where the diversions occur, so Denver Water's ability to manage for environmental flows in the Fraser River watershed is restricted to a binary choice of diverting or not at one or more locations. In 2006, Denver Water began working with TNC to develop a tool for objectively evaluating potential changes in operations that could lead to greater ecosystem health. The tool was designed specifically to provide guidance on this question: if Denver Water were to change operations or to leave a next increment of water in the river, where should they focus? A particular emphasis was placed on the potential to change diversion patterns during high-runoff years, because this could conceivably be done with no loss of yield. High-gradient and elevation streams in the Fraser Watershed were the focus of the project.

The first step in the project was to convene a workshop of scientists, project directors, and outside experts to list the key ecological attributes associated with streamflow in the Fraser River watershed, to identify the key components of the flow regime that sustain those values, and to set preliminary, quantifiable criteria that could be used to make informed management decisions. Results of the workshop were summarized and expressed as explicit relationships between ecological status and flow status for flow metrics (Arthington et al. 2006; Poff et al. 2009). Based on the needs of five biological components of the system (cutthroat trout, amphibians, riparian plant communities, beaver, and aquatic macroinvertebrates) and two abiotic characteristics (sediment and water quality), six streamflow parameters were selected as essential for maintaining the health of the system (Table 8.1). Statistical parameters such as those listed in Table 8.1 are common tools for quantifying the various characteristics of natural and altered flow regimes and setting environmental flow targets, as described in Richter et al. (1996) and Mathews and Richter (2007). In general, flood flow parameters indicate condition of channel maintenance and sediment transport functions, and low flow parameters indicate total habitat availability.

Next the level of hydrologic alteration of the different characteristics of flow listed in Table 8.1 was evaluated. Four condition classes were defined (Table 8.2): natural, minimally altered, moderately altered, and strongly altered. Choosing specific criteria to define condition classes was challenging, because few empirical data were found to link specific flow conditions to specific ecological responses. Criteria used in the template came from a variety of sources. For several parameters, 10% was used to define a minimally altered mean, based on Arthington and Pusey (2003), who suggest that 80–90% of natural flows may be needed to maintain a low risk of environmental degradation. Results of a study by Ryan (1997) in the

Table 8.1 Flow parameters with key functions in support of ecological targets in subalpine, high-gradient streams

Flow parameter	Relationship to ecological targets
1-day minimum flow for the year	Short-term minimum can limit aquatic organisms that require flowing water.
Extreme low flow duration (number of days in the lowest 10th percentile of the natural flow regime)	Many organisms (e.g., fish) can withstand short-duration but not long-duration extreme low flows.
1-day maximum flow for the year	Indicates the magnitude of annual floods, which transport sediment and maintain active channel width.
Small flood Frequency (2–10 year return interval)	Over time, most sediment is transported under bankfull conditions (Andrews 1980; Troendle and Olsen 1994); small flood frequency indicates the regularity of this condition.
Mean small flood duration (from beginning of rising limb to end of falling limb)	Over time, most sediment is transported under bankfull conditions (Andrews 1980; Troendle and Olsen 1994); small flood duration indicates the period of which this condition is sustained.
Mean daily flow for each month	Monthly mean flows relate to total habitat availability and ecosystem productivity (Annear et al. 2004).

Table 8.2 Flow parameter departures from natural that define condition class

	Departure from natural			
Flow parameter (units)	Natural	Minimally altered	Moderately altered	Strongly altered
Mean 1-day minimum flow (cms)	<10%	10% to 1 std. dev.	1 std. dev. to 90%	>90%
Mean extreme low flow duration (days)	<10%	10% to 1 std. dev.	1 std. dev. to 2 std. dev.	>2 std. dev.
Mean 1-day maximum flow (cms)	<10%	<45%	45–90%	>90%
Small flood frequency (proportion of years where bankfull is reached)	0.50	0.33	0.14	< 0.10
Mean small flood duration (days)	<10%	10% to 1 std. dev.	1 std. dev. to 2 std. dev.	>2 std. dev.
Mean monthly flows (all months meet criterion)	<10%	<1 std. dev.	Smaller of (<2 s.d. or 90%)	>2 std. dev. or 90%

Fraser watershed were used to define minimally altered peak flows. Alteration of one standard deviation was used as a defining criterion for several parameters, from Richter et al. (1997). Preliminary data on aquatic macroinvertebrate diversity (Albano 2006; McCarthy 2008) were used to define strongly altered conditions.

Using the template in Table 8.2, hydrologic status of key flow parameters was evaluated for six locations (see Fig. 8.6) using the Indicators of Hydrologic Alteration software (IHA, Richter et al. 1996). Inputs to IHA were modeled natural and managed mean daily streamflow for 1947–1991. The degree of alteration among flow parameters varied both within and among streams. Small flood duration was the most impacted across streams, being strongly altered in five of six streams, with reductions from 37% to 72%. One-day maximum flows were the least impacted across streams, being minimally altered in five of six streams (reductions from 15% to nearly 45%), and natural in the sixth (4% reduction). The greatest percent alteration was in extreme low flow duration, which increased from 117% to 482% in three of six streams.

The information generated was used to conduct an evaluation of trade-offs among streams for the management of bypass flows during high runoff, in order to identify which parts of the flow regime it would be most useful to restore, and in which locations the most ecological benefit could be expected (Table 8.3). The conclusion reached was that the greatest ecological benefit would be achieved by improving flood conditions where low flow conditions are either natural or minimally altered, thereby bringing all flow metrics at those streams to minimally altered or better.

Because operations managers cannot be expected to develop an in-depth understanding of the complex concepts and analyses just described, simple operational principles were developed that could be readily implemented by Denver Water, as follows:

1. *Achieve bankfull conditions.* Bankfull defines the flood threshold, and typically is the flow rate at which significant sediment transport and channel maintenance occurs. Bankfull can be achieved readily in wet years simply by ensuring that diversions do not reduce the flow rate below the magnitude of the natural 2-years recurrence interval flood.
2. *Minimize abrupt changes inflow rate; avoid frequent reversals.* In snowmelt dominated streams, organisms may not be capable of the rapid behavior adjustments needed to respond to abrupt rate changes. Hydrologic data indicate that on the ascending limb of the flood, flow rate typically does not increase by more than ~40% from day to day; on the descending limb, it typically does not decrease more than ~20% from day to day. High ecological function of floods is more likely if these rates are not exceeded by diversion management.
3. *Rotate between and among streams over time.* Under natural conditions, by definition, floods occur in 50% of years, but typically not every other year; depending on snowpack, floods may or may not occur several years in a row. Reducing flood frequency to 33% or even once in 7 years would likely provide significantly more benefit than no floods.

While significant conceptual progress has been made in evaluating how to prioritize and implement each increment of water available for environmental flows

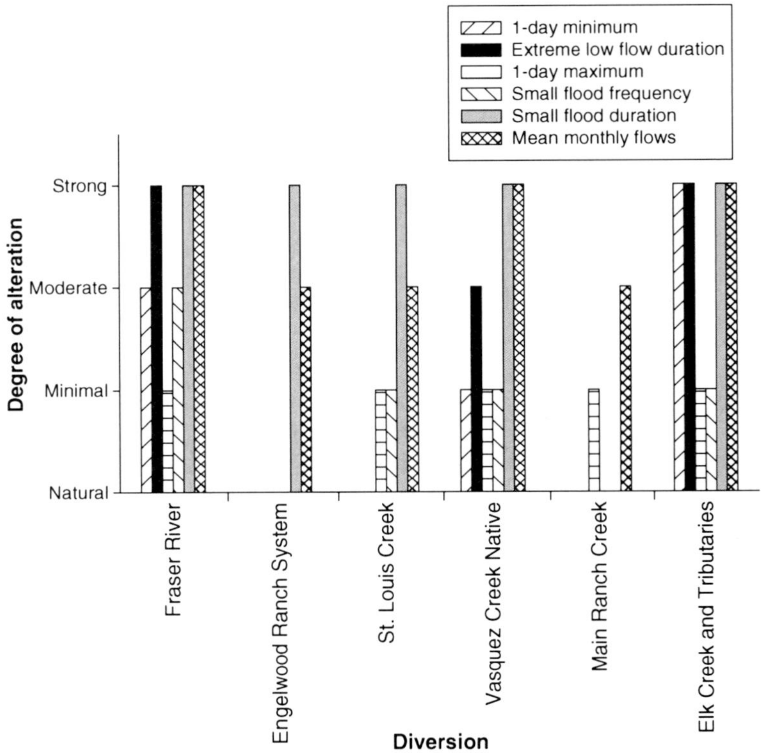

Fig. 8.6 Results of flow parameter analysis at six diversion points in the Moffat Collection system

Table 8.3 An analysis of potential trade-offs between the Fraser River diversion and the Englewood Ranch system

Node	Flow component	Current Status	To achieve moderate[a]		To achieve minimal[a]	
Fraser River Diversion	1d max	Minimal	−2.15	cms	No change	
	Small flood freq	Moderate	No change		dump ~2× as often	
	Small flood duration	Strong	28	days	44	days
Englewood Ranch System	1d max	Natural	−2.7	cms	−1.05	cms
	Small flood freq	Natural	Reduce to 1 in 7 years		Reduce to 1 in 3 years	
	Small flood duration	Strong	9	days	23	days
St. Louis Creek Diversion	1d max	Minimal	−1.8	cms	No change	
	Small flood freq	Minimal	Reduce to 1 in 7 years		No change	
	Small flood duration	Strong	20	days	33	days

[a]Minus (−) sign indicates flow could be reduced

in the Fraser basin, translation of this progress into actual environmental flow releases is still a work in progress. Attempts at ecologically oriented diversion management during 2008 illustrated that system operations can provide substantial barriers to implementation of these principles, as implementation of flow recommendations had to be deferred to permit infrastructure maintenance. Other barriers include: maintaining system flexibility to allow for trans-basin tunnel capacity, legal barriers to instream flows, social impacts from an on-going environmental impact study, and, of course, the need to fulfill water supply needs. An evaluation of these and other potential barriers is on-going. Upcoming discussions with the county and local water users in the Fraser Watershed are expected to present further opportunities for application of these flow–ecology concepts to management of Fraser River diversions.

Summary of Case Studies

FitzHugh and Richter (2004) stated that "achieving ecological sustainable water management [for cities] will require a proactive planning process that examines the economic and ecological trade-offs between water conservation, options for increasing supply, and the integrity of freshwater and estuarine ecosystems." The case studies of the Upper Delaware, Rivanna, and Fraser watersheds show that progress is being made in developing planning processes that factor the needs of ecosystems into decisions about the design and operation of urban water supply systems. The details of how these three projects are being implemented differ, both because of the particular circumstances of each project and because of the lack of a standardized approach to balancing human and ecological needs in urban water management, but some commonalities do exist among the case studies.

First, it is clear that management of flows during droughts is an important component of any successful water supply plan, and the Upper Delaware and Rivanna cases both show that the "share the pain" approach, where drought restrictions on water use are accompanied by reductions in environmental flows, is an acceptable solution from a societal standpoint. Second, another important commonality is the need for quantitative information about ecosystem needs for water and the operations of the water supply system, which can be used in a decision support system or other tool for analyzing trade-offs. None of these projects could have moved forward without such information. Analysis of such information often yields important insights that were not obvious beforehand, but are critical to implementing the project.

A third commonality is the need to represent ecosystem needs and environmental flow goals using operating rules that are ecologically meaningful yet also straightforward enough to be implemented. In the Rivanna case, this need for simplification necessitated a move from a more complicated framework of flow recommendations to the simpler and more elegant use of percent-of-flow standards. In the Fraser watershed case, translating more detailed analyses into simpler operational principles has potential, though more work needs to be done to assess how best to implement such principles in a system of multiple diversions. In the Upper Delaware, seasonal flow releases were arrived at through an iterative process that used outputs of the USGS flow and temperature study and water supply risk metrics to evaluate scenarios of relatively simple release patterns.

Conclusion

Though all of the case studies described in this chapter are in the United States, we feel that the same framework can be fruitfully applied in the developing world. Of course, a key question here is the applicability of this process in places where information and data are limited. The absence of detailed data and information on ecosystem and human needs for water and the trade-offs between them is a serious potential stumbling block in applying the ESWM framework described in Richter et al. (2003). But TNC does have experience working on environmental flow projects outside the developed world (in Honduras, China, and Colombia), and although the projects involved were not with urban water suppliers, or are not advanced enough to describe in this chapter, we have found that data limitations are generally not so severe that the process cannot be implemented.

For example, TNC has used the same process used in the Rivanna project on an environmental flow project in Honduras on the Patuca River, to make recommendations about how a planned hydropower dam should be operated to better preserve downstream ecosystems. In that project indigenous communities that live downstream of the dam site are in danger of having their food sources severely impacted by operation of the dam. Information sources used in this project included scientific data and expert opinion based on knowledge of similar river systems, and a survey of traditional ecological knowledge that was conducted to gather information about river dynamics and life cycles of fish species that live in the river (Esselman and Opperman in press). This survey approach will not be applicable everywhere, but it shows that sometimes in the absence of data there are creative ways to gather information.

The key is that in situations where information is limited, there will have to be a greater emphasis on use of expert opinion, on initiating data collection and research early on in the project, and on adaptive management and experimentation as a way of gathering knowledge as the project is implemented. Use of expert opinion is in fact a common feature of all of TNC's environmental flow projects (including those in the USA), but it would be even more critical in data-poor situations. Planning schedules for new water supply infrastructure projects should give sufficient lead-time to collect baseline and other useful data on streamflow and biota. If information on flow–ecology relationships is totally lacking, then one approach would be to use a purely hydrologic method such as the Range of Variability Approach (Richter et al. 1997) to set preliminary flow standards based on natural ranges of flow variability, which could be improved upon and replaced over time as knowledge improves. This is the method used in the Fraser case study to establish some of the thresholds of alteration. It is essential that use of such preliminary flow thresholds is accompanied by an understanding that these initial standards are only preliminary, and should be improved through a research and adaptive management program.

Table 8.4 Key points from chapter

Benefits of implementing ecologically sustainable water management for cities:
• Maintain ecosystem integrity and ecosystem services while providing for urban water needs
Environmental flow recommendations:
• Should cover all components of the hydrograph (extreme low flows, low flows, high flow pulses, small floods, large floods)
• Need to be more than just constant or minimum flows
• Should take into account seasonal variability
Necessary data and information:
• Streamflow data (on natural and altered hydrology)
• Information on flow–ecology relationships (i.e., flow needs of biota)
• Data or modeling to analyze trade-offs between urban water needs and ecosystem needs
Lessons from US case studies:
• Management of flows during droughts is particularly critical: "Share the pain" approach where drought restrictions on human water use are coupled with reductions in environmental flows seems to be most equitable
• Modeling and analysis of quantitative information about ecosystem needs for water and operations of water supply system helps generate key insights on how best to operate system
• Need to represent ecosystem needs and environmental flow goals using operating rules that are ecologically meaningful yet also straightforward enough to be implemented (i.e., percent of flow rules)
When applying the process in data-limited situations, will put greater emphasis on:
• Expert opinion on flow–ecology relationships
• Immediate and ongoing field data collection
• Adaptive management to refine environmental flow recommendation (learning by doing)
• Setting preliminary standards using purely hydrologic methods (i.e., Richter et al. 1997), which will need to be refined as more ecological information becomes available
• Creative approaches to gathering information (such as survey of traditional ecological knowledge)

Improving the balance between human and ecosystem needs for water is critical to maintaining watershed services, such as fisheries production, wildlife habitat, and mitigation of water and air pollution. A study of the value of ecosystem services globally found that when measured per hectare, the two most valuable biomes were estuaries and floodplains (Costanza et al. 1997), ecosystem types that are both vulnerable to upstream water development. Degradation of watershed services can have particularly negative consequences for poorer communities. For instance, water flowing into Apalachicola Bay, Florida, is diverted for use in Atlanta, potentially affecting oyster fishing communities of Apalachicola, Florida, whose poverty rate is twice the national average (NOAA 2006). These poor communities and their livelihoods are less likely to be taken into account in water management decisions relative to the needs of the residents of Atlanta, without the additional consideration of the need to preserve environmental flows.

There is ample evidence from the three case studies described here that progress is being made on methods for balancing environmental flow and human needs in water supply planning. Table 8.4 summarizes some of the key points covered in this chapter. In a world where climate change will be altering water budgets and affecting the reliability of water supplies, the type of integrated water supply planning described in this chapter will be even more essential. A key to moving forward will be to develop additional examples of how to conduct integrated planning, especially in the developing world, as further experimentation is needed to define methods that can be used in a wider range of settings. This is not to understate the challenges ahead, as pressure on ecosystems from human population growth and increasing water use is certain to increase through time. But with stakeholder involvement, political will, and innovation in the array of information tools at our disposal, it is clearly possible to create a better balance in future urban water supply development between environmental and human needs than has previously existed in most urban water management projects.

References

[ACOE] US Army Corps of Engineers. 1998. Water allocation for the Apalachicola–Chattahoochee–Flint (ACF) River Basin, main report of the Draft Environmental Impact Statement. Federal Register 63:53023–53024.

[ACOE] US Army Corps of Engineers. 2003. Scoping summary: Moffat Collection System project. Wyoming Regulatory Office. Cheyenne, Wyoming.

Albano, C. 2006. Structural and functional responses of aquatic macroinvertebrate communities to streamflow diversion in Rocky Mountain streams. Masters Thesis. Colorado State University. Fort Collins, CO.

Andrews, E. D. 1980. Effective and bankfull discharges of streams in the Yampa River Basin, Colorado and Wyoming. Journal of Hydrology 46:311–330.

Annear, T., I. Chisholm, H. Beecher, A. Locke, P. Aarestad, C. Coomer, C. Estes, J. Hunt, R. Jacobson, G. Jobsis, J. Kauffman, J. Marshall, K. Mayes, G. Smith, C. Styalnaker and R. Wentworth. 2004. Instream flows for riverine resource stewardship, revised edition. Instream Flow Council, Cheyenne, WY.

Arthington, A. H. and Pusey, B. J. 2003. Flow restoration and protection in Australian rivers. River Research and Applications 19:377–395.

Arthington, A. H., S. E. Bunn, N. L. Poff and R. J. Naiman. 2006. The challenge of providing environmental flow rules to sustain river ecosystems. Ecological Applications 16:1311–1318.

Bovee, K.D., Waddle, T.J., Bartholow, J., and Burris, L. 2007. A decision support framework for water management in the upper Delaware River: U.S. Geological Survey Open-File Report 2007–1172, 122 p.

Bunn, S.E. and Arthington, A.H. 2002. Basic principles and ecological consequences of altered flow regimes for aquatic biodiversity. *Environmental Management,* 30:492–507.

[CSWRCB] California State Water Resources Control Board. 1994. Final Environmental Impact Report for the Review of Mono Basin Water Rights of the City of Los Angeles. (12 July 2004; www.monobasinresearch.org/images/mbeir/fititlepage.pdf)

Costanza, R., R. d'Arge, R. de Groot, S. Farber, M. Grasso, B. Hannon, S. Naeem, K. Limburg, J. Paruelo, R.V. O'Neill, R. Raskin, P. Sutton, and M. van den Belt. 1997. The value of the world's ecosystem services and natural capital. *Nature*, 387:253–260.

Denver Board of Water Commissioners. 2002. Water for tomorrow: the history, results, projections and update of the integrated resource plan. Denver, Colorado.

Elliot, W. P. 1998. Delaware River Basin releases from a trout's perspective. Clearwaters., New York Water Environment Association, Inc., Spring 1998, 37–42.

Esselman, P. C. and Opperman, J. J. In press. Overcoming information limitations for the prescription of an environmental flow regime for a Central American river. *Ecology and Society.*

FitzHugh TW, Richter BD. 2004. Quenching urban thirst: growing cities and their impacts on freshwater ecosystems. *BioScience*, 54:741–754.

Gleick, P.H. 2000. The changing water paradigm: A look at twenty-first century water resources development. Water International 25:127–138.

Hogarty, R.A. 1970. The Delaware River Drought Emergency. Bobbs-Merrill Company: Inter-University Case Program. 53 p.

Homa ES, RM Vogel, MP Smith, CD Apse, A Huber-Lee, J Sieber, 2005. An Optimization Approach for Balancing Human and Ecological Needs. Proceedings of the EWRI 005 World Water and Environmental Resources Congress, ASCE, Anchorage, Alaska.

Hundley N Jr. 2001. The Great Thirst: Californians and Water—A History. Berkeley: University of California Press.

Hydrologics, Inc., 2008. What is OASIS with OCL?: Product Fact Sheet (29 August 2008; http://www.hydrologics.net/pdf/oasis.pdf)

Lellis, W. A. 2001. Freshwater mussel survey of the upper Delaware Scenic and Recreation River: Qualitative survey 2000. Wellsboro, PA, U.S. Geological Survey-Northern Appalachian Research Laboratory: 56.

Maharaj, V., McGurrin J., and Carpenter J. 1998. The Economic Impact of Trout Fishing on the Delaware River Tailwaters in New York. Prepared for the American Sportfishing Association and Trout Unlimited. 28 p.

Mathews, R. and Richter, B.D. 2007. Application of the Indicators of Hydrologic Alteration Software in Environmental Flow Setting. Journal of the American Water Resources Association 43(6):1400–1413.

McCarthy, J. M. 2008. Factors influencing ecological recovery downstream of diversion dams in Southern Rocky Mountain streams. Masters Thesis. Colorado State University. Fort Collins, CO.

NOAA. 2006. Fishing Communities of the U.S. 2006. (April 22, 2009; http://www.st.nmfs.noaa.gov/st5/publication/fisheries_communities.html)

[NYCDEP] New York City Department of Environmental Protection. 1998. Drought Management Plan and Rules. (www.nyc.gov/html/dep/pdf/droughtp.pdf)

[NYC DEP] New York City Department of Environmental Protection. 2007. New York City 2007 Drinking Water Supply and Quality Report. (29 August 2008; http://www.nyc.gov/html/dep/pdf/wsstate07.pdf)

[NYC DEP] New York City Department of Environmental Protection. 2008. History of Drought and Water Consumption (29 August 2008; http://www.nyc.gov/html/dep/html/drinking_water/droughthist.shtml

Poff NL, Allan JD, Bain MB, Karr JR, Prestegaard KL, Richter BD, Sparks RE, Stromberg JC. 1997. The natural flow regime: a paradigm for river conservation and restoration. BioScience 47:769–784.

Poff et al. 2009. The Ecological Limits of Hydrologic Alteration (ELOHA): A New Framework for Developing Regional Environmental Flow Standards. *Freshwater Biology* 55(1):147–170.

Postel, S. and Richter, B. 2003. *Rivers for Life: Managing Water for People and Nature,* Island Press, Washington, D.C.

Power M.E., Sun A., Parker G., Dietrich W.E., and Wootton J.T. 1995. Hydraulic Food-Chain Models. *BioScience*, 45:159–167.

Revenga C, Brunner J, Henninger N, Kassem K, and Payne R. 2000. Pilot Analysis of Global Ecosystems: Freshwater Systems. Washington (DC): World Resources Institute.

Richter, B.D., Baumgartner, J.V., Powell, J., and Braun, D.P., 1996. A Method for Assessing Hydrologic Alteration within Ecosystems. *Conservation Biology,* 10:1163–1174.

Richter, B.D., Baumgartner, J.V., Wigington, R., and Braun, D.P. 1997. How much water does a river need? *Freshwater Biology*, 37:231–249.

Richter, B.D. 2007. Meeting Urban Water Demands While Protecting Rivers. A Case Study from the Rivanna River in Virginia. *Journal of the American Water Works Association*, June 2007.

Richter BD, Mathews R, Harrison DL, and Wigington R. 2003. Ecologically sustainable water management: managing river flows for ecological integrity. *Ecol Appl.* 13:206–24.

Richter, B. D., and G. A. Thomas. 2007. Restoring environmental flows by modifying dam operations. *Ecology and Society* 12(1):12. [online] URL: http://www.ecologyandsociety.org/vol12/iss1/art12/

Richter, B. D., A. T. Warner, J. L. Meyer, and K. Lutz. 2006. A collaborative and adaptive process for developing environmental flow recommendations. *River Research and Applications* 22:297–318.

Rosegrant MW, X Cai and SA Cline, 2002. *Global Water Outlook to 2025: Averting an Impending Crisis*, A Report Summary of the 2020 Vision for Food, Agriculture and the Environment Initiative, International Food Policy Research Institute, Washington DC and International Water Management Institute, Colombo.

Ryan, S. 1997. Morphologic response of subalpine streams to transbasin flow diversions. Journal of the American Water Resources Association 33:839–854.

Shen Y, T Oki, N Utsumi, S Kanae, and N Hanasaki, 2008. Projection of future world water resources under SRES scenarios: water withdrawal, *Hydrol. Sci. J.*, 53(1), Feburary, pp.11–33.

Shiau, J. T. and Wu, F. C. 2006. Compromise programming methodology for determining instream flow under multiobjective water allocation criteria, *Journal of the American Water Resources Association*, 42(5), 1179–1191.

Shiau, J. T. and Wu, Fu-Chun, 2007a. "A dynamic corridor-searching algorithm to seek time-varying instream flow releases for optimal weir operation: Comparing three indices of overall hydrologic alteration", *River Research and Applications*, 23:35–53.

Shiau, J. T. and Wu, F. C. 2007b. Pareto-optimal solutions for environmental flow schemes incorporating the intra-annual and interannual variability of the natural flow regime, *Water Resources Research*, 43(6), W06433, doi:10.1029/2006WR005523.

Tharme R.E. 2003. A global perspective on environmental flow assessment: emerging trends in the development and application of environmental flow methodologies for rivers. River Research and Applications 19(5–6):397.

Troendle, C.A. and Olsen, W.K. 1994. Potential effects of timber harvest and water management on streamflow dynamics and sediment transport. pp. 34–41 In: Sustainable Ecological Systems Proceedings, United States Department of Agriculture, Forest Service, Rocky Mountain Forest and Range Experiment Station, GTR RM-247.

[UN] United Nations. 2007. World Urbanization Prospects: The 2007 Revision Population Database. http://esa.un.org/unup/

[UN] United Nations. 2006. World Population Prospects: The 2006 Revision. Population Database. http://esa.un.org/unpp/

[USGS] United States Geological Survey. 2008. Office of the Delaware River Master. History of reservoir releases program in the Upper Delaware Basin. (29 August 2008; http://water.usgs.gov/osw/odrm/releases.html)

2030 Water Resources Group. 2009. Charting our water future: Economic frameworks to inform decision –making. (11 July 2011; http://www.2030waterresourcesgroup.com/water_full/Charting_Our_Water_Future_Final.pdf)

Vogel R., Sieber J., Archfield S., Smith M., Apse C. and Huber-Lee A. 2007. Relations among storage, yield and instream flow. *Water Resources Research*, 43, W05403, doi:10.1029/2006WR005226.

Weiskel, P. K., R. M. Vogel, P. A. Steeves, P. J. Zarriello, L. A. DeSimone, and K. G. Ries III. 2007. Water use regimes: characterizing direct human interaction with hydrologic systems, *Water Resources Research,* 43, W04402, doi:10.1029/2006WR005062.

Wiens JA, Patten DT, and Botkin DB. 1993. Assessing ecological impact assessment: lessons from Mono Lake, California. Ecological Applications 3:595–609.

Chapter 9
Water, Ecosystems, and Poverty: Roadmap for the Coming Challenge

Casey Brown

Introduction

The next century will be a time of major transitions for those who manage and depend on water. With population increases and economic development, the demand for water will grow and likely exceed available supply (2030 Water Resources Group 2009). Competition for water will also grow and, consequently, the economic value of water will increase. In addition, changes in climate are likely to increase the variability of water resources and complicate the challenge of providing water services. As has been the case, historically, those who can pay for water will be able to gain access or be able to pay for alternatives, such as water-saving technologies. However, those without such means will become increasingly vulnerable to the variability of water resources and the hazards that accompany that variability. Already, economic development in the least developed nations is impeded by their inability to manage hydroclimatic extremes (Brown et al. 2010). The dual stressors of climate and growing water demand may be especially difficult for aquatic ecosystems that are likely to be threatened by increased need for hydraulic control structures and to lose in the competition for water. To meet this coming challenge, innovations in how we manage our water resources are required. In this chapter, we review the major challenges that growing water scarcity and climate variability pose to aquatic ecosystems, the poor and those who would manage water for them. While those topics have been addressed individually before, here we explore the nexus of their implications. We then review some possibilities for improving our ability to manage water resources for the benefit of poverty reduction and ecosystem sustainment, including water

C. Brown (✉)
Department of Civil and Environmental Engineering,
University of Massachusetts at Amherst, 37 Mather Drive, Amherst, MA 01003, USA
e-mail: cbrown@ecs.umass.edu

J.C. Ingram et al. (eds.), *Integrating Ecology and Poverty Reduction: Ecological Dimensions*, DOI 10.1007/978-1-4419-0633-5_9,

trading mechanisms and hydrologic forecasting. Finally, we present a roadmap for achieving the innovations that are needed to insure a sustainable future for ecosystems and communities.

Consequences of Water Scarcity

There is growing concern that the world will run out of water. The logic is that water is a finite and relatively scarce resource and as population grows, the proportion of available water inevitably decreases. Supporting this view is the seemingly increasing occurrence of droughts that affect urban populations, farmers, and pastoralists in developed countries and developing countries alike. The well-documented depletion of groundwater in as varied locations as the American southwest and the High Plains, peninsular India, and the north plain of China is further evidence of the growing scarcity of water.

Still, the question of water scarcity is not as simple as it is typically portrayed. Scarcity results from a mismatch between supply and demand that may persist over short or long time periods. Water supply is naturally variable in time and space, and water infrastructure is largely an effort to reduce that variability. Water demand is a complex function of societal values that remains difficult to predict. A number of studies have attempted to estimate the total available water resources and the total use of water resources by the earth's population. Postel et al. (1996) estimated that humans currently use about 26% of freshwater resources, or 54% of what they consider to be "geographically and temporally accessible." Oki et al. (2003), using a slightly different methodology, arrived at an estimate of less than 10% for the proportion of renewable water resources currently being used for human demands. Given the nature of the task, the fact that the results are within reasonable explanation of each other provides some confidence that they are meaningful. We can probably say that we are currently using a significant percent of the renewable water resources, but well less than half. On a global basis, at least, there appears to be plenty of water. Not included in these calculations, of course, is the essentially infinite supply of sea water available from the oceans. Many urban centers around the world are already looking to desalination as a source for additional drinking water. Ultimately, given an energy source to purify saline water and the ability to transport it (or its services), we will never run out of freshwater supplies.

These estimates and others in the literature focus on quantifying the current use of water and comparing it to demand. However, as noted above, water use is a function of many factors. Typical studies consider demand to be static or a function of factors that increase it, namely population growth and economic development. This results in metrics such as the water scarcity index, which is water use divided by renewable available water, with an arbitrary value of 0.4 or above selected to indicate water scarcity. The problem with such an index is that it cannot distinguish between excessive water use that results from subsidies, water use that is simply exploitation of available resources and actual water scarcity where legitimate water demands cannot be met due to limited supply. For example, the American High Plains are an area that appears as water scarce on most global maps (e.g., see Oki et al. 2006) due to high water use (largely groundwater extraction) and low renewable resources (rainfall). However, this is hardly water scarcity. This is the outcome of a decision to extract groundwater for economic gain through the production of agricultural products that are highly subsidized. A set of policies that did not encourage the exploitation of the available water resources would largely solve the water "scarcity" at this location.

Estimating water scarcity based on current use of water is equivalent to measuring food needs based on consumption at an all-you-can-eat buffet. The price that water users face is typically much less than the economic value of water (Rogers et al. 1998). The difference is usually a result of the opportunity value of water and the environmental value of water. These are often not considered in the pricing of water. For example, withdrawing water from a river or controlling the flows for human use can have negative impacts on the ecological health of the river. The costs associated with this environmental damage are difficult to quantify and easy to ignore and, thus, frequently remain unconsidered. Since the water user is not faced with the cost of repairing the damage, either directly or in the price of the water, more water is used than is beneficial to society. In places where there are few protections for maintaining environmental flows of water and where there are financial subsidies for water use, the water consumed and the economically beneficial demand for water will be different. These areas may appear to suffer water scarcity, but in fact are symptomatic of a scarcity of good policy. At present it is impossible to estimate how much water demand falls into this category.

Despite the low global proportion of water consumption for human use, and the existence of localized areas of policy driven over-consumption, there are real and growing examples of water scarcity around the world. This is because a global surplus of water does very little to help resolve local imbalances between supply and demand. There is no global trade in water. There is a global trade in so-called "virtual water," which is the accounting of water used in the production of goods. But this does very little to alleviate water shortages. In fact, in many cases it probably exacerbates water shortages. This is because virtual water does not necessarily flow from places with water excess. Instead, it flows from places where water use is subsidized or where water has very low prices leading to overuse of the resource. Even if water were appropriately priced (this occurs rarely but is increasing), the virtual water trade only alleviates shortages of products, not shortages of water. It can provide food where there is hunger such as through temporary food aid during acute drought, but it is not clear nations that are willing to export food if they fear possible future shortages. But, this trade will not provide farmers with water for their crops and, thus, they will still be without livelihood security and income to buy available food. It will not provide water for domestic uses, water for hydroelectricity production or water for ecosystems. It is unlikely that we can solve water scarcity issues through trade from surplus regions to dryer areas for the reasons described above. Instead, society will need to find ways to simultaneously manage the temporal and spatial variability of water as well as potential growth in water use and demand.

Managing Hydroclimatic Variability

A second major forcing that affects the relationship between water, poverty, and ecosystems is climate. The negative impacts of climate change on the availability of water resources have garnered the attention of scientists and policy makers around the world. The impacts are among the most significant attributed to climate change. Global warming is expected to increase the demand for water, especially for expanded irrigation in rain-fed areas. Acceleration of the hydrologic cycle is an anticipated effect of climate change that could lead to even greater variability in water availability. According to theory, rainfall may become more intense and yet less frequent. Several studies of long rainfall records report increases in extreme events that seem to support the theory (see Alexander et al. 2006). The same acceleration of the hydrologic cycle may lead to increases in the occurrence of droughts, especially in the lower latitudes, where the vast majority of the world's poorest live. It is in these regions, where the climatic changes are expected to be among the most challenging and where the populations have the least capacity to manage them, that the issues at the nexus of water, poverty, and ecosystems will be most critical to resolve.

In the developing world, economic development is already strongly impacted by climate. Several studies have found consistent negative consequences of climate variability on economic growth. Hydroclimatic extremes, particularly drought, have been found to present the most significant climate impacts on economic growth. In an econometric study of economic growth and climate in Sub-Saharan Africa, Brown et al. (2009) found that drought had a significant and negative impact on economic growth, more important even than temperature, which is the climate variable that is typically used for projecting climate change impacts on economic growth. The analysis was repeated at the global level in a report for the 2009 World Development Report (Brown et al. 2009), finding again that hydroclimatic extremes were a significant impediment to economic growth and more important than changes in temperature.

Evidence of the strong effects of hydrologic extremes, such as floods and droughts, has significant implications for the practice of achieving poverty reduction and, consequently, for ecosystems in developing countries. It's widely accepted that vulnerability to extreme hydrologic events should be reduced if poverty alleviation is to be achieved. In addition, the prospect that climate change will increase the occurrence of hydrologic extremes increases the need to reduce the vulnerability of poor people. Without doing so, other development gains achieved through improved education and health services may be wiped out in a single flood or drought. Given the need to reduce the vulnerability of the poorest of the world to climate extremes, can this be achieved while protecting vulnerable ecosystems? The answer is critical for the benefit of the poor and for the future health of those ecosystems.

The historical model in the industrialized world for reducing vulnerability to climate extremes is through investment in infrastructure to control the variability of natural streamflows. Reservoirs have been built to store variable flows to provide irrigation water in times of drought and to store excess water during flood events. Grey and Sadoff (2007) review the status of nations in terms of their ability to control hydroclimate extremes and the investments, largely infrastructural, that were used. They describe countries such as Ethiopia and Yemen where economic growth is vulnerable to climate fluctuations as "hostage to hydrology." It is widely accepted that these countries have a deficient inventory of water infrastructure, especially relative to the variability they experience (Brown and Lall 2006), and that this contributes to the strong negative influence of climate.

Simple examples illustrate the stark differences in infrastructure levels between developed and developing countries. The Connecticut River Basin in the Northeast USA, which has an area of about 10,000 square miles, has over 1,000 dams within the basin. It has been long been known as a "working river," with the water used primarily for power (originally hydraulic power for sawmills and grist mills and eventually hydroelectricity) and for navigation when overland travel was difficult. After major floods ravished the cities of the basin in the twentieth century, 12 flood control dams were built and now floods are largely forgotten. The ecological impact was substantial and contributed to the decline of Atlantic salmon, among other species. However, the salmon are now being re-introduced, many other native species are recovering and dam operators are altering operations to attempt to restore some measure of the natural variability. In contrast, the Niger River Basin in West Africa covers an area of 817,600 square miles and there are estimated to be less than 40 dams. The flows of the Niger and its tributaries are extremely variable both within an average year and between years. The 110 million people living there are among the poorest in the world.

Despite this evidence, there is resistance in the donor community to making investments in infrastructure that control water on a large scale. This is a result of the documentation of negative impacts and poor economic returns on investments made in large dams in the past, which were summarized in the World Commission on Dams (WCD 2000). Are there alternatives to large dam projects? Are there examples of success, especially in regard to flood risk reduction? There are wide hopes for measures such as rainfall harvesting to provide poor farmers additional control of their water resources. However, these methods are primarily effective at relieving the effects of dry spells while providing little protection from prolonged droughts and no protection from floods. Major investments have been made in observation and modeling of the earth's climate system and it is hoped that these can provide the basis for early warning systems to give people time to prepare for adverse climate events. Unfortunately, the vast investments made in the technology are rarely matched with investments in the "software" of tailoring the information for use by vulnerable populations and creating options for them to respond to climate extremes (Brown and Hansen 2008). Such investments are sorely needed to improve the performance of the infrastructure that is built and to provide protection from hydroclimatic variability where it is not built.

Conserving Water Through Incentives

It is likely that a major response to growing water scarcity will be increasing acceptance of the concept of water as an economic good. This has been long recognized as a needed step to increase the productivity of water and was adopted as one of the four "Dublin Principles" of water management (Dublin International Conference on

Water and the Environment 1992). There are a variety of mechanisms by which water can be managed economically. In general, this entails the water users paying a price for water that is representative of the cost of supplying it. In theory, this would include the infrastructure and operational costs of water, as well as the opportunity costs and environmental costs. In practice, pricing water in some proportion of those actual costs would be a good start compared to current pricing (or lack thereof) policies. A major benefit of this approach is that it creates incentives for water users to conserve. In addition, water would be made available for its most valuable uses and not available where the cost of supply exceeds the benefits from water use. Water suppliers can do this directly by changing prices to reflect the actual cost of supply and increase them according to the amount used. Other methods that have been employed include water markets, water banks and water contracts.

The water allocation scheme that embraces economic principles in their purest form is establishing water markets based on tradable or leasable water rights. There are a variety of implementation techniques, but the basic tenet is that the right to use water is treated as a commodity (for in depth reviews, see Anderson and Hill 1997; Easter et al. 1998; Marino and Kemper 1999). This right can be traded or leased temporarily, or permanently sold, without any changing of hands of the land with which the water right was originally associated. For example, a farmer with the right to withdraw a specific volume of water from an irrigation canal could sell that water right to another farmer who could then use the water. The impetus for water markets is that, in theory, they allow water to be utilized in the most economically productive uses. As each individual decision maker attempts to maximize profit (a necessary assumption) through selling or buying water, the "invisible hand" of the market allocates water to its most productive use. The value of water is determined in the free market through many transactions. In theory, society as a whole benefits because the productivity of water is maximized, there are economic incentives for increasing water use efficiency and allocation is based on voluntary decisions. Some form of tradable water rights has been established in California, Chile, Mexico, Spain, South Africa, Texas, and Australia.

Objections to water markets stem from the failure to meet the assumptions necessary to achieve the benefits of perfect market conditions, such as profit maximizing decision makers, perfect information and minimal transaction costs, and the impacts of these market failures. In practice, water markets are often only appropriate where the external costs of water use, such as the cost of damages to the environment when water is withdrawn, are minimal. In places where changes in water withdrawals have strong negative impacts on the environment, a water market is less appropriate because the associated costs would be difficult to include in the transactions. Environmental damages could accrue as a result of the water rights exchanges leading to opposition to future transactions.

Water banks offer a more moderate approach to the marketing of water. Water banks are an attempt to reform water allocation through market incentives while maintaining the central authority and power of water agencies. Banks have typically been established in response to drought or expected drought, as in the case of the California Water Bank (CWB), which was established after recommendation by a state Drought Action Team in response to a multiyear drought (California DWR 1992). Water banks now exist in most US western states. They operate as "facilitators" in transactions that enable surface water to be transferred temporarily to areas of critical need from areas that are able and willing to do without water deliveries during times of drought or prior to drought. To secure water, banks offer a price that they estimate to be attractive to those with water rights. The CWB has used consultation with farmers and economic modeling to determine their price offer (Jercich 1997). On the delivery side, the bank advertises available water at a price it chooses. Price setting provides the bank the opportunity to account for the costs of operation and administration. The CWB includes these costs in its prices and so the sale price exceeds their buying price (Howitt 1998). However, the Snake River water bank of Idaho considers the banked water to be a drought relief measure and buys and sells water at low prices that include a smaller administrative fee, irrespective of supply and demand (Green and O'Connor 2001). This tends to keep the activity of the bank low; few are willing to sell water for low prices during a drought.

Water contracts are a third mechanism for incentive-based water management. Water contracts involve a legal agreement between an individual user or association of users and a water authority. This method of water allocation is typically the result of a government body creating a water storage and distribution system and then contracting with water recipients. The authority is typically governmental and owns storage and conveyance infrastructure that the users rely on, or may be a nongovernmental organization that owns water rights that it allocates. Those seeking water contracts typically apply for a specific quantity and may be limited by local norms based on land area and use. Contract payments are nominally based on the capital, operations, and maintenance costs for the water system infrastructure, although it is common for governments to subsidize these costs. The Bureau of Reclamation implemented this approach in the US West; the Central Valley Project in California and the Colorado-Big Thompson project in Colorado are examples.

Treating water as an economic good provides an avenue for maximizing the productivity of water. This can be attractive to policy makers. Due to the increasing demands for water that often accompany population growth and economic development, there is likely to be subtle, but consistent movement in this direction. The result is that growing water scarcity will likely cause a transition from a world in which water is not valued to a world in which water is valuable and allocated to valuable uses. This will likely have many benefits in terms of increasing incentives for conservation and increasing the productivity of water.

A transition to a world of valuable water will only be effective as a way to manage the resource to the extent that we know how to quantify the value of all water uses. Difficult to value uses, such as water for the environment and water for those who are not able to pay for it, are not readily preserved in an incentive-based water allocation scheme. Instead, direct intervention by government is required to set aside water for purposes that are not easy to quantify in monetary terms. This has been achieved in South Africa, where a right to water was written into the new constitution. And this has been accomplished in the USA and other countries where water for the environment has been preserved through legally enforced instream flow requirements. These examples show that economic allocation of water and preservation of water for uses with unquantified financial values is possible. The question is whether they will spread to other countries and whether they will be upheld when water is scarce. Already there were calls for the cessation of instream requirements in the Appalachicola-Chattahoochie-Flint River system during the drought that struck the US city of Atlanta in 2007.

Roadmap to Meet the Coming Challenge

At present, there is no consensus approach for managing hydroclimatic variability to promote economic development while simultaneously protecting ecosystems, which are further stressed by climate change. Here, we propose a path forward. It begins with the premise that more investment in infrastructure is needed for much of the developing world to manage current and future climate variability. Next, the design and operation of infrastructure must be coupled with sufficient decision support systems to allow these investments to best satisfy multiple objectives that include water for ecosystems. While trade-offs are often unavoidable, often small compromises in operations can achieve significant environmental benefits. Decision support systems that incorporate forecast information are an underutilized tool that can reduce the costs of tradeoffs and make providing water for ecosystems more easily achieved. Finally, training is required for water managers to be able to utilize these tools and to understand the need for an integrated approach to water management. Here our attention focuses on the development of decision support systems to improve the ability of water infrastructure to achieve poverty reduction while mitigating ecosystem impacts.

The field of water resources engineering is at a crossroads due to the growing recognition of the implications of climate change and nonstationarity of the hydrologic record (Milly et al. 2008). Since water resources systems have been designed on the principal of "stationarity," meaning that the historical statistics of streamflow were assumed to be representative of future flows, systems designed under this assumption are now vulnerable to changes in streamflow regimes or trends. Nonstationarity, meaning that the hydrologic record is not stationary, implies that our ways of designing and managing water resources systems may not be valid in the future. This realization is a topic of much debate and consternation among water managers and policy makers. The response to nonstationarity will have significant implications on our ability to provide hydrologic services, while reducing poverty and maintaining ecosystems.

Traditionally, the engineering response to uncertainty might be more and larger infrastructure. Larger infrastructure typically entails more inundated area, more relocated poor people, and more lost ecosystems. Alternative approaches to managing hydroclimatic variability may be possible. However, if they are to be viable with planners and decision makers, a concerted research effort is needed to establish the advantages of these innovations over traditional approaches (Brown 2010).

There is great potential for water innovations that rely on flexibility and forecasts to adapt to changing climate conditions. Hydrologic forecasts make use of the vast investments that have been made to observe, monitor, and model the earth's climate system. Forecasts provide water managers the foresight that was previously taken for granted when stationarity was assumed. They have the potential to provide the foresight needed to manage water resources under conditions of nonstationarity. Currently, the state of streamflow forecasting is highly variable, depending on location and season. In temperate climates, streamflow forecasts are often dependant on weather forecasts, and thus provide less than 1 week of lead-time. The exception is snowmelt fed basins, where monitoring of the snowpack often provides forecasts of water availability up to 3 months in advance. In the tropics, the large influence of the El Niño/Southern Oscillation provides the basis for streamflow forecasts at up to 3 months in advance in certain locations and during certain seasons.

Billions of dollars of investments have been made in our ability to observe, monitor, and model the earth's climate system with the hopes of improving natural resources management. Unfortunately, only minimal investments have been made in the hard work of supporting the use of this information in decision making (Lettenmaier 2008). Water managers are ill-equipped to make decisions based on the uncertain, but potentially useful information that scientific research yields. As a result, a potential resource for the improvement of water management under climate change is underutilized. Without sustained funding for the development of decision support systems and the capacity to use climate information in water resources decision making, the ability to monitor and model climate will not benefit water resources management or the ecosystems that compete for water in controlled environments.

With the kind of support that the earth observation systems receive, the use of climate information and development of flexibility in water resource systems can progress to a point where the stark choices between poverty, development, and ecosystems can be relieved. With a scientific effort in the vein of sustainability science, which is described by Clark (2007) as "the quest for advancing both useful knowledge and informed action by creating a dynamic bridge between the two," innovations can move from the literature to have positive impacts in the field. The water sector, with its outdated operating policies in need of rebooting and its key role in poverty reduction and ecosystem sustainment, is an ideal setting for putting the theory of sustainability science to practice.

In this chapter, we advance the argument that in the next century ecosystems and the poor will be increasingly vulnerable to water scarcity as a result of increasing water demand and climate variability. The traditional responses to scarcity, variability, and uncertainty involve investments in static infrastructure that often have disproportionate negative impacts on the poor and the environment. However, technologies and tools exist that can mitigate these impacts.

There are underutilized tools and technology available that can improve our ability to manage hydroclimatic variability with less impact on the vulnerable. These include economic mechanisms for the allocation of scarce water resources that improve the efficiency of water use, while preserving some fraction for the environment and the poor. Some incentive-based mechanisms, such as water markets, have been implemented in some locations around the world with some success. But there is the potential to do much more, especially using approaches such as water banks and option contracts that facilitate better management of externalities and may be more politically acceptable than water markets.

In addition to economic tools, there are also technological advances that can help solve the coming water challenges. In particular, our current ability to observe the evolution of the earth's weather and climate is a potential valuable resource that is underutilized. For this potential to be realized, research and learning through case studies with real practitioners of water resources management must be undertaken and repeated. The "sustainability science" of water resources management must be supported to explore the potential benefits of such opportunities. Considering the challenges facing water managers in the next century, the payoffs could be huge.

References

Alexander, L.V., Zhang, X., Peterson, T.C., Caesar, J., Gleason, B., Klein Tank, A., Haylock, M., Collins, D., Trewin, B., Rahimzadeh, F., Tagipour, A., Ambenje, P., Rupa Kumar, K., Revadekar, J., Griffiths, G., Vincent, L., Stephenson, D., Burn, J., Aguilar, E., Brunet, M, Taylor M., New, M., Zhai, P., Rusticucci, M., Vazquez-Aguirre, J.L. 2006. Global observed changes in daily climate extremes of temperature and precipitation. Journal of Geophysical Research 111:D05109. doi:10.1029/2005JD006290

Anderson, T.L., and P.J. Hill. 1997. Water Marketing: The Next Generation. Rowman and Littlefield, Lanham, MD.

2030 Water Resources Group, 2009. The Barilla Group, The Coca-Cola Company, The International Finance Corporation, McKinsey & Company, Nestlé S.A., New Holland Agriculture, SABMiller plc, Standard Chartered Bank, and Syngenta AG. http://www.mckinsey.com/clientservice/water/charting_our_water_future.aspx Accessed January 15, 2010.

Brown, C., R. Meeks, K. Hunu, and W. Yu. 2008. Hydroclimatic risk to economic growth in Sub-Saharan Africa. IRI Technical Report No. 08-03. International Research Institute for Climate and Society, Palisades, NY, USA.

Brown, C., Meeks, R., Ghile, Y. and K. Hunu. 2009. An empirical analysis of the effects of climate variables on national level economic growth. Background paper for the World Bank World Development Report. University of Massachusetts, Amherst, USA.

Brown, C., R. Meeks, K. Hunu, and W. Yu. 2010. Hydroclimatic risk to economic growth in Sub-Saharan Africa. Climatic Change (in revision).

Brown, C. 2010. The End of Reliability. ASCE Journal of Water Resources Planning and Management **136 3**.

Brown, C., and U. Lall, U. 2006. Water and Development: The role of variability and a framework for resilience. Natural Resources Forum **30 4:** 306–317.

Brown, C., and J. Hansen. 2008. Agricultural water management and climate risk. IRI Technical Report No. 08-01. International Research Institute for Climate and Society, Palisades, NY, USA.

California Department of Water Resources. 1992. The 1991 Drought Water Bank. California DWR Report.

Clark, W. C. 2007. Sustainability Science: A room of its own. Proceedings of the National Academy of Science **104 6:** 1737–1738.

Dublin International Conference on Water and the Environment. 1992. The Dublin Statement on Water and Sustainable Development. World Meteorogical Organisation. Geneva, Switzerland.

Easter, K.W., M.W. Rosegrant and A. Dinar. 1998. Markets for water: Potential and Performance. Kluwer Academics, Norwell, MA, US.

Green, G. P. and J.P. O'Connor. 2001. Water Banking and Restoration of Endangered Species Habitat: An Application to the Snake River. Contemporary Economic Policy **19 2**: 225–237.

Grey, D., C. Sadoff. 2007. Sink or swim? Water security for growth and development. Water Policy **9**: 545–571.

Howitt, R.E. 1998. Spot prices, option prices, and water markets: An analysis of emerging markets in California, in Easter, K.W., M.W. Rosegrant and A. Dinar, eds. 1998. Markets for water: Potential and Performance. Kluwer Academics, Norwell, MA, USA.

Jercich, S.A. 1997. California's 1995 water bank program: Purchasing water supply options. Journal of Water Resources Planning and Management **123 1**: 59–65.

Lettenmaier, D. P. 2008. "Have We Dropped the Ball on Water Resources Research?" Journal of Water Resources Planning and Management **134 6**: 491–492.

Mariño, M. and K.E. Kemper. 1999. Institutional Frameworks in Successful Water Markets. World Bank Technical Paper #427, Washington, D.C.

Milly, P.C.D., et al. 2008. Stationarity is dead: Whither water management? Science **319**: 573–574.

Oki, T., Y. Agata, S. Kanae, T. Saruhashi and K. Musiake. 2003. Global water resources assessment under climatic change in 2050 using TRIP. Water Resources: Systems water availability and global change, in Franks S., G. Blöschl, M. Kumagai, K. Musiake and D. Rosbjerg, eds., 2003 IAHS Publication **280:** 124–133

Oki, T. and S. Kanae, S. 2006. *Science* **313, 5790**: 1068–1072.

Postel, S., G. Daily, and P. Ehrlich. 1996. Human appropriation of renewable fresh water. Science **271**: 785–788.

Rogers, P., A.Bhatia, and A. Huber. 1998. Water as a social and economic good: how to put the principle into practice. Global Water Partnership/Swedish International Development Cooperation Agency, Stockholm, Sweden.

World Commission on Dams, 2000. Dams and development: A new framework for decision-making. The Report of the World Commission on Dams.

Chapter 10
Introduction to Human Health, Ecosystems, and Poverty Reduction

Samuel S. Myers

Until recently, the term "human-dominated ecosystems" would have elicited images of agricultural fields, pastures, or urban landscapes; now it applies with greater or lesser force to all of Earth. Many ecosystems are dominated directly by humanity, and no ecosystem on Earth's surface is free of pervasive human influence.

Our understanding of the human health impacts of global environmental change resembles ancient maps of the world in which large swaths of the globe were blank with only the words "terra incognita" scrolled across their surfaces and the occasional arrow warning "this way lurk monsters." This is a nascent field, and there remains more mystery than clarity. Our goal in the following chapters cannot be to provide a comprehensive summary of all the health consequences of global environmental change: most of these consequences have yet to be discovered. We will introduce the reader to what is currently known about the consequences of different classes of environmental change. We hope that the reader will be left with a sense of the scope and magnitude of impacts that are well described, as well as an appreciation for how much remains to be studied.

Humanity's ecological footprint is immense and continues to balloon exponentially. That human activity has reconfigured the terrestrial surface of the planet, the composition of its biological species, the function of its ecosystems, and the nature of its climate is indisputable. Each of these classes of human alteration of Earth's natural systems has significant consequences for health, some better understood than others. The chapters which follow are an attempt to explore systematically, though not exhaustively, some of these consequences.

In the chapter on Land Use Change, we explore several mechanisms by which the dominant types of land use change can impact health, primarily through altered exposure to infectious disease, but through a variety of other mechanisms as well.

S.S. Myers (✉)
Department of Environmental Health, Harvard School of Public Health, Harvard Medical School, Landmark Center, 401 Park Drive, Boston, MA 02215, USA
e-mail: Sam_myers@hms.harvard.edu

J.C. Ingram et al. (eds.), *Integrating Ecology and Poverty Reduction: Ecological Dimensions*, DOI 10.1007/978-1-4419-0633-5_10,

The second chapter focuses on climate change and describes a series of causal links, some short and direct and some longer and more indirect, between disruption of climate and health impacts. The Disease Ecology chapter explores the implications of changing the composition of whole biological communities of organisms. This is a rapidly growing field and one that powerfully illustrates the law of unintended consequences: that perturbations of systems that seem quite distant from disease-relevant organisms can have strong impacts on infectious disease exposure. The final chapter explores what is known about the connections between disruption of ecosystem services and health and explores some of the methodological challenges to establishing these linkages.

From these chapters, several principles emerge. The first is that the causal pathways between environmental change and health impacts are often complex and indirect. While the impacts of climate change on heat-related mortality may be the simplest and most direct, the impacts on nutrition, mediated through a complex array of different mechanisms, are likely to cause a much higher burden of disease. A second principle, related to the first, is that health impacts are often unpredictable and fall into the category of unintended consequences. Farmers applying fertilizer to hillside fields in Belize were probably surprised to discover that they were causing increases in malaria exposure well downstream. A third principle is that the health impacts of environmental change are immense. These are not esoteric mechanisms of interest only to academics. Food scarcity, water shortages, changes in the distribution of vector-borne diseases (which already affect nearly half the global population), increasing natural hazards, and widespread population displacement are all expected to result from rapidly accelerating environmental change. Collectively, they represent the greatest public health challenges of the twenty-first century and are expected to affect hundreds of millions or billions of people. A final, and critically important, principle emerging from these chapters is that the health impacts of environmental change are expected to disproportionately impact the poor. As is discussed in depth in the Ecosystem Services chapter, there are several mechanisms by which populations can insulate themselves from the impacts of environmental change: they can trade on international markets for locally scarce resources, they can build infrastructure, apply technology, and generate early warnings based on surveillance systems, for example. But these approaches require access to resources, and those who are resource poor will not be able to afford them. Whether discussing food scarcity resulting from climate change, water-borne disease resulting from breakdowns in ecosystem function, or malaria epidemics resulting from changes in land cover, it will be the poor who are caught without the resources to respond to rapidly changing conditions.

We started out saying that much of the map describing health impacts of environmental change has yet to be filled in. Enough is understood, however, to justify a call for action. Rapidly changing environmental conditions are already generating significant health threats. There is no doubt that current and future change will be the cause of widespread human suffering. There is an urgent need to disconnect current economic growth and development from future ecological impoverishment and degradation. Otherwise, today's development is simply driving tomorrow's suffering.

We will need ecologists to help us understand the implications of ecological change and avoid its worst consequences, while simultaneously helping us to slow the pace of ecological disruption. The great challenge of the twenty-first century will be to address global development goals like poverty alleviation and food security, while nurturing the ecological life support systems that provide the basic building blocks of public health. Ecologists will play a central role in meeting this challenge.

Chapter 11
Land Use Change and Human Health

Samuel S. Myers

Human activity is rapidly transforming our planet. The most pervasive changes to the landscape include deforestation, extension and intensification of agriculture, and livestock management, the construction of dams, irrigation projects, and roads, and rapidly spreading urbanization. In addition to the well-known environmental costs of these changes, each also has important health implications that are often less recognized. However, a growing number of studies that combine ecology and human health are demonstrating how these activities impact the emergence of new infectious diseases and alter the distribution of already recognized diseases.

There are a variety of mechanisms by which land use change may alter exposure to infectious disease (Table 11.1). These mechanisms include the alteration of: (1) bio-physical conditions of habitats that can affect the density or presence of disease-related organisms; (2) exposure pathways, or the way organisms (including humans) interact with each other; (3) the genetics of pathogens; (4) the life cycles of pathogens and vectors; and (5) species composition within a community of organisms (Myers and Patz 2009). Infectious diseases which are transmitted by a vector (usually an arthropod), or have a non-human host or reservoir are particularly sensitive to these types of change (Wilson 2001; Eisenberg et al. 2007). Given that such diseases affect over half the human population, particularly the poor, alterations in their transmission rates can have significant impacts on human health and well-being (Lemon et al. 2008).

In this chapter, we discuss the major types of land use change and how these changes are known to impact exposure to infectious disease. In the chapter by Keesing and Ostfeld (this Volume), we discuss disease ecology, which explores how complex changes in whole communities of organisms are likely to alter disease exposure. Both chapters bring into focus the many ways that changing the natural

S.S. Myers (✉)
Department of Environmental Health, Harvard School of Public Health, Harvard Medical School, Landmark Center, 401 Park Drive, Boston, MA 02215, USA
e-mail: Sam_Myers@hms.harvard.edu

J.C. Ingram et al. (eds.), *Integrating Ecology and Poverty Reduction: Ecological Dimensions*, DOI 10.1007/978-1-4419-0633-5_11,

Table 11.1 Mechanisms of altered infectious disease exposure resulting from environmental change

Mechanism by which environmental change alters disease transmission	Examples	Diseases known to be impacted by this mechanism
Changes in density or identity of disease-related organisms	Deforestation or irrigation projects improve breeding habitat and survival of certain anopheline mosquitoes that transmit malaria in Africa, Latin America, and Asia. Deforestation in Cameroon favors one snail species over another, thereby increasing human exposure to pathogenic schistosomes. Sea surface warming and nutrient loading lead to proliferation of *Vibrio cholerae* and disease outbreaks.	Malaria, schistosomiasis, dengue, Japanese encephalitis, filariasis, trypansomiasis, leishmaniasis, cholera, plague, Rift Valley fever, dracunculosis, onchocerciasis, hantavirus, hemorrhagic viruses, Chagas disease, Oropouche/Mayaro virus harmful algal blooms
Changes in Exposure Pathways	Incursions into wildlife habitat can lead t o new exposure to zoonotic disease as seen in Ebola, simian retroviruses, and, possibly, HIV. Dense urban settlements with poor sanitation, waste disposal, or water treatment can lead to increased exposure to many diseases including diarrheal disease, dengue, and leptospirosis.	Malaria, trypanosomiasis, cryptosporidiosis, giardiasis, Ebola, simian retroviruses, HIV, dengue, filariasis, Chagas disease, plague, leptospirosis, typhus, diarrheal disease, food poisoning
Changing the environment in which organisms live creates genetic alterations which can increase disease transmission	Livestock management relying on extensive use of antibiotics in concentrated animal feeding operations (CAFOs) leads to the emergence of pathogens resistant to numerous antibiotics. Confinement of different animal species in wet markets or pig-duck farms can lead to genetic rearrangements resulting in increased virulence or altered infectivity.	Antibiotic resistant bacteria, influenza, SARS

Changes in life-cycle of vectors or pathogens	Deforestation causes increased ambient temperature in homes and breeding sites which shortens gonotrophic cycles, reduces development time, and increases survivalship of anopheline mosquitoes in Kenya.	Malaria
Changes in species composition of communities of organisms	Biodiversity loss in Northeastern forests of the United States increases exposure to Lyme Disease.	Lyme Disease, West Nile virus, malaria, hantavirus, Guanarito, Junin, Machupo, bartonellosis, Nipah virus, St. Louis encephalitis
	Altered species composition of wetlands in Belize in response to nutrient loading creates favorable habitat for a more effective malaria vector.	

world can impact disease exposure in unexpected ways. We hope it will become clear that understanding these ecological relationships is an important element of improving public health globally.

Tropical Deforestation

Widespread deforestation has been one of the most dramatic and biologically profound changes to our global landscape. Over the past 300 years, we have cut down between 7 and 11 million km^2 of forest—an area the size of the continental United States. Approximately, two million km^2 of natural "forest" in temperate and tropical regions are now highly managed plantations with significantly reduced biological diversity (Foley et al. 2005). Deforestation alters biological composition and complexity, soil dynamics, biogeochemical cycles, surface water chemistry, ambient air temperature, exposure to sunlight, and hydrological cycles. It creates forest edges which provides new habitat for a variety of disease vectors and often creates an active interface between human populations and forest-dwelling vectors and host species. Not surprisingly, it has impacts on infectious disease exposure, particularly to vector-borne and zoonotic diseases. Vector-borne diseases are caused by pathogens that are transmitted from one individual to another by an intermediary organism—usually an arthropod such as a mosquito or a tick. Common examples include malaria and Lyme disease. Zoonoses are diseases that exist in both human and non-human vertebrates and, therefore, have natural host reservoirs in the non-human community. Yellow fever (monkeys) and rabies (dogs, raccoons, foxes, skunks, etc.) are common examples. In this section, we focus on tropical deforestation because very little is known about the infectious disease threats of temperate deforestation, with the exception of the well-described relationship between forest fragmentation and increased risk for Lyme disease in temperate forests (LoGiudice et al. 2003).

Because of its global importance, extensive research has been conducted on the relationship between tropical deforestation and malaria. Here, we describe this research in some detail to illustrate the complexity of land use-disease relationships.

Of an estimated three billion people living in malaria-prone areas, approximately 500 million contract the disease annually, and roughly one million people die each year, mostly in Africa (Hay et al. 2005; Snow et al. 2005; Guerra et al. 2006a). Malaria is transmitted by a number of different species of *Anopheles* mosquito. In order to understand the effect of deforestation on malaria, it is necessary to know which species of mosquito are responsible for transmission in a given location and what their breeding habitats and feeding preferences are. Understanding the ecology of the species that transmit malaria locally, including their breeding habitat, feeding preferences, behavior, and environmental niche is essential to understanding the effects of deforestation on the spread of malaria. This, in turn, is essential to predicting the spread of the disease and developing effective control measures.

In the Americas, 4 out of 5 cases of malaria occur in the Amazon. While there are over 50 different *Anopheles* mosquito species in this region, only one appears to be an important vector of malaria: *Anopheles darlingi* (Guerra et al. 2006b). *A. darlingi* prefers to breed in partly shaded pools with slightly acidic pH. Slash and burn land clearing and road building change the chemical composition of the soil, reduce shading, and often create small pools of water with ideal conditions for *A. darlingi*. As a result, deforestation favors breeding of this vector (Singer and de Castro 2001) and has been shown to increase malaria exposure in the Amazon (Tadei et al. 1998). In the Peruvian Amazon, investigators found that biting rates of *A. darlingi* in deforested areas were 278 times higher than biting rates in forested areas (Vittor et al. 2006). These findings supported earlier work showing strong associations between deforestation and malaria and epidemiological evidence of malaria surges following periods of deforestation.

In Africa, rather than a single malaria vector, there are four primary vectors: *A. gambiae*, *A. funestus*, *A. moucheti*, and *A. nili*. *Anopheles nili* and *A. moucheti* have more localized distributions and are considered "subsidiary" vectors. Of these four, *A. nili* is considered a localized forest species with a relatively limited role in transmission (Carnevale et al. 1992). As in the Amazon, deforestation increases habitat for the primary African malaria vectors; it is, therefore, not surprising that a number of studies have found associations between deforestation and increased malaria exposure in sub-Saharan Africa (Coluzzi et al. 1979; Coluzzi 1984, 1994; Cohuet et al. 2004). A recent review of these and other studies of the relationships between deforestation and malaria concluded that deforestation in sub-Saharan Africa tends to increase malaria transmission (Guerra et al. 2006b). Given the breeding preferences of the primary malaria vectors in Africa, this generalization is likely to hold, although exceptions may well be identified in specialized breeding areas.

Increasing habitat or breeding sites for mosquito species is not the only way that deforestation can increase malaria exposure. An elegant series of investigations by Afrane and colleagues (Afrane et al. 2005, 2006) has gone beyond counting the density of anopheline vectors in forested versus deforested areas. In experiments performed in the western Kenya highlands, they evaluated how deforestation may affect the lifecycle of *Anopheles* species through microclimatic change and showed that, by reducing shading, deforestation raises the average temperature in homes by 1.8°C and in nearby aquatic habitats by 4.8–6.1°C. These ambient temperature changes are associated with much shorter reproductive cycles (nearly 60% shorter), reduced larva-to-adult developmental time, and increased larval and adult survivorship all of which improve the vectorial capacity of the mosquitoes and increase exposure to malaria (Afrane et al. 2005, 2006). For example, as a result of increased ambient temperatures in deforested areas, *A. arabiensis* has a 49–55% higher adult life span and a reproductive rate about twice that in forested areas.

In addition, the alteration of microclimate through deforestation can also increase the geographic range of less abundant vectors. In the case of *A. arabiensis*, deforestation has been shown to facilitate the migration to higher elevations. Afrane and colleagues argue that the combination of deforestation and climate change may facilitate the establishment of *A. arabiensis* as an important malaria vector in the Kenya highlands (Afrane et al. 2007).

In Asia, the relationship between deforestation and malaria appears to be more complex. In part, this is because there are a wider variety of *Anopheles* species that are effective malaria vectors and that have more variable habitat preferences.

Unlike the Americas and Africa where higher rates of transmission occur after deforestation, much of the malaria transmission in Asia occurs in intact forested areas. Almost all of the malaria transmission in Bangladesh in 1989 occurred in forests (Sharma et al. 1991). In India in 1987, 7% of the human population lived in forested areas but contributed to 30% of the malaria cases in the country (Narasimham 1991). In fact, deforestation has driven down the density of important malaria vectors including *A. dirus*, *A. minimus*, and *A. barbirostris* in Thailand, Nepal, India, and Sri Lanka. However, it has also caused an increase in the density of alternative vectors including *A. fluviatilis*, *A. annularis*, *A. jamesii*, *A. nigeririmus*, *A. subpictus*, and *A. peditaeniatus* (Amerasinghe and Ariyasena 1990; Konradsen et al. 1990; Karla 1991; Taylor 1997). In a comprehensive review of over 60 studies of land use change and malaria, Yasuoka and Levins describe the unpredictable nature of the impacts of these complex changes. In Kanchanaburi, Thailand, widespread deforestation from 1986 to 1995 eliminated breeding sites for *A. dirus* and decreased malaria incidence. However, in northeast India, deforestation increased malaria transmission by replacing the historical vector *A. minimus* with *A. fluviatilis* that has since become more abundant (Yasuoka and Levins 2007). In Sri Lanka, deforestation has driven major malaria epidemics (Konradsen et al. 1990). While anopheline ecology is particularly complex in Asia, it conforms, nonetheless, to a general trend throughout the world: deforestation tends to reduce mosquito diversity and the surviving dominant species, for reasons that are not well understood, are almost always more effective vectors for malaria than earlier vectors (Molyneux et al. 2008).

Exposure to another disease, schistosomiasis, also appears to be related to deforestation. Schistosomiasis is a disease caused by parasitic worms (Schisotoma spp.) that spend part of their life cycle in freshwater snails and then leave the snails to penetrate the skin of people who enter contaminated water. The disease can damage liver, lungs, intestines, and bladder and infects roughly 200 million people. Deforestation changes the ecology of freshwater snail populations by increasing sunlight penetration, encouraging growth of vegetation, and changing water levels and flow rates. Many snail species do not survive these changes, but those which do tend to be better hosts for the parasitic worms (schistosomes) that cause this disease (Molyneux et al. 2008). In Cameroon, the upsurge in schistosomiasis following deforestation has been well described. Deforestation there led to the displacement of one type of freshwater snail, *Bulinus forskalii*, by another, *Bulinus truncatus*, better suited to cleared habitats. While *B. forskalii* hosted a type of schistosome that causes little illness in humans, *B. truncatus* is an effective host for *Schistosoma haematobium*, a primary cause of urinary tract Schistosomiasis (Southgate et al. 1976).

The effects of deforestation on other vector-borne diseases have been less well characterized. Deforestation in West Africa has expanded the range of both onchocerciasis (river blindness), which is spread by the bite of the black fly and yellow fever, a mosquito-borne virus (Cordellier 1991; Wilson et al. 2002; Patz and Confalonieri 2005). There is also evidence that deforestation has increased the incidence of cutaneous leishmaniasis (transmitted by the bite of the sandfly) in Latin America (Weigle et al. 1993; Desjeux 2001). However, further research is necessary to understand the full impacts of deforestation on exposure to a variety of vector-borne diseases.

Since nearly half the human population suffers from one or more vector-borne diseases this area represents a rich field for research with important consequences for improving human health (Lemon et al. 2008). Deforestation is one of the most pervasive features of global change driving dramatic biological, geochemical, and hydrological changes that will impact the dynamics of many insect-transmitted diseases. In the majority of diseases that have been studied, impacts associated with deforestation have increased disease exposure. However, regardless of whether deforestation has a positive or negative impact on disease exposure, understanding these relationships is essential in guiding disease surveillance, control, and mitigation efforts. The importance of understanding these relationships has been recently emphasized by the World Health Organization (WHO) as part of their emphasis on integrated vector management (2008b).

Not all health consequences of deforestation are related to infectious diseases however. Over the past decade, investigators have uncovered an important association between deforestation in the Amazon basin and "natural" mercury contamination, particularly of rivers. In watersheds that are hundreds of kilometers removed from the nearest gold mining operations, areas disturbed by deforestation have significantly higher mercury loads than upstream areas that remain intact (Veiga et al. 1994; Fostier et al. 2000). This "natural" contamination is thought to be caused by the release of mercury from burned trees and shrubs and increased soil erosion. The contamination has led to elevated mercury levels in fish and in the local people who eat them. Investigators have demonstrated neurological deficits in Amazonian forest-dwellers even at very low levels of mercury contamination (Lebel et al. 1996, 1998).

There are other, less direct, effects of deforestation on human health. At global scales, deforestation makes a significant contribution to climate change through the release of carbon from soils and forest biomass to the atmosphere and by decreasing carbon uptake from growing trees. Deforestation has contributed roughly one-quarter of the total rise in green house gases (GHG). As we will see in the chapter by Hess and Myers (this volume), the health impacts of climate change are quite significant.

In addition to climate regulation, forests play an important role in maintaining a reliable supply of clean fresh water. Forest litter absorbs water, filters it, and releases it slowly over time. Deforestation causes more rapid runoff, increased sediment loads, and poorer water quality. It can lead to flooding and landslides as well as increased incidence of water-borne disease. In 1998, upstream deforestation played an important role in the Yellow River flood disaster which killed more than 3,500 people, damaged over seven million houses, submerged 25 million hectares of farmlands and caused US$30 billion worth of damage (August 28 2001). In the same year, Hurricane Mitch killed nearly 10,000 people in Central America and left roughly one million people homeless. Areas with deforested hillsides and floodplains suffered disproportionate morbidity and mortality (Environmental Impacts of Hurricane Mitch 1999; Cockburn et al. 1999).

The complex and diverse pathways by which deforestation can impact human health provide an excellent example of the importance of ecology to public health. Without an understanding of ecology, it would be difficult to anticipate that changes in forest cover could have such profound impacts on disease exposure.

Crop Cultivation

Rapid human population growth coupled with economic development and increasing adoption of a western-style diet has driven dramatic increases in global grain and meat production. Since 1960, global food production has risen by roughly two and a half times (2005). Achieving these increases has required bringing more land into cultivation and pasturage. Roughly 40% of the planet's ice-free land surface has been converted to croplands or pasture (Foley et al. 2005). Pressure to increase yields-per-acre has also driven agricultural intensification with industrial fertilizers and pesticides, widespread irrigation, new crop varieties, and mechanization. As with deforestation, there are a wide variety of potential health impacts arising from these practices most of which are not well studied. However, an overview of some of the relationships that have been documented suggests the breadth and variety of health consequences resulting from these types of land use changes.

One way that crop cultivation and livestock management impacts health is by creating new ecological niches that favor disease vectors or hosts. In Trinidad, in the 1940s, the development of cacao plantations caused a major malaria epidemic. The mechanism driving this epidemic was the use of anagroforestry system where species of the *Erythrina* tree were used to provide shade and nitrogen to the cacao. However, the shade provided by the *Erythrina* trees also provided ideal habitat for epiphytic bromeliads which, in turn, created excellent breeding sites for *A. bellator*, the principal local malaria vector. The epidemic was not controlled until the *Erythrina* trees were reduced in number and plantation techniques were changed (Downs and Pittendrigh 1946; Yasuoka and Levins 2007). This example also serves to show the complexity of managing ecological relationships: agroforestry systems, such as the *Erythina* and cacao systems described above, have largely been promoted for their contribution to biodiversity conservation, soil conservation, and carbon sequestration functions amongst others, yet, these case studies demonstrate that there also can be unforeseen health consequences associated with these practices. Thus, conserving and managing a multi-functional landscape will require knowledge of the tradeoffs and synergies among ecosystem services and will require adaptive management to respond to unforeseen consequences resulting from land use decisions.

Numerous other relationships between agricultural cultivation and human health impacts have been documented. In Thailand, both cassava and sugarcane cultivation led to reductions in the density of *A. dirus* but created widespread breeding grounds for *A. minimus* with a resulting surge in malaria (Yasuoka and Levins 2007). In Cote d'Ivoire, cultivation of coffee and cacao plantations has been associated with exposure to African trypanosomiasis (sleeping sickness). The plantations create habitat for the tsetse fly vector and cause exposure by bringing agricultural workers into contact with the vector (Fournet et al. 2000). In a final example, the drainage and cultivation of papyrus swamps in highland Uganda appears to increase the risk of malaria. Households located near drained and cultivated swamps have higher ambient temperatures and more *A. gambiae* mosquitoes per household than households in villages surrounding undisturbed papyrus swamps (Lindblade et al. 2000). As with deforestation, ecological changes associated with crop cultivation can create new habitat for disease vectors and new modes of human exposure, both of which can increase overall infectious disease exposure.

Agricultural practices can also impact health through contamination of waterways with pathogens and excess nutrients. In 1993, the largest water-borne disease outbreak ever recorded occurred in Milwaukee where over 400,000 people were infected with *Cryptosporidium parvum*, a protozoan that can cause severe diarrhea. The outbreak followed a period of heavy rainfall and runoff that contaminated Milwaukee's water supply despite new filtration and disinfection facilities and killed 54 people (Mac Kenzie et al. 1994). Heavy rainfall and runoff has been associated with other cryptosporidiosis outbreaks. *Cryptosporidium* oocysts are shed in the feces of many animals including ruminants like cows and sheep. They are very small (roughly 3 μm), pervasive, and are not easily filtered from water. Investigators found that 64% of farms in Pennsylvania had at least one cow infected with *Cryptosporidium* and on 44% of the farms, all bovine stool samples were positive. On these farms, the cattle had full access to waterways that could be contaminated by their feces (Graczyk et al. 2000). This combination of land clearing and grazing ruminants with no buffer zones to protect waterways provides the ideal ecological conditions for human infection by this parasite.

In addition to pathogens, agricultural runoff contains high concentrations of nutrients, particularly fixed nitrogen which is naturally limiting in most terrestrial environments. Overall, human activity, adds at least as much fixed nitrogen to the terrestrial environment as all natural sources combined (Vitousek and Mooney 1997). Because fixed nitrogen is a critical and rate-limiting nutrient in many ecosystems, its widespread addition has profound impacts on these systems. In marine and freshwater environments, nutrient enrichment is responsible for a rapidly increasing number of harmful algal blooms. These blooms are caused by a wide variety of algae, causing massive fish kills, shellfish poisonings, disease and death of marine mammals, and human morbidity and mortality. In the United States alone, roughly 60,000 individual cases and clusters of human intoxication caused by algal blooms occur annually. Health impacts range from acute neurotoxic disorders and death to subacute and chronic disease. The cost of HABs in the US over a 15-year period was estimated to exceed $400 million (2001). Nutrient enrichment of coastal waters is also likely to play a role in cholera outbreaks. Plankton blooms stimulated by warm temperatures and increased nutrient levels help to transform the cholera bacteria, *Vibrio cholera*, from a quiescent to an infectious state (Ezzell 1999; Colwell and Huq 2001). Unfortunately, this problem is likely to increase since agricultural ecologists anticipate a global increase in nitrogen and phosphorus application of roughly 250% from current levels in order to meet projected food demand by 2050 (Tilman 2001).

Nutrient enrichment of waterways from agricultural runoff can lead to increased risk of parasitic and infectious diseases through other types of ecological change as well. A recent review of the literature included 34 studies involving 41 different species of pathogens on 6 continents. The authors concluded that in 95% of observations

(51 of 55), nutrient enrichment *increased* exposure to pathogens (McKenzie and Townsend 2007). In India, for example, extensive use of synthetic fertilizers in rice fields has been associated with increased exposure to Japanese encephalitis. Elevated nitrogen in these rice fields is associated with increases in the density of mosquito larvae, presumably because of increased growth of the microorganisms which are the primary food source for these larvae (Victor and Reuben 2000; Sunish and Reuben 2001). Similar associations between nutrient loading of surface water and increased concentrations of mosquito larvae have been shown for malaria vectors in Mexico (Rejmankova et al. 1991), Belize (Rejmankova et al. 2006) and Taiwan (Teng et al. 1998) and for *Culex* and *Aedes* species of mosquito which transmit La Crosse encephalitis, Japanese encephalitis and West Nile virus (Walker et al. 1991; Sunish and Reuben 2001). As with deforestation, this is a rich field for future research given the pervasiveness of nutrient loading, the importance of the diseases likely to be impacted, and the fact that most of these relationships have not yet been described.

A final human health impact of nutrient loading is direct exposure to nitrogenous compounds in air and water. Nitrogen from synthetic fertilizers contributes to the formation of nitrogen oxides, which in turn lead to the production of ground-level ozone (O_3). Nitrogen oxides in the atmosphere are also an important driver of particulate air pollution. Both O_3 and nitrogen oxides contribute to respiratory disease (both chronic and acute) and cardiovascular disease. In addition, agricultural application of nitrogen to land surfaces leads to contamination of groundwater with nitrates. The World Health Organization (WHO) maximum standard of nitrate in safe drinking water is 10 ppm. Globally, this standard is often exceeded. Even in the USA where strict drinking water legislation applies, 10–20% of groundwater sources may exceed 10 ppm. The potential health effects of excess nitrate in drinking water include reproductive problems, methemoglobinemia (blue-baby syndrome), and cancer (Townsend et al. 2003).

Livestock Management

Livestock and wild animal management practices also have a series of important health consequences. In these cases, the mechanism for altered infectious disease exposure is often genetic change in pathogens resulting from a variety of livestock management practices. The widespread use of antibiotics for livestock has contributed to rapidly increasing microbial resistance. Resistant strains of *Campylobacter*, *Salmonella*, and *Escherichia coli* which can cause serious human infections have all been traced to the use of antibiotics in intensive livestock agriculture (Patz and Confalonieri 2005). A variety of other emerging or resurging infectious diseases are associated with livestock management practices as well. Pandemic influenza in humans is thought to result from genetic exchange among the strains of influenza virus in wild and domestic birds, and pigs. Close confinement of these animals in proximity to each other, for example in Asian "wet markets," and in pig-duck farms in China, fosters this type of genetic exchange (Daily and Ehrlich 1996).

The SARS epidemic is likely to have resulted from similar crowding of animals in live-animal markets in China. In this case, the species at the center of the epidemic were horseshoe bats and palm civet cats as amplifying hosts with a possible role for raccoon dogs, and Chinese ferret badgers as well. Most of the early cases of SARS were among people who worked with the sale or handling of these animals (Shi and Hu 2008).

A more complete understanding of the ecology of this type of zoonotic disease transmission could lead to less risky livestock management practices. Most of the infectious diseases that are now endemic in human populations originated in non-human populations. This includes the major killers of humanity—smallpox, tuberculosis, influenza, malaria, measles, cholera, and plague (Diamond 1999). Practices which bring new populations of wild or domestic animals into close proximity with each other or human populations are likely to be a major source of new emerging diseases in the future (Weiss and McMichael 2004).

A final example of livestock management practices leading to resurgence of infectious disease comes from the mountainous regions of Yunnan Province, China. There, an economic development project tried to raise local incomes by giving villagers cows. Cattle are an important reservoir of *Schistosoma japonicum*, the agent responsible for schistosomiasis. As they spread throughout the region, they shed schistosome eggs into waterways where they infect snails that serve as the intermediate host. As a result, schistosomiasis rates surged, infecting up to 30% of some villages particularly impacting those villagers that owned and managed cattle (Jiang et al. 1997).

Dams and Irrigation Projects

Large dams and irrigation projects have become another pervasive feature of the human-dominated landscape. By the end of the twentieth century, there were over 45,000 large dams and more than 800,000 small dams in over 150 countries. About half of these were built exclusively or primarily for irrigation, and about one-third of the world's irrigated cropland relies on dams (World Commission on Dams 2000; Keiser et al. 2005).

There is no doubt that these dams make an essential contribution to global food production, provide reliable water supplies, and are a massive source of clean power generation. Irrigated croplands represent only one-fifth of total agricultural lands, but produce roughly 40% of total global agricultural yield. Dams are estimated to contribute 12–16% of world food production (2000). However, the ecological and health impacts of dams and irrigation can be both devastating and far reaching.

Dams and irrigation projects also change local ecology and create new favorable habitat for the transmission of a variety of vector-borne diseases. As it is with deforestation, the association with malaria is one of the best documented health impacts of dams and irrigation projects. A study in northern Ethiopia has shown a sevenfold increase in malaria in villages within 3 km of microdams compared with control villages 8–10 km distant (Ghebreyesus et al. 1999). In India, there was a surge of "irrigation malaria" in the 1990s after poorly evaluated irrigation projects improved breeding conditions for the dominant malaria vector, *Anopheles culcifacies*. Malaria became endemic and widespread in a population of roughly 200 million people as a result (Sharma 1996). In general, dams and irrigation projects constructed in endemic areas tend to increase breeding habitat and transmission of malaria (World Commission on Dams 2000; Keiser et al. 2005). Agricultural projects that combine deforestation with dams and irrigation can be a particularly potent combination for increasing malaria exposure given their combined effects on breeding habitat and microclimate.

In the Nile delta area of Egypt, prevalence of lymphatic filariasis (elephantiasis) rose from <1% in 1965 to >20% after construction of the Aswan High Dam and subsequent irrigation projects. This surge resulted from increased surface and sub-surface moisture that created improved breeding sites for *Culex pipiens*, the mosquito vector of this disease (Harb et al. 1993; Thompson et al. 1996). In Ghana, a different vector is primarily responsible for transmission of filariasis, *Anopheles gambiae*. However, similar dynamics are observed. Rates of infection, worm load, annual bites per person, and annual transmission rates were all found to be higher in irrigated areas than in communities without irrigation (Appawu et al. 2001).

Filariasis was not the only disease to surge with the construction of the Aswan Dam. Because of the creation of extensive new habitat for *B. truncatus*, a freshwater snail which is an excellent intermediate host for *Schistosoma haematobium*, prevalence of *S. haematobium* infection in Upper and Middle Egypt rose from about 6% before construction of the dam to nearly 20% in the 1980s. In Lower Egypt, intestinal schistosomiasis rose to an even greater extent (Malek 1975; Cline et al. 1989; Molyneux et al. 2008). In the Tana river region of Kenya, the Hola irrigation project led to the introduction of snail vectors where they had never been before. Between 1956, when the project began, and 1966, the prevalence of urinary schistosomiasis in children in the region went from 0 to 70%. By 1982, it was 90% (Mutero 2002). Around the world, the rapid proliferation of dams and irrigation projects has generated new habitat for freshwater snails well adapted to these environments and to hosting schistosomes resulting in a global surge in schistosomiasis infection. Propagation of Rift Valley Fever, leishmaniasis, dracunculosis, onchocerciasis, and Japanese encephalitis has also been associated with these projects (Jobin 1999; Patz and Confalonieri 2005).

Roads

Construction of roads can provide edge habitat for disease vectors as well as create pools of water that can be excellent breeding sites for mosquitoes. Roads built for transportation, access to mines, or construction of pipelines can become entry points for settlers. In these situations, the combination of deforestation and exposure of a non-immune population to local vector-borne or zoonotic disease can lead to new epidemics. With malaria, this phenomenon is referred to as "frontier malaria."

Road building has also been implicated as an important factor in the penetration of human populations into previously undisturbed wildlife habitat. This mixing of previously isolated human and non-human populations can lead to exposure to new zoonoses. The bushmeat trade which leads to handling, slaughtering, and consuming wild animal species, particularly in Africa and Asia, further increases the risk of human exposure to new pathogens.

In particular, bushmeat hunting may provide opportunities for exchange of pathogens between humans and non-human primates. In Central Africa, 1–3.4 million tons of bushmeat are harvested annually (Fa and Peres 2001). In West Africa, a large share of protein in the diet comes from bushmeat. In West Africa, the bushmeat harvest includes a large numbers of primates, facilitating interspecies disease transfer. The "Taxonomic Transmission Rule" states that the probability of successful cross-species infection increases the more closely hosts are genetically related, since related hosts are more likely to share susceptibility to the same range of potential pathogens (Wolfe et al. 2000).

Recently, infection with simian foamy virus, a retrovirus that is endemic in most Old World primates, was demonstrated in hunters who reported direct contact with blood or body fluid of non-human primates. This finding provides additional support for already compelling hypothesis that the retrovirus causing HIV/AIDS was likely a mutated simian virus contracted through bushmeat hunting (Hahn et al. 2000; Wolfe et al. 2004). It is likely that Ebola virus infection in human populations also had its origin in bushmeat hunting.

Of the wave of emerging infectious diseases over the past several decades, three quarters are zoonotic (Taylor et al. 2001). This fact is consistent with the theory that new incursions of human populations into previously isolated wildlife habitat, and animal markets that bring live wildlife into close proximity with other human and non-human species may represent important sources of new human infectious diseases.

Urbanization

A final pervasive form of land use change has been the rapid, global development and expansion of cities. Over the past two centuries, the proportion of people living in cities or large towns has grown from approximately 5–50% and continues to climb (Global Environmental Change and Human Health 2007). Between 1960 and 1980, the urban population in developing countries more than doubled. By 2025, urban population in developing countries is expected to account for well over half the global population (Knudsen and Slooff 1992).

Widespread urbanization can result in numerous direct ecological consequences that affect human health. As with deforestation, agriculture, dams and irrigation systems, urbanization creates new habitat for disease hosts and vectors while changing human exposure patterns. Much of the rapid urbanization occurring today is taking place in urban or periurban slums with few services for clean water provision, sewage disposal, solid waste management, or quality housing. Pools of contaminated water, water containers kept in homes, tires and other refuse capable of holding water, and piles of municipal waste all create excellent habitat for a variety of rodent hosts and arthropod vectors, particularly those which transmit dengue, malaria, filariasis, Chagas disease, plague, and typhus. In addition, rural to urban migration brings people from different disease endemic regions together in high density providing a source for new infection as well as non-immune hosts. A final contributing factor to the spread of vector-borne disease in slums is the poor quality housing which does not provide an effective barrier to mosquitoes, rodents, or fleas.

Dengue fever provides a good example of a vector that is well adapted to urban habitat. Over recent decades there has been a tremendous surge in dengue cases as it has spread out of South-east Asia and the Pacific and become endemic throughout the tropics. Dengue is the most common mosquito-borne viral disease in the world with roughly 50 million cases in over 100 countries each year (2008a). It is transmitted by the bite of infected *Aedes* mosquitoes, primarily *Aedes aegypti*. These mosquitoes prefer to feed on humans over other animals and usually live close to human dwellings. They breed in man-made containers like earthenware jugs, tires, metal drums, discarded plastic food containers, and other items that collect rainwater. These characteristics make them highly effective at adapting to urban areas (Daily and Ehrlich 1996; Mackenzie et al. 2004). Relatively simple, ecologically based solutions, such as eliminating rainwater holding containers that serve as the breeding habitat for this species have proven effective control measures.

A second consequence of rapid urbanization in slums and squatter settlements is lack of access to safe drinking water or sanitation. Current projections are that, if efforts to provide water and sanitation to the underserved continue at the current rate, more than 692 million people will live without basic sanitation and 240 million without improved sources of drinking water, in urban areas in 2015 (Norstrom 2007). Particularly in crowded urban conditions, the exposure to infectious disease resulting from both contaminated drinking water and inadequate sanitation is significant. Diarrhea alone, caused primarily by contaminated drinking water, causes around 2.2 million deaths each year, about 4% of all global mortality. Poor sanitation can also lead to outbreaks of leptospirosis as was observed in San Salvador between March and November of 1996, where investigators identified over 300 cases of leptospirosis—a disease that is transmitted by direct exposure to rodents or an environment contaminated by their urine. They found the highest incidence rates among the urban poor where exposure to rats and to flood waters contaminated by rat urine was likely to be highest. The disease had a 15% case fatality rate. Investigators noted an epidemiological shift in leptospirosis transmission from rural to urban areas and postulated that urban slums with large rodent populations create an environment in which heavy rains, which can drive rodents out of burrows and lead to contaminated flood waters, will trigger these epidemics (Ko et al. 1999).

Finally, rapidly growing urban areas cast a footprint far beyond their local boundaries. Wastewater from cities is often poorly treated and can expose downstream communities to infectious disease. It can also lead to outbreaks of harmful algal blooms and shellfish poisoning as it contaminates coastal marine environments (Rose et al. 2001). Urban emissions contribute to local air pollution but also have impacts regionally (acid rain, for example) and globally (climate change). Air pollution, primarily from urban centers in Asia, is causing the formation of atmospheric brown clouds which have become so extensive that they are impacting regional weather patterns, reducing agricultural productivity, and increasing glacial melting in addition to causing extensive deaths from cardiorespiratory disease (Ramanathan et al. 2008). Urban demand for food, building materials, fiber, and other ecosystem services drives many of the land use trends discussed earlier. Whether these demands are reduced by concentrating people in urban environments and creating efficiencies of scale or increased because of transportation requirements and waste has not been well studied.

Conclusion

Human activity has changed the face of the global landscape. These changes in land use and cover have, in turn, altered the dynamics of infectious disease transmission in numerous ways. They have created new habitat and breeding sites for disease vectors which, in many cases, favor disease transmission. They have altered exposure pathways by changing the way organisms (including humans) interact with each other. They have led to changes in the genetics, and thereby, the virulence, and infectiousness of pathogens. They have also changed the lifecycle of pathogens and vectors and the species composition of whole communities of disease-relevant organisms. Not surprisingly, these changes have occurred coincident with a rise in new or reemerging infectious diseases. These are not merely academic concerns; the diseases impacted by these changes represent a large percentage of the total global disease burden.

For practitioners working to improve human wellbeing, understanding how large-scale changes in the landscape can create new dynamics in disease transmission is important. This type of understanding can help us anticipate that, despite their potential economic benefits, there may be significant health effects from environmental change such as dam building, irrigation projects, the use of fertilizers, and the clearing of forests. Knowledge of these interactions can and should help guide surveillance and mitigation efforts. The goal of the practitioner must be to maximize the benefits of these projects while minimizing the negative health impacts. Accomplishing that goal requires an understanding of the underlying ecological relationships between the disease, its vectors, and the environmental niches of both. It also requires active use of health impact assessments (HIAs) as discussed in our chapter on climate change and health.

Finally, it is important, wherever possible, to develop a fine-grained understanding of local field conditions and of the complex relationships between communities of organisms, (including pathogens), habitat, and human populations.

Ecological understanding of the community interactions between diseases, hosts, vectors and the afflicted not only serve to help us understand how our alteration of the environment impacts the spread of these diseases, but can be a powerful tool in preventing, immobilizing, and controlling these diseases as will be discussed in Chap. 13 (this volume) on disease ecology.

References

1999. Environmental Impacts of Hurricane Mitch. Environ Health Perspect.

2000. Dams and Development: A New Framework for Decision-Making. World Commission on Dams, London.

2001. Oceans and Human Health Roundtable Report. National Institute of Environmental Health Sciences and National Science Foundation, Research Triangle Park, North Carolina.

Ecosystems and Human Well-being: Synthesis. Millenium Ecosystem Assessment. Island Press, Washington, DC.

2007. Global Environmental Change and Human Health (2007) Science Plan and Implementation Strategy. Earth System Science Partnership (Diversitas, IGBP, IHDP, and WCRP) Report No.4; Global Environmental Change and Human Health Report No.1.

2008a. Fact Sheet No. 117: Dengue and dengue hemorhagic fever. Page WHO Fact Sheet *in* W. H. O. M. Centre, editor. Fact Sheet No. 117. World Health Organization, Geneva.

2008b. Malaria control: the power of integrated action. Health and Environment Linkages Policy Series. World Health Organization, Geneva, Switzerland.

August 28, 2001. Asian Development Bank News Release 091/01: Novel Approach To Managing Floods On People's Republic Of China's Yellow River.*in* A. D. Bank, editor. New Release 091/01.

Afrane, Y. A., B. W. Lawson, A. K. Githeko, and G. Yan. 2005. Effects of microclimatic changes caused by land use and land cover on duration of gonotrophic cycles of Anopheles gambiae (Diptera: Culicidae) in western Kenya highlands. J Med Entomol **42**:974–980.

Afrane, Y. A., G. Zhou, B. W. Lawson, A. K. Githeko, and G. Yan. 2006. Effects of microclimatic changes caused by deforestation on the survivorship and reproductive fitness of Anopheles gambiae in western Kenya highlands. Am J Trop Med Hyg **74**:772–778.

Afrane, Y. A., G. Zhou, B. W. Lawson, A. K. Githeko, and G. Yan. 2007. Life-table analysis of Anopheles arabiensis in western Kenya highlands: effects of land covers on larval and adult survivorship. Am J Trop Med Hyg **77**:660–666.

Amerasinghe, F. P. and T. G. Ariyasena. 1990. Larval survey of surface water-breeding mosquitoes during irrigation development in the Mahaweli Project, Sri Lanka. J Med Entomol **27**:789–802.

Appawu, M. A., S. K. Dadzie, A. Baffoe-Wilmot, and M. D. Wilson. 2001. Lymphatic filariasis in Ghana: entomological investigation of transmission dynamics and intensity in communities served by irrigation systems in the Upper East Region of Ghana. Tropical Medicine & International Health **6**:511–516.

Carnevale, P., G. Le Goff, J. C. Toto, and V. Robert. 1992. Anopheles nili as the main vector of human malaria in villages of southern Cameroon. Med Vet Entomol **6**:135–138.

Cline, B. L., F. O. Richards, M. A. el Alamy, S. el Hak, E. Ruiz-Tiben, J. M. Hughes, and D. F. McNeeley. 1989. 1983 Nile Delta schistosomiasis survey: 48 years after Scott. Am J Trop Med Hyg **41**:56–62.

Cockburn, A., J. St Clair, and K. Silverstein. 1999. The Politics of "Natural" Disaster: Who Made Mitch So Bad? International Journal of Health Services **29**:459–462.

Cohuet, A., F. Simard, C. S. Wondji, C. Antonio-Nkondjio, P. Awono-Ambene, and D. Fontenille. 2004. High malaria transmission intensity due to Anopheles funestus (Diptera: Culicidae) in a village of savannah-forest transition area in Cameroon. J Med Entomol **41**:901–905.

Coluzzi, M. 1984. Heterogeneities of the malaria vectorial system in tropical Africa and their significance in malaria epidemiology and control. Bull World Health Organ **62 Suppl**:107–113.

Coluzzi, M. 1994. Malaria and the Afrotropical ecosystems: impact of man-made environmental changes. Parassitologia **36**:223–227.

Coluzzi, M., A. Sabatini, V. Petrarca, and M. A. Di Deco. 1979. Chromosomal differentiation and adaptation to human environments in the Anopheles gambiae complex. Trans R Soc Trop Med Hyg **73**:483–497.

Colwell, R. and A. Huq. 2001. Marine Ecosystems and Cholera. Hydrobiologia **460**:141–145.

Cordellier, R. 1991. [The epidemiology of yellow fever in Western Africa]. Bull World Health Organ **69**:73–84.

Daily, G., C. and P. Ehrlich, R. 1996. Global Change and Human Susceptibility to Disease. Annual Review of Energy and the Environment **21**:125–144.

Desjeux, P. 2001. The increase in risk factors for leishmaniasis worldwide. Trans R Soc Trop Med Hyg **95**:239–243.

Diamond, J. 1999. Guns, Germs, and Steel. W.W.Norton & Co., New York, NY.

Downs, W., G. and C. Pittendrigh, S. 1946. Bromeliad malaria in Trinidad, British West Indies. American Journal of Tropical Medicine **26**:47–66.

Eisenberg, J. N., M. A. Desai, K. Levy, S. J. Bates, S. Liang, K. Naumoff, and J. C. Scott. 2007. Environmental determinants of infectious disease: a framework for tracking causal links and guiding public health research. Environ Health Perspect **115**:1216–1223.

Ezzell, C. 1999. It Came from the Deep. Scientific American **280**:22.

Fa, J., E and C. Peres, A. 2001. Game vertebrate extraction in African and Neotropical forests: an intercontinental comparison. Pages 203–241 *in* J. Reynolds, D, G. Mace, M, K. Redford, H, and J. Robinson, G, editors. Conservation of exploited species. Cambridge University Press, Cambridge, U.K.

Foley, J. A., R. Defries, G. P. Asner, C. Barford, G. Bonan, S. R. Carpenter, F. S. Chapin, M. T. Coe, G. C. Daily, H. K. Gibbs, J. H. Helkowski, T. Holloway, E. A. Howard, C. J. Kucharik, C. Monfreda, J. A. Patz, I. C. Prentice, N. Ramankutty, and P. K. Snyder. 2005. Global consequences of land use. Science **309**:570–574.

Fostier, A. H., M. C. Forti, J. R. Guimaraes, A. J. Melfi, R. Boulet, C. M. Espirito Santo, and F. J. Krug. 2000. Mercury fluxes in a natural forested Amazonian catchment (Serra do Navio, Amapa State, Brazil). Sci Total Environ **260**:201–211.

Fournet, F., S. Traore, A. Prost, E. Cadot, and J. P. Hervouet. 2000. Impact of the development of agricultural land on the transmission of sleeping sickness in Daloa, Cote d'Ivoire. Ann Trop Med Parasitol **94**:113–121.

Ghebreyesus, T. A., M. Haile, K. H. Witten, A. Getachew, A. M. Yohannes, M. Yohannes, H. D. Teklehaimanot, S. W. Lindsay, and P. Byass. 1999. Incidence of malaria among children living near dams in northern Ethiopia: community based incidence survey. BMJ **319**:663–666.

Graczyk, T. K., B. M. Evans, C. J. Shiff, H. J. Karreman, and J. A. Patz. 2000. Environmental and Geographical Factors Contributing to Watershed Contamination with Cryptosporidium parvum Oocysts. Environmental Research **82**:263–271.

Guerra, C. A., R. W. Snow, and S. I. Hay. 2006a. Defining the global spatial limits of malaria transmission in 2005. Adv Parasitol **62**:157–179.

Guerra, C. A., R. W. Snow, and S. I. Hay. 2006b. A global assessment of closed forests, deforestation and malaria risk. Ann Trop Med Parasitol **100**:189–204.

Hahn, B. H., G. M. Shaw, K. M. De Cock, and P. M. Sharp. 2000. AIDS as a zoonosis: scientific and public health implications. Science **287**:607–614.

Harb, M., R. Faris, A. M. Gad, O. N. Hafez, R. Ramzy, and A. A. Buck. 1993. The resurgence of lymphatic filariasis in the Nile delta. Bull World Health Organ **71**:49–54.

Hay, S. I., C. A. Guerra, A. J. Tatem, P. M. Atkinson, and R. W. Snow. 2005. Urbanization, malaria transmission and disease burden in Africa. Nat Rev Microbiol **3**:81–90.

Jiang, Z., Q. S. Zheng, X. F. Wang, and Z. H. Hua. 1997. Influence of livestock husbandry on schistosomiasis transmission in mountainous regions of Yunnan Province. Southeast Asian J Trop Med Public Health **28**:291–295.

Jobin, W. 1999. Dams and Disease: Ecological Design and Health Impacts of Large Dams, Canals, and Irrigation Systems. E&FN Spon, London.

Karla, N., L. 1991. Forest Malaria Vectors in India: Ecological Charcteristics and Epidemiological Implications. Pages 93–114 *in* V. Sharma, P. and A. Kondrashin, V, editors. Forest Malaria in Southeast Asia. World Health Organization/Medical Research Council, New Delhi, India.

Keiser, J., M. C. De Castro, M. F. Maltese, R. Bos, M. Tanner, B. H. Singer, and J. Utzinger. 2005. Effect of irrigation and large dams on the burden of malaria on a global and regional scale. Am J Trop Med Hyg **72**:392–406.

Knudsen, A. B. and R. Slooff. 1992. Vector-borne disease problems in rapid urbanization: new approaches to vector control. Bull World Health Organ **70**:1–6.

Ko, A. I., M. G. Reis, C. M. R. Dourado, J. W. D. Johnson, and L. W. Riley. 1999. Urban epidemic of severe leptospirosis in Brazil. The Lancet **354**:820–825.

Konradsen, F., F. Amerasinghe, P., W. van der Hoek, and P. Amerasinghe, H. 1990. Malaria in Sri Lanka, Current Knowledge on Transmission and Control. International Water Management Institute, Battaramulla, Sri Lanka.

Lebel, J., D. Mergler, F. Branches, M. Lucotte, M. Amorim, F. Larribe, and J. Dolbec. 1998. Neurotoxic effects of low-level methylmercury contamination in the Amazonian Basin. Environ Res **79**:20–32.

Lebel, J., D. Mergler, M. Lucotte, M. Amorim, J. Dolbec, D. Miranda, G. Arantes, I. Rheault, and P. Pichet. 1996. Evidence of early nervous system dysfunction in Amazonian populations exposed to low-levels of methylmercury. Neurotoxicology **17**:157–167.

Lemon, S., M, F. Sparling, P, M. Hamburg, A, D. Relman, A, E. Choffnes, R, and A. Mack, editors. 2008. Vector-Borne Diseases: Understanding the Environmental, Human Health, and Ecological Connections. National Academies Press, Washington, DC.

Lindblade, K. A., E. D. Walker, A. W. Onapa, J. Katungu, and M. L. Wilson. 2000. Land use change alters malaria transmission parameters by modifying temperature in a highland area of Uganda. Trop Med Int Health **5**:263–274.

LoGiudice, K., R. S. Ostfeld, K. A. Schmidt, and F. Keesing. 2003. The ecology of infectious disease: effects of host diversity and community composition on Lyme disease risk. Proc Natl Acad Sci USA **100**:567–571.

Mac Kenzie, W. R., N. J. Hoxie, M. E. Proctor, M. S. Gradus, K. A. Blair, D. E. Peterson, J. J. Kazmierczak, D. G. Addiss, K. R. Fox, J. B. Rose, and et al. 1994. A massive outbreak in Milwaukee of cryptosporidium infection transmitted through the public water supply. N Engl J Med **331**:161–167.

Mackenzie, J. S., D. J. Gubler, and L. R. Petersen. 2004. Emerging flaviviruses: the spread and resurgence of Japanese encephalitis, West Nile and dengue viruses. Nat Med **10**:S98-109.

Malek, E., A. 1975. Effect of Aswan High Dam on prevalence of schistosomiasis in Egypt Tropical and Geographical Medicine **27**:359–364.

McKenzie, V., J and A. Townsend, R. 2007. Parasitic and Infectious Disease Responses to Changing Global Nutrient Cycles. EcoHealth **4**:384–396.

Molyneux, D., R. S. Ostfeld, A. Bernstein, and E. Chivian. 2008. Ecosystem Disturbance, Biodiversity Loss, and Human Infectious Disease. Pages 287–323 *in* E. Chivian and A. Bernstein, editors. Sustaining Life: How Human Health Depends on Biodiversity. Oxford University Press, Oxford; New York.

Mutero, C. M. 2002. Health impact assessment of increased irrigation in the Tana River Basin, Kenya.*in* I. W. M. Institute, editor. The Changing Face of Irrigation in Kenya: Opportunities for anticipating change in Eastern and Southern Africa. International Water Management Institute, Colombo.

Myers, S. S. and J. Patz. 2009. Emerging threats to human health from global environmental change. Annual Review of Environment & Resources **34**:223–252.

Narasimham, M., V,V,L. 1991. Perspective of forest malaria in India. World Health Organization, New Delhi, India.

Norstrom, A. 2007. Planning for drinking water and sanitation in peri-urban areas. 21.

Patz, J. A. and U. E. C. Confalonieri. 2005. Human Health: Ecosystem Regulation of Infectious Diseases. Pages 391–415 *in* R. M. Hassan, R. Scholes, and N. Ash, editors. Millennium

Ecosystem Assessment (Program). Condition and Trends Working Group. Ecosystems and human well-being : current state and trends : findings of the Condition and Trends Working Group of the Millennium Ecosystem Assessment. Island Press, Washington, DC.
Ramanathan, V., M., H. Agrawal, M. Akimoto, S. Aufhammer, and L. Devotta. 2008. Atmospheric Brown Clouds: Regional Assessment Report with Focus on Asia., United Nations Environment Program, Nairobi, Kenya.
Rejmankova, E., J. Grieco, N. Achee, P. Masuoka, K. Pope, D. Roberts, and R. Higashi. 2006. Freshwater community interactions and malaria. Page 227 *in* S. Collinge, K and C. Ray, editors. Disease ecology. Oxford University Press, Oxford.
Rejmankova, E., H. M. Savage, M. Rejmanek, J. I. Arredondo-Jimenez, and D. R. Roberts. 1991. MULTIVARIATE ANALYSIS OF RELATIONSHIPS BETWEEN HABITATS, ENVIRONMENTAL FACTORS AND OCURENCE OF ANOPHELINE MOSQUITO LARVAE ANOPHELES ALBIMANUS AND A. PSEUDOPUNCTIPENNIS IN SOUTHERN CHIAPAS, MEXICO. Journal of Applied Ecology **28**:827–841.
Rose, J. B., P. R. Epstein, E. K. Lipp, B. H. Sherman, S. M. Bernard, and J. A. Patz. 2001. Climate variability and change in the United States: potential impacts on water- and foodborne diseases caused by microbiologic agents. Environ Health Perspect **109 Suppl 2**:211–221.
Sharma, V., P., C. Prasittisuk, and A. Kondrashin, V. 1991. Magnitude of forest related malaria in the WHO Southeast Asia region., World Health Organization, New Delhi, India.
Sharma, V. P. 1996. Re-emergence of malaria in India. Indian J Med Res **103**:26–45.
Shi, Z. and Z. Hu. 2008. A review of studies on animal reservoirs of the SARS coronavirus. Virus Research **133**:74–87.
Singer, B. H. and M. C. de Castro. 2001. Agricultural colonization and malaria on the Amazon frontier. Ann N Y Acad Sci **954**:184–222.
Snow, R. W., C. A. Guerra, A. M. Noor, H. Y. Myint, and S. I. Hay. 2005. The global distribution of clinical episodes of Plasmodium falciparum malaria. Nature **434**:214–217.
Southgate, V., H. Wijk, and C. Wright. 1976. Schistosomiasis in Loum, Cameroun: Schistosoma haematobium, S. intercalatum, and their natural hybrid. Zeitschrift fur Parasitenkund **49**:149–159.
Sunish, I. P. and R. Reuben. 2001. Factors influencing the abundance of Japanese encephalitis vectors in ricefields in India - I. Abiotic. Medical and Veterinary Entomology **15**:381–392.
Tadei, W. P., B. D. Thatcher, J. M. Santos, V. M. Scarpassa, I. B. Rodrigues, and M. S. Rafael. 1998. Ecologic observations on anopheline vectors of malaria in the Brazilian Amazon. Am J Trop Med Hyg **59**:325–335.
Taylor, D. 1997. Seeing the forests for the more than the trees. Environ Health Perspect **105**:1186–1191.
Taylor, L. H., S. M. Latham, and M. E. Woolhouse. 2001. Risk factors for human disease emergence. Philos Trans R Soc Lond B Biol Sci **356**:983–989.
Teng, H., J, Y. Wu, L, S. Wang, J, and C. Lin. 1998. Effects of environmental factors on abundance of Anopheles minimus larvae and their seasonal fluctuations in Taiwan. Environmental Entomology **27**:324–328.
Thompson, D. F., J. B. Malone, M. Harb, R. Faris, O. K. Huh, A. A. Buck, and B. L. Cline. 1996. Bancroftian filariasis distribution and diurnal temperature differences in the southern Nile delta. Emerg Infect Dis **2**:234–235.
Tilman, D. 2001. Forecasting Agriculturally Driven Global Environmental Change. Science **292**:281–284.
Townsend, A., R, R. Howarth, W, F. Bazzaz, A, M. Booth, S, C. Cleveland, C, S. Collinge, K, A. P. Dobson, P. Epstein, R, E. Holland, A, D. Keeney, R, M. Mallin, A, C. Rogers, A, P. Wayne, and A. Wolfe, H. 2003. Human health effects of a changing global nitrogen cycle. Frontiers in Ecology and the Environment **1**:240–246.
Veiga, M. M., J. A. Meech, and N. Onate. 1994. Mercury pollution from deforestation. Nature **368**:816–817.
Victor, T. J. and R. Reuben. 2000. Effects of organic and inorganic fertilisers on mosquito populations in rice fields of southern India. Medical and Veterinary Entomology **14**:361–368.
Vitousek, P. M. and H. A. Mooney. 1997. Human domination of Earth's ecosystems. Page 494 Science.

Vittor, A. Y., R. H. Gilman, J. Tielsch, G. Glass, T. Shields, W. S. Lozano, V. Pinedo-Cancino, and J. A. Patz. 2006. The effect of deforestation on the human-biting rate of Anopheles darlingi, the primary vector of Falciparum malaria in the Peruvian Amazon. Am J Trop Med Hyg **74**:3–11.

Walker, E. D., D. Lawson, L, and R. Merritt, W. 1991. Nutrient dynamics, bacterial populations and mosquito productivity in tree-hole ecosystems and microcosms. Ecology **72**:1529–1546.

Weigle, K. A., C. Santrich, F. Martinez, L. Valderrama, and N. G. Saravia. 1993. Epidemiology of cutaneous leishmaniasis in Colombia: environmental and behavioral risk factors for infection, clinical manifestations, and pathogenicity. J Infect Dis **168**:709–714.

Weiss, R. A. and A. J. McMichael. 2004. Social and environmental risk factors in the emergence of infectious diseases. Nat Med **10**:S70–76.

Wilson, M. D., R. A. Cheke, S. P. J. Fiasse, S. Grist, M. Y. Osei-Ateweneboana, A. Tetteh-Kumah, G. K. Fiasorgbor, F. R. Jolliffe, D. A. Boakye, J. M. Hougard, L. Yameogo, and R. J. Post. 2002. Deforestation and the spatio-temporal distribution of savannah and forest members of the Simulium damnosum complex in southern Ghana and south-western Togo. Transactions of the Royal Society of Tropical Medicine and Hygiene **96**:632–639.

Wilson, M. L. 2001. Ecology and Infectious Disease. Pages 283–324 *in* J. L. Aron and J. Patz, editors. Ecosystem Change and Public Health. Johns Hopkins University Press, Baltimore, MD.

Wolfe, N. D., M. Eitel, N, J. Gockowski, P. Muchaal, K, C. Nolte, A. Prosser, T, J. Torimiro, N, S. Weise, F, and D. Burke, S. 2000. Deforestation, hunting, and the ecology of microbial emergence. Global Change and Human Health **1**:10–25.

Wolfe, N. D., W. M. Switzer, J. K. Carr, V. B. Bhullar, V. Shanmugam, U. Tamoufe, A. T. Prosser, J. N. Torimiro, A. Wright, E. Mpoudi-Ngole, F. E. McCutchan, D. L. Birx, T. M. Folks, D. S. Burke, and W. Heneine. 2004. Naturally acquired simian retrovirus infections in central African hunters. Lancet **363**:932–937.

Yasuoka, J. and R. Levins. 2007. Impact of deforestation and agricultural development on anopheline ecology and malaria epidemiology. Am J Trop Med Hyg **76**:450–460.

Chapter 12
The Health Impacts of Climate Change and Ecological Diagnosis and Treatment

Jeremy Hess and Samuel S. Myers

Introduction

As humans evolved, we were actors on the global ecological stage no different from any other organism with whom we shared the planet. However, over the last two to three hundred years, we have increasingly monopolized the stage, becoming the playwright as well as the dominant actor. Today, we are fundamentally altering the stage itself, often in ways that disrupt the planet's ecological systems with deleterious consequences for our own health and well-being.

The previous chapter on the human health effects of land use change discussed some of the profound changes we have wrought on the terrestrial landscape. In this chapter, we discuss the health implications of a second major component of accelerating ecological change: the disruption of our planet's climate. Although both are treated in isolation here, there is a strong interaction between the two. Land use change alters the resistance and resilience of ecosystems which, in turn, changes the stability and quality of ecosystem services relied on by humans. Climate change acts to further disrupt and destabilize many of these ecological systems. In many instances, including infectious disease transmission and coastal vulnerability, land use and climate change act synergistically to multiply threats to human health. In this chapter, we briefly summarize the science of climate change and discuss the impact of these physical changes on human health, emphasizing that the greatest burdens will be borne by those least responsible and least able to cope. Finally, we discuss approaches to reducing vulnerability to climate change, highlighting the value of an ecological lens to evaluate the impacts of climate change. We hope this analysis will help to guide decisions about vulnerability reduction; and to propose

J. Hess (✉)
Department of Emergency Medicine, Department of Environmental Health,
Rollins School of Public Health, Emory University, Steiner Building, Room 120,
49 Jesse Hill Jr Drive, Atlanta, GA 30303, USA
e-mail: jhess@emory.edu

J.C. Ingram et al. (eds.), *Integrating Ecology and Poverty Reduction: Ecological Dimensions*, DOI 10.1007/978-1-4419-0633-5_12,

integrated solutions that include ecological considerations. We do not pretend that ecological tools in isolation will be able to relieve the impacts of climate change on human health, but do suggest that integrating both ecology and medical knowledge amongst other disciplines will elucidate more sustainable and effective mitigation and adaptation strategies.

Climate Change

Our disruption of the climate system began with the age-old human pursuit of energy that was dramatically altered during the Industrial Revolution when we tapped the enormous new energy reserve of fossil fuels, liberating stored energy (and carbon) and applying it to industrial activities. This moment ushered in a host of societal changes that continue to this day, including the Green Revolution in agriculture, increased urbanization, and urban sprawl. Collectively, these changes allowed dramatic increases in food production, life expectancy, and economic prosperity, which in turn have driven an explosion in the human population. At the same time, these changes have reinforced our widespread dependence on fossil fuels. As our numbers have grown and our prosperity increased, our global ecological footprint has expanded dramatically. Our liberation of carbon and other greenhouse gases into the atmosphere represents a critical component of that global ecological footprint (Wackernagel et al. 2002).

The earth's climate is regulated within a fairly narrow range by the atmosphere, which has among its constituents several "greenhouse gases," including carbon dioxide (CO_2), methane, nitrous oxide, and water vapor. Carbon is stored in four distinct reservoirs: the atmosphere, the terrestrial biosphere, the oceans, and sediments. The Industrial Revolution began a rapidly accelerating redistribution of carbon from a sedimentary reservoir (in the form of fossil fuels such as coal, natural gas, and petroleum) into the atmosphere and oceans. Land use changes, particularly deforestation and agricultural practices, have liberated carbon from soils and plants into the atmosphere, contributing roughly one quarter of the carbon humans have added. The result of these combined activities has been an accelerating, dramatic increase in atmospheric greenhouse gas concentrations (Fig. 12.1) and significant increases in oceanic carbon concentrations.

These trends are expected to continue and accelerate, resulting in projected temperature rise in the range of 1.1–6.4°C on average globally through the end of the century (Figs. 12.2 and 12.3). Variability in estimates arises from differences between models and among emissions scenarios, some areas, particularly northern latitudes, are expected to experience a much more dramatic rise (Solomon et al. 2007). With increased temperatures, there will be greater heat energy in weather systems, resulting in more water evaporation into the atmosphere and stronger storms. Less obvious is the effect of droughts that are predicted to be both more frequent and severe. Finally, the combination of thermal expansion of sea water and melting of land-based ice is driving sea level rise (IPCC 2007a) with recent projections predicting rise between 0.8 and 2.0 m by 2100 (Pfeffer et al. 2008).

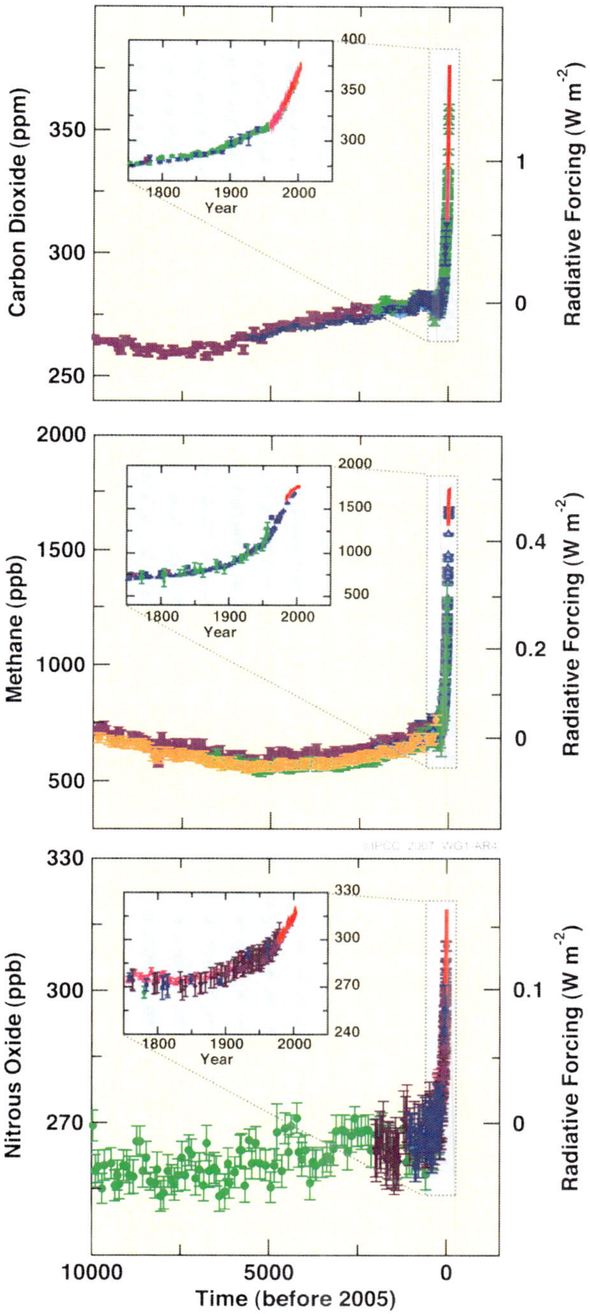

Fig. 12.1 Atmospheric concentrations of carbon dioxide, methane and nitrous oxide over the last 10,000 years (*large panels*) and since 1750 (*inset panels*). Measurements are shown from *ice cores* (symbols with different colours for different studies) and atmospheric samples (*red lines*). The corresponding radiative forcings are shown on the *right hand axes* of the *large panels* (From IPCC 2007a)

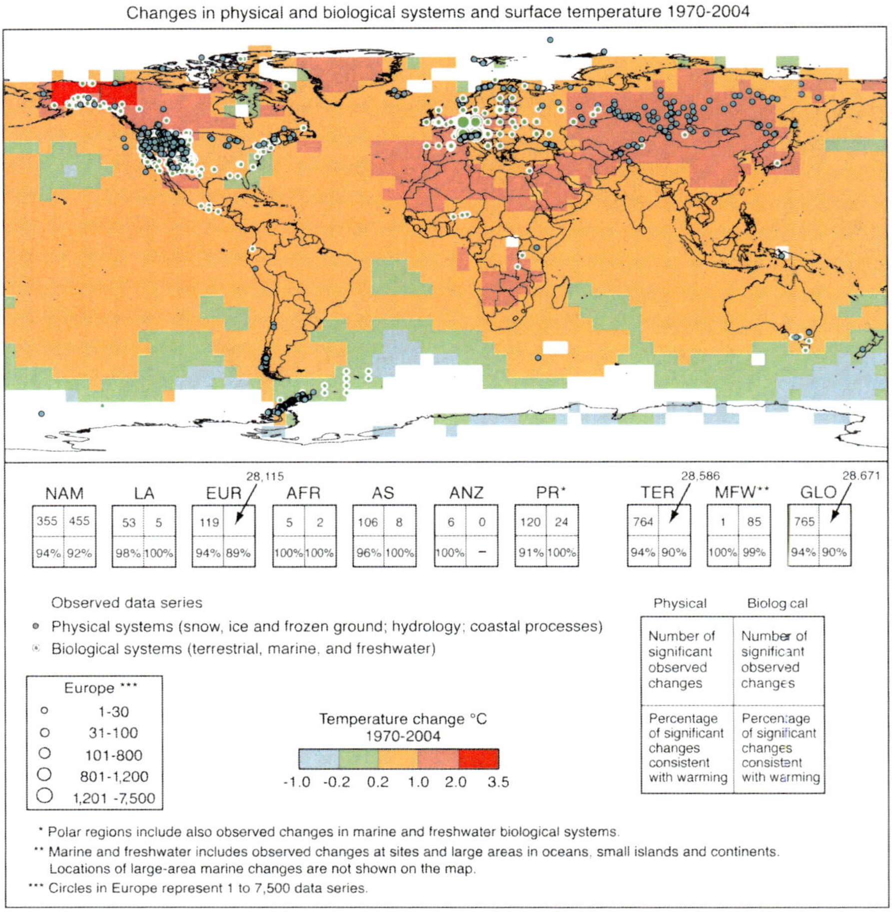

Fig. 12.2 Evidence of changes in physical and biological systems consistent with increased warming, increased precipitation, and sea level rise, overlaid with data on global temperature changes (From IPCC 2007a)

Ultimately, many of these physical changes will translate into adverse outcomes for human health, both direct and indirect. The direct effects result from hazardous exposures such as increased temperatures, increasingly frequent natural disasters, and worsening air quality. Indirect effects are the product of more complex causal pathways and include changes in the range of vector born diseases, reduced access to food and water, and large-scale population displacement. Collectively, these health impacts have become a growing cause for concern. In 2007, the WHO Director General identified climate change as one of the world's most urgent public health priorities (Chan 2007). A critical understanding of these ecological processes is essential to mitigating the impacts of climate change on human health.

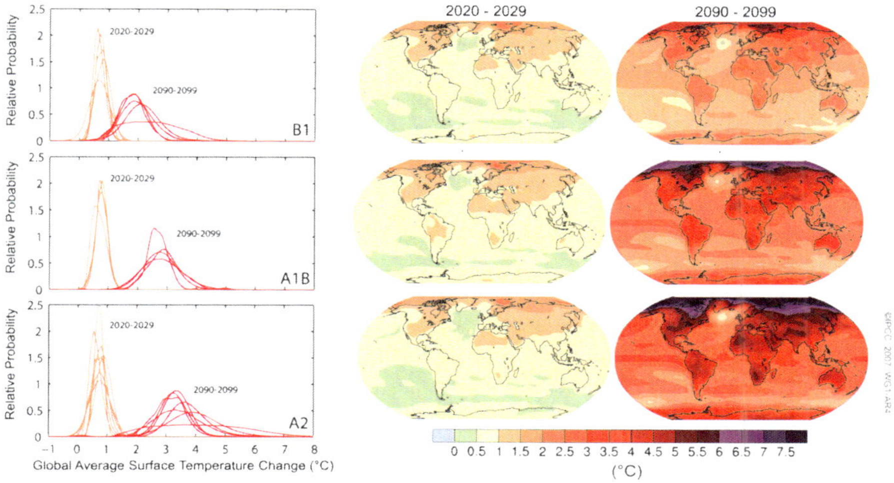

Fig. 12.3 Projected changes in temperature for the periods 2020–2029 and 2090–2099, compared with 1990–1999 (*left side*). Projections of global surface temperature rise for several models and graphic projection for 2020–2029 and 2090–2099 compared with 1990–1999 (*right side*) (From IPCC 2007a)

Health Effects of Climate Change

The health effects of climate change are "location specific and path dependent" (Yohe and Tol 2002), and certain regions are likely to be particularly affected (Patz and Kovats 2007; Hess et al. 2008). The physical changes in climate will vary by region, and particular populations have specific health burdens, vulnerabilities, and adaptive capacities. Climate change will thus have different impacts in the Arctic than in sub-Saharan Africa. Paradoxically, the health hazards from climate change will most dramatically affect those least responsible: the developed world will be relatively insulated, though by no means immune, and the developing world will likely suffer devastating effects (Fig. 12.4a–d) (Patz et al. 2007).

Climate change will impact human health through several parallel mechanisms (Fig. 12.5, Table 12.1). Several direct exposures associated with climate change have known human health effects, including increased temperatures, reduced air quality, and more frequent natural disasters. However, a number of indirect exposures will also have significant health impacts (Table 12.1). Indeed, the greatest burdens of disease may be associated with the more complex, indirect effects. Several of the most important exposures are further discussed below, with particular attention to the ways in which local environments enhance or dissipate hazardous exposures associated with climate change.

a

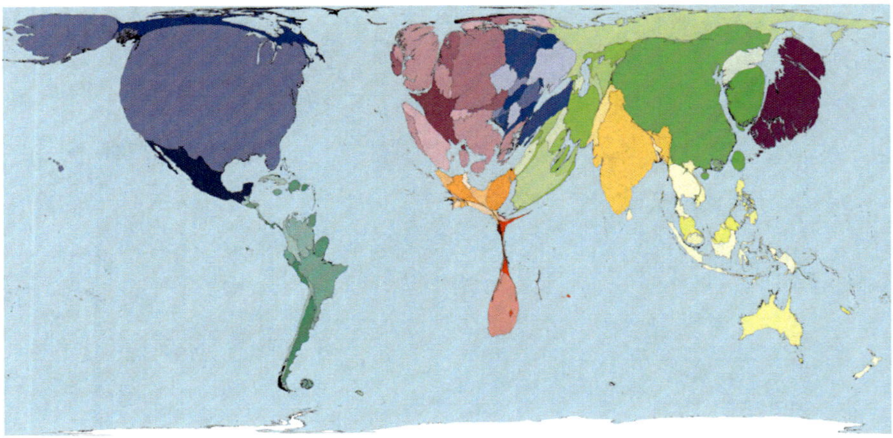

b

Fig. 12.4 (**a**–**d**) Maps that distort the area of a country in proportion to distinct metrics utilized, in this case: (**a**) CO_2 emissions by country as a proportion of the total for the year 2000. (**b**–**d**) The proportion of total global population killed by extreme weather related disasters from 1975 to 2000 for the following hazards: (**b**) drought-related disasters, estimated total of 560,000 deaths, the majority of which were in Ethiopia, Mozambique, and Sudan; (**c**) floods, estimated total of 170,000 deaths; and (**d**) storms, estimated deaths 276,000 (Available at http://www.worldmapper.org, accessed 15 December 2008. © Copyright 2006 SASI Group (University of Sheffield) and Mark Newman (University of Michigan))

c

d

Fig. 12.4 (continued)

Heat

Human beings, and other species, have evolved to live within the historic temperature range they have previously experienced on evolutionary time scales. People living in hot climates are less sensitive to high temperatures than those in more temperate regions, but regardless of the average ambient temperatures, steep increases above the norm will result in heat stress. Climate change will have two impacts on the distribution of extreme heat by both increasing average temperature, and its variability (Fig. 12.6). Combined, these shifts will result in a large increase in extreme heat events with several impacts on human health. Hyperthermia from environmental

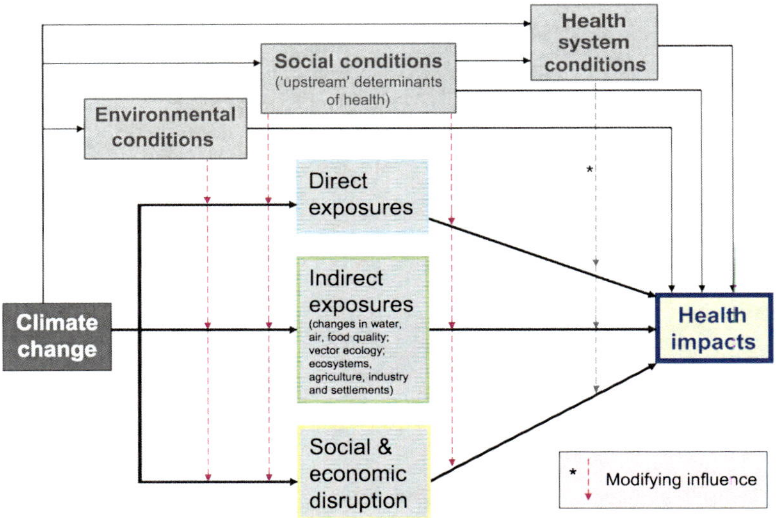

Fig. 12.5 Schematic diagram of pathways by which climate change affects health, and concurrent direct-acting and modifying (conditioning) influences of environmental, social, and health-system factors (From IPCC 2007b)

exposure causes a spectrum of illness ranging from relatively benign heat rash, exhaustion, or heat syncope (fainting), to heat stroke, which is often fatal. For those who do not die from heat stroke, there is a significant increase in illness and mortality in the years after the extreme heat event (Argaud et al. 2007; Wallace et al. 2007). Deaths from heat are the leading weather-related cause of mortality in the USA, despite the existence of high heat warning systems, and are likely significantly underestimated as a consequence of how heat-related deaths are reported (CDC 2003; Luber and McGeehin 2008). Extreme heat events are associated with increased hospital admissions for cardiovascular disease and increases in overall mortality, though certain risk groups experience particularly high risks, as outlined in Table 12.1 (Kovats et al. 2008).

In addition to these risk factors, several socioecological components affect heat exposure and health outcomes. The built environment can concentrate or dissipate heat burden markedly, and neighborhood microclimates have been shown to impact residents' heat burden (Harlan et al. 2006). On a larger scale, the urban heat island effect intensifies heat exposure for residents in the center of a city, both during daytime and overnight, as the city center retains heat that would otherwise be dissipated (Clarke 1972). Urban sprawl is also significantly correlated with an increased incidence of extreme heat events (Stone et al. 2010). Green roofs comprised of living vegetation, and urban forests can make significant contributions to reducing this heat island effect through shading and transpiration (in the same way that the human body utilizes sweating). These can be particularly effective when combined with technological solutions such as high-albedo (highly reflective) roofing, compact urban design, and other green strategies that reduce heat exposure (Younger et al. 2008).

Table 12.1 Major health effects of climate change

	Exposure	Health outcomes	Mechanism(s)	Majorrisk groups	Major affected regions
Primarily direct	Heat	Heat exhaustion, heat stroke (direct); exacerbations of chronic cardiovascular and pulmonary disease (indirect)	Dehydration, direct tissue injury, cardiovascular strain	Extremes of age, athletes, outdoor workers, socially isolated, people with certain psychiatric and chronic diseases	Urban areas, temperate northern-latitude cities
	Extreme precipitation	Morbidity and mortality from injuries (direct); waterborne disease (direct); depression and suicide (indirect); exacerbations of chronic disease (indirect)	Direct tissue injury from flying debris, drowning, injury during preparation and cleanup (direct); separation from social and medical support, economic loss, loss of place (indirect)	Island dwellers, floodplain residents, lower socioeconomic status, patients with mental health and chronic medical problems	Islands, floodplains, river delta megacities
	Drought	Water scarcity, increased incidence of water-washed diseases, morbidity and mortality from armed conflict over arable land and water	Changes in hydrologic cycle lead to increased drought in certain areas; water scarcity compromises hygiene and increases armed conflict	Children, marginalized populations; sub-Saharan Africans	Sub-Saharan Africa
	Sea level rise	Injuries (see extreme precipitation); water stress; malnutrition; displacement; depression	Drowning; water table and soil salinization; loss of place, infrastructure damage	Island dwellers, coastal residents, low socioeconomic status	Islands, coasts, river deltas
	Increased ozone	Exacerbations of asthma, chronic obstructive pulmonary disease (COPD)	Increased temperature catalyzes ozone formation, causes reversible lung injury and airway irritation	Asthmatics, COPD patients, children, people with chronic disease, smokers	Regulatory nonattainment areas; developing country megacity residents

(continued)

Table 12.1 (continued)

	Exposure	Health outcomes	Mechanism(s)	Majorrisk groups	Major affected regions
Primarily indirect	Increased particulates	Exacerbations of asthma and COPD; cardiovascular dysrhythmias; deep venous thrombus formation	Airway irritation; cardiac irritation; abnormal activation of the blood's clotting cascade	Asthmatics, COPD patients, patients with cardiovascular disease and blood clotting disorders	Regulatory nonattainment areas, developing country megacity residents
	Vector-borne disease	Depends on disease; likely increased incidence and shifts in the range of several vector-borne diseases, including malaria, dengue, and Lyme	Ecosystem disruption and shifting alters vector distribution and predator-prey relationships; precipitation changes allow for epizootics resulting in human epidemics	Depends on disease of concern. For malaria, sub-Saharan Africans, those with lower socioeconomic status	Depends on the disease of concern. Higher altitude areas for malaria.
	Water-borne disease	Increased incidence of bacterial and parasitic disease; increased incidence of water-borne disease outbreaks	Waterborne disease prevalence increases with temperature; mechanisms for increased outbreaks unclear	Extremes of age; developing country residents; well-water users	Developing countries; areas with endemic and epidemic cholera; areas with combined sewer overflows
	Forest fires	Respiratory and mucus membrane irritation; displacement and associated mental and physical health effects	Increasingly frequent and severe droughts, earlier spring and snowmelt, later winter, land use changes lead to increased forest fire fuel and human exposures	People with respiratory and cardiovascular disease	Arid regions including the western and southwestern USA
	Malnutrition	Protein-energy malnutrition, stunting, micronutrient deficiencies	Changes in CO_2 and temperature change yields of staple grains; change in rainfall and soil quality compromise yields in marginal areas; increasingly variable weather threatens crops	Children, subsistence farmers, sub-Saharan Africans, Pacific island residents	Sub-Saharan Africa, Pacific Islands, developing countries
	Population displacement	Injuries, sexual violence, waterborne disease, malnutrition, depression, exacerbations of chronic disease	Disasters, drought, conflict over scarce resources	Children, elderly, chronically ill, sub-Saharan Africans, coastal residents, island residents	Sub-Saharan Africa, Southern and Southeast Asia

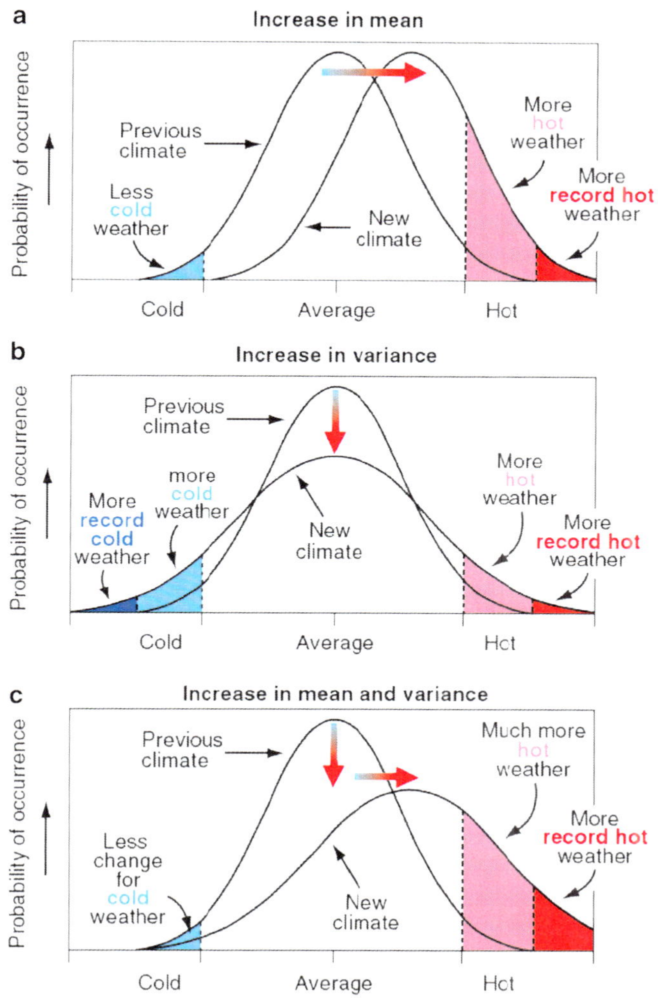

Fig. 12.6 Changes in the mean and variance of extreme heat events (From IPCC 2007a)

Disasters

The Fourth Assessment of the Intergovernmental Panel on Climate Change (IPCC) expressed high confidence that a warming of up to 2°C above 1,900–2,000 levels (on the lower end of most projections for 2,100) would increase the risk of many extreme events, including severe tropical cyclones, floods, droughts, heat waves, and fires (Schneider et al. 2007). In addition to increasing the frequency of natural disasters, climate change amplifies their impact in two other important ways: by

Table 12.2 Hydrometeorological disasters and regions at greatest risk

Disaster type	Socioeconomic risk factors	Population at risk globally	Deaths 1980–2000
Tropical cyclone	Low human development indices, large proportions of arable land, and high exposure (large populations in coastal plans; Small Island Developing States)	119,000,000	251,000
Flood	Low GDP, low population density, and high levels of flood exposure (populations in river floodplains and coastal areas)	196,000,000	170,000
Drought	Low rates of improved water supply access, high exposure, and in sub-Saharan Africa	220,000,000	833,000

Data source: UNDP 2004

increasing the intensity of storms, particularly tropical cyclones, and increasing sea level. Coastal areas are much more densely settled than other parts of the planet. More than a third of the human population lives within 100 km of the shore and less than 50 m above sea level (McGranahan et al. 2007; Barbier et al. 2008). Human settlement has been associated with degradation of coastal barriers – wetlands, vegetated dunes, mangrove forests, and coral reefs – which buffer coasts from extreme storms.

Hurricane Katrina, a category five hurricane that developed as a result of very warm sea surface temperatures, provides a cautionary tale. Though Katrina was downgraded to a category three by the time it hit the Mississippi Delta, years of development had degraded natural coastal buffers including wetlands and barrier islands, and the storm proved to be the most devastating natural disasters in US history (U.S. House of Representatives 2006). Scientific and anecdotal evidence both suggest that the preservation of natural coastal ecosystems, including mangroves, wetlands and coastal forests could have greatly reduced the impact of Hurricane Katrina. Kareiva and Marvier (2007) provide one of the best case studies of the use of ecological restoration to develop a multifunctional landscape. They use the Florida panhandle as an example to suggest that the combination of flood control management and conservation planning can lead to a win-win situation, reducing the vulnerability of poor coastal communities to the expected increases in coastal storms. By identifying areas containing rare and threatened biodiversity (of interest to conservationists), plus areas prone to flooding (of interests to engineers), and finally, the location of poor communities (of interest to development and health practitioners) and overlaying these three maps, we can quickly identify specific locations where conservation efforts would not only protect biodiversity, but would also protect poor communities from the impacts of flooding driven by severe storm events (see Chapters 19, 20 and 22, this volume, for further discussion on these issues).

The impacts of natural disasters are not limited to storm events or injuries however. There are other significant longer-term health effects as well. Severe storms can result in pollution or biological contamination of water supplies in addition to the physical damage to human life. Flooding due to these storms can increase the prevalence of molds and mildews with associated increases in respiratory ailments. Increased drought associated with climate change may increase the intensity, frequency and duration of the fire season in temperate environments as evidenced by the numerous, catastrophic urban fires of 2008–2009 in California, Australia, and the Canary Islands. As a result of these fires, air quality may suffer, again, increasing the incidence of respiratory ailments. Such particulates have been associated with increased morbidity and mortality from respiratory disease (Schwartz 1994; Johnston et al. 2002; Kunzli et al. 2006; CDC 2008). Dust storms associated with droughts in the southwestern USA, coupled with increased soil disruption from building activities, are associated with outbreaks of coccidioidomycosis, or Valley Fever, a fungal pneumonia prevalent in the southwest (Williams et al. 1979) that has increased in incidence in recent years (CDC 2009). The overall increased burden of disease from these harmful exposures has not yet been quantified, but is likely to be substantial (Kinney 2008). Ecological knowledge of the interaction between fuel loading, vegetation composition, structure and dynamics, particularly in the urban interface, is essential to preventing, reducing, and mitigating the impacts of fire on human health.

Air Contaminants

Climate change is expected to reduce air quality, both by increasing ground level ozone and increasing aeroallergen exposure. Ground level ozone is strongly associated with increased morbidity and mortality from cardiorespiratory disease (Eilers and Groot 1997). Ozone formation from primary pollutants such as nitrous oxide and other hydrocarbon combustion byproducts is highly temperature sensitive, particularly above 32°C (90°F). Modeling studies project increased ground level ozone concentrations from higher temperatures, with consequent increases in respiratory morbidity and mortality from diseases such as asthma and chronic obstructive pulmonary disease (COPD) (Knowlton et al. 2004; Bell et al. 2007). Urban forests can likewise make significant contributions to improving air quality, both by absorbing and intercepting airborne pollutants. In a study by the University of Florida Extension Service, the Gainsville urban forest was estimated to have removed more than 200 tons of ground level ozone in one year. They estimate that in 2000, trees removed a total of 360 tons of atmospheric pollutants, equal to a $2 million saving in health care benefits. Similarly, a 1994 study determined that trees in New York City removed an estimated 1,821 metric tons of air pollution at an estimated value to society of $9.5 million.

Warmer temperatures and higher CO_2 concentrations are associated with longer pollen seasons and increased pollen production for many allergenic plants. This trend is likely to cause additional allergic respiratory disease, particularly asthma. Worldwide, 300 million people suffer from asthma, and there are a quarter of a million deaths annually (World Health Organization 2008b). While this may conflict with our suggestion that urban forests would improve air quality (removing pollutants, but adding pollen), combined ecological and medical knowledge can be used to select species that have low pollen loads or non-allergenic pollens. For example, the female individuals of dioecious species (those with female flowers only) would make no contribution to pollen loads while contributing to reducing air pollution. This is a simple example of how ecological knowledge of plant traits, can be manipulated to optimize desired ecosystem services while minimizing negative impacts of biodiversity.

Malnutrition

Adequate nutrition – protein, calories, and micronutrients – is vital to preventing and fighting infectious disease, cognitive development and learning, metabolic and endocrine functioning, reproductive health, and overall vigor. It has been estimated that at least one-third of the burden of disease in poor countries is due to malnutrition (Mason et al. 2003), and roughly 16% of the global burden of disease is attributable to childhood malnutrition (Murray and Lopez 1997). In 2009, over 1 billion people suffered from chronic hunger (FAO 2009).

As the human population grows by roughly another 2.5 billion people by 2050, and people in the developing world with greater prosperity strive to add more meat to their diets, global agricultural production will need to roughly double over the next 50 years to keep up (Alexandratos 1999). One of the central public health questions of this century is whether we can meet the overall caloric requirements of this burgeoning global population, or if ecological constraints will stymie adequate production (Myers and Patz 2009). Meeting this caloric requirement is even more dubious if we expect to be able to sustainably increase food *and* biofuel production on existing arable land, as most estimates suggest that all available arable land is already under cultivation with only marginal land remaining.

Climate change will impact agricultural production in a variety of ways depending on location and approaches to adaptation. Increased concentrations of CO_2 may create a fertilization effect that can increase plant growth (Tubiello et al. 2007); however, it may also reduce both the protein and micronutrient concentration of the most important grain crops (Penuelas and Matamala 1993; Loladze 2002; Lieffering et al. 2004; Högy and Fangmeier 2008; Taub et al. 2008). Warming may increase the amount of available arable land through the thawing of the massive boreal and tundra regions in Canada and Russia, but may also make portions of the tropics, especially the Sahelian belt, unarable (Battisti and Naylor 2009). It has been widely estimated across a variety of different grains and in different regions that a 1°C rise in temperature corresponds to roughly a 10% decline in agricultural yields (Lobell et al. 2008). This temperature effect, alone, is likely to have very significant impacts on global food production.

The impacts of extreme weather and sea level rise may include direct destruction of crops as well as inundation of coastal lands with salt water and net arable land loss, particularly in low-lying areas supporting large populations and in small island states (Patz et al. 2005; Ebi et al. 2006; Nicholls and Tol 2006). For example, Cyclone Nargis hit the Irrawaddy delta in Burma in 2008 and destroyed 20% of the country's rice crop (FAO, 2009). Some have suggested that deforestation may have amplified the effects of this cyclone, and though the ecological consensus is still out on the impact of forested landscapes on mitigating extreme events, there is evidence that these landscapes can mitigate the effects of "normal" events. As an aside, Nargis also provided a lesson in political economy and disease, illustrating the disastrous consequences that can befall the citizens subject to an isolationist totalitarian regime: the storm resulted in over 100,000 deaths, many of which could have been prevented with appropriate early warning systems, evacuation, and government intervention, had the government of Myanmar not ignored warnings and mounted an appropriate response (Webster 2008).

In addition, there will be important impacts of climate change on agriculture mediated through increased water scarcity, changes in exposure to pests and pathogens, and changes in ozone levels. Through a variety of mechanisms, climate change is expected to constrain the quantity and timing of water flow for irrigation. It is also expected to elevate ground-level ozone concentrations which are not only a potent cardiorespiratory toxin, but also a plant toxin with strong impacts on crop yields. The impacts of climate change on agricultural pests and pathogens are not well understood, but already shifts are being observed in the distribution of pests and pathogens and their control by natural predators as a result of climate signals. The pine bark beetle is one example.

From these numerous different mechanisms, it becomes clear that global agriculture is sensitive to a vast array of environmental conditions, many of them in flux as a result of climate change. It will take sophisticated ecological understanding to address these changes in ways that maximize the resilience of crop systems to ongoing environmental change while maximizing the benefits to human nutrition. These types of interventions will be the basis for the rapidly growing movement toward sustainable agriculture which has the potential to simultaneously improve health, biodiversity conservation, and carbon sequestration while also improving livelihoods for many of the world's poor.

Vectorborne and Zoonotic Disease

Vulnerability to environmentally sensitive infectious diseases results from a combination of land use, climate, and sociocultural factors that are complex and interrelated. Alterations in climatic conditions are likely to change the distribution of many vector-borne and zoonotic diseases. With roughly 500 million cases and 1 million fatalities each year, malaria is possibly the vector-borne disease of greatest concern to global health and there is an ongoing debate as to the ultimate impact of climate change on malaria incidence (Rogers and Randolph 2000; Reiter 2001; Ostfeld in press). In theory there are several ecological reasons that malaria's incidence may increase: warmer temperatures decrease the time required for the parasite to reproduce and increase the biting frequency of mosquitoes that serve as malaria vectors. Warmer temperatures are likely to extend vector ranges into higher elevations and latitude, and more extreme precipitation events and flooding may also favor disease outbreaks. The ecological understanding of how organisms move at landscape scales in response to climate conditions, and how organisms perceive habitat are all critical components to understanding the spread of these vector borne diseases. Of particular concern is the extent to which epidemic malaria, or malaria outbreaks in populations with little ongoing immunity, will increase. Current consensus holds that epidemic malaria is likely to increase its range in sub-Saharan Africa, particularly in higher elevations, though malaria's range will also contract in other regions. In some regions with endemic malaria, the transmission seasons may lengthen, increasing the overall disease burden. Similar changes in seasonality associated with altered climate have been observed for a variety of other infectious diseases.

Although many variables in addition to the vector species' ecological niches (temperature, rainfall, humidity, predator-prey relationships, resource concentrations, land use changes, population concentration, and reproductive requirements) affect malaria's prevalence (Reiter 2001), there are several well-documented instances where malaria's incidence is significantly dependent on climate. In the highlands of East Africa, a warming trend from 1950 to 2002 coincided with increases in malaria incidence (Pascual et al. 2006). This relationship does not appear to be linear: just a half degree centigrade increase in temperature translates to a 30–100% increase in mosquito abundance allowing successful breeding and survival of the vector (Patz et al. 2008b). In the Punjab region of India, malaria epidemics are strongly associated with precipitation where incidence increases approximately fivefold following an El Nino event when monsoons are particularly extreme (Bouma and Van der Kay 1996). Similar associations have been shown between malaria outbreaks and El Nino related climate variability in Botswana (Thomson et al. 2006).

The relationships are not limited to malaria, clear climate associations have been established for cutaneous leishmaniasis (Chaves and Pascual 2006), cholera (Koelle et al. 2005), plague (Stenseth et al. 2006; SnÃ¤ll et al. 2008), and particularly for dengue fever (Cazelles et al. 2005). A clearer understanding of the ecological dimensions of vector-borne diseases that elucidates general trends, particularly the interaction between climate change, land use change, and the disease ecology of both vectors and hosts is critically needed (also see Chapters 11 and 13, this volume).

Water-Related Disease

We depend on water for drinking, sanitation, hygiene, and food preparation. People need roughly 50 l of uncontaminated fresh water per day to meet these needs

(Gleick 1996). Inadequate access to water, sanitation, and hygiene is estimated to cause 1.7 million deaths annually and the loss of at least 50 million healthy life years (Vorosmarty et al. 2005). Half of the urban population of Africa, Asia, and Latin America and the Caribbean suffers from one or more diseases associated with inadequate water and sanitation (Vorosmarty et al. 2005).

Water is already scarce, and getting scarcer. Richard Leakey, the chairman of Wildlife Direct, effectively argues that the real threat of climate change is not in changing temperature, but in changing water regimes (personal communication). Roughly 40–50% of renewable, accessible fresh water supplies are already being used (Postel et al. 1996; Vorosmarty et al. 2005). From 1960 to the present, water use relative to accessible supplies has increased between 15% and 30% per decade. In many parts of the world, water is being mined unsustainably from fossil aquifers or withdrawn faster than the rates of replenishment. In the Middle East and North Africa, for example, current rates of fresh water use are equivalent to 115% of total renewable runoff (Vorosmarty et al. 2005).

In addition to water scarcity, changes in extreme precipitation and sea surface temperatures are also expected to increase the incidence of water-related diseases. In the IPCC Fourth Assessment Report, four distinct concerns were articulated: (1) an increase in diarrheal disease as a result of decreased water availability and unimproved water supplies; (2) outbreaks of waterborne disease associated with extreme precipitation events and piped water supplies; (3) amplification of microbial contamination of coastal and recreational waters as a result of increased temperatures and rainfall; and (4) direct effects of increased temperatures on the virulence of waterborne disease pathogens (Confalonieri et al. 2007). Each is discussed in turn below.

Diarrheal disease already causes a significant burden of disease worldwide, with 1.81 million deaths in 2004, the last year for which global statistics are available (World Health Organization 2008a). The vast majority of these deaths is in the developing world, and most are in children under age 5. General studies linking diarrheal disease with various environmental exposures project a 2–5% increase in the global burden of diarrheal disease by 2020 from climate change (Campbell-Lendrum et al. 2003; McMichael 2004).

In countries with improved water supplies, waterborne disease outbreaks related to distribution systems are likely to increase as a result of heavy precipitation events made more frequent as a result of climate change. A study of all-cause waterborne disease outbreaks in the USA found a strong association with heavy precipitation where two thirds of outbreaks followed exceptionally heavy rainfall months (Curriero et al. 2001). While these data need to be confirmed with additional studies, a review of significant outbreaks implicates extreme precipitation as a primary cause of these outbreaks.

Sea surface warming can also drive increased exposure to water-related disease. A surge in the number of harmful algal blooms (HABs) has resulted from the combination of rising sea surface temperatures and increased application and runoff of fertilizers that cause nutrient enrichment of fresh water and coastal systems (Hoagland et al. 2002). HABs can lead to massive fish kills, shellfish poisonings, disease and death of marine mammals, and human morbidity and mortality. Health impacts range from acute neurotoxic disorders and death to subacute and chronic disease. Worldwide, roughly 60,000 individual cases and clusters of human intoxication occur annually (Van Dolah et al. 2001).

Sea surface temperature, along with changes in rainfall patterns and nutrient loading from agricultural runoff, has also been associated with cholera outbreaks. In 2006, the last year for which global figures are available, there were 131,943 cases of cholera with 2,272 deaths (World Health Organization 2006). Cholera outbreaks in Asia and South America have been associated with increased sea surface temperature, rainfall patterns and nutrient loading from agricultural runoff. High nutrient loads and warm water temperatures cause blooms of *Vibrio cholerae* zooplankton and can lead to the transformation of *V. cholerae* from a quiescent to a virulent form (Colwell 1996; Ezzell 1999; Colwell and Huq 2001; Koelle et al. 2005).

Increased water temperature is also associated with increased disease rates from certain pathogens known to cause intestinal disease, particularly bacterial pathogens. Food-borne diseases show a similar pattern (Bentham and Langford 1995; Kovats et al. 2004). Associated increases may be partially offset by declines in certain viral pathogens with higher virulence at colder temperatures, but overall the burden is likely to increase (Ebi et al. 2008). Specific projections of disease burdens linked with climate projections are awaiting further information regarding the climate sensitivity of specific pathogens.

As with the impacts of climate change on air, food, natural disasters, and infectious disease, the impacts of climate change on water-related disease will be mediated by other types of environmental change as well as a host of sociocultural factors that determine vulnerability. The interconnectedness of these factors serves to underline the critical importance of using an ecological lens in trying to understand and address these types of impacts. One of the most successful ecologically based interventions for reducing the impacts of land use change on these HABs is the construction and restoration of riparian buffers along streams and rivers in agricultural landscapes. Riparian buffers simply consist of strips of vegetation that serve as an ecotone between agricultural fields and waterways. Such buffers have been demonstrated to intercept and store excess nutrients from agro-chemical application, preventing it from entering the waterway. While these systems are quite useful during "normal" precipitations events, additional research is needed to understand how they perform during "extreme" precipitation events. Ecological tools combined with environmental engineering may be useful in creating natural infrastructure, rather than built infrastructure, capable of mitigating the combined impacts of climate and landuse change on HABs. Of particular interest is the intersection of landscape ecology, which focuses on the spatial relationships between natural and semi-natural elements of a landscape, and community ecology, which focuses on the interactions between populations of species. Much of the focus of biodiversity and ecosystem function research has demonstrated that diversified systems have greater resource use efficiency (Tilman et al. 2005), however this finding has yet to be applied within the context of riparian ecosystems at the farm scale, and the impacts on HAB's at landscape scales.

Mass Population Movements

Violent conflict over scarce resources and population displacement may represent final common pathways as large, vulnerable populations suffer amplified exposure to water scarcity, hunger, and natural disasters. Sea level rise and more extreme storms will make some low-lying coastal areas untenable for habitation. Research reveals that, although coastal areas less than ten meters above sea level only represent 2% of the world's land area, they house 10% of the world's population (McGranahan et al. 2007). Local scarcities of food and water may drive populations out of resource poor regions. These forces may drive hundreds of millions of people, with few resources and many needs, to seek new homes (Sachs 2007). The UN High Commission for Refugees estimates that between 250 million and one billion people would be displaced by climate change alone between 2008 and 2050 (Johnstone 2008). The impacts of migration on the environment and human well-being are discussed further in the chapter on population mobility by Adamo and Curran in Volume 2.

Displacement on such a large scale can have dramatic health impacts. Non-immune populations migrating into endemic areas are more susceptible to a variety of infectious diseases (Molyneux 1997). Poor housing, sanitation, and waste management infrastructure combined with inadequate drinking water and poor nutrition lead to epidemics of infectious disease, particularly diarrheal diseases, measles, and acute respiratory infections. Protein malnutrition increases mortality from these communicable diseases and contributes independently to morbidity and mortality. Prevalence rates of acute malnutrition have reached up to 50% in refugee populations in Africa (Toole and Waldman 1993). In addition, people displaced by conflict and disasters suffer high levels of violence, sexual abuse, and mental illness. One study found symptoms and signs of post traumatic stress disorder (PTSD) in 30–75% of resettled refugee children and adolescents (McCloskey and Southwick 1996). Overall, crude mortality rates as high as 30 times baseline are not unusual following an acute movement of refugees, with much of the mortality occurring in children under the age of 5 (Toole and Waldman 1997).

The Role of Ecology

With climate change humans have become the world's predominant ecosystem managers. Recognizing this role, and that most of the important impacts discussed above are by ecological processes, efforts to reduce vulnerability to health impacts of climate change will rely on bolstering ecological systems and the services that they provide. We will need to develop agricultural systems that are more heat, salt, and drought tolerant if we are going to improve grain yields at the rate required to forestall widespread malnutrition. We will need to improve the resilience of coastal zones to storm surge and flooding by strengthening mangrove forests, coral reef systems, vegetated dunes, and wetlands (Kareiva and Marvier 2007). Ecologists may provide important breakthroughs in developing and managing new wetland systems that can provide water filtration services under altered environmental conditions.

Some have argued that natural systems can be considered as natural infrastructure, replacing built infrastructure such as dams and levees with the added advantage that natural systems, by their nature, are adaptable, self-repairing, and can be designed to be multifunctional. These issues are discussed further in the chapters by Rumbaitis del Rio, this volume, and Ingram and Khazai, this volume.

Controlling the spread of infectious disease will require strong ecological understanding to avoid the mistakes of the past when new agricultural and development projects altered habitat for disease vectors and hosts and led to epidemics of vector-borne disease. Fundamentally, climate change threatens ecosystem functions which in turn underpin humanity's health and wellbeing. Ecologists will need to play a central role in developing new approaches to support, strengthen, or recreate those ecosystem functions that are being degraded by anthropogenic changes.

Climate Change, Public Health, and Sustainability

The regions of the world most vulnerable to climate change are those with the fewest resources to adapt. Wealthy nations may be able to buy their way out of constraints on food and water and create infrastructural and technological solutions to many of the hazards we have described. Those in the least developed countries, in contrast, will suffer disproportionately, a fact that is painfully ironic given that they have contributed the least to changing climate. Climate change thus brings two concerns into sharp relief: how to protect human health in the face of increasingly frequent and severe hazardous exposures, and how to shape future development so that it mitigates the dangerous impacts of climate change while preserving its associated health gains.

These two concerns are related but distinct (McMichael 2006). While there are effective public health interventions for many hazardous exposures associated with climate change, these interventions are often expensive and require significant built infrastructure. An effective response is less a matter of developing new interventions than of vigorous, widespread implementation of existing interventions with greater interdisciplinary attention to synergies and unintended consequences. For instance, ecologically based interventions to facilitate adaptation to extreme heat events, such as green roofs and urban forests, can be combined with social and technological interventions including social network development such as "buddy programs" which pair neighbors together to check on one another during a heat wave, central air conditioning, and cooling centers to significantly reduce morbidity and mortality among vulnerable populations. Green roofs have been used to dramatically reduce building temperatures during heat waves, as have urban forests. For instance, temperature measurements of the green roof on Chicago's City Hall, and the adjacent unvegetated roof show a temperature difference of more than 10°F (City of Chicago Climate Action Plan 2008). In such cases the ecologically based interventions serve to increase the resistance of the system by increasing the threshold for extreme heat events, whereas the social interventions are essential for when the threshold is crossed.

There are effective interventions for other extreme weather events, including floods and severe storms, ranging from appropriate land-use policies that preserve natural barriers to storm surge to physical flood control measures to early warning systems. In areas of the developed world where such strategies have been implemented, severe storms tend to cause significant infrastructure damage and property loss, but relatively little direct morbidity and mortality. In the developing world, where resources are scarce and infrastructure less extensive, morbidity and mortality from hydrometeorological disasters is much more extreme. Adaptation to climate change requires a global investment in risk management.

A novel approach to human and economic development is also required. Climate change is one of several major disruptions of global ecosystems, including extensive land use change, marine degradation, ozone depletion, biosphere nitrification, and biodiversity loss, occurring as a result of the remarkable growth of the past two centuries. None of these major disruptions can be considered in isolation however, as they are all interacting. Collectively, these disruptions are overwhelming the production of ecosystem services (Raven 2002). In past decades, environmentalists and others have warned about resource depletion, degradation and the dire ecological consequences of such overindulgence. Of increasing importance is the recognition that ecosystem health, and human health in the broadest sense, are inextricably linked. Healthy ecosystems, capable of providing critical ecosystem services should not be taken for granted, nor should the stability of these systems be overestimated in the face of climate change. It is becoming increasingly clear that human health and survival are the ultimate "bottom line" concern (McMichael 2006). McMichael argues that we need to apply a public health lens to sectors that have traditionally been well removed from public health:

> Much discussion about sustainability treats the economy, livelihoods, environmental conditions, our cities and infrastructure, and social relations as if they were ends in themselves; as if they are the reason we seek sustainability. Yet their prime value is as the foundations upon which our longer-term health and survival depend.

Economic and human development are, ultimately, expressions of health, and thus development activities should be explicitly structured with health in mind. Conversely, public health must extend its reach conceptually and practically to include other sectors involved in human and economic development. Public health practitioners must play an active role in traditionally non-health related sectors. The ecological disruptions associated with these sectors – energy production, transportation, urban planning, agriculture, natural resource management, etc. – have become so globally profound and pervasive, it is now clear that they are driving some of the world's greatest public health threats.

The process of incorporating health outcomes into human and economic development activities and applying public health science to processes in other sectors is in its infancy. The epidemiological study of global environmental change serves as an early example of this type of work, though the field is still young. The results of viewing climate change through an ecosystem lens were evident in the preceding sections on the health effects. The following sections outline initial steps toward incorporating health outcomes into human and economic development.

Pursuing Resilience in Development Activities

The immediate and proximate health impacts of most large development projects are already evaluated prior to implementation. In addition to this routine evaluation, however, future development activities will need to consider more remote and indirect health impacts as well as co-benefits to maximize community resilience to dangerous climate change. These indirect impacts are often mediated by complex ecological processes and their full evaluation will require input from ecologists. Co-benefits are indirect consequences of climate mitigation and adaptation strategies, many of which result in improved health (or reductions in harmful exposures). Decreased reliance on motorized transport, for instance, not only reduces greenhouse gas emissions but also decreases air pollution, improves cardiovascular fitness, and reduces morbidity and mortality from road traffic injuries (Woodcock et al. 2007), all of which are unintended but happy consequences of a climate mitigation strategy. Health Impact Assessments (HIAs) are a useful tool – patterned after environmental impact assessments (EIAs) – for identifying and arraying such impacts and introducing these concerns into the development process. The time may be ripe for combining HIAs with EIA with particular emphasis on the impacts of interventions on both human and environmental health.

HIAs have several components: screening, scoping, appraisal, reporting, and monitoring (World Health Organization 2009).

- Screening is a systematic process to determine when an HIA should occur. For instance, funders can consider whether to make HIAs a routine part of all their projects, or a subset depending on predefined criteria. HIAs can be incorporated systematically or on an *ad hoc* basis, though systematic use of HIAs enables early consideration of health impacts and encourages a focus on healthy outcomes at all levels.
- Scoping is the process by which the parameters of the HIA are set. Scoping determines the project's focus and boundaries, identifies stakeholders and facilitators, deadlines, and provides a natural entrée for considerations such as how long-term monitoring will be arranged.
- The reporting process includes presentation of findings to relevant stakeholders. In the case of climate-related impacts, reporting will need to focus not only on immediate and proximate impacts, but also on longer term, potentially delayed and geographically remote impacts on other populations. Co-benefits of certain strategies in terms of health impacts should be highlighted. Reporting necessarily includes presentation of decision options, options analysis, and recommendations for action, with acknowledgement of conflicting impacts.
- Finally, monitoring is the process of evaluating the extent to which the HIA actually influenced the decision making process, and evaluating the degree to which the projected impacts occurred. Monitoring can include both process and outcome indicators; for instance, for an HIA related to switching from petroleum-based diesel to biodiesel in a car fleet, monitoring might include process decisions such as distribution of findings to relevant stakeholders and fuel costs as well as outcome indicators such as particulate air pollution and local rates of

Emergency Department visits for cardiovascular disease. Longer-term monitoring is sometimes, but not necessarily, a part of HIAs.

HIAs allow for leveraging health co-benefits in both climate change mitigation and adaptation activities. With attention to transparency in the assessment and decision-making process, broad conceptualization of policy options, and novel assessment methods, HIAs enable identification of co-benefits along with more conventional disease-reduction opportunities, and allow a health equity discussion to enter the decision making process (Patz et al. 2008a). Smith and Haigler recently used HIA methodology to evaluate co-benefits from novel household energy production methods in the developing world (Smith et al. 2008). HIAs are a relatively new decision support tool (Joffe et al. 2005; Dannenberg et al. 2006), and absent of a centralized policy regarding their application, use of HIAs has been dependent on local leadership (Ahmad et al. 2008), but their potential is significant for a range of decisions and projects with potential health impacts (Davenport et al. 2006; Fielding et al. 2006; Cole et al. 2007; Lee et al. 2007; Scott-Samuel et al. 2007). As more HIAs of climate-related policies and programs are completed, a database of impacts can be constructed and methods can be further revised.

From an ecological and health perspective (sometimes termed eco-health), there are abundant opportunities for co-benefits in climate change mitigation and adaptation. For instance, as deforestation accounts for 18–25% of all global carbon emissions historically, avoiding deforestation is important for reducing emissions (Stern 2006). Avoiding deforestation also preserves ecosystem integrity, important in reducing the emergence of infectious disease (Patz et al. 2004; Foley et al. 2005; Vittor et al. 2006) and in maintaining other ecosystem services important to human health (see the chapter by Levy et al., this volume). A host of other co-benefits become apparent when decisions concerning urban development, public transportation, energy production, land conservation, and other issues related to both mitigation and adaptation are considered using an HIA framework.

From the public health perspective, incorporating HIAs into major development activities has the potential to dramatically enhance decision-making and improve health equity while reorienting future development activities toward health maximization. This has not been undertaken as yet; in fact, there is relatively little literature on the intersection of health impact assessments and development activities, with the notable exception of a Bulletin of the World Health Organization in 2003 devoted to the topic. While the World Bank has included environmental impact analyses (EIAs) in its projects since the 1980s, but these assessments have not formally included sustainability or health co-benefits (Mercier 2003).

Conclusion

Applying an ecological framework to climate change brings several insights for both public health and development. Climate change results from a redistribution of carbon and a market failure in pricing of energy from fossil fuels. These failures have resulted in a transfer of wealth – and consequently of health – to the industrialized world in a sustained global environmental subsidy. Having been left out from this subsidy, the less industrialized world is woefully underprepared for a host of hazardous environmental exposures with potentially devastating health consequences, from direct exposures such as heat, reduced air quality, and extreme weather to indirect exposures such as ecosystem disruption, malnutrition, water related disease, and mass population movements. While the situation is bleak, an ecological lens that identifies humans as managers of the world's complex socioecological systems highlights the importance of resilience and adaptability in the climate change response. Climate change will likely overwhelm the services provided by degraded ecosystems compromising the resilience in certain communities. Socioecological transformation will be required in particularly vulnerable areas. Overall, the extent of human suffering that will result from climate change and other types of large-scale anthropogenic environmental change is ours to influence. Rapid and dramatic efforts to reduce greenhouse gas emissions coupled with a large-scale, well-funded mobilization to reduce vulnerability of those in harm's way would reduce suffering a great deal. An important component of these efforts will be increasing our understanding of the way that ecosystems and human health interact and identifying how they can be managed to reduce vulnerability of ecosystem services as diverse as food production to water filtration to protection from natural disasters. Continuing business as usual and allowing the poor in climate change "hotspots" to suffer as a result, will lead to far greater suffering and may be considered one of the great ethical lapses of human history.

References

Ahmad, B., D. Chappel, T. Pless-Mulloli, M. White, B. Ahmad, D. Chappel, T. Pless-Mulloli, and M. White. 2008. Enabling factors and barriers for the use of health impact assessment in decision-making processes. Public Health **122**:452–457.

Alexandratos, N. 1999. World food and agriculture: Outlook for the medium and longer term. Proc Natl Acad Sci USA **96**:5908–5914.

Argaud, L., T. Ferry, Q. Le, A. Marfisi, D. Ciorba, P. Achache, R. Ducluzeau, and D. Robert. 2007. Short- and long-term outcomes of heatstroke following the 2003 heat wave in Lyon, France. Arch Intern Med **167**:2177–2183.

Barbier, E. B., E. W. Koch, B. R. Silliman, D. D. Hacker, E. Wolanski, J. H. Primavera, E. F. Granek, S. Polasky, S. Aswani, L. A. Cramer, D. M. Storms, C. Kennedy, J, D. Bael, C. V. Kappel, G. M. E. Perillo, and D. J. Reed. 2008. Coastal ecosystem-based management with nonlinear ecological functions and values. Science **319**:321–323.

Battisti, D. S. and R. L. Naylor. 2009. Historical warnings of future food insecurity with unprecedented seasonal heat. Science **323**(5911):240–244.

Bell, M. L., R. Goldberg, C. Hogrefe, P. Kinney, K. Knowlton, B. Lynn, J. Rosenthal, C. Rosenzweig, and J. A. Patz. 2007. Climate change, ambient ozone, and health in 50 U.S. cities. Climatic Change **82**:61–76.

Bentham, G. and I. H. Langford. 1995. Climate change and the incidence of food poisoning in England and Wales. International Journal of Biometeorology **39**:81–86.

Bouma, M. and H. Van der Kay. 1996. The El Nino Southern Oscillation and the historic malaria epidemics on the Indian subcontinent and Sri Lanka: an early warning system for futre epidemics? Tropical Medicine & International Health **1**:86–96.

Campbell-Lendrum, D., A. Pruss-Ustan, and C. Corvalan. 2003. How much disease could climate change cause? Pages 133–159 *in* A. McMichael, D. Campbell-Lendrum, C. Corvalan, K. Ebi, A. Githeko, J. Scheraga, and A. Woodward, editors. Climate Change and Human Health: Risks and Responses. World Health Organization/World Meteorological Organization/United Nations Environmental Programme, Geneva.

Cazelles, B., M. Chavez, A. J. McMichael, and S. Hales. 2005. Nonstationary Influence of El NiÃ ± o on the Synchronous Dengue Epidemics in Thailand. PLoS Medicine **2**:313–318.

CDC. 2003. Heat-related deaths—United States, 1999–2003. MMWR - Morbidity & Mortality Weekly Report **55**:796–798.

CDC. 2008. Monitoring Health Effects of Wildfires Using the BioSense System — San Diego County, California, October 2007. MMWR - Morbidity & Mortality Weekly Report **57**:741–747.

CDC. 2009. Increase in Coccidioidomycosis - California, 2000–2007. MMWR - Morbidity & Mortality Weekly Report. 58(5):105–9.

Chan, M. 2007. Climate change and health: preparing for unprecedented challenges. 2007 David E. Barmes Global Health Lecture, Bethesda MD.

Chaves, L. F. and M. Pascual. 2006. Climate Cycles and Forecasts of Cutaneous Leishmaniasis, a Nonstationary Vector-Borne Disease. PLoS Medicine **3**:e295.

Chicago Climate Task Force. 2008. Chicago Climate Action Plan. Available: http://www.chicago-climateaction.org/filebin/pdf/finalreport/CCAPREPORTFINALv2.pdf. Accessed 10 July 2011.

Clarke, J. 1972. Some Effects of the Urban Structure on Heat Mortality. Environmental Research **5**:93–104.

Cole, B. L., J. E. Fielding, B. L. Cole, and J. E. Fielding. 2007. Health impact assessment: a tool to help policy makers understand health beyond health care. Annual Review of Public Health **28**:393–412.

Colwell, R. and A. Huq. 2001. Marine Ecosystems and Cholera. Hydrobiologia **460**:141–145.

Colwell, R. R. 1996. Global climate and infectious disease: the cholera paradigm. Science **274**:2025–2031.

Confalonieri, U., B. Menne, et al. 2007. Human Health. Climate Change 2007: Impacts, Adaptation and Vulnerability. Contribution of Working Group II to the Fourth Assessment Report of the Intergovernmental Panel on Climate Change, M. L. Parry, O. F. Canziani, J. P. Palutikof, P. J. Van der Linden and C. E. Hanson, Eds., Cambridge University Press Cambridge, pp. 391–431.

Curriero, F. C., J. A. Patz, J. B. Rose, and S. Lele. 2001. The association between extreme precipitation and waterborne disease outbreaks in the United States, 1948–1994. Am J Public Health **91**:1194–1199.

Dannenberg, A. L., R. Bhatia, B. L. Cole, C. Dora, J. E. Fielding, K. Kraft, D. McClymont-Peace, J. Mindell, C. Onyekere, J. A. Roberts, C. L. Ross, C. D. Rutt, A. Scott-Samuel, H. H. Tilson, A. L. Dannenberg, R. Bhatia, B. L. Cole, C. Dora, J. E. Fielding, K. Kraft, D. McClymont-Peace, J. Mindell, C. Onyekere, J. A. Roberts, C. L. Ross, C. D. Rutt, A. Scott-Samuel, and H. H. Tilson. 2006. Growing the field of health impact assessment in the United States: an agenda for research and practice. American Journal of Public Health **96**:262–270.

Davenport, C., J. Mathers, J. Parry, C. Davenport, J. Mathers, and J. Parry. 2006. Use of health impact assessment in incorporating health considerations in decision making. Journal of Epidemiology & Community Health **60**:196–201.

Ebi, K., J. Balbus, P. Kinney, E. Lipp, D. Mills, M. O'Neill, and M. Wilson. 2008. Synthesis and Assessment Product 4.6: Chapter 2, Effects of Global Change on Human Health. CCSP.

Ebi, K. L., N. D. Lewis, and C. Corvalan. 2006. Climate variability and change and their potential health effects in small island states: information for adaptation planning in the health sector. Environ Health Perspect **114**:1957–1963.

Eilers, P. and B. Groot. 1997. Effects of ambient particulate matter and ozone on daily mortality in Rotterdam, the Netherlands. Archives of Environmental Health **52**:455.

Ezzell, C. 1999. It Came from the Deep. Scientific American **280**:22.

FAO. 2009. 1.02 billion people hungry: one sixth of humanity undernourished, more than ever before. Page 1. United Nations Food and Agriculture Organization, Rome.

Fielding, J. E., P. A. Briss, J. E. Fielding, and P. A. Briss. 2006. Promoting evidence-based public health policy: can we have better evidence and more action? Health Affairs **25**:969–978.

Foley, J. A., R. Defries, G. P. Asner, C. Barford, G. Bonan, S. R. Carpenter, F. S. Chapin, M. T. Coe, G. C. Daily, H. K. Gibbs, J. H. Helkowski, T. Holloway, E. A. Howard, C. J. Kucharik, C. Monfreda, J. A. Patz, I. C. Prentice, N. Ramankutty, and P. K. Snyder. 2005. Global consequences of land use. Science **309**:570–574.

Gleick, P. H. 1996. Basic water requirements for human activities: Meeting basic needs. Water International.

Harlan SL, Brazel AJ, Prashad L, Stefanov WL, and L. L. 2006. Neighborhood microclimates and vulnerability to heat stress. Soc Sci Med **63**:2847.

Hess, J., J. Malilay, and A. J. Parkinson. 2008. Climate Change: The Importance of Place. American Journal of Preventive Medicine **35**:468–478.

Hoagland, P., D. Anderson, M, Y. Kaoru, and A. White, W. 2002. The economic effects of harmful algal blooms in the united states: estimates, assessment issues, and information needs. Estuaries **25**:819–837.

Högy, P. and A. Fangmeier. 2008. Effects of elevated atmospheric CO2 on grain quality of wheat. Journal of Cereal Science **48**:580–591.

IPCC, 2007a. Summary for Policymakers. In: Climate Change 2007: The Physical Science Basis. Contribution of Working Group I to the Fourth Assessment Report of the Intergovernmental Panel on Climate Change [Solomon, S., D. Qin, M. Manning, Z. Chen, M. Marquis, K.B. Averyt, M.Tignor and H.L. Miller (eds.)]. Cambridge University Press, Cambridge, United Kingdom and New York, NY, USA.

IPCC, 2007b. Climate Change 2007: Impacts, Adaptation and Vulnerability. Contribution of Working Group II to the Fourth Assessment Report of the Intergovernmental Panel on Climate Change, M.L. Parry, O.F. Canziani, J.P. Palutikof, P.J. van der Linden and C.E.Hanson, Eds., Cambridge University Press, Cambridge, UK, 976 pp.

Joffe, M., J. Mindell, M. Joffe, and J. Mindell. 2005. Health impact assessment. Occupational & Environmental Medicine **62**:907–912.

Johnston, F., A. Kavanagh, D. Bowman, and R. Scott. 2002. Exposure to bushfire smoke and asthma: an ecological study. Med J Aust **176**.

Johnstone, L. C. 2008. Planning for the Inevitable, the Humanitarian Consequences of Climate Change. "Linking Climate Change Negotiations and Disaster Risk Reduction" Copenhagen.

Kareiva, P. and M. Marvier. 2007. Conservation for the people. Scientific American **297**(4):50–57.

Kinney, P. L. 2008. Climate change, air quality, and human health. American Journal of Preventive Medicine **35**:459–467.

Knowlton, K., J. E. Rosenthal, C. Hogrefe, B. Lynn, S. Gaffin, R. Goldberg, C. Rosenzweig, K. Civerolo, J.-v. Ku, and P. L. Kinney. 2004. Assessing Ozone-Related Health Impacts under a Changing Climate. Environmental Health Perspectives **112**:1557–1563.

Koelle, K., X. RodÃ³, M. Pascual, M. Yunus, and G. Mostafa. 2005. Refractory periods and climate forcing in cholera dynamics. Nature **436**:696–700.

Kovats, R. S., S. Edwards, S. Hajat, B. Armstrong, K. L. Ebi, B. Menne, and T. C. Group. 2004. The effect of temperature on food poisoning: a time-series analysis of salmonellosis in ten European countries. Epidemiology of Infection **132**:443.

Kovats, R. S., S. Hajat, R. S. Kovats, and S. Hajat. 2008. Heat stress and public health: a critical review. Annual Review of Public Health **29**:41–55.

Kunzli, N., E. Avol, J. Wu, W. J. Gauderman, E. Rappaport, J. Millstein, J. Bennion, R. McConnell, F. D. Gilliland, K. Berhane, F. Lurmann, A. Winer, and J. M. Peters. 2006. Health effects of the 2003 Southern California wildfires on children. American Journal of Respiratory & Critical Care Medicine **174**:1221–1228.

Lee, K., A. Ingram, K. Lock, C. McInnes, K. Lee, A. Ingram, K. Lock, and C. McInnes. 2007. Bridging health and foreign policy: the role of health impact assessments. Bulletin of the World Health Organization **85**:207–211.

Lieffering, M., H.-Y. Kim, K. Kobayashi, and M. Okada. 2004. The impact of elevated CO2 on the elemental concentrations of field-grown rice grains. Field Crops Research **88**:279–286.

Lobell, D., B, M. Burke, B, C. Tebaldi, M. Mastrandrea, D, W. Falcon, P, and R. Naylor, L. 2008. Prioritizing climate change adaptation needs for food security in 2030. Science **319**:607–610.

Loladze, I. 2002. Rising atmospheric CO2 and human nutrition: toward globally imbalanced plant stoichiometry? Trends in Ecology & Evolution **17**:457–461.

Luber, G. and M. McGeehin. 2008. Climate Change and Extreme Heat Events. Am J Prev Med **35**:429–436.

Mason, J., B, P. Musgrove, and J. Habicht. 2003. At least one-third of poor countries' disease burden is due to malnutrition. Disease Control Priorities Project, Bethesda, Md.

McCloskey, L. A. and K. Southwick. 1996. Psychosocial problems in refugee children exposed to war. Pediatrics **97**:394.

McGranahan, G., D. Balk, and B. Anderson. 2007. The rising tide: assessing the risks of climate change and human settlements in low elevation coastal zones. Environment and Urbanization **19**:17–37.

McMichael, A. 2004. Climate Change. Pages 1543–1649 *in* M. Ezzati, A. Lopez, A. Rodgers, and C. Murray, editors. Comparative Quantification of Health Risks: Global and Regional Burden of Disease due to Selected Major Risk Factors, Vol. 2. World Health Organization, Geneva.

McMichael, A. J. 2006. Population health as the 'bottom line' of sustainability: a contemporary challenge for public health researchers. Eur J Public Health **16**:579–581.

Mercier, J-R. 2003. Health impact assessment in international development assistance: the World Bank experience. Bulletin of the World Health Organization **81**(6):461-462.

Molyneux, D. H. 1997. Patterns of change in vector borne diseases. Annals of Tropical Medicine & Parasitology **91**:827–839.

Murray, C. J. L. and A. D. Lopez. 1997. Global mortality, disability, and the contribution of risk factors: Global Burden of Disease Study. The Lancet **349**:1436–1442.

Myers, S. S. and J. Patz. 2009. The human health effects of global environmental change. Annual Review of Environment & Resources **34**(1): 223–252.

Nicholls, R. J. and R. S. Tol. 2006. Impacts and responses to sea-level rise: a global analysis of the SRES scenarios over the twenty-first century. Philos Transact A Math Phys Eng Sci **364**:1073–1095.

Ostfeld, R. S. 2009. The ecology of climate change and infectious disease. Ecology.

Pascual, M., J. A. Ahumada, L. F. Chaves, X. Rodo, and M. Bouma. 2006. Malaria resurgence in the East African highlands: temperature trends revisited. Proc Natl Acad Sci USA **103**:5829–5834.

Patz, J., D. Campbell-Lendrum, H. Gibbs, and R. Woodruff. 2008a. Health Impact Assessment of Global Climate Change: Expanding on Comparative Risk Assessment Approaches for Policy Making. Annual Review of Public Health **29**.

Patz, J., H. Gibbs, J. Foley, J. Rogers, and K. Smith. 2007. Climate change and global health: quantifying a growing ethical crisis. EcoHealth **4**:397–405.

Patz, J. and R. S. Kovats. 2007. Hotspots in climate change and human health BMJ **325**:1094–1098.

Patz, J., S. H. Olson, C. Uejio, and H. Gibbs. 2008b. Disease Emergence from Global Climate and land use change. Medical Clinics of North America **92**:18.

Patz, J. A., D. Campbell-Lendrum, T. Holloway, and J. A. Foley. 2005. Impact of regional climate change on human health. Nature **438**:310–317.

Patz, J. A., P. Daszak, G. M. Tabor, A. A. Aguirre, M. Pearl, J. Epstein, N. D. Wolfe, A. M. Kilpatrick, J. Foufopoulos, D. Molyneux, D. J. Bradley, and E. Working Group on Land Use Change and Disease. 2004. Unhealthy landscapes: Policy recommendations on land use change and infectious disease emergence. Environmental Health Perspectives **112**:1092–1098.

Penuelas, J. and R. Matamala. 1993. Variations in the Mineral Composition of Herbarium Plant Species Collected During the Last Three Centuries. J. Exp. Bot. **44**:1523–1525.

Pfeffer, W. T., J. T. Harper, and S. O'Neel. 2008. Kinematic constraints on glacier contributions to 21st-century sea-level rise. Science **321**:1340–1343.

Postel, S. L., G. C. Daily, and P. R. Ehrlich. 1996. Human Appropriation of Renewable Fresh Water. Science **271**:785–788.

Raven, P. 2002. Science, sustainability, and the human prospect. Science **297**:954–957.

Reiter, P. 2001. Climate change and mosquito-borne disease. Environmental Health Perspectives **109 Suppl 1**:141–161.

Rogers, D. J. and S. E. Randolph. 2000. The global spread of malaria in a future, warmer world. Science **289**:1763–1766.

Sachs, J. D. 2007. Climate Change Refugees. Scientific American **296**:43–43.

Schneider, S. H., S. Semenov, and A. Patwardhan. 2007. Assessing key vulnerabilities and the risk from climate change. *in* M. Parry, L, O. Canziani, F, J. Palutikof, P, P. Van der Linden, J, and C. Hanson, E, editors. Climate Change 2007: Impacts, Adaptation and Vulnerability. Contribution of Working Group II to the Fourth Assessment Report of the Intergovernmental Panel on Climate Change. Cambridge University Press, Cambridge, UK.

Schwartz, J. 1994. Air pollution and daily mortality: a review and meta-analysis. Environ Res **4**.

Scott-Samuel, A., E. O'Keefe, A. Scott-Samuel, and E. O'Keefe. 2007. Health impact assessment, human rights and global public policy: a critical appraisal. Bulletin of the World Health Organization **85**:212–217.

Smith, K. R., E. Haigler, K. R. Smith, and E. Haigler. 2008. Co-benefits of climate mitigation and health protection in energy systems: scoping methods. Annual Review of Public Health **29**:11–25.

Snäll, T., R. B. O'Hara, C. Ray, and S. K. Collinge. 2008. Climate-Driven Spatial Dynamics of Plague among Prairie Dog Colonies. American Naturalist **171**:238–248.

Solomon, S., D. Qin, M. Manning, Z. Chen, M. Marquis, K. Averyt, M. Tignor, and H. Miller, editors. 2007. Climate Change 2007: The Physical Science Basis. Contribution of Working Group I to the Fourth Assessment Report of the Intergovernmental Panel on Climate Change. Cambridge University Press, Cambridge, UK.

Stenseth, N. C., N. I. Samia, H. Viljugrein, K. L. Kausrud, M. Begon, S. Davis, H. Leirs, V. M. Dubyanskiy, J. Esper, V. S. Ageyev, N. L. Klassovskiy, S. B. Pole, and K. S. Chan. 2006. Plague dynamics are driven by climate variation. Proc Natl Acad Sci USA **103**:13110–13115.

Stern, N. 2006. The Stern Review on the Economics of Climate Change. UK Government Economic Service, Cambridge.

Stone, B., Jr. and J. Hess. 2010. Urban form and extreme heat events: are sprawling cities more vulnerable to climate change than compact cities? Environ Health Perspect. doi 10.1289/ehp.0901879: 2010.

Taub, D. R., B. Miller, and H. Allen. 2008. Effects of elevated CO2 on the protein concentration of food crops: a meta–analysis. Global Change Biology **14**:565–575.

Thomson, M. C., F. J. Doblas-Reyes, S. J. Mason, R. Hagedorn, S. J. Connor, T. Phindela, A. P. Morse, and T. N. Palmer. 2006. Malaria early warnings based on seasonal climate forecasts from multi-model ensembles. Nature **439**:576–579.

Tilman, D., S. Polasky, and C. Lehman. 2005. Diversity, productivity and temporal stability in the economies of humans and nature. Journal of Environmental Economics and Management **49**:405–426.

Toole, M. and R. Waldman. 1997. The public health aspects of complex emergencies and refugee situations. Annual Review of Public Health **18**:283–312.

Toole, M. J. and R. J. Waldman. 1993. Refugees and displaced persons. War, hunger, and public health. JAMA **270**:600–605.

Tubiello F.N., J.F. Soussana, and S.M. Howden. Crop and pasture response to climate change. Proc Natl Acad Sci USA 104(50):19686–90, 2007.

United Nations Development Programme (2004). Reducing Disaster Risk: A Challenge for Development. *in* A. Maskrey, A. Arenas, M. Bhatt et al., editors. United Nations Development Programme Bureau for Crisis Prevention and Recovery, New York.

U.S. House of Representatives. 2006. A Failure of Initiative: Final Report of the Select Bipartisan Committee to Investigate the Preparation for and Response to Hurricane Katrina. U.S. House of Representatives Select Bipartisan Committee to Investigate the Preparation for and Response to Hurricane Katrina, Washington, DC.

Van Dolah, F. M., D. Roelke, and R. M. Greene. 2001. Health and Ecological Impacts of Harmful Algal Blooms: Risk Assessment Needs. Human & Ecological Risk Assessment **7**:1329.

Vittor, A. Y., R. H. Gilman, J. Tielsch, G. Glass, T. Shields, W. S. Lozano, V. Pinedo-Cancino, and J. A. Patz. 2006. The effect of deforestation on the human-biting rate of Anopheles darlingi, the primary vector of Falciparum malaria in the Peruvian Amazon. American Journal of Tropical Medicine & Hygiene **74**:3–11.

Vorosmarty, C., J, C. Leveque, and C. Revenga. 2005. Fresh Water. Pages 165–207 *in* F. Rijsberman, R. Costanza, and P. Jacobi, editors. Millennium Ecosystem Assessment (Program) Condition and Trends Working Group Ecosystems and human well-being : current state and trends : findings of the Condition and Trends Working Group of the Millennium Ecosystem Assessment. Island Press, Washington, DC.

Wackernagel, M., N. B. Schulz, D. Deumling, A. C. Linares, M. Jenkins, V. Kapos, C. Monfreda, J. Loh, N. Myers, R. Norgaard, and J. r. Randers. 2002. Tracking the ecological overshoot of the human economy. Proc Natl Acad Sci USA **99**:9266–9271.

Wallace, R. F., D. Kriebel, L. Punnett, D. H. Wegman, and P. J. Amoroso. 2007. Prior heat illness hospitalization and risk of early death. Environmental Research **104**:290–295.

Webster, P. 2008. Myanmar's deadly daffodil. Nature Geoscience **doi:** 10.1038/ngeo257.

Williams, P. L., D. L. Sable, P. Mendez, and L. T. Smyth. 1979. Symptomatic coccidioidomycosis following a severe natural dust storm. An outbreak at the Naval Air Station, Lemoore, Calif. Chest **76**:566–570.

Woodcock, J., D. Banister, P. Edwards, A. M. Prentice, and I. Roberts. 2007. Energy and transport. Lancet **370**:1078–1088.

World Health Organization. 2006. Cholera 2005. Weekly Epidemiologic Record **81**:297–308.

World Health Organization. 2008a. The global burden of disease: 2004 update. World Health Organization, Geneva.

World Health Organization. 2008b. World Health Statistics. World Health Organization, Geneva.

World Health Organization. 2009. Health Impact Assessment. The World Health Organization, Geneva.

Yohe, G. and R. Tol. 2002. Indicators for social and economic coping capacity—moving toward a working definition of adaptive capacity. Global Environmental Change **12**:25–40.

Younger, M., H. Morrow-Almeida, S. Vindigni, and A. Dannenberg. 2008. The Built Environment, Climate Change, and Health: Opportunities for Co-Benefits. Am J Prev Med **35**:517–527.

Chapter 13
Disease Ecology

Felicia Keesing and Richard S. Ostfeld

Introduction

In his 2008 strategic vision for the National Institutes of Health of the United States, Director Elias Zerhouni argued that medical care in the twenty-first century needs to be redirected (Zerhouni 2008). Despite progress in the practice of preventive medicine, he stated, most treatment still focuses on late intervention, once a patient's symptoms are already apparent. We need, he argued, to move instead toward "pre-emptive medicine" in which the development of symptoms is prevented altogether. Zerhouni illustrated this new approach by describing the discovery of genes that predispose certain individuals to particular diseases. With this knowledge, he argued, public health care providers could focus on prevention and early diagnosis in high-risk patients, rather than just on treatment.

We suggest that the full development of preemptive medicine must incorporate another type of strategy as well: the mitigation of disease by environmental monitoring and management. The pathogens that cause infectious diseases interact not just with their hosts to cause disease; they are also embedded within a web of interactions among organisms in ecological communities. Recent research in the ecology of disease has demonstrated that knowledge of these interactions can be used to predict, prevent, and mitigate the transmission of infectious diseases.

In this chapter, we describe this emerging understanding by defining four key principles of disease ecology. We begin by describing the important role of the density of disease organisms such as pathogens, hosts, and vectors (see Box 13.1) in the transmission of infectious diseases. But we also describe how in some situations, the behavior of organisms, including humans, can override density's importance. Further, we describe how disease outbreaks can often be anticipated well in advance

F. Keesing (✉)
Department of Biology, Bard College, PO Box 5000, Annandale, NY 12504, USA
e-mail: keesing@bard.edu

J.C. Ingram et al. (eds.), *Integrating Ecology and Poverty Reduction: Ecological Dimensions*, DOI 10.1007/978-1-4419-0633-5_13,

by looking at the chain of ecological interactions that precede them. We then explore two examples of types of cryptic, indirect interactions that may occur in a wide variety of diseases of humans, their livestock, and their crops. In conclusion, we argue that recent research in disease ecology tightly links human health with environmental conservation and that a critical understanding of disease ecology is essential for establishing a preemptive medicine. This need is particularly strong for people in impoverished rural communities who are more vulnerable to infectious diseases, and who may have greater interaction with the natural environment.

Box 13.1 Disease Terms

Diseases are typically grouped into two categories: infectious and non-infectious. Non-infectious diseases are those caused by genetic disorders, by food, or by the environment (e.g. ultraviolet radiation which can lead to skin cancer). Infectious diseases are caused by infection of a *host* with an organism – a *pathogen*. Pathogens have traditionally been divided into two groups – *microparasites*, which include the smallest organisms (viruses, bacteria, fungi, and some protists); and *macroparasites*, which include organisms that can be seen with the unaided eye (e.g. helminths). Pathogens can be transmitted from one host to another in a variety of ways. Some pathogens are transmitted through sexual contact, others through contact with respiratory droplets, blood, or feces, and still more in other ways. Some pathogens require another organism (e.g. a mosquito, a tick, a fly – a *vector*) to transport them from host to host; these are the *vector-borne diseases*.

The Importance of Density

The chapter on land use by Myers in this volume has described myriad ways in which the environment can influence the transmission of infectious diseases. Conceptually, the examples so far have generally been quite straightforward. For example, the clearing of a forest for agriculture can increase habitat quality for a mosquito that serves as a disease vector (Singer and DeCastro 2001). Implicitly, these types of examples recognize the importance of population density in disease transmission – the more abundant a pathogen or vector or host is, the more frequently infectious diseases should be transmitted. For the past 30 years, the study of the ecology of diseases has been similarly focused on population density as a determinant of disease risk or severity.

Density is at the core of mathematical models that ecologists use to explore the effects of different factors on disease transmission. The simplest, and best-known, of these models is conceptually very simple. Imagine a pathogen, like a cold virus, that's transmitted directly between hosts. Within this model, hosts can be either infected or susceptible, meaning that they're capable of being infected.

Transmission between infected and susceptible hosts occurs at a particular rate, called the rate of transmission. In such simple models, the number of individuals who get infected during a given time is determined by the product of the rate of transmission, how many susceptible hosts there are (because they're the ones available to become infected), and how many infected hosts there are (because they're the ones who can transmit the infection). So in this model, the densities of susceptible and infected hosts, combined with the rate of transmission, entirely determine how many hosts will be infected in the future.

Models such as this simple susceptible-infected (or S-I) one and its extensions have been very useful. For example, derivations from this model can be used to determine what fraction of a population needs to be immunized to prevent the spread of an infectious disease, such as measles or mumps, and this information can then be used to guide public health efforts. Models can also estimate how small a host population needs to be to prevent a disease from spreading, and this information, too, can be used to determine effective public health strategies (for an excellent introduction to basic epidemiological models, see Allman and Rhodes 2004). A recent study using sophisticated density-based models identified the ecological causes of ongoing epidemics of measles in Niger, and suggested that a combination of sustained and reactive vaccination, coupled with stringent surveillance early during seasonal outbreak periods, could reduce mortality and morbidity substantially (Ferrari et al. 2008).

Despite the enormous utility of density-based models, however, the densities of hosts, pathogens, and vectors do not always tell the whole story. In the following sections, we focus on more complex ways in which organisms, including people, interact to affect human health directly, by influencing transmission of human pathogens, or indirectly, by affecting the health of crops or livestock. In many of these examples, ecological interactions can lead to surprising effects on health that could not be predicted from the density of hosts or vectors.

Density Isn't Everything

Bovine tuberculosis (BTb) is a serious disease that causes progressive emaciation in cattle and other mammals, including humans. The disease is caused by infection with the bacterium *Mycobacterium bovis*, a close relative of the bacterium that causes the more familiar human Tb. *M. bovis* is passed from host to host primarily through respiratory secretions and through milk (Cosivi et al. 1999). Humans can become infected from drinking infected milk that has not been sterilized.

To control BTb, public health professionals typically concentrate on reducing the density of infected cattle by closely monitoring herds and euthanizing cattle that show symptoms (Woodroffe et al. 2006). In some areas of the world, this approach has been effective, because the density of infected cattle is the critical factor in transmission. In Great Britain, however, BTb has repeatedly been a problem, despite efforts to solve the problem by culling infected cattle. One reason BTb has been hard to control in Britain is that cattle are not the only host species that can transmit the disease. In the 1970s, scientists discovered that European badgers, *Meles meles*, are also effective hosts for *M. bovis* and they often live in proximity to cattle (Krebs et al. 1997). When this wildlife reservoir for the bacterium was discovered, farmers were given license to cull badgers to reduce badger densities and thus make them less likely to transmit infection to susceptible cattle (Griffin et al. 2005).

But reducing badger density through culling did not reduce transmission either. In fact, BTb incidence was 27% *greater* in areas that had been culled to reduce badger density compared to areas that hadn't been culled (Donnelly et al. 2003). Why? Because badgers are social animals that defend group territories. Culling disrupts their tightly knit social groups, causing them to increase movement distances and the sizes of their home ranges (Woodroffe et al. 2006). The result of this increased level of movement is to increase badger contact rates with cattle, and thus the transmission of tuberculosis.

Culling of badgers in Britain has pitted conservationists, who want to protect badger populations, against farmers, who want to protect their cattle (Krebs et al. 1998). But it turns out that both groups should be on the same side, at least on this issue. It's important to note that while BTb remains a huge *economic* issue in Britain, it is not a prominent issue for human health directly because of the widespread sterilization of milk for human consumption. But in other parts of the world, particularly in Africa, BTb is a major human health issue (Cosivi et al. 1999); we return to BTb in Africa later in this chapter.

The key underlying feature of the badger example is that social behaviors affect the number of interactions that can cause BTb; behavior trumps density. This might be an unusual, or even a unique example that makes an interesting story but isn't broadly applicable. But it turns out that the critical importance of behavior is a feature of many diseases, including a number that are particularly relevant to human health, as we describe below.

For certain kinds of diseases, the number of contacts between susceptible and infected hosts is relatively constant; it doesn't increase as the density of hosts increases. The classic examples are sexually transmitted diseases like HIV/AIDS (Lloyd-Smith et al. 2004). The number of sexual contacts that an individual person has is limited at a relatively constant level; people generally don't have a greater number of sexual contacts just because they live among a higher-density population. The transmission of sexually transmitted diseases depends, then, on the *percentage* of hosts that are infected, because this represents the chance that a particular sexual encounter includes an infected host (Fig. 13.1). A similar situation is thought to exist for diseases transmitted by arthropod vectors such as mosquitoes and ticks. The number of blood meals taken by a vector on a host species tends to be fixed at a relatively constant level. For example, ticks tend to feed once per life stage and female mosquitoes tend to feed once per egg-laying event, with roughly constant intervals between such events[1]. Therefore, the probability that a particular vector

[1] Important exceptions to this general rule occur, for example, when environmental factors such as climate warming accelerate biting rates of some vectors.

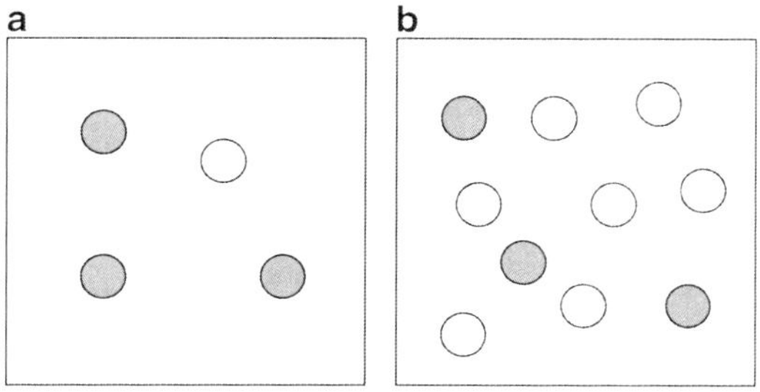

Fig. 13.1 Frequency-dependent versus density-dependent transmission for a sexually-transmitted infectious disease. In density-dependent transmission, the density of infected hosts (*gray circles*) is equal in both panels. Thus, an uninfected host is at equal risk in both cases. However, the situation changes for frequency dependent transmission. If people have relatively constant rates of sexual encounters regardless of density, a particular person will have a sexual encounter with, for example, one other person in a certain time interval, regardless of how many people there are in total. So if this person has a sexual encounter with a person in panel (**a**), he or she is more likely to encounter an infected host (*gray circles*) than an uninfected one (*white circles*), because infected hosts represent 75% of the population. In panel (**b**), this person is much less likely to encounter an infected host, even though there are many more total hosts, because infected hosts represent only 30% of the population. This is an example of frequency-dependent transmission, and it is thought to be common for sexually-transmitted diseases of humans and other animals

individual will acquire an infection depends less on the population density of hosts than on the percentage of hosts that are infected (the frequency).[2] In disease ecology jargon, sexually transmitted and many vector-borne diseases are said to have *frequency-dependent* rather than *density-dependent* transmission.

One interesting side note is that in some situations, density might be correlated with the frequency of infection. In the case of HIV/AIDS, infection is more likely among promiscuous people, intravenous drug users, and men who have sex with men (Fan et al. 2007). If people with those characteristics tend to choose to live in more dense populations, as in urban areas, then there will be a correlation between high density and high frequency of incidence of HIV/AIDS.

Look Upstream

In the previous section, we emphasized that density doesn't always predict infectious disease transmission. But of course in many cases, it does, and mitigation efforts are frequently aimed at reducing the density of hosts or vectors. Disease ecologists are increasingly recognizing that it is often possible to predict when and where hosts or vectors will be at high density, often well in advance.

[2]The probability that a human being will become infected by this vector depends on the likelihood that the vector is infected (which is frequency dependent) and the number of times infected vectors feed on the human (which is related to the density of vectors). In this way, transmission of vector-borne diseases can be both frequency and density dependent.

This potentially allows disease transmission to be dramatically reduced. The trick is to look upstream, sometimes literally.

In Belize, the growth of plants in wetland habitats is limited by the availability of phosphorus. Where phosphorus is in short supply, the wetlands are dominated by sparse, short vegetation interspersed with floating mats of cyanobacteria (Rejkmankova et al. 2006). But when phosphorus is abundant, this sparse vegetation is replaced over time with dense growth of cattails (*Typha* spp.) and other large, dense plants (Rejkmankova et al. 2006). Both types of plant communities harbor larvae of mosquitoes that serve as vectors of malaria. The female of a particular mosquito species, *Anopheles albimanus*, prefers to lay eggs in sparsely vegetated marshes with cyanobacterial mats. The other species, *A. vestitipennis*, prefers the densely vegetated marshes.

When phosphorus runs off from agricultural areas, it fertilizes sparsely vegetated areas, resulting in increased growth. This in turn appears to turn habitat for *A. albimanus* into habitat for *A. vestitipennis*, the species more likely to transmit malaria (Grieco et al. 2002). So the over-use of phosphorus in agricultural fertilizers in Belize leads to a delayed increase in malaria risk because it causes an increase in the density of malaria vectors through a chain of interactions. This discovery leads to several obvious interventions. Increasing the local use of bednets would deal directly with the problem of higher densities of highly competent malaria vectors. Managers could also try to reduce the abundance of *Typha* and other dense vegetation in wetlands (Rejkmankova et al. 2006) to prevent the mosquito increases earlier in the chain. And finally, farmers could reduce phosphorus use, or runoff into wetlands, to prevent the problem in the first place. None of these solutions is easy, but all would have valuable, pre-emptive benefits for human health.

Another example of a chain of direct interactions, although perhaps a more surprising one, comes from the northeastern USA, which is an area of high incidence of Lyme disease. Lyme disease is caused by a bacterium, *Borrelia burgdorferi*, that is passed from host to host by a tick vector. In the northeastern USA, the vector is the blacklegged tick, *Ixodes scapularis*. These ticks feed on a variety of mammals, birds, and reptiles (Keirans et al. 1996). When uninfected ticks feed on infected hosts, they can pick up the infection; once ticks become infected, they can pass the infection on to humans (Ostfeld 1997). But it turns out that not all host species are equally likely to transmit infection to ticks. If ticks feed on skunks or opossums or squirrels, the ticks are relatively unlikely to acquire an infection, even if the hosts are infected with the bacterium (LoGiudice et al. 2003). But if ticks feed on white-footed mice (*Peromyscus leucopus*), they have more than a 90% chance of becoming infected. For this reason, the white-footed mouse is called the most competent reservoir for the Lyme bacterium.

Given that mice are much more likely to infect ticks than any other species is, it is not surprising that the abundance of white-footed mice in a habitat is a good predictor of the number of infected ticks (Ostfeld et al. 2006). If we could predict the number of mice in a habitat, then, we should be able to predict when Lyme disease risk will be high. But what predicts when mouse abundance will be high? In some areas, the answer turns out to be simple: acorns (the fruit of oak trees of the genus *Quercus*). Acorns are rich in protein and lipids, and, perhaps most importantly, some species of acorns store well over the winter when other food is scarce.

In years when mice have acorns to eat, they survive the winter at much higher numbers and begin spring breeding earlier than in years when they don't. In fact, acorns are such a good food for mice that the number of acorns in the fall predicts the number of mice the following summer (Ostfeld et al. 2006). Acorns also predict the number of infected ticks 2 years later, with a time lag that results from the long life cycle of the tick. So based on acorn abundance in 1 year, we can predict with fairly high accuracy what Lyme disease risk will be like almost 2 years later. Unusually high acorn crops occur only every few years and they affect large regions in synchrony (Ostfeld et al. 2006), so there are distinct times and areas of high Lyme disease risk. Two years is plenty of lead-time to inform local health care providers and the public when to be most vigilant, and education has been shown to be effective at reducing the severity of Lyme disease cases (A. Evans, unpublished data).

In summary, in situations in which density is an important determinant of disease risk, knowing the ecology of the organisms involved can often make it possible to predict well in advance the times and locations when disease risk will be greatest. The challenge is to identify the chain of events leading to high abundance. In both the wetland and acorn examples, what leads to high abundance of a vector or host is, not surprisingly, a flush of resources, although the resources in both cases were several steps removed from the disease organisms (Ostfeld and Keesing 2000).

Interestingly, a recent review found that such links between resource pulses and increases in disease transmission are strikingly common (McKenzie and Townsend 2007), occurring in 54 of 55 examples surveyed. Why disease transmission should be exacerbated, rather than inhibited or unaffected by nutrient pulses is not immediately evident. One possibility is that negative or neutral results are less likely to lead to publication, resulting in a publication bias that does not reflect reality (McKenzie and Townsend 2007). Another possibility is that there is a biological reason why organisms that are involved in disease transmission are likely to respond positively to pulses of nutrients. Perhaps the very traits that tend to make organisms good hosts or vectors for pathogens (e.g. being widespread and able to achieve high abundance in a diversity of habitats) make them capable of responding positively to nutrient or other resource availability. Indeed, for a number of diseases, the most competent reservoir for pathogens is highly likely to be a habitat generalist, based on our observations of Lyme disease, West Nile virus encephalitis, babesiosis, various hantaviruses, Junin virus, and others. Habitat generalists are likely to be able to respond opportunistically to resources, and they are also ideal habitats themselves for pathogens because they are widespread, abundant, and resilient to disturbance. These hypotheses remain to be fully explored.

Cryptic Interactions

The examples in the preceding section illustrated chains of direct interactions that could influence the abundance of hosts or vectors for diseases. But interactions that affect transmission can also be much less direct; in fact, they can be so indirect that they are almost hidden.

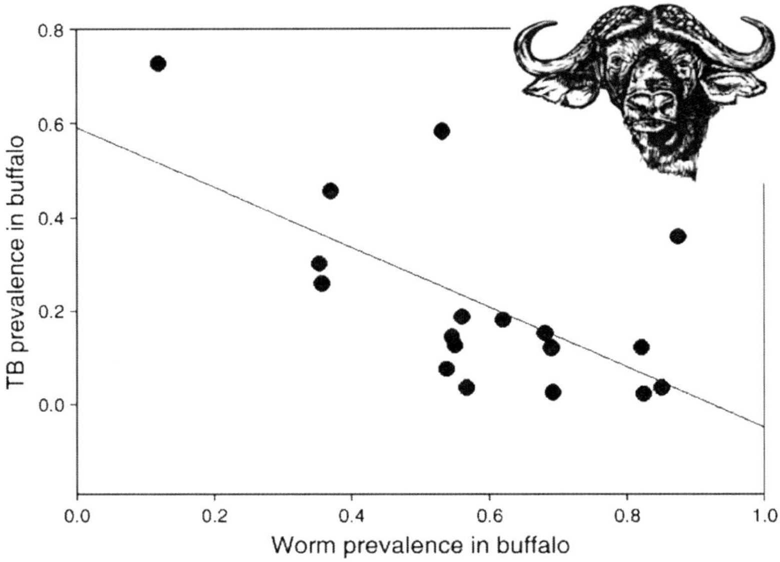

Fig. 13.2 Negative correlation between TB prevalence and worm infection prevalence in African buffalo (Adapted and reprinted from Jolles et al. 2008)

One recent example of a disease system with complex, cryptic, and indirect interactions comes from studies of bovine tuberculosis again, but this time in South Africa. Wild African buffalos (*Syncerus caffer*) can be infected with *Mycobacterium bovis*, just as domestic cattle can be. They can also be infected with other organisms, including gastrointestinal roundworms (Jolles et al. 2008). Jolles, Ezenwa, and colleagues asked whether infection with *M. bovis* is affected by infection with roundworms. They found that buffalos with high numbers of roundworms were unlikely to be infected with *M. bovis*, whereas those with low numbers of roundworms were likely to be infected (Fig. 13.2). What could cause this negative correlation? One possibility is that buffalo that are infected with both pathogens die at a higher rate, and they found evidence that this is indeed the case. For example, buffalos infected with Tb had significantly worse body condition only if they were also infected with worms; infection with worms alone had no effect on body condition.

Another possible explanation for the pattern in Fig. 13.2 is that the animals' immune systems might not be able to simultaneously mount effective immune responses to both pathogens. The immune system of mammals uses two distinct pathways to combat pathogens, depending on whether the pathogens are intracellular, like *M. bovis*, or extracellular, like roundworms (Jolles et al. 2008). The immune system also cross-regulates the two responses: when one pathway is activated, the other is actively suppressed (Jolles et al. 2008). Jolles et al. found that worm-free buffalos had the highest levels of activity of the immunological pathway that combats extracellular pathogens. Thus, buffalos that mount a strong immune response to worm infection may be more susceptible to infection with Tb because their immune systems are suppressing the intracellular response in favor of the extracellular response.

Jolles and her colleagues developed analytical models to assess whether either increased mortality or cross-reactivity could account for the patterns of co-infection that they observed in buffalo herds. Their analyses suggest that neither effect alone is sufficient to cause the pattern they observed; both increased mortality and immunological cross-reactivity must be occurring.

Of course, the health of buffalos is not directly connected to human health, but in many parts of Africa, buffalos and domestic cattle can come into close enough contact for transmission to occur (Cosivi et al. 1999). And, of course, the health of cattle is intimately connected to human well-being as is discussed in the chapters on hunger in this volume. But perhaps more importantly, the buffalo Tb system may serve as a model of how interactions between pathogens occur within hosts, and suggest ways that this knowledge could be used in effective treatment. For example, the buffalo example suggests that treating gastrointestinal worms could reduce Tb infection: if buffalos don't have to activate an immune response to worms, their immune system is freer to tackle Tb infection. Similar patterns have already been observed in a number of human diseases and treatment of parasitic infections is recommended for reducing the severity of Tb and HIV/AIDS in humans (Hotez et al. 2006).

Complex interactions can also occur outside of hosts. We return to our discussion of Lyme disease for an example. We mentioned previously that different host species have different probabilities of infecting ticks that are feeding on them, with the white-footed mouse infecting 92% of feeding ticks with the Lyme bacterium. Recent research has demonstrated that mice are also the hosts from which ticks are most likely to successfully take a blood meal (Keesing et al. 2009). More than half of ticks that are experimentally placed on mice manage to feed successfully, and almost all of those become infected with the Lyme bacterium (Fig. 13.3). In contrast, ticks are much less likely to be able to feed successfully on other host species, and those that do feed successfully are much less likely to become infected with the Lyme bacterium. Of 100 ticks that are placed on an opossum to elicit feeding, for example, only about four manage to feed successfully; the rest are groomed off and killed (Keesing et al. 2009). And of those that feed successfully, only 3% are likely to become infected with the Lyme bacterium (LoGiudice et al. 2003). Remarkably, when we find opossums in the forest, they have hundreds of ticks feeding successfully on them at any one time (LoGiudice et al. 2003). That means that opossums serve as an ecological trap for ticks: it must be the case that thousands of ticks attempt to feed, but 96% of those are killed by the opossum. And then of those few hundred that survive, only a handful become infected, because opossums are also poor reservoirs for the Lyme bacterium (LoGiudice et al. 2003). Having opossums, and other similar hosts, around is a good way to reduce Lyme disease risk, because they remove a huge number of ticks from the environment, while only infecting a small number of them. In contrast, having a lot of white-footed mice around increases risk, because they successfully feed lots of ticks and the ones they feed are highly likely to become infected.

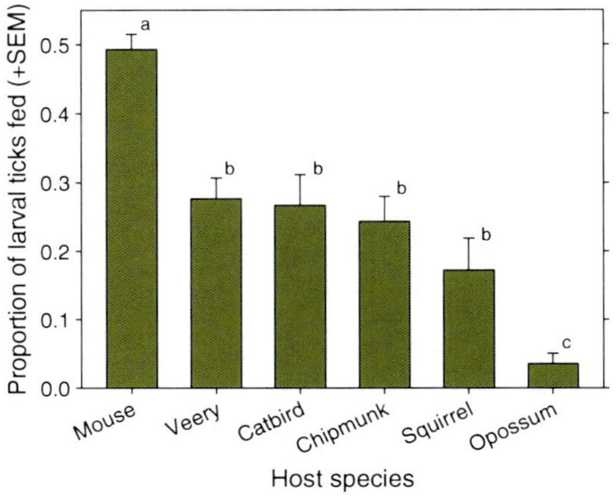

Fig. 13.3 The proportion of larval ticks that fed successfully (+ standard error of the mean) on six species that are common hosts for larval blacklegged ticks (*Ixodes scapularis*) in upstate New York, USA. Hosts were captured in the field and held in the laboratory until ticks naturally feeding on them had fed to repletion and dropped off. Hosts were then reinfested with 100 larval ticks and monitored to determine the proportion of those ticks that fed successfully. Lowercase letters indicate results that were significantly different (Data from Keesing et al. 2009)

Creating habitats with few mice is actually not that difficult, at least conceptually. As described earlier, mice are more abundant after big acorn crops, so acorns largely determine abundance through time. But from place to place, mice are found at high densities in habitats that have lost most of their vertebrate biodiversity (Nupp and Swihart 1998; LoGiudice et al. 2008). This is apparently because in these habitats, mice have lost both competitors, like squirrels and chipmunks, and predators, like foxes and weasels (Rosenblatt et al. 1999). Not surprisingly, habitats with higher predator diversity have lower mouse densities, and lower Lyme disease risk (Logiudice et al. 1998) because of an abundance of alternative hosts like opossums that siphon tick meals away from mice and then don't infect many of those ticks. This is an example of a phenomenon called the "dilution effect" in which high diversity reduces disease risk. The dilution effect has been described for a variety of diseases of plants and animals (Keesing et al. 2006; Keesing et al. 2010) including a number of human diseases such as West Nile virus encephalitis (Allan et al. 2008; Swaddle and Calos 2008; Ezenwa et al. 2006, 2007), hantavirus pulmonary syndrome (Dizney and Ruedas, in litt.), bartonellosis (Telfer et al. 2005), and of course Lyme disease (LoGiudice et al. 2003).

To protect human health, then, we should, it seems, create or preserve habitats with high vertebrate biodiversity. But how is that done? The only way we know of, at least in temperate forests, is to prevent forest fragmentation. Diversity is highest in continuous forest and large forest patches, presumably because bigger animals have larger requirements for space to maintain viable populations and can't survive in small fragments. In fact, small patches of fragmented forests in the eastern USA have the highest abundance of mice (Nupp and Swihart 1998), the lowest vertebrate diversity (Rosenblatt et al. 1999), and the highest Lyme disease risk (Allan et al. 2003).

Higher diversity is correlated with larger habitat areas in ecosystems throughout the world, an ecological phenomenon called the species-area relationship (Lomolino et al. 2006). If the dilution effect is as widespread as it appears to be, then preserving large, intact habitats would be a good strategy for conserving wildlife and, not coincidentally, might also be a good strategy for protecting human health, at least in some disease systems. Of course, the relationship between disease and habitat fragments occurs for Lyme disease because the best host for the pathogen is also an ecologically resilient host that is present in – and reaches high abundance in – degraded habitats such as forest fragments. If this is unique to the Lyme disease system, then the dilution effect might not occur for other diseases. But the same phenomenon does occur for a number of other diseases, as we have already noted. The most competent reservoirs for West Nile virus, for example, are birds such as house sparrows, house finches, American robins, and bluejays that live in degraded human habitats. Perhaps, then, it's not only a coincidence that the most competent reservoirs for pathogens are frequently those generalist species that survive best in a wide range of habitats. Perhaps successful pathogens have evolved to be best adapted to hosts that are ecologically resilient, and therefore most likely to respond positively to human-caused disturbances. An alternative, but not mutually exclusive, hypothesis is that ecologically resilient hosts have evolved life history strategies in which allocation of resources to particular types of immune defense is minimized. Such a strategy could be advantageous if disease rarely kills these animals before predators do. The result would be that these resilient species that are permissive to vectors and pathogens tend to dominate in human-degraded environments. These questions remain to be explored both theoretically and empirically.

Both the Lyme disease and buffalo tuberculosis examples illustrate that complex and indirect interactions among pathogens, hosts, and, in some cases, vectors, can lead to surprising outcomes. Just because interactions are complex, however, does not mean they are intractable. Both examples suggest general principles that might occur in a wide variety of disease systems, allowing us to make educated guesses about complex interactions in previously unstudied disease systems.

Summary and Conclusions

Throughout this chapter, we have emphasized features of disease transmission that appear to be broadly applicable. The densities of pathogens, hosts, and vectors are clearly important in many disease systems. So, too, though, are other things. Social behavior, in some circumstances, can override the effects of density, as in the badger example. So when social interactions are changing, for humans or for other animals involved in transmission of infectious diseases, we should not be surprised if disease transmission is affected. Diseases that are sexually-transmitted or vector-borne, and those that involve animals with a highly structured social organization, like that of badgers, would be most likely to be affected this way.

When density is the major factor determining disease transmission, we might be able to predict disease risk well in advance by keeping in mind the chain reactions that can lead to high densities. If we can, then we should be able to initiate public health interventions, including education, in time to reduce transmission. We must also be mindful of less direct interactions that might nevertheless be crucial. As the buffalo Tb example illustrates so well, when hosts are infected with both extracellular and intracellular pathogens, mitigation focused on one pathogen might reduce infection or severity of the other. Not surprisingly, given their shared mammalian ancestry, humans and buffalos appear to share responses to co-infection, so this strategy could apply to human health both directly and indirectly, though much on this topic remains to be explored.

Finally, lessons from Lyme disease suggest that the preservation of natural habitats might have direct bearing on human health because those habitats are likely to harbor high diversity and this diversity protects humans from disease transmission. The generality of the dilution effect is currently being explored by a number of research groups and their results should provide a great deal of insight into this intriguing possibility that aligns the goals of conservation with those of public health.

Collectively, the examples in this chapter demonstrate the many ways in which ecological interactions among organisms can influence the transmission of infectious diseases. The infectious diseases whose ecological complexities we understand in the most detail are not those that inflict the highest burden of disease on humans, such as tuberculosis, malaria, and dengue. However, there is no reason to believe that other pathogens should be any less sensitive to the types of interactions we have described in this chapter. As these examples have demonstrated, the integration of ecology with public health and sustainable development could improve the lives of people throughout the world.

References

Allan, B. F., F. Keesing, and R. S. Ostfeld. 2003. Effects of habitat fragmentation on Lyme disease risk. Conservation Biology 17:267–272.

Allan, B. F., W. A. Ryberg, R. B. Langerhans, W. J. Landesman, N. W. Griffin, R. S. Katz, K. N. Smyth, B. J. Oberle, M. R. Schultzenhofer, D. E. Hernandez, A. de St. Maurice, L. Clark, R. G. McLean, K. R. Crooks, R. S. Ostfeld, and J. M. Chase. 2008. Ecological correlates of risk and incidence of West Nile Virus in the United States. Oecologia 158:699–708.

Allman, E. S. and J. A. Rhodes. 2004. *Mathematical models in biology: an introduction*. Cambridge University Press.

Cossive, O., J. M. Grange, C. J. Daborn, M. C. Raviglione, T. Pujikura, and D. Cousins et al. 1999. Zoonotic tuberculosis due to Mycobacterium bovis in developing countries. Emerging infectious diseases 4(1):59–70.

Donnelly, C. A., R. Woodroffe, D. R. Cox, J. Bourne, G. Gettinby, A. M. Le Fevre, J. P. McInerney, and W. I. Morrison. 2003. Impact of localized badger culling on TB incidence in British cattle. Nature 426:834–837.

Ezenwa, V. O., M. S, Godsey, R. J, King, and S. C. Guptill. 2006. Avian diversity and West Nile virus: Testing associations between biodiversity and infectious disease risk. Proceedings of the Royal Society of London Series B-Biological Sciences 273:109–117.

Ezenwa, V. O., L. E. Milheim, M. F. Coffey, M. S. Godsey, R. J. King, and S. C. Guptill. 2007. Land cover variation and West Nile virus prevalence: Patterns, processes, and implications for disease control. Vector-Borne and Zoonotic Diseases 7:173–180.

Fan, H., R. F. Conner, and L. P. Villarreal. 2007. AIDS: science and society. Jones and Bartlett Publishers, Sudbury, MA, USA.

Ferrari, M. J., R. F. Grais, N. Bharti, A. J. K. Conlan, O. N. Bjørnstad, L. J. Wolfson, P. J. Guerin, A. Djibo, and B. T. Grenfell. 2008. The dynamics of measles in sub-Saharan Africa. Nature 451:679–684.

Grieco, J. P., N. L. Achee, R. G. Andre, and D. R. Roberts. 2002. Host feeding preferences of *Anopheles* species collected by manual aspiration, mechanical aspiration, and from a vehichle-mounted trap in the Toledo District, Belize, Central America. Journal of the American Mosquito Control Association 18:307–315.

Griffin, J. M., D. H. Williams, G. E. Kelly, T. A. Clegg, I. O'Boyle, J. D. Collins, and S. J. More. 2005. The impact of badger removal on the control of tuberculosis in cattle herds in Ireland. Preventive Veterinary Medicine 67:237–266.

Hotez, P. J., D. H. Molyneux, A. Fenwick, E. Ottesen, S. E. Sachs S, and J. D. Sachs. 2006. Incorporating a Rapid-Impact Package for Neglected Tropical Diseases with Programs for HIV/AIDS, Tuberculosis, and Malaria. PLoS Medicine 3:No. 5, e102.

Jolles, A. E., V. O. Ezenwa, R. S. Etienne, W. C. Turner, and H. Olff. 2008. Interactions between macroparasites and microparasites drive infection patterns in free ranging African Buffalo. Ecology 89:2239–2250.

Keesing, F., R. D. Holt, and R. S. Ostfeld. 2006. Effects of species diversity on disease risk. Ecology Letters 9:485–498.

Keesing, F, Brunner, J., Duerr, S., Killilea, M., Logiudice, K., Schmidt, K., et al. (2009). Hosts as ecological traps for the vector of Lyme disease. Proceedings. Biological sciences / The Royal Society 276(1675):3911–3919.

Keesing, F., L. K. Belden, P. Daszak, A. Dobson, C. D. Harvell, and R. D. Holt et al. 2010. Impacts of biodiversity on the emergence and transmission of infectious diseases. Nature 468(7324):647–652.

Keirans, J. E., H. J. Hutcheson, L. A. Durden, and J. S. H. Klompen. 1996. *Ixodes (Ixodes) scapularis* (Acari: Ixodidae): Redescription of all active stages, distribution, hosts, geographical variation, and medical and veterinary importance. Journal of Medical Entomology 33: 297–318.

Krebs, J. R., R. M. Anderson, T. Clutton-Brock, C. A. Donnelly, S. Frost, W. I. Morrison, R. Woodroffe, and D. Young. 1998. Badgers and bovine TB: Conflicts between conservation and health. Science 279:817–818.

Krebs, J. R., R. Anderson, T. Clutton-Brock, I. Morrison, D. Young, C. Donnelly, S. Frost, and R. Woodroffe. 1997. Bovine tuberculosis in cattle and badgers. HMSO, London, UK.

Lloyd-Smith, J. O., W. M. Getz, and H. V. Westerhoff. 2004. Frequency-dependent incidence in models of sexually transmitted diseases: portrayal of pair-based transmission and effects of illness on contact behaviour. Proceedings of the Royal Society of London Series B-Biological Sciences 271:625–635.

LoGiudice, K., S. Duerr, M. Newhouse, K. A. Schmidt, M. Killilea, and R. S. Ostfeld. 2008. Impact of community composition on Lyme disease risk. Ecology 89:2841–2849.

LoGiudice, K., R. S. Ostfeld, K. A. Schmidt, and F. Keesing. 2003. The ecology of infectious disease: effects of host diversity and community composition on Lyme disease risk. Proceedings of the National Academy of Sciences 100:567–571.

Lomolino, M., B. Riddle, J. H. Brown. 2006. Biogeography, 3rd edition. Sinauer Associates, Sunderland, MA, USA.

McKenzie, V. J. and A. R. Townsend. 2007. Parasitic and infectious disease responses to changing global nutrient cycles. EcoHealth 4:384–396.

Nupp, T. E. and R. K. Swihart. 1998. Effects of forest fragmentation on population attributes of white-footed mice and eastern chipmunks. Journal of Mammalogy 79:1234–1243.

Ostfeld, R. S. 1997. The ecology of Lyme-disease risk. American Scientist 85:338–346.

Ostfeld, R. S., C. D. Canham, K. Oggenfuss, R. J. Winchcombe, and F. Keesing. 2006. Climate, deer, rodents, and acorns as determinants of variation in Lyme-Disease risk. PLoS Biology 4:No. 6, e145.

Ostfeld, R. S. and F. Keesing. 2000. Pulsed resources and community dynamics of consumers in terrestrial ecosystems. Trends in Ecology and Evolution 15:232–237.

Rejmánková, E., J. Grieco, N. Achee, P. Masuoka, K. Pope, D. Roberts, and R. M. Higashi. 2006. Freshwater community interactions and malaria. Pages 90–104 in S. K. Collinge and C. Ray editors. Disease ecology: Community structure and pathogen dynamics. Oxford University Press, New York, NY, USA.

Rosenblatt, D. L., E. J. Heske, S. L. Nelson, D. M. Barber, M. A. Miller, and B. MacAllister. 1999. Forest fragments in East-central Illinois: Islands or habitat patches for mammals? The American Midland Naturalist 141:115–123.

Singer, B. H., M. C. de Castro. 2001. Agricultural colonization and malaria on the Amazon frontier. Ann N Y Acad Sci. 954:184–222.

Swaddle, J. P. and S. E. Calos. 2008. Increased avian diversity is associated with lower incidence of human West Nile infection: Observation of the dilution effect. PLoS ONE 3:e2488

Telfer, S., K. J. Bown, R. Sekules, M. Begon, T. Hayden, and R. Birtles. 2005. Disruption of a host-parasite system following the introduction of an exotic host species. Parasitology. 130:661–668.

Woodroffe, R., C. A. Donelly, D. R. Cox, F. Johnbourne, C. L. Cheeseman, R. J. Delahay, G. Gettinby, J. P. McInerney, and W. I. Morrison. 2006. Effects of culling on badgerMeles meles spatial organization: implications for the control of bovine tuberculosis. Journal of Applied Ecology 643:1–10.

Zerhouni, E. 2008. NIH Strategic Vision for the Future: From Curative to Preemptive Medicine Available at http://www.nih.gov/strategicvision.htm. Accessed Sept 24 2008.

Chapter 14
Human Health as an Ecosystem Service: A Conceptual Framework

Karen Levy, Gretchen Daily, and Samuel S. Myers

Introduction

To live in good health and, in many ways, to live at all, people need a wide array of life-support benefits that derive from ecosystems. Collectively these are called *ecosystem services*, a term referring to the conditions and processes through which ecosystems, and the species that make them up, sustain and fulfill human life (Ehrlich and Ehrlich 1981; Daily 1997; Millennium Ecosystem Assessment 2005). These processes underpin the production of goods (such as seafood and timber), life-support functions (water purification and flood control), and life-fulfilling conditions (beauty and inspiration), as well as the preservation of options (such as genetic diversity for future use).

Ecosystems and human health are thus intimately interlinked. The preceding three chapters illustrate how changes in land use and climate can impact health directly, and how numerous indirect impacts on human health are mediated through changes in the composition of species in a given ecosystem. Here, we explore more directly the ways in which the condition of ecosystems and the health of human populations are linked, and we explore prospects for illuminating these linkages to advance scientific understanding and inform management options and decisions. The kinds of questions that stand out include,

1. Are there practical and reliable indicators of ecosystem condition/function that signal levels of risk to human health?
2. Can change in certain ecosystem attributes (size, configuration, and composition) be reliably translated into changes in health risks?

K. Levy (✉)
Department of Environmental Health, Rollins School of Public Health, Emory University, 1518 Clifton Rd NE, Atlanta, GA 30322 USA
e-mail: karen.levy@emory.edu

J.C. Ingram et al. (eds.), *Integrating Ecology and Poverty Reduction: Ecological Dimensions*, DOI 10.1007/978-1-4419-0633-5_14,

The relationships between biophysical attributes of ecosystems and human communities are complex. Destruction of ecosystems can improve aspects of community health. For example, draining swamps can reduce habitat for the mosquito vector that transmits the parasite that causes malaria, as discussed by Myers, this volume. At the same time, ecosystems provide many services that sustain human health, for which substitutes are not available at the required scale, such as purification and regulation of drinking water flow.

To date, there is little rigorous research establishing the links between ecosystem conditions and human health. In order to understand the complexities of these relationships, there is a need to clarify the factors that confound them, and to establish a common lexicon for ecologists and health scientists to discuss them. In this chapter, we describe some of the evidence that exists to indicate important adverse health impacts from deteriorating ecosystem services, and also outline the reasons that epidemiological evidence for these relationships remains difficult to establish. We also discuss how to move forward not only to establish more clear evidence, but also to help set policy agendas for addressing these relationships.

Background

Ecosystem services are the conditions and processes through which ecosystems, and their biodiversity, sustain and fulfill human life. Ecosystem services are generated by a complex of natural cycles, powered by solar energy. These cycles operate on a wide spectrum of temporal and spatial scales, from protracted and global biogeochemical cycles to comparatively instantaneous life cycles of tiny bacteria (Daily 1997).

Ecosystem services can be classified by four different types (Millennium Ecosystem Assessment 2005):

Provisioning services include the products obtained from ecosystems, such as food, freshwater, building materials, fuels, and precursors to pharmaceutical and industrial products;

Regulating services are the benefits obtained from regulation of ecosystems, including flood and storm control, climate regulation, water purification, disease regulation, and carbon sequestration;

Supporting services are defined as services needed for the production of all other ecosystem services, and include nutrient dispersal and cycling, soil formation, waste decomposition and detoxification, primary production, crop pollination, and seed dispersal; and

Cultural services include all non-material benefits obtained from ecosystems, such as cultural heritage, intellectual and spiritual inspiration, recreational experiences, educational opportunities, and aesthetic value.

The ecosystem services framework allows the benefits that human societies obtain from ecosystems to be explicit in policy considerations (Millennium

Table 14.1 Some relationships between ecosystem conditions or processes and human well-being

Ecosystem condition or process	Intermediate ecosystem service	Final ecosystem service	Dimension of human well-being
Biodiversity in oceans		Production of a wide array of seafood	Nutrition
Primary production and herbivory	Herbivory	Production of animal biomass	
Predation	Control of agricultural pests	Production of plant biomass for use as food, fiber, timber, and fuel	
Pollination	Pollination		
Nutrient cycling	Generation and renewal of soil fertility		Clothing Shelter Energy
	Decomposition of waste	Protection from pathogens and toxins	Health
	Purification of water		
Photosynthesis	Carbon sequestration	Climate stabilization	Protection from climate variability (storms, floods, droughts, heat waves)
Seed dispersal	Replenishment of natural vegetation	Landscape stabilization	
		Seed dispersal	Economic flexibility and security
Ecological stability		Relatively constant production	
Generation and maintenance of biodiversity	Preservation of options	Possibility of using a good or service (e.g., a natural medicinal product or crop pollination) in the future	
		Beauty	Aesthetic inspiration
		Complexity	Intellectual stimulation
			Diverse cultures
		Serenity	Peace of mind

Ecosystem Assessment 2005). It also provides a way to consider the losses we might be incurring when we lose well-functioning ecosystems.

From the inception of the concept of ecosystem services in the 1970s and 1980s (Mooney and Ehrlich 1997), health has been widely cited as a main service that ecosystems provide. Table 14.1 illustrates one of the many possible perspectives relating the conditions and processes occurring in ecosystems to key elements of human well-being. Although very simple, this table reveals several important observations.

First, relatively few ecosystem conditions and processes confer direct benefits on humanity, such as in the way some ocean biodiversity contributes, via seafood, to human nutrition. Rather, most ecosystem conditions and processes confer numerous indirect benefits, or "intermediate services." The second column in the table presents a variety of such intermediate services, such as pollination, agricultural pest control, and renewal of soil fertility, all of which contribute indirectly to human nutrition and health. One could easily envision additional columns that would illuminate more intermediate services, revealing more details of pollination, predation, and soil processes.

Second, individual ecosystem conditions and processes contribute to more than one final ecosystem service, and ultimately all lead to aspects of human well-being. This observation holds true in virtually any classification, at any level of detail. Thus, nutrient cycling contributes to nutrition, clothing, shelter, energy, and health. Similarly, predation contributes not only to control of agricultural pests, but also to control of reservoirs and vectors of human pathogens, and thereby to health (this link is not shown in the table).

Third, the inverse of the second point is also true, meaning that maintaining any single aspect of well-being, such as food security, requires attention to a great many aspects of the ecosystems supporting it. Protection from climate variability underscores this point, depending on strikingly different ecological conditions and processes. The two shown in Table 14.1 are photosynthesis and seed dispersal.

Fourth, biodiversity is involved in virtually every part of the table. Indeed, if one were to create a relational table such as this at a very fine level of detail, one could list as intermediate services the conditions and processes required to support each individual population, of each species, involved in each final service (Luck et al. 2003, 2009). Thus, ecosystem services contribute to making human life both possible and worth living, in complex and interesting ways. Because of their basis in cycles, operating over such varying scales, the classification of ecosystem services is inherently arbitrary, a function of context and the most useful point of entry into the cycles, and the appropriate level of detail of analysis.

In some situations, the services provided by ecosystems can be replaced by physical infrastructure (such as water treatment facilities). But in many situations, intact ecosystems provide services more effectively or efficiently than engineered alternatives; sometimes they are irreplaceable. For example, animal-pollinated crops provide one-third of the calories in the human diet (Klein et al. 2007) and pollinator-provided calories (in the form of nuts, seeded vegetables, and fruits) are especially rich in nutrients that support human health. Ongoing declines in populations of pollinators may threaten the crops upon which human communities depend to meet their nutritional needs. While bees have been managed for honey production for thousands of years, only in the past century have people managed bees for pollination services – to crops in highly intensified systems from which natural sources of pollinators have been eliminated (B. Brosi, personal communication, 4/5/09). Refer to the chapter on ecosystem services in agricultural landscapes by Smukler et al., this volume, for additional discussions related to these issues.

Natural watersheds provide a filtering mechanism for improving water quality for human consumption. While water treatment plants can replace this service, the quality of source water entering into water treatment plants can affect the quality of water consumed by human communities even in areas with high investment in water purification systems. There is some evidence to suggest that higher turbidity levels of source water are associated with higher rates of hospital visits for gastrointestinal illness (Schwartz et al. 2000; Mann et al. 2007; Tinker et al. in press). In some cases, it can be more cost-effective to maintain watershed functioning than to build a water treatment plant, and there are examples of this from many cities in the United States and other parts of the world (Postel and Thompson 2005). For example, the city of New York recognized this in their decision to restore the Catskill watershed to provide the city with water purification rather than investing in a water filtration plan (Chichilnisky and Heal 1998). Other municipalities are using managed wetlands as tertiary water treatment facilities (Humboldt State University 2009).

Natural barriers, such as vegetated dunes, reefs, mangrove forests, and wetland systems, can aid in controlling natural hazards, and intact ecosystems in some cases eliminate the need for extensive human engineering to control the forces of nature. In the 2004 Southeast Asian tsunami, the coastal areas of Thailand flanked by several hundred meters of mangrove forests withstood the tsunami's impacts far better than did areas where mangroves had been cut down, such as in Sri Lanka. Coral reefs also had a wave-buffering effect. Recognizing this, in the wake of the tsunami, the Thai government is allowing only a few of the impacted and displaced shrimp farmers to return and instead allocating more land for mangrove forests (Englande 2008). In India, mangroves forest cover has been associated with lower cyclone-associated death tolls (Das and Vincent 2009). Similar management issues concern natural wetlands and barrier islands that once protected the Gulf Coast of the United States from storms like Hurricane Katrina (see chapter by Ingram and Khazai, this volume, for further discussion on coastal disasters).

The threat that the depletion of ecosystem services represents to human health has been acknowledged in various global policy arenas. Policy papers produced by major international undertakings like the Millennium Ecosystem Assessment (MA) (Millennium Ecosystem Assessment 2005) and Global Environmental Outlook (GEO4) (United Nations Environment Programme (UNEP) 2007) contain numerous statements that ecosystem service degradation will have significant impact on human health and well-being and threatens to reverse progress on the Millennium Development Goals (MDGs) (United Nations 2009). The MA is the broadest review to date of ecosystem services research, synthesizing information from scientific, governmental, private, and local sources. According to the MA, "The degradation of ecosystem services is harming many of the world's poorest people and is sometimes the principal factor causing poverty." At the same time the authors acknowledge that "the information available to assess the consequences of changes in ecosystem services for human well-being is relatively limited," (Millennium Ecosystem Assessment 2005). The Health Synthesis of the MA, which summarizes the findings of the MA's global and sub-global assessments of how ecosystem changes specifically affect human health and well-being, concludes that, while "ecosystem services are indispensable to the well-being of people everywhere", "limited information [exists] on the details of linkages between human well-being and the provision of ecosystem services, except in the case of food and water," (Corvalán et al. 2005). Thus we know that human communities depend on nature to provide services and that reduced access to these services below some threshold "should" impact health and well-being. But while we may accept the logic that humans depend on their local environments to diversify their food supply, provide safe drinking water and sanitation, or additional sources of income, how much direct evidence is there that ecosystem service degradation is causing poor health outcomes? Or that good health outcomes are attributable, in part, to good supply of ecosystem services?

Efforts to quantify the relationships between ecosystem services and human health on a large scale have tended to underestimate the complexity of these relationships. Two studies have explicitly explored the association between ecological conditions and human health outcomes (Sieswerda et al. 2001; Huynen et al. 2004). Both studies report the results of linear regression analysis on a global scale, using aggregated country-wide datasets from sources including the World Resources Institute (WRI), World Bank, and World Health Organization (WHO). Neither study's conclusions support the hypothesis that loss of ecosystem services leads to a decline in the health and well-being of human communities. Any relationship that they did find disappeared once indicators of socioeconomic status were controlled for in the models. However, several methodological problems limit the conclusions drawn from this work, many of which were acknowledged by the authors.

Challenges to Linking Ecosystem and Human Health

The design failures of the aforementioned studies illustrate some of the problems that have plagued this field of inquiry. To move toward a better understanding of the role of ecosystems in supporting human health, we highlight some of these methodological issues, in an effort to move beyond them.

Scale of Inquiry

Both Sieswerda et al. (2001) and Huynen et al. (2004) undertook an analysis of the relationship between ecosystem status and human health at the global scale, using countries as the unit of analysis. This scale of analysis misses many of the complexities of the relationship between ecosystem integrity and human health. Many different human and ecological conditions exist within any given country, urban and rural, wealthy and poor. Rural populations rely heavily on local ecosystem services to support their livelihoods (Gadgil 1998). Communities that rely directly on local or regional ecosystem services will experience the impacts of loss of these services that may not be felt country-wide. In addition, many different biophysical realities exist within any given country. Effects will vary depending on ecosystem type, and even within a given ecosystem, relationships between ecosystem services and health may be scale-dependent. For example, hydrological services are largely regional (Brauman et al. 2007). Thus regional level analyses are needed before data can be aggregated on a global scale. Appropriate variables should be used to stratify data analysis, including characterizations of the ecosystem (e.g., arid vs. wet; high vs. low altitude; tropical vs. temperate) and of the population in question (e.g., urban vs. rural).

Measures of Ecosystem Services

Sieswerda et al. were interested in the broad effects of "ecological integrity" on human well-being and health, whereas Huynen et al. had a more specific focus on biodiversity loss and its effects. Indicators of ecosystem function used by the authors included such variables as percentage of threatened species per 10,000 km^2, current forest as a percentage of original forest, percentage of land highly disturbed by human activity, percentage of the country's land mass totally or partially protected, percentage of forest remaining since pre-agricultural times, and average annual change in forest cover.

These measures of ecosystem health address land use and species composition broadly, but do not capture ecosystem services, *per se*. Again, the country-wide scale at which these indicators are measured do not capture regional and local effects. Country-wide metrics of forest cover and species counts provide no information about the spatial distribution of these metrics in relation to the location of human populations.

Many habitat types other than forests provide ecosystem services, and ecosystem services are also provided in working landscapes (Daily 1997), and local sociological, economic, and other human factors are also important factors to consider. More appropriate metrics of ecosystem health with respect to the provisioning of ecosystem services might include measures of water quality and quantity, air quality, food resource availability, or abundance of specific disease vectors or hosts.

Definitions of "Health"

The WHO defines human health as "a state of complete physical, mental and social well-being" (World Health Organization (WHO) 2006). Several different proxies can be used as indicators of overall health. Sieswerda et al. (2001) used life expectancy as the outcome of interest, whereas Huynen et al. (2004) used several different indicators of health: life expectancy, disability adjusted life expectancy (DALE), infant mortality rate, and percentage of low birth weight babies.

While these are useful overall measures of health as an integrated measure of well-being, further insights about the relationship between health and ecosystem services will be gained by stratifying the outcome in question. Lumping all diseases into a broad category of "health" can obfuscate our understanding, because changes to ecosystems will have drastically different effects on diseases of different etiologies. For example, insecticide application for mosquito abatement might reduce the incidence of vector-borne diseases while at the same time increasing the incidence of cancer. Draining swamps might reduce exposure to malaria or West Nile virus but reduce water quality and increase exposure to water-related disease.

The WHO has adopted the global burden of disease (GBD) approach to measuring burden of disease using the disability-adjusted life years (DALYs) metric. DALYs are a time-based measure that combines mortality (years of life lost from premature death) and morbidity (years of life lost due to time lived in states of less than full health). This approach allows for a consistent assessment of the burden of disease across diseases, risk factors, and regions (World Health Organization (WHO) 2009).

Several efforts have quantified the global burden of disease attributable to environmental factors, with estimates of 23–33% (World Health Organization 1997; Smith et al. 1999; Prüss-Ustün and Corvalán 2007). However, the definitions of "environment" used in these assessments were not specific to ecosystem-level change.

A systematic inventory of the burden of disease attributable to changes in particular ecosystem services, stratified by regions and ecosystem types, would help elucidate the varied relationships between ecosystem health and human health (Corvalán et al. 2005). To be sure, burden of disease assessment cannot fully account for complex causal pathways, long timescales, and potential irreversibility of alterations to ecosystems (Corvalán et al. 2005). However, this type of analysis would provide more insight into specific relationships in a way that an aggregated analysis cannot.

Data Availability

Both Sieswerda and Huynen relied on large publically available databases for their analyses; both utilized WRI data and in addition Huynen included data from World Bank and WHO sources. Sieswerda rightly points out that the availability of information depends on a country's ability to collect or willingness to provide data, which is not unrelated to health status. The scale of data used must match the question of interest. Analysis of large datasets can provide important insights if appropriate questions are asked. But in many cases, finer grained data will be needed to address the relationship between ecosystem health and human health at an appropriate scale. Where data exist, a series of analyses at smaller scales can be aggregated to provide insight into more generalizable phenomena. Where such data are lacking, research should be designed and carried out to fill in these gaps. We next discuss approaches to collecting such data.

Analytical Approach

Linear regression, as applied by Sieswerda and Huynen, represents a convenient analytic approach to investigating relationships between explanatory variables and outcomes of interest. However, these techniques cannot account for many of the complexities of the relationships between ecosystem condition and human health. The relationship between different types of resource scarcity and negative health outcomes is not likely to be linear. Effects of ecosystem change on community health are not likely to be felt immediately, but rather experienced gradually over

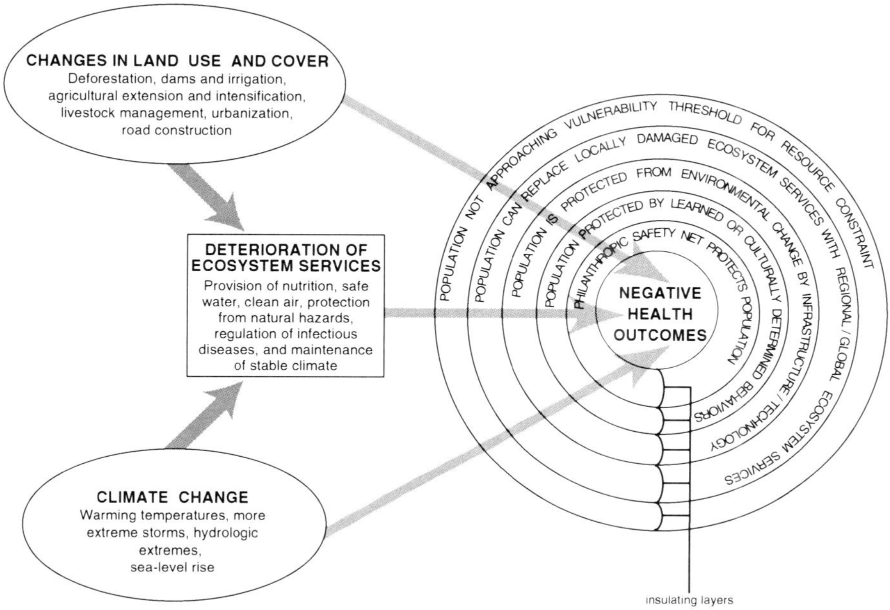

Fig. 14.1 *A schematic of the complex relationships between altered environmental conditions and human health.* Drivers of global environmental change (e.g., Land use change or climate change) can directly pose health risks, or impair ecosystem services that subsequently influence health. For hazards that affect human health, however, exposures will be modified by multiple layers of social or infrastructure barriers that can buffer or eliminate risk. Together, all components must be considered to achieve realistic assessments of population vulnerability (Reprinted, with permission, from Myers and Patz (2009). ©2009 by Annual Reviews www.annualreviews.org)

time. Huynen notes this effect, explaining that "only when a threshold in the losses of biodiversity is reached, the provision of ecosystem goods and services gets compromised" (Huynen et al. 2004). A strong correlation with health will be reached only when resources are very constrained. Until this "threshold" is reached, depletion of ecosystem services might have little impact on health (Fig. 14.1). Because complex and interdependent causal pathways introduce non-linearity into the relationship, appropriate analytic techniques will be necessary to account for time lags in effect, non-linearity of responses, and threshold effects. We discuss alternative analytical approaches further below.

Insulating Factors

Ultimately, direct relationships between measures of ecosystem health and measures of human health may be difficult or impossible to establish because human populations tend to be insulated from direct impacts of ecosystem service degradation by a

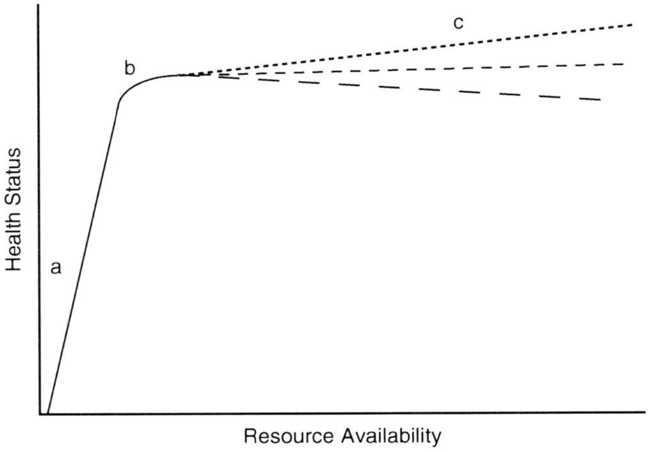

Fig. 14.2 *A schematic diagram of a proposed relationship between resource scarcity and human health.* When resources are tightly constrained (**a**), increases or reductions in access to them can have significant health consequences. Once access to adequate food, water, fuel building materials, etc. has been achieved (**b**), the relationship between increased access and health gains becomes much less pronounced. Further increases in resource access (**c**) may lead to marginal improvements in health status, but overuse may also lead to reduced health status, for example, excess food consumption and obesity (Reprinted, with permission, from Myers and Patz (2009). ©2009 by Annual Reviews www.annualreviews.org)

variety of mitigating factors (Fig. 14.2). According to the WHO Director General, "Nature's goods and services are the ultimate foundations of life and health, even though in modern societies this fundamental dependency may be indirect, displaced in space and time, and therefore poorly recognized" (Corvalán et al. 2005).

The ability to "trade" ecosystem services through access to local or global markets or philanthropic safety nets, and the ability to mitigate loss of ecosystem services through infrastructure or behavioral practices confound the relationship between ecosystem health and human health. Dasmann contrasts "ecosystem people" and "biosphere people" in terms of the radius over which they have access to resources, and over which they are vulnerable to ecosystem disruption (Dasmann 1988; Gadgil 1998).

"Trading" ecosystem services can occur when locally damaged ecosystem services are replaced by regional or global ecosystem services. In essence, richer countries have been able to buy their way out of the health effects of ecosystem destruction. This is seen with the effect of socioeconomic indicators, such as the Gross Domestic Product, that overwhelm the effect of loss of ecosystem integrity for predicting human health outcomes in both Sieswerda and Huynen's models. Higher income countries are buffered from the effects of loss of ecosystem services with their ability to import these services by accessing goods produced on "ghost acreage" elsewhere (Borgstrom 1965).

Philanthropic efforts represent another sort of "trading" of ecosystem services. For example, the health impacts of a hurricane that hits an area with a heavily degraded ecosystem might be ameliorated by humanitarian relief efforts. Or regions affected by famine or natural hazards may be supported by importing goods and services from outside the affected region. This type of humanitarian assistance may come from national or international organizations.

Technology and infrastructure can also confound the relationship between loss of ecosystem services and subsequent health impacts. Technology such as the introduction of fertilizers for increased food production can insulate populations from altered environmental conditions. Likewise, the development of infrastructure such as water treatment plants to protect against deteriorating water quality or seawalls to protect against storm surges can reduce the impacts of changing ecological systems on human populations. Such measures can replace some of the services provided by ecosystems, or otherwise buffer communities from their loss. Technology interacts with human behavior which represents another type of protective measure. For example, even if microbial water quality has deteriorated, people can treat their water to prevent the ingestion of pathogenic organisms. The use of bednets can also protect against the spread of malaria vectors. To the extent that these behaviors are culturally mediated, however, they may have less ability to adapt to rapidly changing environmental conditions, as these behaviors often evolve over many generations. Technology, infrastructure, and trade are all highly dependent on access to resources, so more impoverished populations will tend to be less buffered to ecosystem degradation than wealthier ones.

Vulnerability, as a concept, includes not only exposure to the health risks associated with changing environmental conditions but also the population-level conditions we have discussed above that make such exposure more or less safe (Turner et al. 2003).

Relationships Between Specific Ecosystem Services and Health

The ways in which provisioning, regulating, supporting, and cultural ecosystem services support human health and well-being have been reviewed elsewhere (Corvalán et al. 2005). The most direct relationships between ecosystem services and health exist for provisioning of food and safe water and the regulation of infectious disease, climate, and natural hazards. However, even with these apparently direct relationships, there is often a paucity of data showing a direct correlation between deteriorating ecosystem services and adverse health outcomes. Here, we highlight three classes of ecosystem services that have strong and clear relationships with human health. We summarize the salient evidence supporting these relationships, and also describe where more information is needed to determine the strength and direction of the relationship in various different circumstances. In order to illustrate the methodological issues described above, we discuss why it has been difficult to establish the evidence, key confounders, and suggestions for lines of inquiry and approaches to these topics. We focus on the ecosystem services of food production, fresh water supply, and protection from natural hazards. While other ecosystem services, such as regulation of climate and infectious diseases, also have direct relation to human health, they have been covered extensively in the previous three chapters on land use, climate change, and disease ecology.

Food Production

Food production is a key provisioning service as is outlined in the four chapters on ecology and hunger, this volume. Food production as a provisioning service is supported by other services, such as soil formation, biodiversity as a source of new crop varieties, pollination, pest control, climate regulation, water supply, and nitrogen fixation. An adequate supply of food supports human nutritional requirements and overall bodily function, including cognitive development, metabolic and endocrine functioning, reproductive health, immune status, and overall vigor. Food shortages lead to malnutrition, which causes stunting, cognitive impairment, diarrhea, and can ultimately lead to starvation. Worldwide, roughly 16% of the global burden of disease is attributable to childhood malnutrition (Murray and Lopez 1997) and, in 2008, the FAO estimated that 923 million people worldwide are suffering from malnutrition (2008).

During the next 50 years, global demand for food is projected to double (Alexandratos 1999). In certain parts of the world, particularly sub-Saharan Africa and parts of South Asia, rapidly growing populations are already encountering ecological constraints to local food production. Soil degradation and water scarcity have prevented yields from rising over the past 35 years, and, in some areas, they have been falling. Poor access to fertilizers, new crop varieties and irrigation have increased local vulnerabilities to these environmental trends. Future threats to food production include further land degradation, increasing water scarcity, accelerating climate change, and growing demand caused by population growth and increased meat consumption. Loss of wildlife habitat and fisheries depletion also decrease protein intake from hunting and fishing. Given the apparent strength of this relationship between the ecosystem service of food production and nutrition that supports human life, what are the challenges in providing evidence of a link between the depletion of this service and adverse health outcomes?

First, the scale of analysis is critical. Global food production can be a misleading indicator of food scarcity. For the time being, global food production exceeds global demand, and yet, nearly a billion people are chronically hungry. For many poor populations, global food supply is irrelevant because they do not have the resources to access global grain markets. For them, meeting nutritional requirements is based on locally productive ecosystems providing sources of basic nutrition. When these systems become less productive as a result of increasing water scarcity, land degradation, or nutrient depletion, the impacts may be quite immediate. Poor, rural, populations are unable to insulate themselves from these changing environmental conditions by accessing global grain markets or importing water or nutrients in the form of irrigation and fertilizer. As a result, these people feel the loss of ecosystem services most acutely. Thus, the scale of inquiry is important to consider with respect to assessing this relationship. Global or national agricultural data may not reflect a particular region's vulnerability to malnutrition as a result of local environmental change. In addition, other factors may come into play. For example, the relationship between degradation of arable land, food supply, and malnutrition may not be linear. The effect may be lagged in time, and subject to a threshold phenomenon: only after agricultural productivity falls below a certain level do communities experience the health effects of declines in productivity. Thus, the analytical approach is important to consider in order to take these distortions into account.

In the more developed regions of the world, access to global markets plays the role of an insulating factor preventing communities from experiencing health impacts of the destruction of ecosystem services. In the words of Sieswerda et al., "Our results suggest that there is a separation of consumption from consequence," (Sieswerda et al. 2001). Of course, if enough pressure is placed on productive landscapes at a global scale then people from all regions will feel the impact.

Fresh Water Supply

The fresh water that humans depend on flows directly from ecosystems, which provide water for extractive and in-stream use, for water-related cultural services, and for other water-related supporting services (also see chapters 6–9, this volume). "Extractive water supply" is that water available for municipal, agricultural, commercial, industrial, and thermoelectric power use. Ecosystems can act as natural water purification plants, filtering out the chemicals, microbes, nutrients, salts, and sediments that contaminate surface and groundwater. They can also buffer extreme water flow events. Intact forests and riparian buffers promote the transfer of surface water to groundwater by infiltration, which reduces flood peaks and can increase base flow, generally increasing the predictability of water availability. Floodplain wetlands also reduce flooding by absorbing and slowing floodwaters (Brauman et al. 2007). Riparian forests, upland forests, wetlands, and mangroves, all of which play a disproportionate role in the provisioning of these hydrological services, are particularly vulnerable to human interventions.

Water availability is a function of factors such as regional climate patterns and natural hydrological processes, and is increasingly affected by anthropogenic impacts, such as climate change, loss of vegetation, and increased demand. Because of these stresses, one-third of the world's population now lives in countries experiencing moderate to high water stress (Corvalán et al. 2005). During the next 50 years, water demand for irrigation – which accounts for roughly 70% of total fresh water use – is expected to triple while household and manufacturing uses are also expected to increase significantly (Postel 1998).

Over a billion people do not have access to adequate safe water and 2.6 billion people do not have access to adequate sanitation (United Nations Development

Programme (UNDP) 2006). Water supply is known to be an important factor in reducing the incidence of waterborne disease (Fewtrell et al. 2005), and inadequate access to water, sanitation, and hygiene is estimated to cause 1.7 million deaths annually and the loss of at least 50 million healthy life years. Half of the urban population in Africa, Asia, Latin America, and the Caribbean suffers from one or more diseases associated with inadequate water and sanitation (Vorosmarty et al. 2005). Inadequate access to uncontaminated fresh water is likely to increase exposure to waterborne disease by reducing access to sanitation and increasing direct exposure to pathogens. In addition, as we have discussed, water scarcity is a major threat to agricultural production and is already reducing local food production in certain regions. Altered flow regimes can also lead to injuries and other effects of flooding. But what are the challenges of providing evidence to support these relationships?

Scale of inquiry is, once again, a key factor to consider. Water timing, quality, and availability are all highly regional phenomena, and often depend on land management of highly localized watersheds. Extrapolations of local and short-term effects of hydrologic services to larger scales may be flawed because effects observed on small scales are not always seen within an entire basin (Brauman et al. 2007). Additionally, many aspects of hydrologic response are dominated by extreme but infrequent events (Brauman et al. 2007). Thus, care must be taken in defining the *measure of ecosystem services* when considering hydrologic services. Health effects may not be a function of average flows but rather of extreme flows. For example, reviewing almost 50 years of data from the USA, (Curriero et al. 2001) found that 51% of waterborne disease outbreaks were preceded by precipitation events above the 90th percentile, and 68% by events above the 80th percentile.

Several insulating factors confound the relationship between fresh water supply and human health, especially in higher income countries (Fig. 14.1). Infrastructure plays a critical role in insulating populations from declining quality and quantities of fresh water. Highly efficient irrigation technology, water-free sanitation systems, and water filtration plants can all reduce dependence on large amounts of uncontaminated fresh water. Flood control infrastructure can reduce vulnerability to more extreme runoff patterns. Human behavior related to water treatment (boiling, filtering, etc.) can mask the effects of degraded water quality. And, increasingly, water is essentially being imported in the form of grain grown elsewhere (it takes roughly 1,000 tons of water to grow 1 ton of grain). Affluent countries are often net importers of water, which means they may be less dependent on local services but consume more ecosystem services overall than less affluent countries (Brauman et al. 2007). People who lack the resources to engage these different mechanisms are the ones who will suffer the most direct health impacts from deteriorating access to safe water.

Protection from Natural Hazards

Natural hazards can have immediate impacts on human health in the short term, through injuries, drowning, and heat stress (also see the chapters by Rumbaitis del Rio; Ingram and Khazai; and March, this volume). In addition, human communities may experience several longer term health effects from the loss of living shelters, population displacement, chemical and biological pollution of water supplies, degradation of air quality due to fires, exposure to mold as a result of flooding, and mental health impacts from the trauma of the experience. The specific impacts of floods on human health can be related to injuries in the short term and to outbreaks of water-borne, vector-borne, and rodent-borne diseases as well as mental health disorders over the medium and longer terms (Ahern et al. 2005). Ecosystems can reduce human vulnerability to natural hazards in several ways. For example, intact wetlands provide natural water filtering capacity. Coral reefs, vegetated dunes, mangroves, and wetlands buffer the effects of storms on coastal areas. Standing forests can mitigate flooding associated with extreme rainfall events. Environmental degradation, therefore, reduces the capacity of certain ecosystems to serve as a buffer against climate extremes (Corvalán et al. 2005). For example, the 2004 tsunami in Southeast Asia disproportionately affected regions with degraded coral reefs and mangrove forests (Dahdouh-Guebas et al. 2005; Danielsen et al. 2005; Marris 2005; Kunkel et al. 2006), and, in 1998, Hurricane Mitch disproportionately caused landslides in settings of non-terraced farming on steep slopes in Central America (Cockburn et al. 1999). Bradshaw et al. (2007) found loss of natural forests to be correlated with flood risk and severity in developing countries.

In addition, vulnerability to disasters has increased in recent decades. This may be, in part, a result of the growth of human populations in areas that are at greater risk from extreme weather or natural hazards, such as settlements in low-lying coastal areas or floodplains, or in dryland ecosystems at risk of drought. It is also likely a reflection of more frequent weather-related hazards as projected by the Intergovernmental Panel on Climate Change (IPCC) (Schneider et al. 2007). Globally, the annual absolute number of people killed, injured or made homeless by disasters is increasing (Corvalán et al. 2005).

What are some of the barriers to establishing evidence for the ways that intact ecosystems provide protection from the health impacts of disasters? Many insulating factors obfuscate the ways in which environmental degradation affects human communities. Human vulnerability to natural hazards is mediated by a wide variety of factors including where people live, the quality of their housing, disaster preparedness, early warning systems, and environmental conditions (Adger et al. 2005). Infrastructure is of primary importance. Flood control systems can reduce the impact of hydrological peaks. Disaster preparedness measures, such as early warning systems, can also minimize the impact of natural hazards. Housing quality also affects the ability of a community to withstand extreme events. In addition, philanthropic response in the form of disaster relief efforts can conceal the health impacts that might otherwise be experienced as a result of disasters. These types of organizations assuage the short- and medium-term impacts of disasters through the provisioning of food, fresh water, medical supplies, temporary housing structures, and other goods and services. National and international disaster relief organizations can also enforce positive human practices such as water treatment.

Data availability is another big issue. There are limited data available to evaluate the contribution that environmental change has played in increasing vulnerability to fires, floods, storms, tidal waves, landslides or other hazards. For example, the evidence about the health effects of floods is dominated by studies of slow-onset floods in high-income countries that may have little relevance to flash floods and floods in low-income settings. Yet floods with the largest mortality impacts have occurred where infrastructure is poor and the population at risk has limited economic resources (Ahern et al. 2005). The relationship between forest cover and risk of flooding is also debated with respect to extreme events, such as cyclones and typhoons (Laurance 2007). More data are needed to fully understand the relationships between ecosystem degradation, incidence of disasters, and their associated health outcomes.

Research Agenda

These examples of the relationships between three different classes of ecosystem services and human health illustrate the ways that ecosystem services are interrelated in dynamic and complex ways, and how the causal pathways between ecosystem service production and human health are complex, difficult to quantify, and mediated by a variety of non-environmental influences. These complexities highlight the importance of interdisciplinary partnerships to improve our understanding. If we can better understand the health impacts of the loss of ecosystem services, we will be able to apply this information to guide policy, and to help measure health improvements following the implementation of a new ecosystem services management approach. The MA concludes that efforts "will require an unusual level of interdisciplinary analysis and synthesis in which the population health sciences are central, especially epidemiology."

More clarity is needed on the different intellectual paradigms that characterize epidemiology and ecology, in order to move toward integrating the two fields. In order to be more explicit about what defines health and identify confounding factors in the relationship, ecologists and public health researchers must start to speak a common language. For example, public health scientists should understand that ecologists take offense when "ecological" is used to describe a study's limitations as purely correlational, since much research by ecologists establishes causal mechanism. Ecologists and environmental scientists should take into consideration that "health" comprises many outcomes, with many distinct and multifactorial etiologies. Only when this type of understanding is built can we begin to move forward not only in talking about these issues, but toward building data in support of the relationships in question, and ultimately addressing them with calls to action.

Understanding the complex causal webs and establishing causal inference about the relationship between ecosystem health and human health will require collaborations between ecologists and health scientists, to produce more data, and to carry out more sophisticated analyses that recognize the complexity of the problem. Large country-wide datasets can be used to ask questions about when ecosystems do or do not support community health, but appropriate stratification should be used, as discussed above. For example, a global burden of disease approach can be used to investigate the impact of changes in particular ecosystem services on the incidence of specific diseases in a series of different types of ecosystems and human communities. Several large databases on both human health conditions and ecosystem conditions are available for such an effort. In addition, collaborations between ecologists and health scientists at the time of data collection will improve the ability to provide causal inference. Rather than tacking on health outcomes or environmental conditions as an afterthought, ecologists and epidemiologists can work together to design studies that incorporate aspects of both from the outset. Investigators from different disciplines can work together to develop a conceptual model for how they believe environmental conditions are linked to health outcomes at a specific location before beginning any data gathering activities. The causal pathways that make up such a model can then be tested as hypotheses by gathering the appropriate targeted data.

Ecological data can be incorporated into different epidemiological study designs. *Active surveillance* can be used to monitor both disease incidence and ecological parameters at the same time. *Prospective studies* could be employed to look for expected health outcomes in a population affected by a particular ecosystem management approach (e.g., places where a service will be significantly degraded as a result of planned activity where the consequences of this degradation can be tracked.) *Case–control* epidemiological studies can be used to investigate the particular ecological conditions surrounding a health condition (e.g., places with very similar populations and histories but strongly different ecosystem health (neighboring watersheds) that are managed very differently). Where good historical data exist about both health and ecosystem service changes, *retrospective* studies can explore their relationship in time. Where significant efforts have been made to restore services *intervention trials* can track the health impacts of a restoration project aimed at restoring ecosystem services.

Where traditional epidemiological methods fail, newer approaches have been developed to address the broader contexts that determine population and ecosystem-level health risks. In recent years, many epidemiologists have argued for public health research to move beyond traditional risk factor analysis at the individual level and toward analysis concerned with multiple levels and types of causation. Several more sophisticated approaches have been proposed, such as environmental epidemiology (Pekkanen and Pearce 2001), ecoepidemiology (Susser and Susser 1996), social-ecologic systems perspectives (McMichael 1999), and ecosocial theory (Krieger 2001). These efforts all use a systems theory-based approach to extend the purview of causation across axes of space, time, and organizational level and propose to inter-relate research at different scales through feedbacks and interactions (Eisenberg et al. 2007; Plowright et al. 2008). Multilevel statistical models, and dynamic mathematical models, time-series analysis, panel studies, and risk analysis are all examples of these newer approaches that can be used toward the goal of understanding ecosystem health–human health linkages.

The MA Health Synthesis concludes that "the level of uncertainties and the unsuitability of standard approaches lead many scientists to avoid attempting to answer some questions posed directly by decision-makers. ... Scientists tend to respond with a scientifically more rigorous and less uncertain answer to a small part of the equation." More extensive collaborations between ecologists and epidemiologists can help provide rigorous data to fill gaps in knowledge and, at the same time, produce science that addresses policy-makers' immediate concerns.

Conclusion

In this section, we have seen that large-scale, anthropogenic, environmental changes can cause significant threats to human health. The preceding chapters on the impacts of land use change, disease ecology, and climate change describe how accelerating climate and land use change are likely to impact human health. This chapter describes how many other impacts may be mediated through deterioration of ecosystem services (see Fig. 14.2). In combination, this deterioration is increasing the exposure of hundreds of millions of people to food scarcity, water scarcity, natural hazards, infectious diseases, and population displacement.

For each of these risks, different populations around the world have dramatically different vulnerabilities. In part, this is because the biophysical changes human activity is causing around the planet are not uniform. Rapid glacial melting on the Tibetan plateau threatens dry season water supply for over a billion people living and growing irrigated crops in the river basins of Asia's great rivers. Droughts and increased temperatures caused by climate change in sub-Saharan Africa will interact with existing water scarcity, soil degradation, and nutrient depletion to reduce crop yields and constrain already tight food supplies. The triple threat of more severe storms, rising sea levels, and degraded coastal barriers will pose significant risks to low-lying coastal populations (10% of the human population lives in coastal areas at less than 10 m elevation).

But differential exposure to the biophysical changes associated with human activity is not the only reason why vulnerabilities to these threats will vary across different populations. Vulnerability, as a concept, includes not only exposure to health risks associated with changing environmental conditions but also characteristics of a population that determine its ability to adapt to such conditions. Many of the threats associated with global change can be reduced by means of trade, technology, infrastructure, behavior change, philanthropy, and governance. Populations with the resources (economic and sociocultural) to engage these mechanisms to reduce vulnerability will suffer less than those without such resources (see Fig. 14.2).

There is an urgent need to characterize and quantify these growing threats more accurately. We need to begin modeling each of these types of vulnerability to accelerating environmental change and mapping out which populations are at greatest risk. To do so, will require collaboration from a wide variety of scientific disciplines, and central among them, will be ecology.

Acknowledgments Much of the content of this chapter was discussed in a workshop held at Stanford University in December, 2008. During this workshop, the authors received very constructive input from K. Arkema, B. Brosi, J. Davis, P Ehrlich, D. Ennaanay, R. Gould, A. Luers, H. Mooney, and G. Schoolnik.

References

Adger, W. N., T. P. Hughes, C. Folke, S. R. Carpenter, and J. Rockstrom. 2005. Social-Ecological Resilience to Coastal Disasters. Science **309**:1036–1039.

Ahern, M., R. S. Kovats, P. Wilkinson, R. Few, and F. Matthies. 2005. Global Health Impacts of Floods: Epidemiologic Evidence. Epidemiologic Reviews **27**:36–46.

Alexandratos, N. 1999. World food and agriculture: Outlook for the medium and longer term. Proceedings of the National Academy of Sciences of the United States of America **96**: 5908–5914.

Borgstrom, G. 1965. The hungry planet: the modern world at the edge of famine. Collier-MacMillan, London.

Bradshaw, C. J. A., N. S. Sodhi, K. S.-H. Peh, and B. W. Brook. 2007. Global evidence that deforestation amplifies flood risk and severity in the developing world. Global Change Biology **13**: 2379–2395.

Brauman, K. A., G. C. Daily, T. K. e. Duarte, and H. A. Mooney. 2007. The Nature and Value of Ecosystem Services: An Overview Highlighting Hydrologic Services. Annual Review of Environment and Resources **32**:67–98.

Chichilnisky, G. and G. Heal. 1998. Economic returns from the biosphere. Nature **391**:629–630.

Cockburn, A., J. St Clair, and K. Silverstein. 1999. The Politics of "Natural" Disaster: Who Made Mitch So Bad? International Journal of Health Services **29**:459–462.

Corvalán, C., S. Hales, and A. J. McMichael. 2005. Ecosystems and human well-being : health synthesis. Millennium Ecosystem Assessment, World Health Organization, Geneva, Switzerland.

Curriero, F. C., J. A. Patz, J. B. Rose, and S. Lele. 2001. The Association Between Extreme Precipitation and Waterborne Disease Outbreaks in the United States, 1948–1994. American Journal of Public Health **91**(8):1194–1199.

Dahdouh-Guebas, F., L. P. Jayatissa, D. Di Nitto, J. O. Bosire, D. Lo Seen, and N. Koedam. 2005. How effective were mangroves as a defence against the recent tsunami? Current Biology **15**:R443–R447.

Daily, G. C., editor. 1997. Nature's Services. Island Press, Washington, DC.

Danielsen, F., M. K. Sorensen, M. F. Olwig, V. Selvam, F. Parish, N. D. Burgess, T. Hiraishi, V. M. Karunagaran, M. S. Rasmussen, L. B. Hansen, A. Quarto, and N. Suryadiputra. 2005. The Asian Tsunami: A Protective Role for Coastal Vegetation. Science **310**:643.

Das, S. and J. R. Vincent. 2009. Mangroves protected villages and reduced death toll during Indian super cyclone. Proceedings of the National Academy of Sciences of the United States of America **106**:7357–7360.

Dasmann, R. F. 1988. Towards a biosphere consciousness. Pages 177–188 *in* D. Worster, editor. The Ends of the Earth: Perspective on Modern Environmental History. Cambridge University Press, Cambridge.

Ehrlich, P. and A. Ehrlich. 1981. Extinction: The Causes and Consequences of the Disappearance of Species. Ballantine Books, New York.

Eisenberg, J. N., M. A. Desai, K. Levy, S. J. Bates, S. Liang, K. Naumoff, and J. C. Scott. 2007. Environmental determinants of infectious disease: a framework for tracking causal links and guiding public health research. Environ Health Perspect **115**:1216–1223.

Englande, A. J., Jr. 2008. Katrina and the Thai Tsunami - Water Quality and Public Health Aspects Mitigation and Research Needs. International Journal of Environmental Research and Public Health **5**:384–393.

Fewtrell, L., R. B. Kaufmann, D. Kay, W. Enanoria, L. Haller, and J. M. Colford, Jr. 2005. Water, sanitation, and hygiene interventions to reduce diarrhoea in less developed countries: a systematic review and meta-analysis. Lancet Infectious Diseases **5**:42–52.

Food and Agricultural Organization (FAO). 2008. Assessment of the World Food Security and Nutrition Situation. Page 20 Committee on World Food Security, 34th Session. FAO, Rome.

Gadgil, M. 1998. Prudence and profligacy: a human ecological perspective. Pages 99–110 *in* T. M. Swanson, editor. The Economics and Ecology of Biodiversity Decline. Cambridge University Press, Cambridge.

Humboldt State University. 2009. Arcata's Wastewater Treatment Plant & The Arcata Marsh and Wildlife Sanctuary. http://www.humboldt.edu/~ere_dept/marsh/. Accessed on: April 5, 2009.

Huynen, M., P. Martens, and R. S. De Groot. 2004. Linkages between biodiversity loss and human health: a global indicator analysis. International Journal of Environmental Health Research **14**:13–30.

Klein, A., B. E. Vaissiere, J. H. Cane, I. Steffan-Dewenter, S. A. Cunningham, C. Kremen, and T. Tscharntke. 2007. Importance of pollinators in changing landscapes for world crops. Proceedings of the Royal Society B **274**:303–313.

Krieger, N. 2001. Theories for social epidemiology in the 21st century: an ecosocial perspective. Int J Epidemiol **30**:668–677.

Kunkel, C. M., R. W. Hallberg, and M. Oppenheimer. 2006. Coral reefs reduce tsunami impact in model simulations. Geophysical Research Letters **33**.

Laurance, W. F. 2007. Forests and floods. Nature **449**:409–410.

Luck, G. W., G. C. Daily, and P. R. Ehrlich. 2003. Population diversity and ecosystem services. Trends in Ecology and Evolution **18**:331–336.

Luck, G. W., R. Harrington, P. A. Harrison, C. Kremen, P. M. Berry, R. Bugter, T. R. Dawson, F. d. Bello, S. Díaz, C. K. Feld, J. R. Haslett, D. Hering, A. Kontogianni, S. Lavorel, M. Rounsevell, M. J. Samways, L. Sandin, J. Settele, M. T. Sykes, S. V. D. Hove, M. Vandewalle, and M. Zobel. 2009. Quantifying the contribution of organisms to the provision of ecosystem services. BioScience **59**:223–235.

Mann, A. G., C. C. Tam, C. D. Higgins, and L. C. Rodrigues. 2007. The association between drinking water turbidity and gastrointestinal illness: a systematic review. BMC Public Health **21**.

Marris, E. 2005. Tsunami damage was enhanced by coral theft. Nature **436**:1071.

McMichael, A. J. 1999. Prisoners of the proximate: loosening the constraints on epidemiology in an age of change. American Journal of Epidemiology **149**:1–11.

Millennium Ecosystem Assessment. 2005. Ecosystems and Human Well-being : Synthesis. Island Press, Washington, DC.

Mooney, H. A. and P. R. Ehrlich. 1997. Ecosystem services: a fragmentary history.*in* G. C. Daily, editor. Nature's Services. Island Press, Washington, DC.

Murray, C. J. L. and A. D. Lopez. 1997. Global mortality, disability, and the contribution of risk factors: Global Burden of Disease Study. The Lancet **349**:1436–1442.

Myers, S. S. and J. Patz. 2009. Emerging threats to human health from global environmental change. Annual Review of Environment & Resources **34**:223–252.

Pekkanen, J. and N. Pearce. 2001. Environmental epidemiology: challenges and opportunities. Environ Health Perspect **109**:1–5.

Plowright, R. K., S. H. Sokolow, M. E. Gorman, P. Daszak, and J. E. Foley. 2008. Causal inference in disease ecology: investigating ecological drivers of disease emergence. Frontiers in Ecology & the Environment **6**:420–429.

Postel, S. L. 1998. Water for food production: Will there be enough in 2025? . BioScience **48**:629–637.

Postel, S. L. and B. H. Thompson, Jr. 2005. Watershed protection: Capturing the benefits of nature's water supply services. Natural Resources Forum **29**:98–108.

Prüss-Ustün, A. and C. Corvalán. 2007. How much disease burden can be prevented by environmental interventions? Epidemiology **18**:167–178.

Schneider, S. H., S. Semenov, and A. Patwardhan. 2007. Assessing key vulnerabilities and the risk from climate change.*in* M. Parry, L, O. Canziani, F, J. Palutikof, P, P. Van der Linden, J, and C. Hanson, E, editors. Climate Change 2007: Impacts, Adaptation and Vulnerability. Contribution of Working Group II to the Fourth Assessment Report of the Intergovernmental Panel on Climate Change. Cambridge University Press, Cambridge, UK.

Schwartz, J., R. Levin, and R. Goldstein. 2000. Drinking water turbidity and gastrointestinal illness in the elderly of Philadelphia. Journal of Epidemiology and Community Health **54**:45–51.

Sieswerda, L. E., C. L. Soskolne, S. C. Newman, D. Schopflocher, and K. E. Smoyer. 2001. Toward Measuring the Impact of Ecological Disintegrity on Human Health. Epidemiology **12**:28–32.

Smith, K. R., C. F. Corvalan, and T. Kjellstrom. 1999. How much global ill health is attributable to environmental factors? Epidemiology **10**:573–584.

Susser, M. and E. Susser. 1996. Choosing a future for epidemiology: II. From black box to Chinese boxes and eco-epidemiology. American Journal of Public Health **86**:674–677.

Tinker, S. C., C. L. Moe, M. Klein, W. D. Flanders, J. Uber, A. Amirtharajah, P. Singer, and P. E. Tolbert. In Press. Drinking water turbidity and emergency department visits for gastrointestinal illness in Atlanta, 1993–2004. Journal of Exposure Science and Environmental Epidemiology.

Turner, B. L., R. E. Kasperson, P. A. Matson, J. J. McCarthy, R. W. Corell, L. Christensen, N. Eckley, J. Kasperson, X, A. Luers, M. L. Martellow, C. Polsky, A. Pulsipher, and A. Schiler. 2003. A framework for vulnerability analysis in sustainability science. Proc Natl Acad Sci U S A **100**:8074–8079.

United Nations. Millennium Development Goals 2015. http://www.un.org/millenniumgoals/. Accessed on: May 19, 2009.

United Nations Development Programme (UNDP). 2006. Human Development Report 2006. Beyond scarcity: Power, poverty and the global water crisis. New York.

United Nations Environment Programme (UNEP). 2007. Global Environment Outlook: environment for development (GEO-4) Malta.

Vorosmarty, C., J, C. Leveque, and C. Revenga. 2005. Fresh Water. Pages 165–207 *in* F. Rijsberman, R. Costanza, and P. Jacobi, editors. Millennium Ecosystem Assessment (Program) Condition and Trends Working Group Ecosystems and human well-being : current state and trends : findings of the Condition and Trends Working Group of the Millennium Ecosystem Assessment. Island Press, Washington, DC.

World Health Organization. 1997. Health and Environment in Sustainable Development. World Health Organization, Geneva.

World Health Organization (WHO). 2006. Constitution of the World Health Organization- Basic Documents, Forty-fifth edition, Supplement. Geneva.

World Health Organization (WHO). 2009. Global Burden of Disease. http://www.who.int/topics/global_burden_of_disease/en/. Accessed on: April 5, 2009.

Chapter 15
Introduction to Ecological Dimensions of Global Energy Poverty

Cristina Rumbaitis del Rio

Globally, 1.6 billion people, 16% of the world's population, lack access to electricity and more than three billion people rely on traditional fuels such as wood, dung, agricultural waste, or charcoal, to meet everyday domestic needs (IEA 2010). This condition of energy poverty constrains multiple aspects of human development and growth. Women and children must spend time, often hours, collecting wood-fuel instead of investing time in other productive pursuits, such as attending school or producing value-added commodities for sale or trade. When fuel is in short supply, water is not boiled to kill pathogens, contributing to debilitating ill-health, and mortality of infants. Smoke from indoor combustion of biomass fuels contributes to acute respiratory infections and chronic obstructive pulmonary disease, which kills an estimated two million people a year, mostly young children and women (World Bank 2006). The collection of these traditional fuels puts additional stresses on forest and agricultural ecosystems, undermining rural livelihoods over the long term. Trees are cut for wood-fuel, forests are cleared for charcoal production, and agricultural residues are burned instead of returning nutrients to the soil. While the energy poor are predominantly concentrated in rural areas, access to electricity in urban and peri-urban areas of developing countries is often limited to wealthier communities, forcing the poor to go without or pay comparatively higher prices for domestic fuels. Billions of dollars worth of energy subsidies are rarely targeted to benefit the poor.

Investment in reducing energy poverty has been significant through the past several decades, but has yielded negligible advancements. Rural access to electricity has remained stagnant at 6% over the past decade (World Economic Forum 2010). Only China has made significant progress in improving rural energy access, now achieving 98% electrification of the country through large-scale investments in

C. Rumbaitis del Rio (✉)
The Rockefeller Foundation, 420 Fifth Ave., New York, NY 10018-2702, USA
e-mail: crumbaitisdelrio@rockfound.org

J.C. Ingram et al. (eds.), *Integrating Ecology and Poverty Reduction: Ecological Dimensions*, DOI 10.1007/978-1-4419-0633-5_15,

electrification and multiple phases of electric industry policy reforms (Global Energy Network Institute 2007).

Progress in other geographies has been impeded in part by a large funding gap. It has been estimated that US$ 435 billion is required to provide electricity to all of the world's presently un-served population (Practical Action 2009). The World Bank estimates that less than half of this amount is currently available to meet those needs (World Bank 2006). Other barriers include a focus on large-scale interventions that do not necessarily meet the needs or capacities of the poor, lack of private sector investment in energy provision models that benefit the poor, and multiple policy and regulatory barriers.

An additional complicating factor is the global imperative to provide energy services to the under-served without producing the same level of greenhouse gases that developed nations have emitted in meeting their own energy needs. Thus, there is a need to provide clean energy services for the poor, while grappling with the equity implications of asking the poor, who have contributed the least to global warming, to adopt clean energy solutions, which may be less affordable than more polluting energy intervention alternatives. Indeed, by some calculations, meeting the needs of the energy poor with fossil fuels would only increase greenhouse gas emissions by 2% (Practical Action 2009), a small amount compared to the energy needs of other sectors in society (transportation, industry, commerce, etc.). Thus, it is important to ensure that measures taken to reduce global greenhouse gas emissions, such as greater biofuel production and use, do not inadvertently further marginalize or impoverish poor communities.

Despite these challenges, sustainable solutions, promising technologies and viable business models do exist to meet the energy needs of the poor. Extension of electricity grids is most viable in urban and peri-urban areas. In rural areas, decentralized energy production systems are most relevant in the near term, and can be powered by renewable energy sources such as micro-hydro, biomass, wind, and solar energy. Improved stoves are available to improve the efficiency and reduce the health impacts of indoor fuel combustion, and cleaner burning fuels are increasingly also available. The social, economic, and health benefits of improving energy access are now better understood and are helping to create greater demand for energy interventions.

The chapters in this section provide frameworks, approaches, and examples to illustrate how the energy needs of the poor can be met sustainably, affordably, and equitably, with a particular focus on the ecological considerations relevant to reducing energy poverty. The chapter by Doll, *Ecological Context for Sustainable Energy Solutions*, provides a framework to assist in the analysis of appropriate energy sources to meet the energy needs of a given community. Assessment criteria are by necessity multi-dimensional – including physical, social, economic, political, and operational constraints. The ecological contribution to this framework includes an environmental assessment of fuel source options, as well as the conceptualization of the human – energy production system as a dynamic complex system that evolves over time. The assessment framework is illustrated with examples from the author's experience implementing rural energy interventions as a development program in Rwanda.

The chapter by Ganz et al., *Ecology-Poverty Considerations for Developing Sustainable Biomass Energy Options*, focuses on biomass fuels specifically and applies an ecosystem services approach toward assessing biomass energy options for energy poverty reduction. This approach looks at all the components of the biomass energy supply chain from producers to end users, and makes the interconnected social, environmental, and economic costs and benefits of the various biomass options explicit. Ecological modeling tools help assess the tradeoffs and aid in decision-making. The authors provide a highly replicable phased approach to evaluate sustainable biomass options through a series of guiding questions aimed to help practitioners develop and implement biomass energy alternatives that benefit both poor communities and biological diversity.

The final chapter in the section, *Ecological Sustainability of Woodfuel as an Energy Source in Rural Communities*, by Bailis et al., looks at the ecological implications of wood-fuel use at both local and global scales, and evaluates different approaches toward improving ecological sustainability of wood-fuel use. A case study based upon a spatial analysis of wood-fuel supply–demand imbalances illustrates how spatial analysis tools, biomass inventories, and multi-criteria analysis can be used to identify hotspots where local demand for wood-fuel exceeds production capacity, providing an evidence base with respect to where the most critical locations are for focused actions. They conclude that while wood-fuel dependence is unlikely to decrease in developing countries in the near-term, there is reason for optimism in the increased understanding of the ecological dynamics of wood-fuel use, and the availability of multiple tools to better assess the benefits of more sustainable wood-fuel production and use.

Taken together, these chapters illustrate that despite the challenge of closing the gap on energy poverty, and, concurrently, in meeting related or derived development goals, there is a multiplicity of sustainable energy options, and distinct contributions that ecologists can make in assessing those options, prioritizing areas for intervention, and implementing ecologically sound solutions.

References

International Energy Agency (IEA). 2010. Comparative Study on Rural Electrification Policies. In Emerging Economies: Keys To Successful Policies. International Energy Agency. Paris, France. *http://www.iea.org/papers/2010/rural_elect.pdf*

Meisen, P. and N. Cavino. 2007. Rural Electrification, Human Development and the Renewable Energy Potential of China. Global Energy Network Institute, San Diego, United States. http://www.geni.org/globalenergy/research/china-development/china-development.pdf

Practical Action. 2009. Energy poverty: The hidden energy crisis. Practical Action. Warwickshire, United Kingdom. http://practicalaction.org/advocacy/docs/advocacy/energy_poverty_hidden_crisis.pdf

The World Bank. 2006. Energy Access, Security, Key to Reducing Poverty. The World Bank. Washington DC, United States. http://go.worldbank.org/HT2WKGTT20

World Economic Forum. 2010. Energy Poverty: Impeding Sustainable Development and Growth. World Economic Forum. Geneva, Switzerland. http://www.weforum.org/en/initiatives/EnergyPovertyAction/index.htm

Chapter 16
Ecological Context for Sustainable Energy Solutions

Susan C. Doll

Introduction

The objectives of this chapter are to explore the broad context of energy sources; how this information can be used to provide energy services; the role of energy in alleviating poverty; the types of energy sources available to reduce poverty; and, finally, to provide a framework that considers local conditions that ultimately influence which energy solutions will be sustainable. Ecology is relevant in this endeavor not only in terms of considering the ecological impacts of different potential energy interventions, but also in constructing an integrated framework to consider the interactions between the physical environment and the socio-cultural and economic components on the functioning of the system as a whole (ecology, *sensu* Odum 1977). As such, no matter the status of the natural environment, which is highly degraded in many developing countries, an integrative definition of "ecology" rather than an environmentally focused interpretation can be applied to evaluate the applicability and long-term sustainability of energy solutions.

Understanding the role of energy in poverty reduction begins with understanding that the earth is a materially closed but energetically open system. There is no material being added to or removed from the earth, save the occasional meteorite crashing into the surface and the material launched into orbit or left on the moon by the space program. The main implication of the earth as a closed system is that our material resources are finite, and unless we understand the conditions for sustainability, we are destined to run out of valuable resources and be poisoned by our own waste (Cairns 1977). It is the continuous and relatively constant input of energy from our sun, the

S.C. Doll (✉)
Department of Technology and Environmental Design,
Appalachian State University, Boone, North Carolina, USA
e-mail: scdoll2@gmail.com

J.C. Ingram et al. (eds.), *Integrating Ecology and Poverty Reduction: Ecological Dimensions*, DOI 10.1007/978-1-4419-0633-5_16,

origin of essentially all of the earth's energy sources, that is the "engine" of the biosphere, powering the cycles of water and materials. Solar radiation heats the earth and creates temperature gradients that generate air movement (i.e. wind) and evaporates water from the oceans that falls back to the surface and makes its way again to the ocean in a never-ending cycle. Plants capture and convert sunlight to chemical energy in the form of complex molecules via photosynthesis, thus providing the base of the food chain. Under the right conditions, when plants die, they are sequestered underground and over time, under high temperature and pressure conditions, become fossil fuels. Animals convert some of the stored chemical energy in the food they eat to create their own complex molecules for growth and repair, and to provide energy for basic metabolic functions and mobility. Physical characteristics of animals determine where they are able to live, how fast they can move, how well they can hear and see, and how strong they are. The amount of physical exertion, or work, that can be performed is related to and constrained by the available energy stores in their bodies.

Through the use of tools and the ability to harness external energy sources that were independent of their food intake, humans were able to overcome their physio-energetic limitations. Fire provided warmth and protection from predators, and expanded the types of foods that could be eaten. Domesticated animals carried heavy loads and pulled plows and wagons, increasing productivity and relieving humans of tiresome work. Wind was used to power boats and turn windmills to pump water, grind grain (all things previously done by manual labor) and eventually to generate electricity; water wheels were used in much the same way. The invention of the steam and combustion engines in the nineteenth century, powered by fossil fuels, revolutionized farming, manufacturing, and service industries. All of these energy sources can be used to produce electricity, enabling the use of electrical devices for lighting, communication, and medical care. In order to quantify the magnitude of the impact of these external energy sources on the quality of human life, the amount of mechanical work that can be performed by a single person in a given period is used as a standard unit for comparison. An equivalent amount of work done using a non-human energy source is expressed in units of man-equivalents or "energy slaves" (Buckminster and Marks 1973). One review of per-capita energy use calculated that each person in the USA had the equivalent of 100 energy slaves working for them 24 hours a day, largely in the form of fossil fuels, while in some developing countries the available energy was <1 energy slave per person (Boyden 1987). Fundamentally, this means that the majority of work in developing countries is performed by, and limited to, the energy available from people. When modern tools are not available, the majority of that energy goes into subsistence farming for survival. When people are malnourished or sick, their ability to do work is diminished, further reducing productivity.

Energy and Poverty

The connection between energy and poverty reduction, hunger, education, gender equality, heath, water, and sanitation, and environmental sustainability has been explicitly discussed in the context of the Millennium Development Goals (MDGs)

(Modi et al. 2005, DeClerck et al., Chapter 1, this volume). Although not identified as an MDG target in and of itself, energy has been acknowledged as one of the underlying enablers for escaping the cycle of poverty. Strategic energy interventions can have a positive impact on almost every facet of daily life (EAC 2007). Increased access to mechanical power can improve productivity and relieve the burden of activities that are traditionally done with human labor such as water lifting and delivery, growing and processing agriculture products, and transporting goods to home and market. Increasing the sustainable production of biomass and improving fuel-efficiency of stoves so that they use less fuel, provide the dual benefits of reducing the labor burden of fuelwood collection on women and children, and lessening pressure on stressed ecosystems. Improving cookstove performance to produce less smoke and increasing access to cleaner-burning biomass-derived or modern fuels such as kerosene and LPG (liquefied petroleum gas) can help reduce adverse health effects associated with the use of biomass fuel, primarily in women and infants. Increasing access to electricity for lighting, computers, and specialized equipment in households and public institutions (health clinics/centers, schools, government offices, and community centers) can improve the capacity and quality of basic social services. All of these interventions can in turn support income-generating opportunities (Sachs 2005). It should be noted, however, that access to these different energy sources, progressing from low-quality fuels such as animal dung and crop waste, to higher-quality "modern" fuels, is strongly correlated with increasing prosperity. There is a generally accepted progression of fuel types, in order of increasing cleanliness, convenience, and cost of fuel and end-user appliances: dung and crop waste, wood, charcoal, LPG, and gas, and, finally, electricity (Hosier and Dowd 1987). While the cleanliness and convenience are highly desirable, poverty is an obstacle to modern fuels due to the increasing cost. In some cases, there may be ways to "leap-frog" to higher-quality energy sources using solutions suited to developing countries, such as using animal dung to produce biogas or human power to generate electricity.

Role of Ecology

In a resource-constrained world, where there are many competing needs, the use of ecological concepts can provide the basis for a systematic methodology to identify appropriate energy intervention(s) that not only satisfy targeted, high-priority energy needs and are technically feasible in the environment within which they will be operating, but also take into consideration other factors related to long-term sustainability. Another important dimension of using an ecological lens is the concept that all systems are dynamic and elements of the system change and co-evolve over time. As such, decisions about the implementation of energy intervention(s) should not be seen as the final answer, but rather as steps in a process, requiring a longer-term vision and strategy to see if/how interventions that are implemented today will be compatible with tomorrow and provide a foundation on which to build future improvements.

Evaluation Approaches

Two basic approaches to evaluating energy interventions are (1) to identify a commonly used and/or abundant energy source and evaluate how it can be most efficiently and effectively used to meet energy needs, and (2) to identify high-priority energy needs and evaluate which intervention(s) and energy sources are most appropriate for a specific time and place. These two approaches are not inherently incompatible and, in fact, can be complimentary steps in developing a comprehensive energy strategy. The first is a reasonable approach for locations where resources and income are severely limited, and there may only be a single energy source that makes sense for near-term interventions. The second approach evaluates a wider range of energy options to identify which energy sources are most appropriate for specific energy services and potential barriers to implementing certain types of interventions. The approach taken in this chapter is to describe a general framework for screening a broad range of potential energy sources for specific applications and identifying technological advances and capacity building that may be needed to enable future desirable interventions. This author uses recent experiences in a rural village in Rwanda to (1) serve as a practical case study to demonstrate the methodology and (2) describe site-specific factors that were found to be critical determinants for sustainable interventions. In addition to availability of energy resources, requirements for institutional support and human skills are addressed. Background information includes an overview of the range of energy solutions that are available to provide services that can help alleviate poverty. Other chapters in this book address the implementation of specific types of energy solutions, including biomass fuel (Ganz et al., this volume) and fuelwood (Bailis et al., this volume).

Energy Services and Sources

Although discussions about energy for development are most often focused on biomass and cooking issues, it is important to remember that the broader goal is to provide a range of energy services that will improve the overall quality of life, and that these services can be achieved by a number of different energy sources. Each type of energy source has features which determine its applicability for a given energy service, independent of the location where it is used.

Energy Services

As mentioned in the introduction, energy services that have helped humans move beyond their physical limitations include fuel combustion for cooking and heating, animal power to reduce work burden and provide transportation, and wind and water power to provide energy for grinding, pumping water, and generating electricity that can be used for lighting, computers, and specialized equipment in households and public institutions. To understand the relationship between energy services and poverty, it is necessary to understand how they are related to human needs and the quality of life. There is a hierarchy of human needs that can be broadly categorized as life critical, life enhancing, and luxury, generally corresponding to improving quality of life. Life critical services are those that are necessary for survival, including obtaining water, food, and shelter. Humans have been struggling to secure these since the dawn of time when the only resources they had at their disposal were the products of nature and their own muscle power. In many parts of the world this is still the case, but with added pressure from ever-increasing population. Life-enhancing services increase or improve value, quality, or convenience. Many things in this category reduce the burden associated with satisfying life-critical needs (e.g. pumps for better access to water, machines to assist with growing and processing food) and also loosen the constraints of local ecosystem productivity (e.g. fertilizer and transportation to distribute food, water, building materials). The luxury service category encompasses things that add comfort or pleasure, but are not absolutely necessary. Satisfying luxury needs typically builds on innovations devised for the two previous categories; food transported by airlines so that consumers can have fresh fruit and vegetables any time of the year, air conditioning and heating for maximum thermal comfort, and fine food and wine, for example. One of the keys to moving up the human needs hierarchy is the availability of energy. Using transportation as an example, Fig. 16.1 shows various modes of transportation that illustrate the progression from rudimentary transportation by foot, to increased energy input from animals and bicycle-assisted transport that improves agriculture productivity and carrying capacity, and, finally, to fossil fuel-powered vehicles that extend the limits of distance and carrying capacity and drastically reduce travel time and work burden. For poverty alleviation, providing life-critical services is an absolute minimum with a progression to life-enhancing services as conditions and increasing access to energy resources allow.

Energy Sources

Energy sources are typically classified as *primary* or *secondary* sources. Primary energy is contained in the form of chemical bonds (fossil fuels, biomass, adenosine triphosphate (ATP), and glycogen in humans and animals), nuclear bonds (radioactive molecules for fission reactors), electromagnetic radiation from the sun (solar thermal, solar photovoltaic, and wind), gravitational energy from attraction between two bodies of mass (falling water and tidal energy), and energy from the heat of the earth's core (geothermal). In order to do useful work, primary energy must be converted to secondary energy through one or more conversion steps to create thermal, mechanical, or electrical energy. Thermal energy is used for services that require heat, such as cooking, hot water, and food processing. Mechanical energy makes

Fig. 16.1 Different modes of transportation use different sources of energy and end-use technologies. (**a**) Walking is the most common mode of transportation in developing countries. *Energy = human power*. (**b**) Animals reduce work burden and increase carrying capacity. *Energy = animal power*. (**c**) Bicycles for transport of people and cargo are a common mode of transport in many countries. *Energy = human-assisted power*. (**d**) Motorized vehicles are the most common mode of transportation in industrialized countries. *Energy = gasoline, diesel*

things move, including turbine blades, engine parts, and grinding mills. Electrical energy is the most versatile and can not only power electrical appliances, but can also produce heat and be used to make things move. Energy is lost during each conversion step so single-step processes are the most efficient. Each primary energy source is only able to produce one kind of secondary energy with a single conversion step: fossil fuel and biomass energy is converted to thermal energy through the process of combustion, wind and water energy is converted to mechanical energy using blades that rotate a shaft, and solar radiation is converted to electricity by electron excitation in photovoltaic cells. With additional conversion steps, each primary energy source is able to provide other types of secondary energy: thermal energy from fossil fuel and biomass combustion can be used to generate hot air or steam turn turbine blades and produce mechanical energy, then this mechanical energy or mechanical energy from wind and water can be used to drive a generator that produces electricity, and, to complete the circle, electrical energy from any of the primary energy sources can provide thermal energy via resistive heating or mechanical energy using pumps and motors. From an energy-efficiency stand point, it is best to use a primary energy source that requires the fewest number of conversion steps to get the desired secondary energy form needed, as there are additional energy conversion losses and distribution devices required at each step.

Matching Energy Services and Sources

Table 16.1 cross-references a list of energy services on the left and energy sources across the top. Every cell in the table indicates whether a particular energy source is able to provide each energy service with a single-conversion device. The lists of energy services and energy sources are not intended to be exhaustive, but rather to capture the ones most relevant to poverty reduction. Although coal, oil, and natural gas are the major energy sources used in developed countries and have demonstrably satisfied the entire hierarchy of human needs, on a global level they are not considered a sustainable option and therefore are not included here. Also, because of the complexity and/or cost of the support equipment and operating systems, nuclear, tidal, and geothermal primary energy sources are also not considered as energy options for poverty alleviation.

Biology-based primary energy sources included in Table 16.1 are (1) unprocessed biomass (such as crop residue, dung, and wood) that may be compressed or formed into briquettes, (2) biofuel that is easily extracted from oil plants (preferably from nonedible

Table 16.1 Energy services and how/if they can be provided by different primary energy sources

	Primary energy sources										
Energy services	Biomass	Biofuel -plant oil	Biogas	Kerosene	LPG	*Human labor*	Animal labor	Solar thermal	Solar PV	Wind	Micro-hydro
Cooking/heating	(T)	(t)	(T)	(T)	(T)	–	–	(t)	–	–	–
Food/fuel drying	(t)	(t)	(t)	(t)	(t)	–	–	(T)	–		–
Lighting	(t)	(t)	(T)	(T)	–	M/E	–	–	(E)	(E)	(E)
Electrical hardware IT equipment Medical instruments	–	M/E	M/E	–	–	M/E	–	–	(E)	(E)	(E)
Water lifting/delivery	–	(M)	(M)	–	–	(M)	(M)	–	(E)	(E) (M)	(E) (M)
Agroprocessing	–	(M)	(M)	–	–	(M)	(M)	–	(e)	(E) (M)	(E) (M)
Transportation	–	(M)	–	–	(M)	(M)	(M)	–	–	–	–

The secondary energy sources are color-coded: RED = Thermal (T), BLUE = Electrical (E), PURPLE = Mechanical (M). If cells have a lower case letter, this indicates that the service can be provided by that energy source but it is not an optimum match. Dashes indicate that a particular energy source *cannot* provide that energy service without increased cost, complexity, and inefficiencies of additional conversion device(s) and infrastructure plants such as *Jatropha* sp. to minimize competition with food crops), (3) biogas produced from anaerobic digestion of organic material (the most commonly available feedstock is animal manure or human feces), and (4) people and animal power. Additional primary energy sources included are two petroleum-based fuels (kerosene and LPG) that are currently relatively widely used in developing countries, and four solar-based energy sources (solar thermal, solar PV, wind, and water). For cells in the table where a particular energy source is able to provide an energy service, the kind of secondary energy needed is also indicated. Important observations from this table indicate which energy services each primary energy source *can* and *cannot* provide. Biofuel (plant oil) is the only primary energy source that can provide all seven energy services. Limitations of the other primary energy sources are listed below:

- Solar thermal is the most limited primary energy resource, only efficient for producing thermal energy to provide two of the seven energy services – it cannot provide electrical or mechanical energy at the scale needed for poverty alleviation.
- Biomass and kerosene have slightly greater functionality in terms of the services provided, are a better match for cooking, and add the functionality for lighting.
- Wind and water energy sources are suited for providing both mechanical (rotating shaft) and electrical energy (connect the shaft to a generator) – future innovative designs should consider incorporating interchangeable functionality where the rotating shaft can be disengaged from the generator when mechanical energy is needed and re-engaged when electricity is needed.
- Biogas can provide all three secondary energy types, though it has limited application for transportation. Biogas and biofuel (plant oil) provide the most flexibility for meeting a wide range of energy service needs.

Support Equipment Requirements

For each primary energy source to produce useable secondary energy, a system for collection, processing, storing, and distributing must be in place, and specific end-use equipment is generally required to make energy functionality available to an end-user. For example, using solar energy to produce electricity requires manufacturing facilities to produce the photovoltaic array that will collect and convert the sunlight, batteries to store the electricity so that it is available when it is needed (not just when the sun is shining), and internal wiring to distribute the electricity from the panel to the battery and from the battery to the electrical equipment/appliance. Similarly, burning LPG to produce thermal energy for cooking or mechanical energy to power an engine requires facilities to manufacture and fill a pressurized canister, transport networks to get the canister from the filling site to the supplier and then from the supplier to the customer, valves to control the release of the gas, and tubing to transfer the gas from the tank to the end-use gas-burning appliance or engine. Contrast this with the simplicity of using biomass for cooking, which is the main reason it is the dominant energy source in most developing countries. It is available from local sources and can be collected by hand, can meet the most basic human needs for heat and cooking, requires little processing and no special end-use equipment to use. Though other energy solutions will be needed to meet mechanical and electrical energy needs, biomass will continue to be a major energy source, making it all the more critical to implement sustainable biomass production and use practices. Issues related to biomass use are addressed further in the following chapters by Ganz et al. and Bailis et al., this volume.

Human Health and Environmental Issues

Each energy source also has associated human health and environmental impacts that are relatively independent of local conditions, although housing characteristics do play a significant role in adverse health effects from household energy use. The dominant health issue related to energy use worldwide is by far the production of smoke from combustion of lower-quality fuels (dung, crop waste, wood) in confined indoor spaces, particularly among woman and infants (Warwick and Doig 2004). There have been many studies conducted to investigate the correlation between indoor air pollution from biomass use and adverse health effects that indicate indoor particulate concentrations can be orders of magnitude higher than accepted standards (Ezzati and Kammen 2002). However, due to the uncertainty associated with exposure assessment (there are no specific biomarkers for exposure to biomass smoke) and also with attributable health risk (respiratory disease is generally non-specific and may be caused by other factors), a definitive causal effect has not been established. Recommendations for reducing indoor concentrations of smoke, in order of generally increasing efficacy and cost, include: increased household ventilation, improved stoves with more efficient combustion, venting of combustion gases using chimneys or other removal devices, and ideally, replacing biomass fuels with cleaner-burning fuels such as charcoal, biogas, and LPG. There are also labor burden and safety issues associated with collection of biomass fuels, particularly to women and children (Ruhfuess 2006). Additional environmental concerns related to biomass fuel use include the degradation of forests from unsustainable fuelwood harvesting practices, with subsequent soil erosion, and immediate implications for agricultural productivity, soil nutrient cycling, and soil fertility if biomass is burned rather than recycled (Parikh and Ramanathan 1999).

Assessing Local Context

The previous discussion highlights the general limitations of various energy options, independent of where they might be used, that must be considered to determine their potential role in a comprehensive energy strategy, but it does not provide any site-specific information that would indicate the suitability of each energy source under local conditions. The criteria presented next are meant to guide assessments of local suitability of energy options. Through the author's personal experiences, it became apparent that all too often interventions were promoted without adequate understanding of the full context within which they will need to function.

Assessment Criteria

The list of context assessment criteria that is identified below reflects the environmental and organizational elements of a human-dominated system that will apply, to a greater or lesser extent, for any intervention being considered and was largely inspired by the day-to-day challenges of the author's experiences living and working on infrastructure interventions in Rwanda over a 3-year period.

1. Environmental (climate, resources)
   - What natural resources are locally available, especially for wind, water, solar, and biomass?
   - What impact will the use of each energy source have on the health of the ecosystem and community?
2. Physical (support equipment, utilities, structure)
   - Are the physical support structures currently in place at the household and community level?
3. Commercial (vendors, materials, import/export, transport)
   - Is the project, and later the community, going to be able to get hardware and replacement parts through existing supply chains and vendors?
   - What is the role of the private sector in sustainability of the intervention or in transitioning to other energy sources? Are there viable business models?
4. Operational (technical capacity, install, maintain, repair, user training)
   - Is the necessary technical capability available for continued operation?
5. Managerial (planning, budgeting, maintenance, depreciation)
   - Who retains "ownership" of the project or materials in the long term?
   - Is there a structure and knowledge in place to be able to manage the project over time?
6. Socio-economical (acceptance, priority, confounders, general use practices)
   - What are individual/community expectations for the project?
   - Are there any local practices or cultural barriers to people accepting or using the intervention?

- Does it support local priorities?
- Is adoption of this intervention cost-competitive with what is currently being done?
- What is the general direction and trend for increasing user/market sophistication and evolution? Is the project forward compatible and does it support future growth?

7. Political (policy, competing interests)

   - Is the project and expected outcome aligned with local/regional/national agendas?
   - Are there any legal statutes or policies that might affect implementation or long-term operation?

To illustrate how this assessment may be useful to development planners, the context status of each criterion will be described for conditions in Mayange, Rwanda. The assessment of status is based on the author's observations during 2006–2008, while conducting research on local improved stove options, collecting energy use data, and assisting with the coordination of energy, water, communication, and construction interventions for the health, education, business development, and agriculture sectors of the Millennium Villages Project, a development and research project implemented by Columbia University and the United Nations Development Program. It is important to note that any assessment of status is a snapshot that will change over time, hopefully, in a positive direction influenced by the selected intervention. At the end of this section, context assessment criteria for each energy source/energy service combination will be assigned a status "score." This summary of all the context scores for each candidate energy source will provide a preliminary indication of the "fit" of the intervention with current contextual factors and how "prepared" a community is to accommodate a particular intervention.

Environmental Conditions

Two impacts that are central to environmental sustainability are resource depletion and pollution: what resources are available and how are they being used; what is the "health" status of air, water, and soil systems; and what are the negative impacts of human activities? Natural resources harnessed for energy production include resources derived from the ground (fossil fuels, the thermal capacity of the ground itself), the land (trees, grasses, fruit and seeds, wild game, the land itself for producing crops), and from water (water for crops, seafood, the thermal capacity of the water). The energy from the sun in the form of radiation, heat, wind, or falling water is a natural resource that varies depending on location – the available amount, quality, intensity, and duration (seasonal variation) of sunlight, water, and wind determine their use as renewable energy sources. Overuse and abuse of resources can lead to declining game and fish populations, deforestation, ecosystem degradation, depletion, and loss of soil. Pollution of air, water, and soil systems from energy production and use includes: thermal pollution of water systems, release of harmful manmade chemicals and excessive use of fertilizers and growth hormones, and improper disposal of wastes including "natural" materials such as human waste and biodegradable products.

Rwanda is a resource-poor country with few natural energy sources. Located a few degrees south of the equator, sunlight is abundant for the majority of the year. Until recently, 100% of the country's electricity was provided by hydropower. Severe droughts led to an energy crisis and importation of fossil fuel-powered generators and the fuels to operate them. An as yet untapped resource (except on a small, local scale) is the methane in the waters of Lake Kivu on the western border of the country. Destruction of the natural forests in Rwanda is widespread, and although demand for charcoal was one factor that contributed, land clearing for agriculture, habitation, and creating tea plantations are dominant activities leading to deforestation. Currently, virtually all charcoal is being made from planted trees, on private and community-held land, and little to none is made from natural forests (Van der Plas 2008). In Mayange, there is little regular wind and no nearby surface rivers or steams for hydro or wind power. Sunlight and bio-power (plants, human, animals) are the only locally available energy resources. In Mayange, the hilly topography and deforestation have contributed to soil erosion and loss of topsoil that will require water and soil conservation interventions to support opportunities for sustainable biomass production. The use of biomass as the predominant energy source for the country contributes to indoor and outdoor air pollution and the associated disproportional health and labor burden on women and children.

Physical Conditions

Commonly referred to as "infrastructure," the physical buildings, utilities, transportation, and distribution systems that are in place as a starting point will greatly influence the strategic approach for energy solutions and the cost of implementation. A non-energy example that illustrates this point is the donation of computers to local schools. Schools often do not have a reliable source of electricity and the building space is already inadequate for overcrowded classrooms resulting in no secure place to put the computers and no reliable and affordable energy to power computers. To use an energy example, poor transportation infrastructure is a physical barrier to implementing LPG as a clean energy source because it requires transportation of heavy canisters on foot or bicycle to and from a distribution center that could be many kilometers away, over dirt roads that may be impassable during the rainy season. The absence of needed infrastructure does not permanently preclude the future use of a particular energy intervention, but it does require additional resources and commitment to put them in place prior to implementation.

Commercial Conditions

Assuming that the necessary physical conditions are adequate or can be upgraded as part of the project, an unequivocal barrier to sustainability is the lack of a commercial support network for all of the inputs and parts required for the intervention. This may not impede *implementation* but it can be the death knell for long-term operation. For example, all of one particular kind of hand pump for badly needed water wells in Mayange were inoperable at one time or another due to unavailability of the same broken part. The commercial support system is comprised of networks for import/export to get goods into and out of the country, vendors and material providers to make products available to customers, and transportation for the distribution of goods from wholesalers to retailers. This includes supply chains and distributors for anything that is required for installation *and* anything that gets used up, worn out, or broken. Furthermore, products need to be locally affordable in terms of time, effort and money. If they are available in-country, but are too expensive, too far away, or take too much time/effort to obtain, then they are effectively unavailable. Part of the testing of the sustainability and viability of an intervention should include building the intervention from scratch with locally available materials and practices. Difficulties encountered will be a reflection of what the end-user will have to tackle during long-term operation. On a positive note, the absence of supportive commercial conditions may provide opportunities for business development and should be considered as part of the intervention strategy. Using the example of LPG again, if physical and economic conditions were such that there was a demand for LPG, it is highly likely that an entrepreneur could establish a business transporting and delivering canisters. In Mayange, commercial networks are in place for biomass and kerosene, and could be adapted for biogas and biofuel, creating additional income for suppliers. Supplies for simple biogas and solar thermal technologies are currently available through existing networks.

Operational Capacity

There is a critical mix of technical expertise that is required to successfully implement and operate even the simplest energy intervention. Without local technical skills and capacity to install, maintain, and repair the equipment, and to conduct "owner" and user training, projects will soon deteriorate. For example, one of the most common reasons for failure of solar PV systems is not the arrays, but the batteries because the recipients do not know how to replenish the "acid" (i.e. distilled water) when levels get low. With proper assistance, local capacity to install, maintain, and upgrade interventions can be built. However, capacity building takes time and requires dedicated financial support. So if the required skills do not currently exist at the location where the intervention is to be implemented, the project either needs to provide the necessary training, work with a local institution that will provide the training, or wait and come back later when the necessary technical expertise is available. Another possibility is that the presence of an intervention needing technical support will "pull" in technical capability by creating a demand for it.

Operational capacity is an area where phased, long-term strategic planning can introduce initial interventions that build on existing capabilities and provide stimulus for capacity building that can then be used for more advanced interventions down the road. In Mayange, there are trained electricians, plumbers, and masons whose skills could be applied to biofuels, biogas, and solar thermal technologies, but additional skill building would be required for the operation of advanced solar, wind, and water technologies.

Managerial Capacity

Many managerial skills needed for a successful project, such as planning, budgeting, maintenance, and managing depreciation, are not common in rural areas of developing countries. One often neglected aspect of a successful project is a plan for the eventual replacement of system components. To use solar PV interventions as an example again, it is commonly assumed that solar energy is essentially "free" to the end-user once the initial capital costs are covered and installation is complete. However, up to a third of the total cost of the system is for batteries to store the electricity produced during the day so that it can meet usage requirements, particularly at night. Batteries must be maintained and replaced regularly. Without managerial foresight to either budget for battery replacement, develop a business model to produce affordable replacement batteries or find a funding source at the end of battery life, the solar system becomes non-functional. Furthermore, in resource-constrained situations, it is nearly impossible to budget for contingencies that are 5–10 years in the future. Management skills need to be developed to make efficient and effective use of available operational capacity to ensure timely repair and maintenance of equipment. In Mayange, formal management skills are not common, though training for management of cooperatives is being supported by MVP.

Socio-Economic Conditions

Social and economic aspects of a community are intimately inter-twined. There are traditional roles for men and women regarding household chores, work outside the home, money management, and public and private decision making. It is important to understand the economic base of a community, as the introduction of energy interventions may disrupt established income-generating activities, such as the growth, harvest, and distribution of biomass. Also, a clear picture of how household and business resources are expended on energy can provide insight into energy service priorities (where do limited resources get spent first) and what percent of income is currently dedicated to energy. If a different energy source is introduced, the competing cost of the energy must be considered as well as the cost of any different end-use devices, such as a new stove, that are required to use the new energy source. Figure 16.2 shows some typical fuel types used in rural communities in developing countries, progressing from low-quality dung (a) to clean-burning LPG (b). Figure 16.3 shows the corresponding end-user stoves for each fuel type.

Fig. 16.2 Typical fuel types for cooking include (**a**) animal dung, (**b**) fuelwood, (**c**) charcoal, and (**d**) LPG

When energy interventions introduce new devices or methods for securing or using energy, and require change in traditional ways of doing things, non-economic considerations are also critical. New devices may look and operate differently and if they are perceived to not work as well, they will be abandoned. For example, one type of mud stove introduced in Mayange, which when properly installed and used, greatly reduced indoor air pollution. However, because of the thermal mass of the mud, it sometimes took longer for water to boil (though testing showed that time for cooking beans was similar to the traditional three-stone fire). So despite the air-quality improvement, and because improved health benefits may not have been fully understood, many households reverted back to the traditional fires.

In addition, there are often dual functions provided by current energy sources that may not be provided by alternatives. For example, traditional three-stone cooking fires not only provide heat for cooking and hot water, but also provide light in the evening and early morning, and serve as a focal point for social gatherings. In addition, smoke from the fire can help provide pest control, keeping mosquitos away and also drying and fumigating thatch roofs to protect them from decay. If the new energy source/intervention does not provide all of the same functions as the traditional approach that it is replacing, it therefore may not be accepted. There can also sometimes be resistance to change from unexpected quarters. During one of the author's cookstove tests, there was a household in the study where a bachelor lived who did

Fig. 16.3 Typical stoves include (**a**) 3-stone stove suitable for any type of biomass, (**b**) improved stoves made from clay most commonly used with fuelwood, (**c**) metal improved stoves that can accommodate both biomass and charcoal, and (**d**) modern stoves that use LPG

all of his own cooking. It turned out he had a different method for tending fire where he removed ashes part way through the cooking process, which resulted by far in the greatest fuel-efficiency of any household. When the other study participants (all women) were made aware of this apparently beneficial fire-tending method, they flatly refused to accept it despite support of the study data, presumably because cooking is typically a woman's domain.

Typically, interventions are most successful if they can satisfy current household priorities that already require some amount of expenditure, where people may be persuaded to pay slightly higher costs or agree to structured payment schedules if it provides the same or more service and is not too risky. This latter point is often not fully considered, as many of us do not really have much to lose if we make a bad decision, but for families that have almost nothing, losing even a little can tip the balance of survival. In Mayange, economic conditions have been so unpredictable for many years, largely because of the increasingly irregular weather patterns, that households may be very conservative about doing anything that may decrease present security, no matter how enticing the future benefits. Currently, not much money is spent on energy in Mayange. Most households use scavenged fuelwood collected from the ground. Some households purchase fuelwood for $1–$2 a bundle that typically lasts for 2–3 days and spend ten cents to a dollar on small amounts of kerosene for household lighting. Typically, kerosene is sold in small quantities measured with soda or beer bottles, or small tomato paste cans and burned in handmade wick lamps. But the socio-economic situation in Mayange is improving. When the author was last in Mayange in 2008, there were a handful of new community-run cooperatives that provided members, particularly the women in the basket weaver's coop, with cash income for the first time in their lives. There are also indications that the community is receptive to trying new things. An example of this is the receptiveness of households engaged in testing a household lighting intervention and their reluctance to relinquish their prototypes at the end of the study period.

Political Considerations

Energy security is always a critical component of national policy as the instability and scarcity of energy supply can lead to social disruption and conflict within a country or across national boundaries. As a resource-poor, landlocked country, it is important that Rwanda's largest source of energy used is renewable and is as independent of external influences as possible. The structure of national and regional utility management can also play a significant role in the preference for centralized versus decentralized energy sources and conversion facilities. For instance, in Rwanda, there is one electric utility company and other electricity production is unregulated; therefore, unless an area currently without electricity is included in the centralized expansion planning, interventions for decentralized off-grid electricity are needed. In recognition of the fact that biomass is, and will continue to play a very big role in Rwanda's energy future, a national Biomass Energy Strategy is currently being developed and improved stove programs are being pursued. Finally, competing commercial endeavors and NGO efforts will also affect the success of any given intervention.

Summary of Rwanda Context Assessment

Using the narrative description in the previous sections, a relative score is assigned to the assessment criteria for the Energy Sources and Services described earlier in this chapter. The criteria are divided into two categories: (1) performance criteria that are dependent on energy source but not location, including energy source limitations described in the sections of this chapter on "Matching Energy Services and Sources" and "Human Health and Environmental Issues"; and (2) the seven context assessment criteria from the "Assessment Criteria" that are dependent on location and energy source. For the overall performance criteria, a score of "1" indicates good flexibility for the full range of energy services and minimal environment and human health impacts, while a score of "4" indicates serious deficiencies. Scores for the seven context assessment criteria are as follows:

1. A score of "1" indicates that current local conditions and capabilities are essentially in place to support the intervention, though some expansion of capacity may be required.
2. A score of "2" indictates that some elements are in place locally but current conditions and capabilities are limited and if not substantially expanded will make it difficult for that particular intervention to be sustained.
3. A score of "3" indicates that capabilities are available in the region or country but may not be local to all locations where implementation is being considered. Without improvement in local capacity or provisions for bringing in outside personnel and materials when they are needed, the intervention cannot be sustained.
4. A score of "4" indicates that prerequisite conditions and capabilities to support the intervention are essentially not available at this time, missing altogether, only intermittently available, or inaccessible to the local community.

The scores for all criteria are summarized in Table 16.2 in a color-coded "scorecard" that reflects the status for Mayange, Rwanda prior to 2009. The color-coding provides an intuitive green-means-go, red-means-stop visual, with strongly positive scores of "1" shown in green, and strongly negative responses of "4" indicated in red. Intermediate responses of "2" or "3" are shown as yellow and orange, respectively. Though not necessarily of equal importance, averaging the responses across the two categories of assessment criteria, for the individual energy sources, provides a preliminary indication of general suitability for Mayange. It should be noted that the context score will be different for different locations in Rwanda.

Table 16.2 Summary of assessment criteria "scores" for different energy sources in Mayange Rwanda

Assessment criteria	Biomass	Biofuel –plant oil	Biogas	Kerosene	LPG	*Human labor*	Animal labor	Solar thermal	Solar PV	Wind	Micro-hydro
Flexible energy services	3	1	1	3	3	1	3	4	2	2	2
Low env/health impact	4	2	1	2	1	3	2	1	1	1	2
Overall Performance score	*3.5*	*1.5*	*1*	*2.5*	*2*	*2*	*2.5*	*2.5*	*1.5*	*1.5*	*2*
Natural resources	2	3	2	4	4	1	3	1	1	4	3
Physical/infrastructure	1	3	2	2	3	1	2	2	2	3	3
Commercial network	1	3	2	2	3	1	1	2	3	4	3
Operational capability	1	3	2	1	3	1	1	2	3	4	3
Managerial capacity	2	3	2	2	2	1	2	2	3	3	3
Socio-economic	1	2	3	2	3	2	2	2	3	4	3
Political	1	3	2	2	2	1	3	3	2	4	1
Overall Context score	*1.3*	*2.9*	*2.1*	*2.1*	*2.9*	*1.1*	*2.0*	*2.0*	*2.4*	*3.7*	*2.7*

The numerical ranking scores for each energy source indicate which source(s) have the best overall performance and which are most compatible within the local context. The energy sources with the best overall performance scores are biogas and plant oil ranked the highest, followed by solar PV and wind. The poorest performance scores are for biomass, animal labor, and solar thermal energy sources, primarily related to their limited use for certain types of energy services. Of the four energy sources with performance scores <2, biogas has the best context assessment score. In an overall energy strategy, the low-performing energy sources with scores >2 (biomass, kerosene, animal labor, and solar thermal) should always be considered in conjunction with one or more other energy sources that can provide the energy services that they cannot, and should not be considered as stand-alone solutions. The energy sources most compatible with the local context (i.e. scores <2) in Mayange are not surprisingly, *human labor* followed by *biomass* – the dominant sources currently being used.

As discussed in the introduction, output from human labor is constrained by physical limitations and, in poor rural communities, it is already largely committed to day-to-day survival; therefore the results of this context assessment for Mayange indicate that an emphasis on increasing the sustainable supply of biomass and promoting the production of biogas are suitable strategies given the *current* local conditions. The performance scores are primarily related to the nature of the energy source and, therefore, with the exception of biomass, will not change depending on local conditions. In the case of biomass, the poor performance score (currently 3.5) can be increased to a "1" or "2" with improved stove technology and sustainable fuelwood production to reduce impacts on human health and the environment, respectively. Sustainable production of biomass will be a challenge in Mayange due to irregular rainfall and poor soil quality and erosion and will require concurrent implementation of soil and water conservation measures. With regard to the context assessment criteria, all but the natural resource criterion have the potential for improved scores as human and infrastructure capacity develop for the location being evaluated.

While there will likely be debate about the exact ranking of some of the criteria (particularly, the "4" scores), this framework can be used by development planners to gain insight into the likelihood that a specific intervention will be sustainable in a specific location. If the intervention of choice has an overall combined performance and context assessment score greater than a "2," or any single element has scored more than a "3," there are issues that need to be addressed before the intervention has a reasonable chance of being a sustainable solution for the overall alleviation of poverty. Prudent actions taken in response to a poor overall score can be to (1) review the poorly scored area(s) and re-scope or re-evaluate the project approach and objectives, and incorporate necessary activities into the project scope to improve contextual fit; (2) table the project until a later date when local conditions and capabilities are better able to support long-term sustainability; or (3) choose a different intervention. One positive outcome of a failed context assessment is that it can help identify barriers to implementation and can be useful for prioritizing funding to overcome hurdles (primarily capacity building) for the implementation of cleaner, more sustainable energy sources in rural areas of developing countries.

Conclusions for Development Planning

The two main considerations of the proposed framework are energy source performance and site-specific context. The former is important for understanding the role a particular energy source/intervention can play in an *overall* energy strategy and the energy services that it can and cannot provide. Cost analysis and long-term funding allocation should include provisions for energy solutions that will provide the full range of energy services that are critical for poverty alleviation. Understanding the latter, site-specific context, *before* proceeding with a solution is absolutely critical for project success and sustainability; projects that come with predetermined solutions looking for a problem to solve often result in unmet expectations, and rarely have long-term impact, resulting in a waste of limited resources. It is important to be prepared and willing to accept that your primary idea may not be the best answer in all contexts. Paying attention to the following lessons learned will improve the likelihood of selecting a sustainable energy solution.

Natural/environmental – Make sure that the intervention will actually work with respect to the ecological and climate conditions at the site and that projects are informed by adequate field assessments and testing.

Physical – Be sure that all necessary prerequisite support equipment, utilities, and structures are in place or be prepared to add them to the project. Look for synergies with existing systems. For example, battery-powered household lighting should take advantage of the networks and enterprises springing up for charging of cell phone batteries.

Operational – Plan for on-the-job-training for installation, maintenance, repair, and user instruction; start simple, introduce complexity as technical capacity develops. Look for synergies with local skills and capacities and entrepreneurial opportunities.

Commercial – Begin with interventions that can be supported by existing vendors, supply chains, and available materials. Introduction of appropriate advances can often lead to demand for new services that will drive new business. It is generally most effective if interventions can be introduced within the context of small business development – entrepreneurs have a vested interest in the project succeeding. In Mayange, when a new cooperative is formed, people submit applications for membership and there has been no shortage of interest.

Managerial – Work with the individual(s) or entities that will be responsible for continuing function of the intervention, institutionalize processes for planning, budgeting, and maintenance. When possible, use a phased approach for implementation to allow time for issues to surface and adjustments to be made.

Socio-economical– Get buy-in on community priorities, and be prepared for resistance to change and unforeseen confounders to your overall objectives (e.g. the well you have installed may be delivering clean water, but dirty containers and drinking cups are being used). Education is an important component of change – make sure your project does not neglect this element – better yet, plan from the start to team with local educational or community organizations. Work within the existing socio-political conditions and constraints with an eye toward future direction and evolving sophistication of the end-user, evaluate how these may change in the future. For interventions that are shared by community members, it is particularly important to manage expectations and address the issues of risk and mitigation strategies.

Political– Work within the country's development goals, and particularly be sensitive to ongoing, competing programs.

References

Boyden, S. (1987). *Western Civilization in Biological Perspective: Patterns in Biohistory*, (pg. 196). Oxford University Press.

Buckminster, F. and R. Marks. (1973). The Dymaxion World of Buckminster Fuller. (pg. 52). Doubleday Press. New York, NY.

Cairns, J. Jr. (1997). Defining Goals and Conditions for a Sustainable World. *Environ Health Perspect.* 105(11): 1164–1170.

Ezzati, M. and D. Kammen. (2002). Household energy, Indoor air pollution and health in developing countries: Knowledge base for effective interventions. *Annu. Rev. Energy Environ.* 27: 233–270.

Hosier R.H. and J. Dowd. (1987). Household fuel choice in Zimbabwe – an empirical test of the energy ladder hypothesis. *Resources and Energy* 9(4): 347–361.

Modi, V., S. McDade, D. Lallement, and J. Saghir. (2005). Energy and the Millennium Development Goals. Energy Sector Management Assistance Programme, United Nations Development Programme, UN Millennium Project, and World Bank. New York, NY, USA.

Odum, E. P. (1977). The emergence of ecology as a new integrative discipline. *Science* 195: 1289–1293.

Parikh, J.K. and R. Ramanathan. (1999). Linkages among energy, agriculture and environment in rural India. *Energy Economics* 21: 561–585.

Ruhfuess, E. (2006). Fuel for life; household energy and health. World Health Organization, WHO Press, Geneva Switzerland.

Sachs, J. (2005). The end of poverty: economic possibilities for our times. Penguin Press. New York, NY.

Van der Plas, R. (2008). Biomass Energy Strategy (BEST), Rwanda. Volume 1 – Background and analysis. Deutsche Gesellschaft für Technische Zusammenarbeit (GTZ). Eschborn, Germany.

Warwick H. and A. Doig. (2004). Smoke – the Killer in the kitchen. Indoor air pollution in developing countries. ITDG Publishing. London, UK.

Chapter 17
Ecology–Poverty Considerations for Developing Sustainable Biomass Energy Options

David J. Ganz, David S. Saah, Jill Blockhus, and Craig Leisher

Introduction

An estimated 1.6 billion people worldwide, mostly in rural areas, suffer from energy poverty (World Bank 2009), that is, they have insufficient energy sources for cooking, heating, and the 101 small tasks that become big tasks when there are energy shortages. Energy has many forms, as introduced in the chapter by Doll in this section, but for the poor, much of their energy comes from biomass.

Biomass is biological material extracted from living or recently living organisms such as wood or alcohol. It is a sustainable energy source if managed responsibly. However, biomass energy use is rarely well managed. Over use of biomass resources not only compromises biodiversity, it also harms the poor, who are heavily dependent on biomass energy. Moreover, poor households suffer from significant negative health impacts from misuse of biomass energy. Globally, there has been little progress in meeting the challenges of capitalizing on the opportunities presented by biomass energy.

The potential benefits of improved biomass energy production and use are large. Biomass energy can provide a source of light, so a student in a poor household can study after dark. Biomass can be converted to electricity to run water pumps providing clean water and water for productive uses. This same electricity can also yield increased information flows from radio, television, and cell phone. More efficient use of biomass energy can result in improved health for the poor by reducing indoor air pollution.

In this chapter, we focus specifically on biomass and we argue for an "energy–poverty alleviation" approach that involves all components of the biomass energy supply chain from the producer to the manufacturer to end user. We adopt an ecosystem services approach that is driven by a series of guiding questions to help

D.J. Ganz (✉)
Lowering Emissions in Asia's Forests (LEAF),
Winrock International, Bangkok, Thailand
e-mail: Dganz@winrock.org

J.C. Ingram et al. (eds.), *Integrating Ecology and Poverty Reduction: Ecological Dimensions*, DOI 10.1007/978-1-4419-0633-5_17,

practitioners develop and implement biomass energy alternatives that benefit both poor communities and biological diversity.

The approach suggested here is designed to assist people in transitioning to a sustainable livelihood system in which their household and community assets are increased and local resource management is improved. It enables local people to identify and develop potential biomass energy markets that will provide income and benefits without degrading the resource base. The implementation of this approach would result in a climate-friendly outcome while providing immense socio-economic benefits to the world's poor.

Biomass Energy Use in the Developing World

The majority of renewable energy globally is derived from biomass. Renewable energy comprises approximately 18% of total global final energy consumption (Fig. 17.1), of which 72% is biomass. The next largest renewable energy segment is large hydropower constituting 17% of the total share of renewable energy, with the balance of 11% coming from other forms (World Bank 2009).

Developing countries have a high potential for expanding their use of renewable sources.

In sub-Saharan Africa, 70–90% of the primary energy supply comes from biomass sources. Even oil-rich sub-Saharan African countries continue to rely on biomass energy to meet the bulk of their household energy requirements—in Nigeria, it is estimated that about 91% of the household energy needs are met by biomass (Karekezi 1999).

In Asia, biomass accounted for about 11% of all primary energy used in 2000, making biomass the fourth largest source of energy after oil (37.2%), coal (23.5%), and

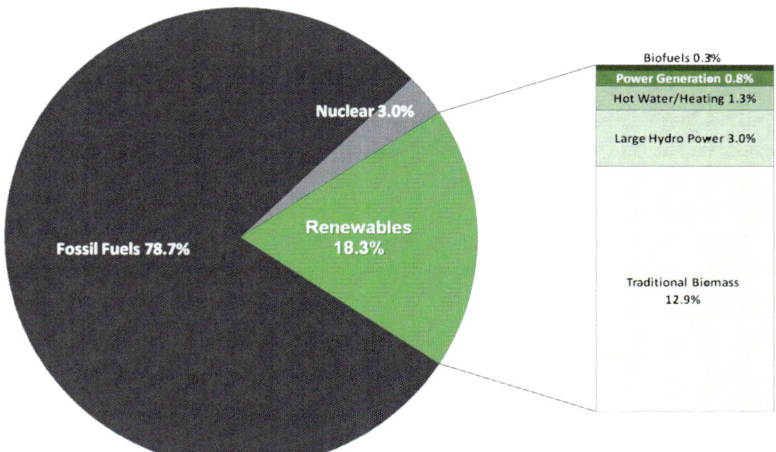

Fig. 17.1 Renewable energy share of global final energy consumption 2006

gas (21.1%). Unfortunately, there is insufficient information available to breakdown which sectors are the largest consumers of biomass energy. Based upon a few small-scale studies, the domestic sector generally accounts for about 70–80%, while the industrial sector consumes the remaining 20–30% of a country's biomass energy production (GreenLineCare 2009; Rural Wood Energy Development Program 1999). In Sri Lanka, 90% of the population consumes 73% of the biomass in domestic uses, mainly for cooking. Nearly 80% of households use inefficient cooking methods such as the open fire or traditional stoves (IDEA 2004). In Lao, PDR, Cambodia, and Myanmar, biomass energy consumption is as high as 80–90%.

More than 65 million people in Latin America do not have access to electricity, including both the rural and peri-urban areas (Sagari 2006). Those that have access in the peri-urban regions often experience erratic electricity supply. Hydrocarbons (oil, natural gas, and coal) comprise 75.7% of Latin America/Caribbean energy supply (UNDP, World Energy Assessment: 2004 update). Although Latin America has surplus energy, the energy markets in the region face the risk of future energy supply shortages. For instance, in the past decade, electricity consumption in Brazil, Argentina, Chile, Paraguay, and Uruguay, which is heavily reliant on hydropower, has at times been restricted due to water shortages. Since rainfall throughout the region varies considerably, electricity supply in years of severe drought may be insufficient to meet the region's growing energy demand (WEC 2008). Latin America and the Caribbean's biomass share is between 14.5% and 15.05% of total primary energy use (TPES) (UNEP 2008). Considering that the primary end uses of biomass are cooking and heating, the expansion of electricity access, used primarily for lighting, is not expected to have a significant affect on biomass use in the future. The International Energy Agency (IEA) projects that Latin American biomass use will still account for 70% of residential energy consumption in 2030 (IEA statistics 2008). The Caribbean region, that is, the nine countries[1] included in the ECLAC/GTZ renewables assessment project, has a heavy dependency on hydrocarbons, which stands at almost 80%. Caribbean renewables, which account for 17%, are basically composed of woodfuel and woodfuel products (7.6%), cane products (almost 9%), and hydroenergy, which is remarkably marginal, at <1% (ECLAC 2003). Given these current levels, the Latin American and Caribbean Initiative for Sustainable Development has met its initial target for the countries of the region to modify their energy structure so that 10% of their TPES comes from renewable sources by 2010 (UNEP 2008).

Yet renewable energy options must be implemented sustainably. As the Intermediate Technology Development Group (ITDG) Power for Poverty Reduction report (2004) states, "it is essential that renewable energy technologies expand the choice of energy for poor people. By focusing on renewables alone, we are in danger of restricting the already very limited choice of poor people. If renewables are promoted to the exclusion or detriment of expanding other options for the poor, then we are limiting development options for the poorest people on earth." Some organizations argue that

[1]Barbados, Cuba, Grenada, Guyana, Haiti, Jamaica, the Dominican Republic, Suriname, and Trinidad and Tobago.

to address energy poverty, there needs to be a shift from burning biomass to burning commercial hydrocarbons. The argument is that this would reduce local ecological impacts and would not materially affect the greenhouse gas balance of the world. Further, this shift in fuel use could make a huge difference in the quality of life and economic prospects of poor nations by helping the poor access safer cooking fuels, electric power, and power for transport and running machinery.

Biomass and Environmental Considerations

The Millennium Development Goals clearly establish a link between poverty alleviation and sustainable development stating that the welfare of people depends, *inter alia*, on the capacity to increase their income through improved access to resources and production factors. The livelihoods of poor rural populations greatly depend on the natural resources they can access from their immediate environment. The benefits that people obtain from the management and use of natural resources are mediated by access to resource and markets, production factors, and how these are maintained over time. For instance, in locations where wood fuels or other forest resources are extracted for commercial sale, local users may find that their own access to energy for subsistence needs is contingent on distant markets, state agents, and powerful business interests (as discussed in Bailis et al., this volume).

Apart from wood, agricultural land produces biomass residues, part of which is available as fuel on an environmentally sustainable basis. At present, the main biomass fuels are crop residues such as bagasse, rice husks and straw, coconut husks and shells, palm oil kernel, shells, and fiber. Wood and other biomass fuels (as well as animal dung) can substitute for each other, though most household consumers have a general preference for wood over other forms of biomass.

Growing concerns for sustainable development and poverty alleviation have focused attention on the potential role of renewable energies. This issue is of particular strategic importance on two different scales. First, in less developed countries, the cost of imported energy is a limiting factor to general growth and development, a constraint further multiplied by the rising prices of oil in international markets. Second, access to energy is a basic need that is usually unmet or under-met for the poor. Rural poor populations often face strong constraints in access to energy. These constraints include:

- A hostile environment (slopes, altitude, and remoteness);
- Poor infrastructure (transportation, adequate roads, communication, and energy) and lack of development options or opportunities;
- Institutional barriers (education, health, and investment systems);
- Poor economic opportunities (limited market development); and
- Limited access to a reliable supply of reasonably priced energy.

As a result, the poor usually rely on resources from their immediate environment, a practice that is considered, in some cases, to be damaging to environmental health

(Earth Institute 2004). The World Energy Council, for instance, argues that the use of traditional energy sources by the poor (mainly wood fuels), combined with the use of inefficient technologies and appliances, results in wastage of wood resources. In addition, the use of crop residue and animal waste as fuels can be detrimental to soil quality and agricultural and livestock productivity, as these resources often have alternative applications as soil conditioners, organic fertilizers, and livestock fodder.

Biomass and Social Considerations

Another constraint to sustainable biomass energy production and use becoming more widely adopted by the poor is that biomass energy sources are frequently common-pool resources. The poor often rely on wood, grass, or dung collected in common areas such as community forests and fields. Wealthier people in the community also rely on the same common-pool resources for income production. Studies from several countries (Narain et al. 2005; Cavendish 2000) suggest that dependence on common-pool resources follows a U-shaped curve (Fig. 17.2). The poor and the wealthier in a community depend considerably on local natural resources than those in the middle economically. The poor collect biomass to cover basic needs like cooking and heating. The wealthier have the capital to invest and exploit the biomass for profit. Biomass resources are in fact often a significant portion of income for those who are relatively wealthy. This means the wealthier people in a community must be a part of any approach that seeks to improve the poor's quantity or quality of biomass collected from common-pool resources. Potential elite capture of benefits from a biomass project must be managed through equitable benefit sharing agreements among community members. These should be negotiated upfront and should include a conflict resolution mechanism in the management structure.

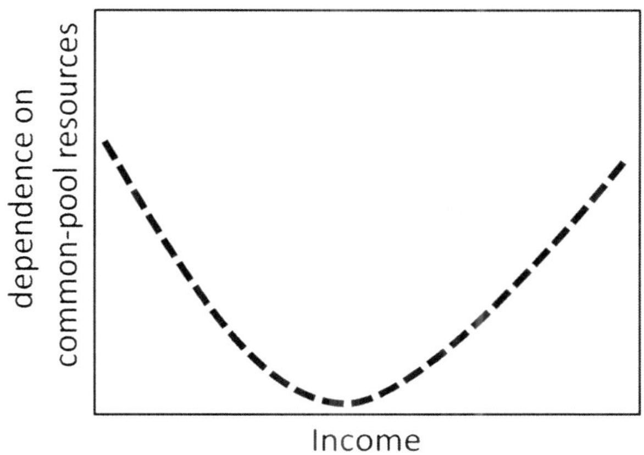

Fig. 17.2 Relationship of household income to dependence on common-pool resources

There also needs to be particular attention paid to gender issues in biomass projects. Women and children are often the collectors of biomass energy, and a biomass project can have significant unintended negative impacts on the resource collectors. Moreover, the most economically marginal households in many communities are often those headed by females. Changing how biomass energy sources are used and managed can have detrimental impacts on the most vulnerable in a community. Dedicated efforts are needed to reach and incorporate the needs of the least-powerful stakeholders, in order to ensure that this segment of society benefits as well.

Biomass and Health Considerations

The rural poor subsist mainly on traditional energy sources such as animal dung, crop residues, and wood (Reddy et al. 1997; Goldemberg et al. 2000). The continuing reliance of poor households on such forms of energy comes with major disadvantages including:

- Substantial, and often increasing, time and effort to procure firewood or other forms of biomass—for example, in rural sub-Saharan Africa, many women have to carry 20 kg of fuel wood an average of 5 km every day (IEA 2002);
- A higher price per unit of energy because subsidies often increase as one goes up the energy ladder (Reddy et al. 1997); and
- Severe and widespread health impacts associated with indoor air pollution resulting from the inefficient combustion of energy sources in poor households, with women and children facing particular risk (Smith et al. 1993).

Indoor air pollution is one of the top preventable health risks in developing countries (Bruce et al. 2000). Estimates provided by the World Health Organization suggest that about 1.6 million premature deaths can be annually attributed to indoor air pollution from biomass and coal use in poor households in developing countries (WHO 2002; Smith et al. 2000). Indoor smoke from these solid fuels is, in fact, responsible for about 38 million disability-adjusted lost years (DALYs)[2] in developing countries with attendant social and economic costs (WHO 2002). The majority of these costs are affiliated with growth defects, impairment of vision, and respiratory illnesses that affect the labor force and community well-being. Supplying biomass requires high labor inputs, which in many places is often a burden for women and small children, thus impacting the ability of these family members to get an education and acquire the necessary skills to advance out of poverty (WEC 2007). For these reasons, the continued dependence on traditional fuels for household energy use presents a particularly troubling aspect of the energy–poverty problem.

[2]One DALY represents one healthy year of life lost by an individual due to disease/adverse health condition.

Investment in Improving Biomass Energy Production

Government agencies, non-governmental organizations (NGOs), development organizations, and other actors have been involved in a variety of programs to enhance—quantitatively and qualitatively—energy services for the poor. But the resources and funds channeled toward rural energy constitute only a small fraction of the total funds targeted for the expansion and transformation of the energy sector in developing countries. The commercial energy segment has received the bulk of investment made in developing countries. Reducing the problems associated with the use of traditional biomass fuels in poor households remains only one of a range of objectives of rural energy programs. As a result, even though efforts such as those promoting cleaner energy-conversion devices (e.g., improved cook stoves), as well as the use of less-polluting fuels such as biogas, have shown some success (Smith et al. 1993), resources and attention devoted to improving household energy use are not at all commensurate with the magnitude of the problem. Perhaps limited investment and attention can be attributed to the fact that many still perceive biomass energy as having substantial environmental and social drawbacks. This need not be the case.

Biomass Resources, Technology, and Sustainability

One of the most promising areas for biomass energy is in growing specific crops for energy—biofuel production. The increasing demand for energy crops can contribute to increased prices of their products. It can also increase job opportunities for rural communities. However, if governments fail to manage biomass resource development appropriately, negative impacts will occur such as forest destruction, conflict with food production (Brown 1980: Ignaciuk et al. 2006; Johansson and Azar 2007), and contamination of natural water systems by excess fertilizer and pesticide inputs (Hall et al. 1982; Pimentel et al. 1992; Pimentel 2005). It is therefore crucial to analyze how the expanding demand of biomass energy will affect rural communities, especially poor small-scale farmers. In this chapter, we suggest analyzing these impacts using an ecosystem services approach. This approach will allow us to put the ecological benefits of biomass fuels in context by also considering other ecosystem services such as water quality, carbon, or biodiversity.

Table 17.1 highlights some of the major advantages and disadvantages of biofuels associated with environmental, economic, and social concerns.

Biofuels can have a neutral or positive effect on the environment if they are managed in a sustainable fashion and the technologies used to convert them to energy are efficient. At present, however, most traditional biomass combustion devices are inefficient, resulting in incomplete combustion, excess greenhouse gas production, and negative health impacts. This has historically made biomass less attractive when compared with other fuels especially fossil fuels. This creates an environmental opportunity to improve the production model by either improving the biomass combustion devices or replacing them with more efficient and/or cleaner biomass technologies.

Table 17.1 Advantages and disadvantages of biofuels (adapted from Hall and Moss 1983)

Advantages	Disadvantages
1. Stores energy and potentially carbon	1. Land and water use competition
2. Renewable	2. Land area required
3. Versatile conversion options and bioenergy and fuel products (sometimes with heat byproducts)	3. Potentially depletes other ecosystem goods and services
4. Dependant on technology already available with minimum capital input	4. Supply uncertain in initial phases
5. Depending on the region, relatively available to all income levels	5. Costs are often uncertain
6. Can be developed with present manpower and material resources	6. Fertilizer, soil, and water requirements
7. Large biological and engineering development potential	7. May be contradictory to existing agricultural, forestry and social practices, and local uses
8. Creates employment and develops skills	8. Bulky resource; transport and storage can be a problem
9. Reasonably priced	9. Subject to climatic variability
10. If scaled appropriately, may be compatible with environmental and social concerns	10. Low conversion efficiencies
11. Does not increase atmospheric CO_2	11. Seasonal production in some geographies

Biomass can be converted into energy by direct combustion or non-combustion means. Direct combustion is how most biomass is put to use for heating, cooking, and industrial processes, or indirectly to produce electricity. Non-combustion methods convert raw biomass into a variety of gaseous, liquid, or solid fuels that can then be used directly in a power plant or home for energy generation. The carbohydrates in biomass, which are comprised of oxygen, carbon, and hydrogen, can be broken down into a variety of chemicals, some of which are useful fuels. This conversion can be done in three ways:

1. *Thermochemical.* Biomass can be changed by heat and time into various gases, liquids, and solids such as methane and alcohol. Another approach is to take these types of fuels and run them through fuel cells, converting the hydrogen-rich fuels into electricity and water, with few or no emissions, although this is unlikely to be technologically suitable with respect to many developing countries' infrastructure.
2. *Biochemical.* Bacteria, yeasts, and enzymes can be employed to breakdown biomass into usable biofuels. Fermentation, the process used to make wine, is used to change biomass liquids into alcohol, a combustible fuel. A similar process is used to turn corn into grain alcohol or ethanol, which is mixed with gasoline to make gasohol. Also, when bacteria break down biomass, methane and carbon dioxide are produced. This methane can be captured—as it is in sewage treatment plants and landfills—and burned for heat and power.

3. *Chemical*. Biomass oils, such as soybean and canola oil, can be chemically converted into a liquid fuel similar to diesel fuel and into gasoline additives. Cooking oil from restaurants, for example, has been used to make "biodiesel" for trucks. Algae have also been used as a source of oil to produce biodiesel.

These gasification approaches will undoubtedly improve rural energy production options and will make biomass a better source of energy.

Of course, the advantages and disadvantages of biomass energy technologies need to be understood within the environmental, political, and economic realities of the potential end users. As demonstrated in the chapter by Doll in this volume, an initial assessment of needs and local context reduces the risk of developing an overly complex biomass solution that is ill-matched to local contexts.

Closer attention needs to be paid to the economics of land use and the competition for land between food and energy crop production under stringent CO_2 control policies. Biodiversity, soil and nature conservation, and carbon sequestration considerations do not imply an explicit land demand. The attractiveness of climate change mitigation options depends on how well they harmonize with other environmental and socioeconomic goals. Future studies should, therefore, avoid assessing the prospects for biomass energy in isolation, but instead should adopt a broader approach where various land-use options are assessed simultaneously with due consideration of other environmental and socioeconomic goals.

An Ecosystem Services Approach to Biomass

The sustainability of biomass energy is fundamentally about using natural resources in a sustainable way. Because biomass resources are interconnected with the wider ecosystem (such as fuel wood and watershed protection), assessing biomass options at the ecosystem scale is a requisite for long-term sustainability. Hence, using an ecosystem services approach to assess biomass energy options is critical. Moreover, this approach makes the interconnected costs and benefits of the various options explicit.

Since the Millennium Ecosystem Assessment (Powledge 2006), much work has been done on modeling ecosystems. Models, such as the InVEST tool (NatCap 2007) and the Natural Assets Information System (NAIS) (Troy and Wilson 2006), allow decision makers to see the tradeoffs between different land-use options and help them find land-use solutions tailored to the characteristics of a specific ecosystem. A modeling approach is particularly useful at a larger scale where the assessment of options community-by-community is impractical.

At the local scale, the assessment of biomass energy options can be achieved through a cost–benefit analysis to explore the short- and long-term costs and expected benefits of a proposed energy solution (Loomis et al. 2008). However, regardless of the scale of analysis, the evaluation can also be integrated with spatially explicit models to weigh potential benefits from different biomass energy development strategies (Troy and Wilson 2006). Spatially explicit models compare a baseline damage probability under a traditional energy scenario with the amount and probability of damages to environmental assets under alternative biomass energy scenarios.

The biomass energy scenarios evaluated could range from small one mega-watt cogeneration units for decentralized electricity production, to large *Jatropha* biodiesel plantations, to community woodlots. Ideally, well-designed biomass energy projects will result in lower damage probabilities, higher income generation opportunities, and a constant flow of ecosystem goods and services.

Evaluating Sustainable Biomass Energy Options

To explore the role of ecology in addressing the many dimensions of poverty reduction, especially health, energy needs, infrastructure development,[3] and livelihood diversification,[4] we propose the following framework to indentify, plan, and develop energy–poverty alleviation approaches that will provide tangible benefits locally without degrading the resource base. This framework is intended to assist the conservation and development communities to systematically identify and develop sustainable biomass energy options. It is also designed to expand the techniques and tools available for planning biomass energy projects that will assist local communities in implementing cost-effective solutions to their energy demands. It is assumed that facilitators will have previous knowledge and expertise in the use of participatory assessment tools, since these will be required to adapt these tools. This framework gives local communities more opportunities to benefit from their biomass resource and greater incentives to better manage and protect those resources. For more information on these participatory approaches, please consult "Participatory Rural Appraisals: Past, Present and Future" by Robert Chambers (1992) or "Shortcut methods in social information gathering for rural development projects" by Chambers (1987) or Grandstaff and Messershmidt (1995) "A Manager's Guide to the use of Rapid Rural Appraisal" and Pretty et al. (1995) "A Trainer's Guide for Participatory Action."

The framework for planning sustainable biomass energy options is a logical sequence of steps arranged into three phases. It uses a series of general tools that have been adapted to achieve specific results in the identification and development of biomass energy resources. The three phases of this framework are:

- *Phase 1: Assess the existing situation.* Develop a profound understanding of the existing biomass uses, the ecosystem services upon which the community depends, the problems associated with biomass energy development, and a shortlist of options for energy–poverty alleviation.
- *Phase 2: Identify potential biomass energy options*, markets, transportation systems, and means of energy distribution. This is the decision-making process for the best alternative energy option utilizing local biomass and the gathering

[3]Infrastructure development includes road building, road upgrading, water and sanitation provision, energy provision, port and other transport development, and urban development.

[4]Livelihood diversification typically refers to seeking a variety of farm and off-farm income and livelihood sources, which might include collecting non-timber forest products or engaging in day labor activities.

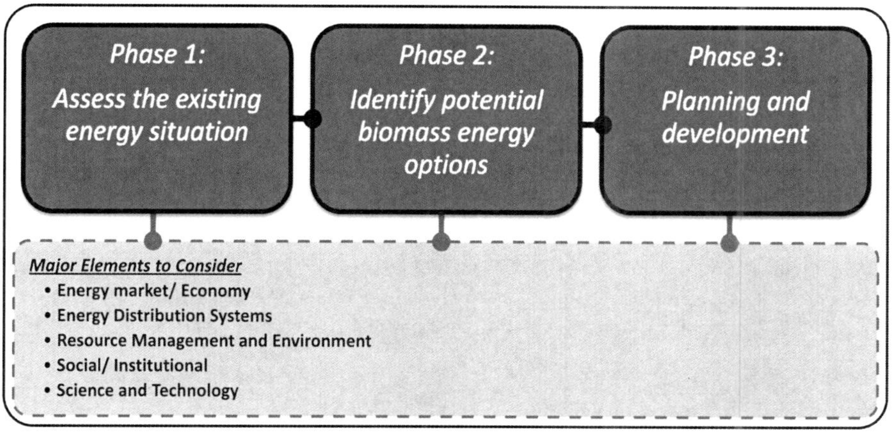

Fig. 17.3 Framework for planning sustainable biomass energy options

of information for their further development. If resources allow, conduct a cost–benefit analysis of various energy–poverty alleviation options.

- *Phase 3: Planning and implementation* of an energy–poverty alleviation solution. Formulate a business plan. Secure political, financial, and institutional backing for the implementation of the plan.

For this framework to adequately meet the challenges of improving energy services for poor households in developing countries, it must take into account environmental, social, and technical factors as well as the commercial and financial aspects of a given form of biomass energy. This focus on social and environmental issues means that the long-term development of energy–poverty alleviation solutions will occur *only* if it meets the needs of the local community, as defined by the community and the development and conservation organizations involved in promoting and, potentially, implementing the energy–poverty alleviation solution.

To gather and analyze the necessary information from the framework, five dimensions of development need to be considered:

1. Energy Market/Economy
2. Energy Distribution Systems
3. Resource Management and Environment
4. Social/Institutional
5. Science and Technology

Information is gathered in these five areas through the three phases of development (Fig. 17.3). The objective is to use this information to screen and adapt sustainable biomass energy options in a systematic way that will allow conservation and development actors to focus activities towards promoting technologies and approaches suited to the local environmental, social, and community infrastructure.

While this process may seem overwhelming to those new to this arena, it is no different than the technology evaluations and due diligence that an investor goes through before financing a new venture.

Box 17.1 provides a series of questions to guide the use of this framework by practitioners. This is not intended to be an exhaustive list of questions, but rather provides the basis for integrated ecological, economic, and socio-cultural evaluation of sustainable biomass energy options. While the focus of these questions is on biomass energy, the same framework could be used to evaluate other energy solutions. General guidance on assessing a range of suitable energy sources, in addition to biomass, for a given community is outlined by Doll, this volume.

Box 17.2 provides examples of successful biomass energy options.

Box 17.1 Framework for Evaluating Sustainable Biomass Energy Options

Phase 1: Assessing the Existing Situation

This is the exploratory phase in which the aim is to understand the boundaries of the existing energy system. This understanding will be used to develop a shortlist of alternative biomass energy options and identify the key issues requiring a resolution to deploy an effective energy–poverty alleviation solution with local[5] biomass resources.

Key questions:

A. Energy Market/Economy

1. How do existing biomass markets operate?
2. What are the current major sources of energy?
3. How much does energy cost to produce?

B. Energy Distribution Systems

1. How does energy get to its various market segments (industrial, residential, commercial, government, manufacturing, etc.)?
2. What is the reliability of energy by market segment?
3. How do existing biomass production and collection systems operate?

C. Resource Management and Environment

1. What are the existing ecosystem goods and services that the community depends upon?
2. Which ecosystem goods and services would be influenced by the collection of biomass resources?
3. How are natural resourses used for energy production currently managed?

[5]Local is to be defined by a group of stakeholders in the energy-poverty alleviation solution. We specifically use the clarification "local" so as to differentiate between these solutions and international trade of biomass resources for meeting the global demands for alternative energy.

(continued)

Box 17.1 (continued)

D. Social/Institutional

1. What are the current energy–poverty alleviation options? What other poverty alleviation efforts are underway and how might they benefit from adding an energy component?
2. What income does the community get from existing energy markets? Which portions of the community benefit from these existing markets? Is it possible to disaggregate the various levels of dependency by various levels of the poor, by socio-demographic group, or by gender?
3. How much energy is collected for subsistence/household use? What are the sources of biomass energy for users? Can these sources be managed differently to maximize community benefits and ecosystem goods and services?
4. What are the current national, regional, and local energy policies and regulations and institutions? Are there any local or customary institutions or policy frameworks that promote or limit the benefit of the poor from their biomass resource? Are there ways to increase incentives to better manage and protect resources?
5. How effective are the current policies and regulations?

E. Science and Technology

1. What are the available biomass resources and existing uses?
2. Are the current systems (harvesting, processing, marketing, and distribution) running in an efficient manner?
3. What scientific or technological infrastructure is available to improve the existing energy system?
4. What next-generation technologies are coming online that if adequately funded and disseminated would improve existing energy systems?

Phase 2: Identify Potential Biomass Energy Options and Future Markets

In this next phase, we investigate the development of potential biomass energy options including the problems and opportunities of the existing situation (examined in Phase 1). Options are then evaluated given potential new markets, technologies, and improved distribution systems. A number of problems can occur if this process is not performed with a high level of rigor. These include: unsustainability of the resource and/or the energy market, erratic

(continued)

Box 17.1 (continued)

supply of biomass, low income-generation, noncompliance with current regulations, lack of awareness of sources for assistance (financial or technical), and degradation of the environment.

Key questions:

A. Energy Market/Economy

1. What are the future demands for energy?
2. What are the predicted future production costs?
3. Which biomass energy options have the best market potential?

B. Energy Distribution Systems

1. How will energy get to its various market segments in the future?
2. What is the future reliability of energy by market segment?
3. What are the future demands to the energy distribution system?
4. How could new biomass collection systems operate?

C. Resource Management and Environment

1. How can the biomass resources be developed to generate a sustainable source of energy while maintaining the flow of ecosystem goods and services?
2. What are the future ecosystem goods and services that the community will depend upon?
3. How will ecosystem goods and services be influenced by harvesting new biomass resources?

D. Social/Institutional

1. What are the potential energy–poverty alleviation options?
2. How can communities generate a sustainable income for the households involved in the energy and/or fuel production both for their own consumption and distribution markets? Are there economies of scale that could be achieved that would enable communities to benefit?
3. How will new proposed biomass energy options affect strategies for subsistence/household energy use?
4. How will new policies and regulation influence biomass energy systems?

E. Science and Technology

1. How can new science and technologies improve biomass energy systems and ecological sustainability?
2. What are the future trends in science and technologies, and how can they be integrated into biomass energy plans?

(continued)

Box 17.1 (continued)

Phase 3: Planning and Development of the Selected Energy–Poverty Alleviation Solution

In this final phase, a local solution has been identified that is both environmentally sustainable and financially promising for the local beneficiaries. The aim of Phase 3 is to plan for the successful implementation of the energy–poverty alleviation solution. The focus of this phase is on developing the procedures and management tools to operationalize the solution. In this phase, proper indicators will be needed to recognize unexpected changes in the biomass energy sector or local income generation such that timely corrective actions can be taken when needed.

Box 17.2 Examples of Successful Biomass Energy Solutions

Example 1: Biogas Support Partnership

In Nepal, an initiative that built on establishing a Biogas Sector Partnership has resulted in one of the best alternative energy examples in the region. The Alternative Energy Promotion Center sold biogas digesters (biogas plants) to households located primarily in the rural areas of Nepal. This project supports the replacement of traditional energy sources used by the rural population, such as firewood and kerosene, with modern biogas plants. Biogas plants use anaerobic decomposition of organic material (mostly animal manure) to produce a flammable gas called biogas, which can be used to meet rural cooking and lighting needs.[6]

Switching to biogas reduces carbon emissions and decreases the frequency of respiratory infections that result from burning solid fuels in poorly ventilated households. Families save approximately 3 hours of labor per day from the conveniences of gas in addition to obtaining financial savings. Women and girls, who are traditionally responsible for collecting firewood and cooking

[6]One component of this project was submitted to the Clean Development Mechanism (CDM) based on greenhouse gas (GHG) emissions reductions from displacing conventionally used fuel sources for cooking, such as fuelwood and kerosene.

(continued)

Box 17.2 (continued)

and cleaning, are among this project's primary beneficiaries. Furthermore, access to biogas will enable families to use gas lanterns after sunset providing light for children's studies or other household activities.

The Biogas Sector Partnership of Nepal (initiated as an affordable energy program with Dutch funding) adopted a multiple-pronged approach including:

- Financial support for end users through microfinance institutions and cooperatives;
- Uniform technical design of biogas plants;
- Thorough quality control and monitoring of the production, installation, and after sales services of the participating biogas companies;
- Continuous research and development efforts to optimize plant operation and to tailor the biogas plants to the needs of the end users;
- Social marketing through outreach, awareness, and training programs;
- Implementation of a fertilizer extension program to maximize the use of bio-slurry, a byproduct of the biogas production;
- Support to institutions servicing various functions of the biogas sector such as financing, construction, maintenance, manufacturing, training, and marketing.

For poor farmers, the initial cost of Rs. 30,000 to install a biogas plant system, including a cooking stove, is a large investment. However, the Biogas Support Program (BSP) provides small-holder farmers who own at least one cow with technical support to build biogas plants that will convert livestock waste into energy. For instance, a typical farmer would receive a subsidy of Rs. 6,500 and a Rs. 12,000 low-interest loan from a Grameen Bank to cover the costs of the digestor and additional inputs needed. The farmer would also be allocated a subsidy of Rs. 1,500 from BSP. Loans were typically repaid in about a year's time through savings from reduced fuelwood purchase costs.

"All we did was to provide guidance and boost the motivation of the villagers so that they could decide how best to use their cattle dung to produce clean gas and improved manure," says Saroj Rai, executive director of BSP-Nepal.

By the end of 2007, BSP had installed 172,858 biogas plants, provided 167 microfinance institutes with wholesale loans from AEPC's Biogas Credit Fund, directly benefited 1,080,000 persons by biogas plants and employed 11,000 people.

This success has led to additional funding for expansion. The World Bank-administered Global Partnership on Output Based Aid (GPOBA) program has provided the Government of Nepal with US$5 million for the verified installation of up to 37,000 new biogas plants in 48 remote districts of Nepal. The program uses an innovative "output-based aid" approach in which subsidy payments are made based on verified results. In 2008, a payment of $592,200 was made following successful installation of 4,772 new biogas plants, and as

(continued)

Box 17.2 (continued)

of July 2009, the program is now in its fourth phase, which aims to support biogas plant installation for over 135,000 new rural households through 2011.

Sources: www.bspnepal.org.np and www.worldbank.org

Example 2: Laguna de Bay Community Carbon Finance Project

With an aggregate area of 91,136 ha and a shoreline of 220 km, Laguna de Bay is the largest lake in the Philippines and the second largest inland freshwater lake in Southeast Asia. Laguna de Bay watershed is a priority watershed for environmental sustainability goals because the area contains 13% of the population of the Philippines and the lake supports fisheries, recreation, and domestic water supply and provides aesthetic value for the many small, historic towns in the area. The objective of the carbon finance project is to implement a set of small-scale waste management and reforestation projects in the watershed, which is heavily degraded due to severe deforestation and pollution from more than ten million people and thousands of industries that discharge largely untreated solid and liquid wastes.

Waste management projects include waste technologies that avoid methane emissions (composting and aerobic wastewater treatment) and those that recover methane (landfill gas collection and wastewater biogas system). These activities will improve waste management practices in the region while reducing methane emissions. Reforestation sub-projects include stream bank rehabilitation, upland reforestation, and agro-forestry. They will increase tree cover in the activity areas while reducing carbon dioxide through its sequestration in the tree biomass during growth. Emission reductions from waste management activities will be sold to the Clean Development Carbon Fund, and those from the reforestation sub-projects to the BioCarbon Fund. An estimated volume of more than 90,000 tons of carbon dioxide equivalent per year from 15 projects over 10 years will be purchased, with additional projects expected to be added to the purchase over time.

(Source: Carbon Finance for Sustainable Development, 2006, World Bank.)

The examples in Box 17.2 represent innovative examples of ways to finance biomass energy projects through greenhouse gas reduction programs. The Clean Development Mechanism, the Community Development Carbon Fund, the BioCarbon Fund, and the Scaling-Up Renewable Energy in low-income countries (SREP) program of the World Bank's Climate Investment Funds (http://go.worldbank.org/2R1V5849O0) all provide financial resources to pilot and demonstrate the economic, social, and environmental viability of low-carbon development pathways in the energy sector .

Conclusions

There is an urgent need to provide "energy–poverty alleviation" solutions for the 1.6 billion people who are energy-deprived. Energy poverty has a negative impact at local to global scales by contributing to deforestation and to climate change (Sagar 2005). In this chapter, we have shown examples where current energy fuel sources and energy technologies in these energy–poverty zones also have a direct negative impact on health (Karekezi 2001; Smith et al. 1993), and how current energy economics create a situation that reinforces these energy–poverty zones, and in some cases make them larger by promoting unsustainable solutions.

The only way out of this cycle of energy poverty is to take into account environmental, social, and technical factors as well as the commercial and financial aspects of plausible sustainable biomass energy solutions. The long-term success of energy–poverty alleviation solutions is dependent on meeting the needs of local communities. Understanding communities' social and environmental issues with systematic analyses, via valuations of the ecosystem services for example, is critical for developing holistic biomass-energy solutions.

Successful energy projects are dependent on several key traits that we have organized into a framework (Box 17.1). The framework's objective is to systematically organize information on factors relevant to planning the development of sustainable biomass energy options. This approach will enable the conservation and development communities to focus their activities of information gathering and analysis and promoting a technology that integrates well with the existing social/community infrastructure and disaggregates the various levels of resource dependency, including the demographics of the poor and the importance of gender. The framework includes the identification of priority target rural areas, socioeconomic assessments of appropriate biomass transformation technologies, stakeholder analysis, policy analysis, and identification of biomass options that do *not* conflict with traditional land tenure and food production systems. Ideally, such a framework will be used to efficiently identify successful energy–poverty solutions. While this process may seem cumbersome to newcomers, it provides the critical due-diligence needed to identify and implement projects that will have a higher chance of being sustainable and successful. Finally, these options should be designed to provide significant health and environmental benefits, maintaining a regular flow of ecosystem goods and services and opportunities for the poor to have a better life with access to a reliable supply of affordable energy.

Acknowledgments We are particularly grateful for the time offered to us by Katherine Warner of New Zealand's Agency for International Development and Cristina Rumbaitis del Rio of the Rockefeller Foundation who provided insightful comments and suggestions on both content and organization of this chapter.

References

America and the Caribbean. http://www.worldenergy.org/documents/lac_executive_summary.pdf

Brown, L. 1980. Food or fuel- new competition for the world's cropland. Worldwatch Institute, Washington, DC.

Bruce, N., Perez-Padilla, R. and Albalak, R. 2000. "Indoor air pollution in developing countries: a major environmental and public health challenge." In *Special Theme: Environment and Health*. World Health Organization: Geneva.

Cavendish, W. 2000. Empirical regularities in the poverty-environment relationship of rural households: evidence from Zimbabwe. World Development, volume 28, issue 11, November 2000.

Chambers, R. 1987. Short cut methods in socila information gathering for rural development projects/Proceedings of the 1985 international conference on rapid rural appraisal (RRA)/ Khon Khen University, Thailand, pp. 66–88.

ECLAC (Economic Commission for Latin America and the Caribbean). 2003. Energy Sustainability in Latin America and the Caribbean: The Share of Renewable Sources. 70 pp. Available at http://www.eclac.org/publicaciones/xml/3/13413/Lcl.1966i.pdf. Accessed on January 5, 2010.

Goldemberg, J., Reddy, A.K.N., Smith, K.R., and Williams, R.H. 2000. Rural energy in developing countries. In: Goldemberg, J. (Ed.), World Energy Assessment: Energy and the Challenge of Sustainability. United Nations Development Program, New York.

Grandstaff, T. and Messerschmidt, D. 1995. A manager's guide to the use of Rapid Rural Appraisal. FARM Program. FAO/ UNDP and Suranaree University of Technology.

GreenLineCare 2009. Biomass is an Important Source of Fuel in Asia. Accessed on November 10, 2009 from: http://www.greenlinecare.com/en/index.php?option=com_content&view=article&id=73%3Abiomass-is-an-important-source-of-fuel-in-asia&catid=1%3Alatest-news&Itemid=1.

Hall, D.O. and Moss, P.A. 1983. Biomass for Energy in Developing Countries. GeoJournal 7.1:5–14.

Hall, D.O., Barnard, G.W., and Moss, P.A. 1982. Biomass for energy in developing countries. Pergamon Press, Oxford.

IDEA, Kundasale, Sri Lanka. Village Hydro Power Development in Sri Lanka. www.inforse.dk/asia/pdf/SL_village_MicroH.pdf.

IEA (International Energy Agency) 2002. Energy and poverty. In: World Energy Outlook, 2002. IEA, Paris.

International Energy Agency (IEA) Statistics Division. 2008. Energy Balances of OECD Countries (2006 edition) and Energy Balances of Non-OECD Countries (2006 edition). Paris: IEA. Available at http://data.iea.org/ieastore/default.asp. Accessed on 5 December 2010.

Ignaciuk A, Vöhringer F, Ruijs A, Van Ireland E.C. 2006. Competition between biomass and food production in the presence of energy policies: a partial equilibrium analysis. Energy Policy 34:1127–1138.

ITDG (Intermediate Technology Development Group) 2004. Powering Pverty Reduction. Position Paper for Renewables 2004 Bonn, June 104, 2004. http://practicalaction.org/docs/advocacy/powering_poverty_reduction%20.pdf.

Johansson, D. and Azar, C. 2007. A scenario based analysis of land competition between food and bioenergy production in the US. Climate Change, 82, 267–291.

Karekezi, S. 1999. Access to modern energy—a view from Africa. In: Energy after the Financial Crisis—Energy and Development Report 1999. World Bank, Washington.

Karekezi, S. 2001. Poverty and energy in Africa—A brief review. Energy Policy 30: 915–919.

Loomis, J., Timm, K., Leslise, R., and Frank, C. 2008. Benefit Transfer Toolkit for Fish, Wildlife, Wetlands and Open Space. Western Economics Forum VII. pp. 33–43.

Narain, U., S. Gupta and K. Van 't Veld. 2005. Poverty and the Environment: Exploring the Relationship between Household Incomes, Private Assets, and Natural Assets. http://www.prem-online.org/archive/7/doc/PREM%20WP%2005-09.pdf. Accessed on 14 November 2009.

NatCap 2007. The Natural Capital Project, Stanford University. http://www.naturalcapitalproject.org/InVEST.html. Accessed on 14 November 2009.

Pimentel, D. 2005. Environmental and Economic Costs of the Application of Pesticides Primarily in the United States. Environment, Development and Sustainability 7: 229–252.

Pimentel, D., Acquay, H., Biltonen, M., Rice, P., Silva, M., Nelson, J., Lipner, V., Giordano, S., Horowitz, A., and D'Amore, M. 1992. Environmental and Economic Costs of Pesticide Use. BioScience, Vol. 42(10); 750–760.

Powledge, F. 2006. The Millennium Assessment. In Bioscience, Vol. 56, No. 11.

Pretty, J.N., Gujit, I., Thomson, J. and Scoones, I. 1995. Partipatory Learning and Action: A Trainer's guide. IIED London.

Reddy, A.K.N., Williams, R.H., Johansson, T.B. 1997. Energy After Rio: Prospects and Challenges. United Nations Development Program, New York.

Rural Wood Energy Development Program 1999. Production, Utilization and Marketing of Woodfuel in Lao PDR – National Workshop (RM 45), FAO RAP Publication 1999/45.

Sagar, A.D. 2005. Alleviating energy poverty for the world's poor. Viewpoint. Energy Policy 33, 1367–1372.

Sagari, S. 2006. Latin America and the Caribbean Region: Selected Energy Sector.

Smith, K.R., Gu, S.H., Huang, K., and Xiu, D.X. 1993. 100 million improved cookstoves in China—how was it done? World Development 21(6), 941–961.

Smith, K.R., Zhang, J., Uma, R., Kishore, V.V.N., Joshi, V., and Khalil, M.A.K. 2000. Greenhouse implications of household fuels: an analysis for India. Annual Review of Energy and the Environment 25, 741–763.

The Earth Institute. State of the Planet 2004. Available at : http://www.earthinstitute.columbia.edu/sop2004/consensus.html.

Troy, A. and Wilson, M.A. 2006. Mapping ecosystem services: Practical challenges and opportunities in linking GIS and value transfer. Ecological Economics 60, 435–449.

UNDP (United Nations Development Program) 2004. World Energy Assessment: 2004 update. http://www.undp.org/energy/weaover2004.htm.

UNEP 2008. Report on the Latin American and Caribbean Initiative for Sustainable Development (ILAC): Five Years after it was adopted. Sixth Meeting of the Foresum of Ministers of Latin America and the Caribbean, held in Santo Domingo, Dominican Republic, January 27th-February 1st, 2008. Available at: http://www.pnuma.org/forumofministers/16-dominicanrep/rde03tre-ILACReport_FiveYearsRev2.doc. Accessed on January 5, 2010.

World Bank 2009. Design Document for the Program on Scaling-up Renewable Energy in Low-Incomes Countries (SREP), a Targeted Program under the Strategic Climate Fund.

WEC (World Energy Council) 2008. Regional Energy Integration in Latin.

WEC (World Energy Council) 2007. The Challenge of Rural Energy Poverty in Developing Countries. http://www.worldenergy.org/publications.

WHO (World Health Organization) 2002. World Health Report: Reducing Risks, Promoting Healthy Life. WHO, Geneva.

Chapter 18
Ecological Sustainability of Woodfuel as an Energy Source in Rural Communities

Rob Bailis, Jeff L. Chatellier, and Adrian Ghilardi

Introduction

Overview of Woodfuel Use in Developing Countries

Between one-third and one-half of the world's population rely on wood and other biomass fuels[1] to meet their energy needs. Table 18.1 shows an estimate of the number of people relying on biomass fuels in 2004 from the International Energy Agency (IEA 2006). The use of wood as a household fuel is overwhelmingly concentrated in less developed countries where alternative fuels like natural gas, kerosene, liquefied petroleum gas (LPG), and electricity are inaccessible. Heavy reliance on woodfuel is associated with a range of social and environmental challenges including health problems resulting from exposure to indoor air pollution (IAP) and environmental change, which ranges from local degradation of forests and woodlands to large-scale changes in land cover and greenhouse gas emissions. In addition, supplying woodfuels requires high labor inputs, which, in many places, is often a burden for women and small children.

The problems associated with biomass use rarely arise as a result of wood consumption alone; rather, they are the result of complex relationships between wood

[1] The terminology used for fuels derived from woody biomass deserves some explanation. In this discussion, we use the term *woodfuel* to encompass minimally processed firewood, as well as charcoal and other solid fuels derived from lingo-cellulosic materials, such as sawdust or wood-waste briquettes (see FAO 2000, for a more detailed explanation). Non-woody forms of biomass, such as crop residues and dried dung are also used for traditional energy applications, and are associated with similar consequences.

R. Bailis (✉)
Yale School of Forestry and Environmental Studies, 195 Prospect Street, New Haven, CT, USA
e-mail: robert.bailis@yale.edu

J.C. Ingram et al. (eds.), *Integrating Ecology and Poverty Reduction: Ecological Dimensions*, DOI 10.1007/978-1-4419-0633-5_18,

Table 18.1 Estimates of the number of people relying on biomass resources as their primary fuel for cooking in 2004 (International Energy Agency 2006b)

	Total population (millions)		Rural (millions)		Urban (millions)	
	%	Total	%	Total	%	Total
Sub-Saharan Africa	76	575	93	413	58	162
North Africa	3	4	6	4	0.2	0.2
India	69	740	87	663	25	77
China	37	480	55	428	10	52
Indonesia	72	156	95	110	45	46
Rest of Asia	65	489	93	455	35	92
Brazil	13	23	53	16	5	8
Rest of Latin	23	60	62	59	9	25
Total	52	2,528	83	2,147	23	461

Source: IEA analysis based on latest available national census and survey data

consumers, the environment in which they live and the larger political economy. Therefore, understanding the environmental challenges associated with woodfuel consumption is only possible by considering the social, political, economic, and environmental context in which they arise. This chapter focuses primarily on the challenges to ecological sustainability that are posed by dependence on biomass, but it also discusses the range of social and political factors that affect household energy choices and their environmental consequences.

Woodfuels and Poverty

Biofuel dependence is closely correlated with income, both among and within countries. Figure 18.1 shows how the prevalence of fuelwood as a *primary* source of household energy declines with increasing income in several African countries.[2] Fuelwood and other solid biofuels are linked to poverty because they are associated with risks, inconveniences, and cultural meanings that people in higher income strata may wish to avoid. The UNDP in its World Energy Assessment defined energy poverty as "the absence of sufficient choice in accessing adequate, affordable, reliable, high quality, safe, and environmentally benign energy services to support economic and human development" (Goldenberg and Johansson 2004). Fuelwood, while affordable for most of the world's population, often has an unreliable supply, it is inconvenient to use as a fuel and requires considerable space for storage, it is unsafe from a health perspective, has limited ability to support economic development and its widespread use can potentially degrade ecosystems and reduce the environmental services they provide.

[2] Fuelwood falls off as a *primary* fuel choice among wealthier households. However, some evidence shows that families do not stop using woodfuels altogether. Instead, they expand their fuel choice as they get wealthier by incorporating additional fuels into their energy mix (Masera et al. 2000; Pfaff et al. 2004).

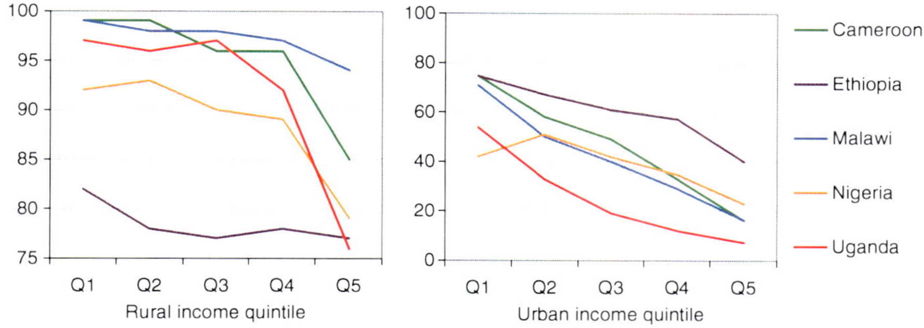

Fig. 18.1 Prevalence of fuelwood as a primary energy source by income quintile in select African countries between 2001 and 2004. The graphs show rural (*left*) and urban (*right*) households. Note the vertical scales in each graph are different (Source: National demographic surveys reported in World Bank (2008))

As it is the main source of fuel of the world's "energy poor," fuelwood consumption is used as an indicator of poverty by development organizations. However, dependence on woodfuel is not just a result of poverty; it can also contribute to factors that reinforce poverty. It is a causal agent of preventable morbidity and mortality and it has a strong association with low educational attainment, both of which make it difficult for families to rise out of poverty (UNDP 2005).

Moreover, lack of modern energy services, which correlates with woodfuel dependence, places a burden on household labor. Economic activities are constrained because the energy to support a variety of income-generating activities is absent. However, it is important not to assume woodfuel reliance is always associated with a complete lack of access to modern energy services. There is wide regional variation. For example, in some developing regions, where biomass is the dominant fuel for subsistence needs, electricity is also widely available (IEA 2004). In such areas, access to electricity may support income generation, but families may continue to depend on woodfuels for the bulk of their cooking and space heating needs.[3]

The Quality and Availability of Data on Woodfuels

Accurate data on fuelwood consumption in developing countries would be a powerful tool for policy makers designing legislation on topics ranging from energy systems to environmental conservation. However, fuelwood consumption data is

[3] This situation is common in parts of Mexico (Masera et al. 2000). Similarly, in rural China, where there has been near universal electrification, the majority of households continue to depend on biomass for their cooking needs (agricultural residues are more common than woodfuels, but the point still holds; see Zerriffi et al. 2008).

difficult to obtain for many reasons. In the developing world, fuelwood is collected and consumed by subsistence users who do not measure the amount or the species composition of their fuelwood stocks. Since fuelwood for domestic consumption is not often traded or sold in formal markets, uniform measurement standards and sales data simply do not exist.

Determining the amount of energy produced from fuelwood consumption is also difficult. Tree species have different calorific values (Harker et al. 1982). In addition, moisture content can vary greatly, which affects the mass of fuel consumed, as well as the useful energy that can be obtained from the fuel. Fuelwood consumption data based on weight must account for the species type and water content in order to give an accurate figure in terms of energy produced, but this is rarely the case. Moreover, the efficiency of energy conversion devices like wood-burning stoves varies enormously, so the estimates based on "typical" consumption rates of a small sample can be unreliable.

Despite the difficulties in obtaining reliable data on fuelwood consumption and energy production, resources are available that can assist policy makers in this realm. Surveys can provide a picture of a households' main source of energy. In recent years, some national censuses and demographic/health surveys have started to include questions about fuel choices. These data can provide a periodic nationally representative snapshot of household fuel choice. Unfortunately, data produced from these large-scale surveys are limited, in that they usually provide the household's primary fuel choice, but offer no insight into multiple fuel use, which is common even among poor populations.[4] In addition, large-scale survey data offers no insight into quantities of energy consumed at the household level. Some countries do conduct targeted surveys specifically exploring household-level energy consumption, but this does not appear to be common practice.

At the national level, aggregate energy consumption data is available from the International Energy Agency. The IEA publishes national and regional energy balances that offer detailed accounts of energy supply and consumption disaggregated by fuel-type and economic sector. Woodfuels are categorized in the "combustible renewables and waste" category (CRW), which includes all biomass fuels: those that are used in traditional applications, as well as feedstock used for modern applications like cogeneration or liquid biofuels.[5] In addition, the IEA disaggregates each energy type by sector, so residential energy may be analyzed separately (International Energy Agency 2008).

---

[4] For example, one nationally representative survey of Kenyan households found that 96% of the rural population used more than one fuel and 45% used three or more types of fuel (Nyang 1999).

[5] For the majority of developing countries, it is safe to assume that CRW consists almost entirely of traditional woodfuels and crop residues. One exception is Brazil, which uses biomass for a number of non-traditional applications: for example, the country produced nearly 19 billion liters of ethanol from sugarcane and generated over 14,000 gigawatt-hours of electricity from biomass feedstocks in 2005 (International Energy Agency 2008).

The FAO provides country and regional level data on woodfuel production in solid volume units (CMU) rather than in energy units. The FAO does not provide data on industries that rely on fuelwood for their energy supply, but it does provide estimates of country and regional level consumption calculated by subtracting exports from total production (FAOSTAT 2008).[6]

(a) Defining sustainability in the context of traditional energy systems

Sustainability has become an important concept in environmental governance, influencing policies ranging from industrial development to environmental conservation. The concept emerged in the 1970s, when it became apparent that the resource base upon which the global economy depended could not support the economy's rapid expansion (Kidd 1992). Economists began to study and model the conditions under which growth could continue in a world of finite resources (Cabeza Gutés 1996). These studies focused on capital, defined broadly as "produced means of production" (Costanza and Daly 1992), and differentiated the total capital stock into three categories: natural, social, and physical (or man-made) (Cabeza Gutés 1996). Natural capital consists of society's endowment from nature, which includes renewable resources like forests and non-renewable resources like fossil fuels. Physical capital includes money, as well as anything produced from natural capital ranging from buildings to machines. Social capital consists of intangible assets derived from interpersonal relationships within social networks and institutions. Social Capital is associated with structured forms of social interaction like formal educational systems, as well as unstructured everyday interactions that build social cohesion, trust, and reciprocity (Bourdieu 1985; Baker 1990).

Two different schools of thought on sustainability, each with different definitions, emerged based on the differentiation of capital. Strong sustainability regards natural capital as a collection of resources that provide functions that are not substitutable by social or physical capital. These functions include a host of ecosystem services ranging from erosion control to genetic diversity. Strong sustainability, therefore, is defined by maintaining the same level of natural capital for future generations. Weak sustainability views natural, social, and man-made capital as interchangeable, which implies that natural capital can be consumed as long as it maintains or increases the stocks of physical and/or social capital (Pearce and Atkinson 1993).

Fuelwood consumption is often portrayed as unsustainable because of its association with deforestation and/or forest degradation. From a strong sustainability perspective, these processes lead to a depletion of natural capital stock, which is not limited to trees, but includes the sum of all forest-related assets. If these assets are reduced for future generations, current extraction is unsustainable. However,

[6] Data from the IEA and FAO originate from different sources and often do not agree (see Bailis et al. 2005, supplemental online material for a discussion of this in the context of African woodfuel data).

unlike fossil fuels, forests are a potentially renewable resource. Forests can recover when wood is harvested for fuel: a form of "natural income" (Costanza and Daly 1992), which maintains the capital stock of the forest. If wood extraction is balanced with the forest's capacity to regenerate, then the capital stock is maintained. Forest management that satisfies strong sustainability criteria should take species composition and growing conditions into account, rather than simply maintaining a standing stock of biomass. This is a complex undertaking. Tropical and subtropical forests can be extremely heterogeneous and a great deal of variation can exist within a small spatial scale (Montagnini 2005).[7]

Challenges to Sustainability in Woodfuel Systems

Ecological Sustainability

Local Environmental Change

In the 1970s, global attention turned to energy issues as a series of price shocks that severely affected the world's economies. At roughly the same time, alarming rates of deforestation began to grab the world's attention (Bajracharya 1983). Analysts merged the two crises into a distinct environmental challenge, dubbed the "other energy crisis" (Eckholm 1975). This crisis revolved around fears that rates of wood harvest were exceeding sustainable yields in many of the world's developing regions. In response, development organizations published alarming reports predicting environmental catastrophe resulting from the so-called " firewood gap" unless drastic measures were taken, including unprecedented levels of tree planting and severe demand reduction programs (de Montalembert and Clement 1983; FAO 1978).

Closer scrutiny soon revealed that other socioeconomic drivers, primarily land clearing for cultivation and timber extraction, both exacerbated by expanding road networks, are more influential drivers of deforestation than wood energy use (Arnold et al. 2003; Geist and Lambin 2002; Kaimowitz and Angelsen 1998). In addition, numerous natural factors like rainfall, soil quality, and wildlife, interact with anthropogenic forces to influence tree cover in complex ways. Box 18.1 provides a description of how human and ecological drivers interact to influence land cover in savannah

[7] When discussing the impacts of woodfuels on forests, it is useful to draw a distinction between deforestation, the "direct human-induced conversion of forested land to non-forested land," and forest degradation, the "direct, human-induced, long-term loss (persisting for X years or more) or at least Y% of forest carbon stocks [and forest values] since time T and not qualifying as deforestation." Both definitions are from the IPCC (2003), but X, Y, and T are left to national or local-level decision makers to define. The distinction is important because woodfuel supply is rarely a sole cause of deforestation (Geist and Lambin 2002), but may be a driver of degradation. The distinction is also important for methodological reasons. Deforestation can be detected by remote sensing methods, but degradation very often cannot.

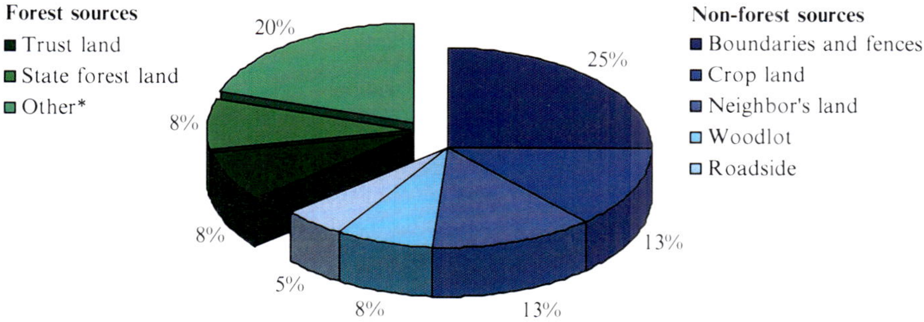

* "Other" includes purchased wood of unknown origin, which may be from forest or non-forest sources

Fig. 18.2 Sources of firewood for rural households in Kenya in 2000

ecosystems, which provide woodfuels and numerous other sources of livelihoods for people across sub-Saharan Africa. These interactions challenge the idea of a direct link between woodfuel demand and deforestation.

In addition, problems arise among subsistence users if access to certain woodlands is denied or when woodlands are cleared as a result of other pressures. This can lead subsistence users to overexploit the little areas that remain accessible to them. Moreover, population and economic pressures can force people to shorten fallow periods or expand the area that they cultivate, which reduces both the time and space in which their home-grown wood accumulates. Hence, while energy demand may not be the primary cause of fuelwood-scarcity, scarcity still affects many who rely on woodfuels for subsistence needs.

Wood for subsistence use rarely comes from mature trees in forests or woodlands; people prefer gathering fallen branches and dead wood (Leach and Mearns 1988). People may also collect wood from their own household compounds or from fallow land to which they have access. Figure 18.2 shows sources of firewood identified by Kenyan households in a national energy survey. The majority report a dependence on non-forest sources of firewood, primarily from their own land (Ministry of Energy 2002). However, when woodfuel becomes commercialized, mature trees are often cut. This is common where woodlands are used to provide fuel to urban markets. For example, charcoal is a common fuel derived from wood that is carbonized (heated in an oxygen-deprived environment, so that full combustion does not occur and the volatile material in the wood is driven off). It is popular in many urban and peri-urban areas across the developing world and is associated with widespread clearance of woodlands (Girard 2002; Ribot 1993). Despite this association, empirical studies have shown that charcoal is not inherently destructive and that under good management, it may be produced sustainably (Chidumayo 1993; Hosier 1993; Young and Francombe 1991). However, charcoal is often produced illicitly; sound management is rare. Box 18.2 discusses charcoal production in Kenya, one of the world's leading charcoal producers, which struggles with the issue of sustainability in the charcoal trade.

Box 18.1 Natural and Human Drivers of Land Cover Change in Woody Savannah

Africa's savannah woodlands provide woodfuel and other subsistence needs, as well as, some sources of commercial production for a large segment of the population. Savannah ecosystems under pressure from herbivory, rainfall variability, and fire can shift from grass-dominant to tree-dominant states or vice versa. Rainfall varies from year to year, and has direct and indirect effects on land cover (Scholes and Hall 1996). Precipitation affects the behavior of people and wildlife. In ecosystems initially dominated by grasses, abundant rains can lead to an increase in the quantity and quality of pasture, temporarily supporting higher concentrations of livestock and/or herbivorous wildlife. Grazing animals can promote the growth of woody biomass by removing the herbaceous layer, which competes for light and nutrients. Browsers, on the other hand, feed on young seedlings, preventing maturation, thereby maintaining the herbaceous layer. Soil types also influence these dynamics, as Breman and Kessler (1995):

> Woody plants vary in their response to grazing...On sandy soils or fluvial landscapes, intensive grazing may lead to an increase in canopy cover, but a strong reduction is also possible. On loamy soils especially in dry zones, canopy cover is reduced by intensive grazing because infiltration [of water] is reduced. However, the highest canopy covers occur on fallowlands and lands near natural or artificial water points (i.e. where grazing pressure is high). Higher rainfalls favor the more positive influences of grazing on woody plants (p. 45).

Fire is a perennial feature of savannah landscapes both as a natural and anthropogenic phenomenon. Fire interacts with rainfall and influences grazing in numerous ways:

> Fire leads to the loss of volatile compounds of nitrogen, carbon and sulfur. It tends to destroy woody seedlings and sensitive species, particularly those lacking seed adaptations, belowground reserves, and the capacity to sprout back. Rangeland systems...,where fire has been a regular feature for centuries, have a correspondingly fire-adapted species composition. In such systems periodic burning enhances the production of good grazing (Homewood and Rodgers 1991, p. 103).

The influence of rainfall and fire also depend on grazing intensity (Homewood and Rodgers 1991). Under moderate grazing, rainfall increases the quantity of herbaceous biomass. Under normal conditions, fires remove the herbaceous layer, but leave established trees and shrubs standing and promote germination of dormant seeds, thereby, reinforcing woody biomass cover. However, after heavy rains, above-normal herbaceous dry matter can cause intense burns, killing extant trees, and destroying the soil seed bank, causing a shift from woody to herbaceous cover. Both grasses and woody species can thrive, but there is a "competitive asymmetry" inherent in these systems such that either may establish dominance on small scales. For example:

(continued)

Box 18.1 (continued)

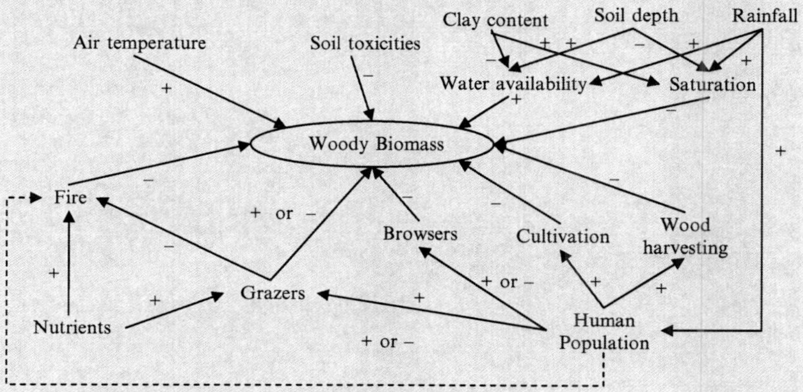

Fig. 18.3 Factors that can increase (+) or decrease (–) woody biomass cover in savanna ecosystems (Source: adapted from Breman and Kessler (1995) and Scholes and Hall (1996))

> …mature trees out-compete grasses for light, water and nutrients, yet grasses out-compete small shrubs and tree seedlings (reducing establishment) and they increase the likelihood of fires which kill small trees…lead[ing] to structural instability. Often some degree of tree clumping takes place adding further complexity with conditions often very different between the under-canopy and inter-canopy areas (House and Hall 2003).

Some of these dynamics are illustrated in Fig. 18.3.

Box 18.2 Charcoal in Kenya

Kenya relies on woodfuels for three-fourths of its primary energy supply (Fig. 18.4). Roughly half of the wood harvested for fuel is converted to charcoal. Despite its widespread use, woodfuel has largely been ignored by policy makers, particularly the supply side of the sector (Bailis et al. 2006). A strong association has been made by the press and the government between charcoal and deforestation (Ecoforum 2002; Okwemba 2003). Unfortunately, little can be said with certainty about the degree to which Kenya's exploitation of wood energy is leading to permanent forest loss. Reliable data is very difficult to obtain. It is certain, however, that the country lacks an effective set of policies to promote or enforce sustainable woodfuel management. This void leads to a great deal of ambiguity in the woodfuel sector. Some charcoal regulations are in

(continued)

Box 18.2 (continued)

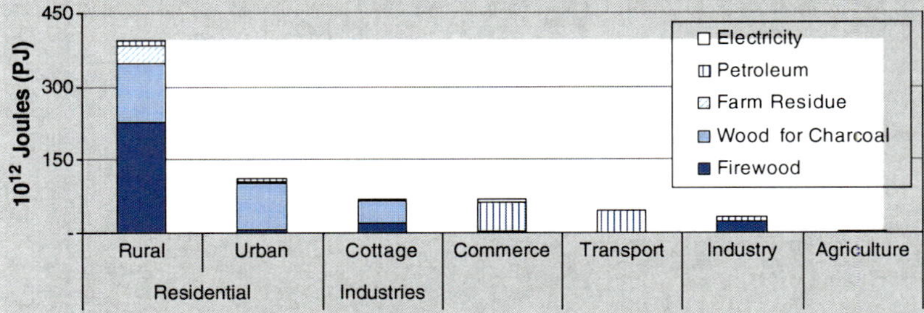

Fig. 18.4 Kenya's energy supply by fuel and sector in 2000 (IEA 2003; Ministry of Energy 2002)

place at the provincial or district level, but these lack transparency and suffer from inconsistent enforcement. Consequently, in many parts of the country, charcoal is illegal to produce and transport, but it is legal to sell, buy, and consume. Such ambiguity discourages investment in the trade, encourages unsustainable practices, and fosters corruption.

For example, in Narok district, a major charcoal production zone, a ban on commercial charcoal transport was in effect between 2003 and 2005. Despite the ban, the district provides as much as 30% of Nairobi's charcoal, with 10–20 lorries ferrying thousands of 40 kg sacks to the city every day (Bailis 2005). The ban, which was meant to protect nearby Mau Forest, a high-value conservation area, was circumvented through bribery, which reached such high levels that as much as 25% of the retail price of each sack of charcoal was fraudulently captured by local officials (Bailis 2005). Ironically, Narok's charcoal does not originate from the forest that the transport ban was meant to protect. Rather, it is harvested from parcels of woody savannah that were formally pastoral, but had been subdivided and allocated to the district's Maasai population throughout the 1990s. This land is private land that would likely be cleared in the absence of charcoal production: charcoal simply facilitates the process.

Woodfuel, particularly the charcoal trade, provides direct employment for as many as 200,000 people across Kenya at different stages of the supply chain (Mutimba and Barasa 2005). For some with little or no land to farm, charcoal provides full-time employment. For others, it presents a source of income when farm production is low or when a bit of extra cash is needed. Thus, woodfuels are a critical part of the economy, not only because of their contribution to household production, but also because of the livelihoods of woodfuel suppliers.

Woodfuel dependence will persist in Kenya; whether it can be managed sustainably is an open question. To promote sustainability, regulations governing the woodfuel trade must be rationalized and clarified to remove the legal ambiguity that currently exists. Investment in woodfuel production must be encouraged, so that the private sector can participate in woodfuel provision.

Woodfuels and Global Change

In addition to local environmental impacts, the scale at which woodfuel consumption occurs has implications for global change. Wood harvesting, fuel processing, and final combustion create a flow of greenhouse gases (GHGs) from terrestrial stocks to the atmosphere. Post-harvest management of woodlands may also result in GHG emissions. On the other hand, carbon is sequestered by the regeneration of harvested trees, thus, the net emissions of GHGs from the wood-energy fuel cycle depend on the degree of sustainability with which the fuel is harvested. Identifying the degree of sustainability of the woodfuel harvest is challenging and few rigorous examples of this research exist. Box 18.3 discusses one effort, which utilized multi-scale spatial analysis of woodfuel supply and demand to identify local-level imbalances.

However, full regeneration of harvested trees does not assure GHG-neutrality. Additional emissions occur, because typical wood combustion devices such as household stoves in the developing world cannot achieve full combustion, which results in emissions of CH_4 and N_2O, as well as GHGs that are not controlled under current climate change policies, but nevertheless have an impact on radiative forcing, such as CO, non-methane hydrocarbons, and aerosols.[8] Furthermore, charcoal production, which utilizes roughly 15% of woodfuel harvest worldwide (FAOSTAT 2008), emits considerable amounts of non-CO_2 GHGs (Kituyi et al. 2001; Pennise et al. 2001; Bertschi et al. 2003; Brocard et al. 1996).[9]

Based on numerous studies of biomass emissions under lab and field conditions, it is estimated that CO_2 contributes roughly 60–70% of the "tree-to-stove" emissions from fuelwood and 30–40% of the tree-to-stove emissions from charcoal (Bond et al. 2004; Bertschi et al. 2003; Brocard et al. 1996; Pennise et al. 2001). Thus, regeneration of harvested trees reduces the impact from CO_2, but can not fully offset the impact from the other GHGs.

The IPCC's Fourth Assessment Report notes that some of the emissions from land use change (LUC) are the result of "traditional biomass use." However, the assessment departs from its usual rigor by assuming that 90% of the traditional biomass harvest is "from sustainable biomass production." The remaining 10% of global harvest is "non-renewable" by default (the IPCC bases this on assumptions made in International Energy Agency 2006a). Based on this assumption, the IEA

[8] CH_4, N_2O, CO, non-methane hydrocarbons, and BC aerosols have a larger warming impact than a molar equivalent quantity of CO_2. OC aerosols have a cooling effect, but these only partially balance the warming impact of BC species (Bond et al. 2004). Each of these compounds has a larger warming impact than a molar equivalent quantity of CO_2 (with the exception of OC aerosols) (IPCC 2007). Therefore, when fuel-bound carbon is emitted in one of these forms rather than CO_2, CO_2 sequestration through future wood growth does not fully counterbalance the warming effect of those pollutants.

[9] Estimates of global woodfuel production and the fraction of woodfuel that is utilized for charcoal vary widely (see Bailis et al. 2005, for a description of limitations in this data).

Box 18.3 Spatial Analysis of Woodfuel Supply-Demand Imbalances in Central Mexico

Historically, studies of woodfuel balance (supply and demand) have utilized either national or regional data or micro-level case studies conducted in specific localities. However, complex relationships between fuelwood supply and demand lead to impacts that are heterogeneously distributed in space and time. Generalized approaches used to determine national or regional wood energy balances are unable to provide information on the spatial distribution of areas suffering from extreme supply–demand imbalances. Localized case studies may be able to identify such "hotspots," but cannot be extrapolated, as fuelwood use and associated impacts can differ substantially, even between neighboring localities.

This study used a spatially explicit method to assess environmental and socio-economic impacts associated with traditional fuelwood use in Mexico in order to identify woodfuel "hotspots": that is, individual localities with high woodfuel consumption and insufficient biomass resources (Fig. 18.5). In the first stage, a multi-criteria analysis was conducted in order to analyze Mexican counties according to seven indicators: number, density, and annual population changes of fuelwood users; percentage of households using fuelwood; resilience of consumption; trends in land use and land cover change; and the balance between supply and demand. The national fuelwood balance – a key value when comparing countries – was extremely positive (165 million tons per year). Compared to Southeast Asia and sub-Saharan African countries, Mexico is not in a crisis situation in terms of fuelwood use and its associated impacts. However, the spatial analysis identified 304 counties (out of a total of 2,424) with negative or close to zero balances. These were grouped into 16 *hot spots*. Approximately 6.3 million fuelwood users live in these counties, which constitutes 25% of the nation's fuelwood users in 2000.

In the second stage, one fuelwood *hot spot* in Michoacan State was selected and a grid-based model was developed in order to identify individual localities with high fuelwood consumption and insufficient supply. The analysis also gave a robust and statistically confident estimate of the non-renewable biomass (NRB) fraction of fuelwood extraction by locality (a critical value to estimate baselines in carbon offset projects) (Fig. 18.6). Ground-truth efforts validated these findings. Importantly, large variations in NRB were found in neighboring communities, which demonstrates that spatial patterns of fuelwood supply and demand are highly site specific. This work shows the value in multi-scale assessments of woodfuel supply and demand in order to focus action on the most critical locations.

(continued)

Box 18.3 (continued)

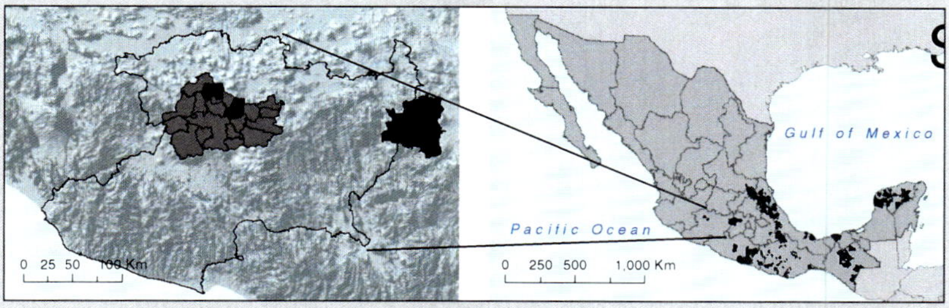

Fig. 18.5 Study area: the Purhepecha Region. Notes: *Black shapes* in each map represent high priority counties following the national-level assessment (Ghilardi et al. 2007). In the *left-hand* map, counties of the Purhepecha Region in the north-west of Michoacán State, are highlighted in *dark gray*

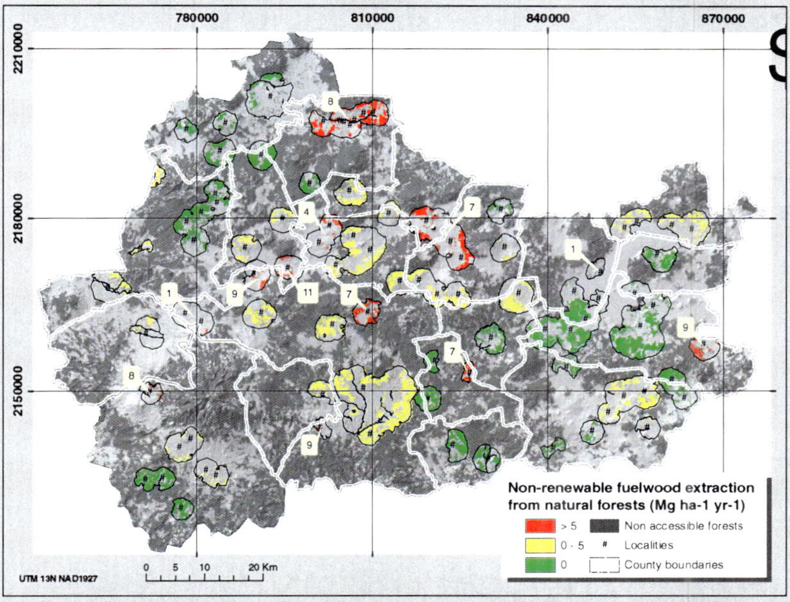

Fig. 18.6 Pressure over natural forests due to fuelwood extraction on a non-renewable basis from accessible areas considering walking fuelwood gatherers. Notes: Highlighted *red, yellow and green* areas within accessible areas correspond to accessible forests. *Dark-gray* areas correspond to non-accessible forests. Labels show the expected time in years for depletion of half the fuelwood stock available from forest areas (Source Ghilardi et al. 2008)

estimates that global woodfuel use contributes to approximately 2% of total global emissions (International Energy Agency 2006a, section III.6).[10] This is roughly equivalent to the emissions from the transport sector in the European Union (World Resources Institute 2008).

Of course, this estimate ignores the interactions between wood harvest for energy and other drivers of LUC discussed above. Moreover, the relationships between proximate causes of land cover change and structural drivers of change like demography, political economy, and technology make it difficult to attribute a specific value to the net emissions from woodfuel demand alone.

Additional aspects of woodfuel sustainability are linked to the social conditions in which production and consumption occur. These include struggles over access among producers and consumers of forest resources for energy and other uses. Social sustainability also extends into the health and well-being of woodfuel users. Each of these are discussed briefly below.

Political Ecology and Resource Access in the Context of Woodfuels in Developing Countries

Woodfuel provision is a critical component of household production, both for domestic use and for provision to the market. Thus, woodfuels are linked very closely to the livelihoods of poor rural populations. In locations where woodfuels or other forest resources are extracted for commercial sale, local users may find that their own access to energy for subsistence needs is contingent on distant markets, state agents, and powerful business interests (Ribot 1999; Bailis 2005).

As was discussed above, woodfuel provision has strong associations with environmental degradation. Environmental impacts, whether real or perceived, often generate attempts by governments to regulate or control access to those resources. For example, charcoal and firewood dealers may be required to obtain permits to ensure that the supply is from a sustainable source. Those attempts may be subverted by poor enforcement, corruption, and/or a failure to incorporate local knowledge and institutions in regulatory design (Ribot 1999; 2004; Dove 1992; Robbins 1998). The illicit nature of woodfuel provision systems tends to make them opaque to outsiders (see the discussion in Box 18.2, which describes charcoal production in Kenya).

The benefits that people obtain by playing a role in these markets are mediated through the degree of access that they maintain. Access is mediated by mechanisms that are both legal and extra-illegal, as well as an array of "structural and relational" factors that include technical capacity, markets for land, labor and capital, as well as social identity and social relations (Ribot and Peluso 2003). These factors actively

[10] The authors checked the IEA's estimate by considering global fuelwood and charcoal consumption as reported by both the FAO and IEA (FAOSTAT 2008; International Energy Agency (IEA) (2007)) and the IEA's assumption that 10% of the global woodfuel harvest was unsustainable. Taken together with published emissions factors for wood combustion and charcoal production/combustion (Bertschi et al. 2003; Brocard et al. 1996; IPCC 1997; Pennise et al. 2001; Smith et al. 2000), we estimate that net GHG emissions from woodfuel combustion ranged from 1 to 2.8% of global GHG emissions

shape dynamic systems of provision that are distributed over space and changing over time (Leslie and Reimer 1999).

Understanding the flow of benefits from woodfuel provision is not only of theoretical importance. It is also of critical relevance. Many interventions attempt to improve rural livelihoods and environmental outcomes by changing the mechanisms of access for certain groups or actors. A lack of understanding of the practices within and between groups of actors in the commodity chain has led to failure of many interventions. For example, in parts of West Africa, devolution of control over state-owned forests has led to increased control by local communities, including management that supplies woodfuel markets. Under such systems, woodfuel dealers typically pay higher prices than when they have access to resources not under community management. However, not all accessible forest has been brought under full community control and influential outsiders, including officials from the Forest Services of numerous countries, still capture disproportionate benefits of the trade, particularly where local community cohesion is weak (Kerkhof 2002).

Thus, interventions may not fail completely. However, they may still have unintended, potentially negative environmental impacts and/or negative outcomes for less powerful economic players (Schroeder 1993). Some of these dynamics are explored in Box 18.4, which gives an example of woodfuel management in Senegal.

Box 18.4 Incorporating Energy Needs in Conservation Policy: The Saloum Delta National Park, Senegal

The Saloum Delta National Park (SDNP) in Senegal has a tradition of conservation dating back to 1935, when it was made a forest reserve under the French colonial forestry code in order to protect it from what the French deemed the "immemorial abusive use by natives" (cited in Ribot 1993). In 1981, the size of the SDNP was increased to include a large portion of the Saloum delta and was later classified as a UNESCO Man and Biosphere Reserve, as well as a RAMSAR site. Despite these conservation efforts, conflict over wood extraction between local people and forestry officials persists (Ribot 1993; Chatellier 2007).

While recent national surveys have documented that Senegal's urban populations use LPG as the dominant cooking fuel (ENDA 2005; Macro International 2008), nearly 99% of the rural population in the area surrounding the SNDP, relies on fuelwood as the main cooking fuel (Chatellier 2007, See Fig. 18.7). Fuelwood is also used in the region for commercial activities, such as shell fish processing, fish smoking, and the production of shell lime, a cement substitute. Regional surveys suggest the presence of a fuelwood shortage; the majority of subsistence fuelwood collectors reported the need to travel longer distances and spend more time searching for fuel than 10 years earlier. With accessible fuelwood far from village centers, fuelwood markets have materialized. Thus, fuelwood is now a commodity with a well-known price, which has

(continued)

Box 18.4 (continued)

Fig. 18.7 Images from communities adjacent to Saloum Delta National Park in Senegal: a woman with purchased fuelwood (*left*) and a shell-lime kiln (*right*) (Source: J. Chatellier)

encouraged professional fuelwood collectors, with access to better tools, transportation, and labor, to enter into the market. Fuelwood for cooking needs, once a subsistence product, has become commercialized; families spend cash instead of labor to meet basic energy needs. Broader economic effects also affect local markets. For example, a recent spike in cement prices led to a boom in shell lime production, which requires vast amounts of fuelwood to process. This added source of demand exacerbates the local fuelwood shortage and drives up prices.

Fuelwood extraction in the SNDP is banned under the Senegalese forestry code, which prohibits resource extraction from national parks even at the subsistence level. Despite the law and a visible presence of park officials, commercial fuelwood extraction, sale, and consumption take place openly. Park officials devote resources to regulating subsistence-level extraction by women who remove mostly deadwood and coppiceable shrubs by requiring that they obtain "free" permits. The permit system dates to the 1930s, when the colonial government used it to assert its ownership of the forest (Ribot 1993; Chatellier 2007). In contrast to the regulations on subsistence collectors, park officials turn a blind eye to the harvest of entire trees for commercial activities, which are often managed by local elites. The criminalization of energy needs in communities adjacent to national parks has created a divide between locals and park officials making future collaboration on conservation difficult. A sustainable silviculture management program, designed to meet the energy needs of local communities, could be one way to meet local fuelwood needs and maintain conservation efforts as it would reduce the current ecologically destructive practice of selective logging (Uhl and Vieira 1989).

Health and Social Welfare

As was mentioned previously, small-scale combustion devices burning solid fuels can not achieve full combustion and, as a result, release numerous pollutants. In addition to having a climate impact, stoves vented directly into the indoor environment result in harmful concentrations of indoor air pollution. Solid fuel use has been shown to cause elevated risks of acute respiratory infection (ARI), chronic obstructive pulmonary disease, and some types of cancer.[11] The WHO estimates that diseases attributable to smoke from solid fuels contribute nearly 3% of the global burden of illness and death (WHO 2002).

These effects are concentrated within particular populations. As a result of the division of labor within most households in the developing world, exposure occurs disproportionately among women and young children (WHO 2002). Other risks from woodfuel use, like burns, can affect young children in particular. In addition, the risk of injury from gathering and transporting heavy loads of fuel over long distances, as well as exposure to possible harassment for girls and women gathering wood far from home can arise. However, evidence of these risks is only collected anecdotally, and, thus, they are not included in official statistics (Diaz et al. 2008).

Interventions in the Traditional Energy Sector

As was discussed above, concern about woodfuels initially focused on the perceived link between woodfuel consumption and deforestation. This dates at least to the 1970s, but the interest in forest conservation gradually subsided and a focus on public health emerged. Much more recently, the link between woodfuels and forest conservation has re-emerged in the context of GHG emission reductions. Each of these types of intervention are discussed below.

Interventions Linking Woodfuels and Forest Conservation

Supply-Side Interventions

For many, the obvious response to woodfuel scarcity is to plant trees. In some cases, this response coincides with the needs of communities that are dependent on wood for energy, but that is not always the case. Planting and maintaining trees can be a

[11] The only conclusive association between cancer and IAP is lung cancer from exposure to coal smoke. Health professionals suspect that other forms of cancer may also be caused by exposure to smoke from solid fuels, but the epidemiological evidence is inconclusive. Similarly, asthma, tuberculosis, cataracts, and low birth weights are suspected, but not yet proven conclusively (Smith and Mehta 2003; Smith et al. 2004).

time consuming, labor-intensive process for local communities. People are unlikely to plant trees for energy if alternative sources exist, such as crop residues. Similarly, if land can be put towards more lucrative uses; planting trees for firewood may be seen as "burning money." Tree planting as a response to wood scarcity is, in any case, complicated by local property institutions. In some places, property rights associated with trees and their products are separable from rights to the land on which the trees grow (Fortmann and Bruce 1988). Planting trees may represent a claim of land ownership, and result in disputes. In addition, in many post-colonial societies there is a long history of land appropriations and forced evictions predicated on real or perceived environmental crises (Leach and Mearns 1996). Thus, any intervention, however, well intended, may be viewed with suspicion (Skutsch 1983).

If tree-planting is introduced as a means of easing the pressure that demand for woodfuels puts on forests, interventions may be either through state-run, community, or farm/ household-level forestry. Many state forestry institutions have a history of antagonistic relations with local communities (Castro and Nielsen 2001; Skutsch 2000). Nevertheless, some governments have successfully established woodlots or managed forests specifically for community wood production (FAO 2003). However, establishing tree plantations is expensive, particularly when state bureaucracies are involved, and highly centralized state-run forestry agencies are not usually an economically feasible way to mitigate woodfuel scarcity. On the other hand, if state-owned forests are already established (for example, in reserves established for timber production), the state can ease wood scarcity by allowing local communities access to dead wood, fallen trees, and pruned branches or by devolving a section of the forest to community control.

Community forestry (CF) contrasts with state forestry in that forest management is partially or wholly vested in the community. Many variations of CF exist. The managed trees may be a section of natural forest, a plantation or a wood-lot. Land may be land held in common, or it may lie on state-owned land with management responsibilities vested in the community. Fuelwood provision is one of many possible dimensions of CF, but energy is rarely the sole purpose of establishing community control. Some CF arrangements limit communities to non-commercial/ non-timber uses: for example, rights to graze livestock, fish, hunt, and extract a variety of forest products like food, medicine, leaves, and thatch. Other community forestry systems vest commercial management rights in communities including the right to sell timber concessions or harvest timber commercially themselves as in Mexico, Laos, and Vietnam (Bray et al. 2003; Sunderlin 2006).

Wood scarcity can also be mitigated by tree planting at the household level. Smallholders throughout the developing world maintain wide varieties of trees on their own land (Chambers and Leach 1989). The majority do so without outside assistance, though outside intervention can help to provide seeds or seedlings, as well as technical advice. As with CF, trees on farms are rarely used only as sources of fuelwood. Agroforestry, which integrates trees with cultivation and livestock systems, is particularly effective for maintaining trees on the homestead (Montagnini 2006).

Demand-Side Interventions

In addition to tree planting, the perceived link between fuelwood consumption and deforestation led to the development and dissemination of fuel conserving cookstoves. Early views presumed that traditional cookstoves were inherently inefficient and attempted to improve upon them by improving combustion efficiency and heat transfer. To date, hundreds of varieties of cookstoves have been developed and hundreds of millions of stoves are said to have been disseminated throughout the developing world. The vast majority of these are in China. Many programs have not succeeded, or have had problems scaling up. Early interventions tended to focus on engineering solutions, but failed to address social issues in which household energy use is situated (Barnes et al. 1994, also discussed in the chapter by Doll, this volume). Few realized that in the hands of an experienced cook, a traditional "three-stone" fire can be as efficient as many heavily engineered stoves. The behavior, perceptions, and motivation of the cook are important determinants of fuel consumption that were largely overlooked (Crewe 1997).

A few programs that were developed during the 1980s have had a lasting impact. In addition to China's massive National Improved Stove Program (NISP) (Sinton et al. 2004; Smith et al. 1993), the Kenyan Ceramic Jiko was also relatively successful (Hyman 1987; Kammen 1995). Some reasons for the success of each program include a slow transition from heavy state or donor support to commercialization so that after a time, stove construction and sale were shifted to the private sector. Importantly, this shift was supported at various stages by substantial research and development, stove marketing, external monitoring, and evaluation, and, in China's case, quality control and certification (Bailis et al. 2008).

Interventions in Household Energy and Health

Both the successes and failures of past projects offer lessons for a new wave of household energy interventions currently underway. These interventions focus on reducing the burden of disease caused by cooking with biomass fuels by improving combustion, venting emissions outdoors, or switching to cleaner fuels. Numerous studies have shown that these strategies can reduce IAP substantially (Chengappa et al. 2007; Dutta et al. 2007; Ezzati et al. 2000; Masera et al. 2007).

The task remains to scale up dissemination of improved stoves. Numerous projects are underway across the developing world with varying levels of donor support.[12] Among the donor community, recent activities are oriented toward

---

[12] There are several web-based sources of information about household energy and health projects including the Household Energy Network (HEDON) at http://www.hedon.info/goto.php/index.htm, SparkNet at http://sparknet.info/home.php and an on-line community of improved stove practitioners at http://www.bioenergylists.org/.

commercialization of stove dissemination. This reflects a shift that has occurred in development practice more generally, where emphasis is placed on business-like approaches rather than models that rely on donors and subsidies (Hoffman et al. 2005; Bailis et al. 2008). Whether this shift will facilitate broader adoption of cleaner household energy technologies remains an open question.

Interventions Linking Woodfuels and GHG Emission Reductions

In addition to the recent attention on public health in household energy interventions, there has also been growing interest in the traditional energy sector as a means to reduce GHG emissions. Academic studies have quantified the differences in emissions between traditional and improved stoves in both lab (Bertschi et al. 2003; Brocard et al. 1996; Pennise et al. 2001; Smith et al. 2000) and field settings (Johnson et al. 2008; Roden et al. 2006). Yet, net emissions reductions depend on forest management, as well as, emissions from stoves (Bailis and Barasa 2008).

Despite the challenges, some carbon markets are accepting carbon offsets generated from substituting traditional stoves with improved ones. The Clean Development Mechanism (CDM) of the Kyoto Protocol has recently accepted two methodologies that cater to this type of project, after a long period in which improved stoves were not considered suited to CDM, because woodfuel themselves were considered to be unsustainable. However, despite this change, very few cookstove projects have yet entered the CDM pipeline (Fenhann 2008). In addition, a voluntary offset methodology has recently been accepted by the CDM Gold Standard, an organization that certifies carbon offset projects that maximize social and environmental co-benefits (ClimateCare 2008). The traditional energy sector has the potential to yield emission reductions with substantial co-benefits and revenue generated from the sale of offsets, which could assist with scale-up of stove projects that often struggle to achieve widespread adoption.

Conclusions

This chapter has discussed multiple dimensions of sustainability relevant to woodfuels. In its most narrow conception, a concern about sustainability in the context of woodfuel use may be limited to local environmental degradation. However, as we argue, sustainability has much broader implications. First, environmental sustainability extends across different scales, including regional impacts and global change. Moreover, the challenge of woodfuel sustainability extends into social and political spheres, including poverty and livelihoods, public health, and social relationships.

In this section, we present policy options relevant for each of the aspects of sustainability discussed above.

Promoting woodfuel sustainability requires an understanding of the drivers of forest degradation and the role of woodfuel harvesting as one of several possible pressures on forest resources. This knowledge is required at a scale that is meaningful to woodfuel users. As Ghilardi explains in his discussion of fuelwood *hotspots* in Mexico (Box 18.3), woodfuel–forest interactions can be very heterogeneous across relatively small scales. Communities with similar social characteristics and woodfuel demand may manage forest resources very differently. Analyses of local heterogeneity can be indispensable in identifying local-level drivers of change, as well as prioritizing areas for intervention.

In addition, in woodfuel-dependent communities where forest degradation is apparent, policy makers should not assume *a priori* that woodfuel demand is the sole or primary driver. Multiple pressures on forest resources can interact with, and supplant, woodfuel extraction as drivers of environmental change. Only local-level research can truly identify causes of forest degradation and lead to solutions.

In cases where woodfuel extraction is identified as a cause of forest degradation, a range of supply and demand-side interventions are possible. To address supply-side challenges, the devolution of forest management to local communities has proven to be effective, provided that communities are sufficiently empowered with strong institutional arrangements and sufficient resources (Ribot 2004). Extension services can support such efforts by providing technical advice and training.

Demand side challenges include technological and behavioral changes that promote cleaner combustion, improve end-use efficiency, and/or shift wood-dependent households toward other sources of energy. These interventions also have the benefit of addressing some of the social sustainability challenges linked to woodfuel-dependence. Efficiency improvements can lower the costs of cooking by reducing the time and/or expense required to procure fuel, and cleaner combustion can reduce exposure to harmful pollutants leading to lower incidence of disease.

Both supply and demand-side interventions also carry global environmental benefits. Supply-side interventions can enhance carbon sinks by promoting afforestation and reforestation or reduce future emissions by avoiding deforestation and degradation. Demand-side interventions can also reduce emissions in multiple ways. First, efficiency improvements reduce the amount of fuel needed for a given cooking task, which reduces pollution. Second, when demand-side interventions promote cleaner combustion, emissions of pollutants with high global warming impacts are reduced.

However, describing potential benefits from interventions that enhance the sustainability of woodfuel use is easy. The true challenge lies in operationalizing interventions at a level that is commensurate with the scale of the challenge. Saving existing forests and planting trees are both universally promoted environmental objectives, but both activities have proven difficult to implement on a grand scale. Similarly, with a few notable exceptions, reducing woodfuel demand by promoting technical and/or behavioral change is proceeding at a very slow pace despite analyses that show such interventions are extremely cost-effective ways of addressing environmental and public health challenges (World Health Organization 2007).

There are numerous barriers to scaling up efforts to promote woodfuel sustainability. For example, many of the negative impacts associated with woodfuel dependence – forest degradation, as well as labor demands and health impacts among highly marginalized populations – often fall outside of the formal economy for which decision makers typically design policy. This stands in contrast to recent approaches in development interventions which have become increasingly market oriented: for example, the commercialization of improved cookstoves and monetization of ecosystem services (see the chapters on Payments for Ecosystem Services, Volume 2). Until the problems associated with dependence on unsustainable woodfuels are fully understood and, to the extent that is possible, quantified, commercialized solutions are unlikely to be effective.

Other barriers arise because the nature of woodfuel sustainability is highly contingent on local circumstances. Locally specific social-environmental factors confound attempts to develop and deploy "best practices", as discussed in the chapter by Ganz et al., this volume. As Ghilardi explains in the case study from Mexico described in Box 18.4, local woodfuel management practices and associated environmental impacts can vary a great deal within a politically defined region that is otherwise culturally and economically similar.

However, despite numerous barriers, there are several reasons to be optimistic about the prospects of sustainability in the woodfuel sector. First, the breadth of tools required to better assess the benefits of more sustainable household energy utilization has expanded a great deal in recent years (Disease Control Priorities Project 2006; Smith et al. 2007; World Health Organization 2007; Ghilardi et al. 2007). Second, many actors in emerging carbon markets have turned their attention to the household energy sector as a promising area to create carbon emission reductions that also carry substantial social benefits. This has raised the profile of household energy among policy makers and created incentives to develop more accurate assessment methodologies to better understand the circumstances in which woodfuel utilization contributes to forest degradation and loss. In a third and related point, the emerging discussions of reducing emissions from deforestation and forest degradation (REDD) in the context of climate change mitigation (see chapters by Rumbaitis del Rio, this volume, Jenkins, Volume 2, and Estrada and Corbera, Volume 2) has also turned attention toward the household energy sector in certain places. However, currently, carbon markets are limited to project-level interventions, which almost always occur at a local scale, while the REDD discussions will require a much needed dialog at the national scale.

Woodfuel dependence in developing countries is unlikely to decrease in the near-term and many barriers to enhancing woodfuel sustainability are still well-entrenched. Nevertheless, new assessment methods and changing approaches to environmental management have shifted the terrain slightly in favor of greater sustainability. Whether these changes translate into real improvements in environmental quality and social welfare for woodfuel-dependent communities remains to be seen.

References

Arnold, M., Kohlin, G., et al. (2003). Fuelwood Revisited: What has Changed in the Last Decade?, CIFOR: Occasional Paper, 39, 37.

Bailis, R. (2005). Fuel from the Savanna: the Social and Environmental Implications of the Charcoal Trade in Sub-Saharan Africa. Energy and Resources Group. Berkeley, CA, University of California. PhD Thesis: 250 pages.

Bailis, R. and Barasa, M. (2008). "Can traditional energy fit the carbon market? Carbon credit accounting from non-renewable biomass." Climate Policy **In review**

Bailis, R., Cowan, A., et al. (2008). "Arresting The Killer In The Kitchen: The Promises and Pitfalls of Commercializing Improved Cookstoves." World Development **In review**

Bailis, R., Ezzati, M., et al. (2005). "Mortality and Greenhouse Gas Impacts of Biomass and Petroleum Energy Futures in Africa." Science **308**(5718): 98-103

Bailis, R., Kirubi, C., et al. (2006). Searching For Sustainability: Kenya's Energy Past And Future, African Centre for Technology Studies (ACTS): Policy Brief, 6.

Bajracharya, D. (1983). "Deforestation in the Food/Fuel Context: Historical and Political Perspectives from Nepal." Mountain Research and Development **3**(3): 227-240. http://www.jstor.org/stable/3673017

Baker, W. E. (1990). "Market Networks and Corporate Behavior." The American Journal of Sociology **96**(3): 589-625. http://www.jstor.org/stable/2781065

Barnes, D., F., Openshaw, K., et al. (1994). What Makes People Cook With Improved Biomass Stoves? A comparative international review of stove programs, World Bank: World Bank Technical Paper, No. 242.

Bertschi, I. T., Yokelson, R. J., et al. (2003). "Trace gas emissions from the production and use of domestic biofuels in Zambia measured by open-path Fourier transform infrared spectroscopy." Journal of Geophysical Research-Atmosphere **108**(D13): 5-1, 5-13

Bond, T., Venkataraman, C., et al. (2004). "Global atmospheric impacts of residential fuels." Energy for Sustainable Development **8**(3): 20-32

Bourdieu, P. (1985). The Forms of Capital. New York, Greenwood.

Bray, D. B., Merino-PÃ©rez, L., et al. (2003). "Mexico's Community-Managed Forests as a Global Model for Sustainable Landscapes." Conservation Biology **17**(3): 672-677. 10.1046/j.1523-1739.2003.01639.x. http://search.ebscohost.com/login.aspx?direct=true&db=eih&AN=9886013&site=ehost-live

Breman, H. and Kessler, J.-J. (1995). Woody Plants in Agro-Ecosystems of Semi-Arid Regions. Berlin, Springer-Verlag, 340.

Brocard, D., Lacaux, C., et al. (1996). Emissions from the combustion of biofuels in Western Africa. In Biomass Burning and Global Change. J. S. Levine Ed. Cambridge, MA, MIT Press. **1**: 350-360.

Cabeza Gutés, M. (1996). "The concept of weak sustainability." Ecological Economics **17**(3): 147-156. http://www.sciencedirect.com/science/article/B6VDY-46BMV06-3/1/b488aa3f7e78c28f163f0c381891136c8

Castro, A. P. and Nielsen, E. (2001). "Indigenous people and co-management: implications for conflict management." Environmental Science & Policy **4**(4-5): 229-239. http://www.sciencedirect.com/science/article/B6VP6-430XPST-7/2/2e7e288bee6c7b27295ccf7e9fb2aa59

Chambers, R. and Leach, M. (1989). "Trees as savings and security for the rural poor." World Development **17**(3): 329-342. http://www.sciencedirect.com/science/article/B6VC6-45BC3F3-8T/2/02375c214532ca3715b0b10d051ede01

Chatellier (2007). Author's Personal Research. Available at.

Chengappa, C., Bajpai, R., et al. (2007). "Impact of Improved Cookstoves on Indoor Air Quality in the Bundelkhand Region in India." Energy for Sustainable Development **In review**

Chidumayo, E. N. (1993). "Zambian Charcoal Production: *Miombo* Woodland Recovery." Energy Policy **21**(5): 586-597

ClimateCare (2008). Indicative Programme, Baseline, and Monitoring Methodology for Improved Cook-Stoves and Kitchen Regimes, 32. Available at: http://www.cdmgoldstandard.org/uploads/file/V01%2010-05-08%20GS%20Cook-stove%20Methodology.pdf.

Costanza, R. and Daly, H. E. (1992). "Natural Capital and Sustainable Development." Conservation Biology **6**(1): 37-46. http://www.jstor.org/stable/2385849

Crewe, E. (1997). The Silent Traditions of Developing Cooks. In Discourses of Development: Anthropological Perspectives. R. D. Grillo and R. L. Stirrat Eds. London, Berg: 59-80.

de Montalembert, M. R. and Clement, J. (1983). Fuelwood Supplies in the Developing Countries, UN Food and Agriculture Organization, 42.

Diaz, E., Bruce, N., et al. (2008). "Self-rated health among Mayan women participating in a randomised intervention trial reducing indoor air pollution in Guatemala." BMC International Health and Human Rights **8**(1): 7. http://www.biomedcentral.com/1472-698X/8/7

Disease Control Priorities Project (2006). Cost-Effective Interventions, The World Bank Group.

Dove, M. (1992). "Foresters' beliefs about farmers: a priority for social science research in social forestry." Agroforestry Systems **17**(1): 13-41

Dutta, K., Shields, K. N., et al. (2007). "Impact of Improved Cookstoves on Indoor Air Quality in rural homes near Pune, India." Energy for Sustainable Development **In review**

Eckholm, E. (1975). The other energy crisis: Firewood, Worldwatch, 22.

Ecoforum (2002). Hot and Dirty: Inside Kenya's 23 Billion Shilling Charcoal IndustrySpecial Issue, 16-22.

ENDA (2005). Energy Use, Energy Supply, Sector Reform and the Poor in Senegal.

Ezzati, M., Kammen, D., et al. (2000). "Comparison of Emissions and Residential Exposure from Traditional and Improved Cookstoves in Kenya." Environmental Science and Technology **34**(4)

FAO (1978). Forestry for Local Community Development, UN Food and Agriculture Organization: FAO forestry paper, 07, 114.

FAO (2000). Unified Wood Energy Terminology: UWET, UN Food and Agriculture Organization: Draft.

FAO (2003). State of the World's Forest.

FAOSTAT. (2008, February 2004). "FAOSTAT Forestry Data." United Nations Food and Agriculture Organization, Accessed on Jan 25, 2008, from http://faostat.fao.org/faostat/collections?subset=forestry.

Fenhann, J. (2008). "CDM Pipeline Overview (June, 2008 edition)." CD4CDM - UNEP Risø Centre, Accessed on July 25, 2008, from http://cd4cdm.org/Publications/CDMpipeline.xls.

Fortmann, L. and Bruce, J., Eds. (1988). Whose Trees? Proprietary Dimensions of Forestry. The Rural Studies Series of the Rural Sociological Society. Boulder and London, Westview Press.

Geist, H. J. and Lambin, E. F. (2002). "Proximate Causes and Underlying Driving Forces of Tropical Deforestation." BioScience **52**(2): 143-151

Ghilardi, A., Guerrero, G., et al. (2007). "Spatial analysis of residential fuelwood supply and demand patterns in Mexico using the WISDOM approach." Biomass and Bioenergy **31**(7): 475-491. http://www.sciencedirect.com/science/article/B6V22-4NJ26HJ-1/2/96a5d691ea561450d9c245299994d58e

Ghilardi, A., Guerrero, G., et al. (2008). "Multi-scale analysis of residential fuelwood supply and demand spatial patterns in Mexico." Journal of Environmental Management **In review**

Girard, P. (2002). "Charcoal production and use in Africa: what future?" Unasylva **53**(4): 30

Goldenberg, J., and Johansson, T. B. (2004). World Energy Assessment: Overview. World Energy Assessment: Overview, 2004 Update, (ISBN: 92-1-126167-8).

Harker, A. P., Sandels, A., et al. (1982). Calorific values for wood and bark and a bibliography for fuelwood, Tropical Products Institute, G162, 20.

Hoffman, K., West, C., et al. (2005). Enterprise Solutions to Poverty: Opportunities and Challenges for the International Development Community and Big Business. Available at: http://www.shell-foundation.org/download/pdfs/SF_Exec_Summary_Enterprise_Solutions_to_Poverty.pdf.

Homewood, K. and Rodgers, W. (1991). Maasailand Ecology: Pastoralist Development and Wildlife Conservation in Ngorongoro, Tanzania. Cambridge, Cambridge University Press, 298.

Hosier, R. H. (1993). "Charcoal Production and Environmental Degradation: environmental history, selective harvesting, and post-harvest management." Energy Policy **21**(5): 491-509

House, J. I. and Hall, D. O. (2003, July 13, 2003). "Savannahs." King's College London, Accessed on July 12, 2005, from http://www.savannas.net/newframe.htm.

Hyman, E. L. (1987). "The strategy of production and distribution of improved charcoal stoves in Kenya." World Development **15**(3): 375-386. http://www.sciencedirect.com/science/article/B6VC6-45DHVWS-J4/2/d07ce0042093cf0aa1cb5537cb3f7886

IEA (2003). Energy Balance of Non-OECD Countries 2000-2001. Paris, The Organization for Economic Cooperation and Development (OECD), 802.

IEA (2004). Energy and Development (Chapter 10 of World Energy Outlook 2004). In World Energy Outlook 2004 Paris, IEA Publications: 329-365.

IEA (2006). World Energy Outlook 2006. Paris, IEA Publications, 586.

International Energy Agency (2006a). CO_2 Emissions from Fuel Combustion: 1971 2004, International Energy Agency, 556. Available at: http://www.ingentaconnect.com/content/oecd/1608019x/2006/00002006/00000022/6106113e

International Energy Agency (2006b). World Energy Outlook 2006. Paris, IEA Publications, 586.

International Energy Agency. (2008). Accessed on 08-29-08, from http://www.iea.org/Textbase/stats/index.asp.

International Energy Agency (IEA) (2007). Energy Statistics of Non-OECD Countries 2004-2005. Paris, The Organization for Economic Cooperation and Development (OECD), 786.

IPCC (1997). Revised 1996 IPCC Guidelines for National Greenhouse Gas Inventories: Reference Manual, Blackwell Press, Vol. 3. Available at: www.ipcc-nggip.iges.jp/public/gl/invs6.htm.

IPCC (2003). Definitions and Methodological Options to Inventory Emissions, Intergovernmental Panel on Climate Change, 32. Available at: http://www.ipcc-nggip.iges.or.jp/public/gpglulucf/degradation.html.

IPCC (2007). Climate Change 2007: The Physical Science Basis: Contribution of Working Group I to the Fourth Assessment Report of the Intergovernmental Panel on Climate Change [Solomon, S., D. Qin, M. Manning, Z. Chen, M. Marquis, K.B. Averyt, M. Tignor and H.L. Miller (eds.)]. New York, NY, Cambridge University Press, 996.

Johnson, M., Edwards, R., et al. (2008). "In-field greenhouse gas emissions from cookstoves in rural Mexican households." Atmospheric Environment **42**(6): 1206-1222. http://www.sciencedirect.com/science/article/B6VH3-4R17V4K-4/2/99f50a8eafd449d4b8838d3c0cfc60a8

Kaimowitz, D. and Angelsen, A. (1998). Economic Models of Tropical Deforestation: A Review, CIFOR, 139. Available at: http://www.cgiar.org/cifor.

Kammen, D. (1995). "Cookstoves for the Developing World." Scientific American **273**: 72-75

Kerkhof, P. (2002). A new social contract for participatory forestry in the Sahel. Second international workshop on participatory forestry in Africa. Arusha, Tanzania, UNFAO.

Kidd, C. (1992). "The evolution of sustainability." Journal of Agricultural and Environmental Ethics **5**(1): 1-26. http://dx.doi.org/10.1007/BF01965413

Kituyi, E., Marufu, L., et al. (2001). "Carbon monoxide and nitric oxide from biofuel fires in Kenya." Energy Conversion and Management **42**: 1517-1542

Leach, G. and Mearns, R., Eds. (1988). Beyond the Woodfuel Crises: people, land, and trees in Africa. London, Earthscan.

Leach, M. and Mearns, R., Eds. (1996). The Lie of the Land: Challenging Received Wisdom on the African Environment. African Issues. London, The International African Institute.

Leslie, D. and Reimer, S. (1999). "Spatializing commodity chains." Progress in Human Geography **23**(3): 401–420

Macro International (2008). MEASURE DHS STAT compiler.

Masera, O., Edwards, R., et al. (2007). "Impact of Patsari improved cookstoves on indoor air quality in Michoacán, Mexico." Energy for Sustainable Development **XI**(2): 45-56

Masera, O., Saatkamp, B., et al. (2000). "From Linear Fuel Switching to Multiple Cooking Strategies: A Critique and Alternative to the Energy Ladder Model." World Development **28**(12): 2083-2103

Ministry of Energy (2002). Study on Kenya's Energy Demand, Supply and Policy Strategy for Households, Small Scale Industries and Service Establishments: Final Report, KAMFOR Company Limited, 158.

Montagnini, F. (2005). (Editor). Environmental Services of Agroforestry Systems. Haworth Press. New York. pp 126.

Montagnini, F. (2006). Homegardens of Mesoamerica: Biodiversity, food security, and nutrient management. In Tropical Homegardens: 61-84.

Mutimba, S. and Barasa, M. (2005). National Charcoal Survey of Kenya, Energy for Sustainable Development - Africa: Final Draft Project Report, 56. Available at: www.esd.co.uk.

Nyang, F. (1999). Household Energy Demand and Environmental Management in Kenya. Amsterdam, Thela Thesis publications, 224.

Okwemba, A. (2003). Charcoal Burning Could Be Legal Again. The Daily Nation, July 17, 2003.

Pearce, D. W. and Atkinson, G. D. (1993). "Capital theory and the measurement of sustainable development: An indicator of "weak" sustainability." Ecological Economics **8**(2): 103-108. http://www.scopus.com/scopus/inward/record.url?eid=2-s2.0-0027749823&partnerID=40&rel=R8.0.0

Pennise, D., Smith, K. R., et al. (2001). "Emissions of Greenhouse Gases and Other Airborne Pollutants from Charcoal-Making in Kenya and Brazil." Journal of Geophysical Research-Atmosphere **106**: 24143-24155

Pfaff, A., Chaudhuri, S., et al. (2004). "Household Production and Environmental Kuznets Curves." Environmental and Resource Economics **27**: 187-200

Ribot, J. (1999). "Decentralisation, Participation and Accountability in Sahelian Forestry: Legal Instruments of Political-Administrative Control." Africa **69**(1): 23-55

Ribot, J. (2004). Waiting for Democracy: the Politics of Choice in Natural Resource Decentralization, World Resources Institute, 140.

Ribot, J. C. (1993). "Forestry policy and charcoal production in Senegal." Energy Policy **21**(5): 559-585. http://www.sciencedirect.com/science/article/B6V2W-4903GSN-59/2/98ad6fe42e3d28f6924a0025b2cb2bb5

Ribot, J. C. and Peluso, N. (2003). "A Theory of Access." Rural Sociology **63**(2): 153-181

Robbins, P. (1998). "Paper Forests: Imagining and Deploying Exogenous Ecologies in Arid India." Geoforum **29**(1): 69-86

Roden, C. A., Bond, T. C., et al. (2006). "Emission Factors and Real-Time Optical Properties of Particles Emitted from Traditional Wood Burning Cookstoves." Environ. Sci. Technol. **40**(21): 6750-6757. http://pubs3.acs.org/acs/journals/doilookup?in_doi=10.1021/es052080i

Scholes, R. J. and Hall, D. O. (1996). The Carbon Budget of Tropical Savannas, Woodlands, and Grasslands. In Global Change: Effects on Coniferous Forests and Grasslands. A. I. Breymeyer, D. O. Hall, *et al.* Eds. Chichester, New York, Wiley: 69-100.

Schroeder, R. A. (1993). "Shady Practice: Gender and the Political Ecology of Resource Stabilization in Gambian Garden/Orchards." Economic Goegraphy **69**(4): 349-365

Sinton, J. E., Smith, K. R., et al. (2004). "An assessment of programs to promote improved household stoves in China." Energy for Sustainable Development **8**(3): 33-52. http://www.ieiglobal.org/ESDVol8No3/chinastoves.pdf

Skutsch, M. (2000). "Conflict management and participation in community forestry." Agroforestry Systems **48**(2): 189-206. http://dx.doi.org/10.1023/A:1006328403023

Skutsch, M. M. (1983). Why People Don't Plant Trees: The Socioeconomic Impacts of Existing Woodfuel Programs (Village Case Studies, Tanzania), Resources for the Future and USAID, 94. Available at: http://www.odi.org.uk/fecc/resources/greyliterature/fuelwood/skutch/skutch.pdf.

Smith, K. and Mehta, S. (2003). "The burden of disease from indoor air pollution in developing countries: comparison of estimates." International Journal of Hygeine and Environmental Health **206**: 279-289

Smith, K., Mehta, S., et al. (2004). Indoor air pollution from household use of solid fuels. In Comparative Quantification of Health Risks: Global and Regional Burden of Disease Attributable to Selected Major Risk Factors. M. Ezzati, A. Lopez, *et al.* Eds. Geneva, World Health Organization: 1435-1493.

Smith, K., Uma, R., et al. (2000). Greenhouse Gases From Small-Scale Combustion Devices In Developing Countries Phase IIa: Household Stoves In India, US Environmental Protection Agency, EPA-600/R-00-052, 98.

Smith, K. R., Edwards, R., et al. (2007). "Monitoring and Evaluation of Improved Biomass Cookstove Programs for Indoor Air Quality and Stove Performance: Conclusions from the Household Energy and Health Project." Energy for Sustainable Development **XI**(2): 5-18

Smith, K. R., Shuhua, G., et al. (1993). "One Hundred Million Improved Cookstoves in China: How was it done?" World Development **21**(6): 941-961

Sunderlin, W. D. (2006). "Poverty alleviation through community forestry in Cambodia, Laos, and Vietnam: An assessment of the potential." Forest Policy and Economics **8**(4): 386-396. http://www.sciencedirect.com/science/article/B6VT4-4H57THV-3/2/81ba02e501bd58255d034aa2eea5d7ef

Uhl, C. and Vieira, I. C. G. (1989). "Ecological Impacts of Selective Logging in the Brazilian Amazon: A Case Study from the Paragominas Region of the State of Para." Biotropica **21**(2): 98-106. http://www.jstor.org/stable/2388700

UNDP (2005). Energizing the Millennium Development Goals: A Guide to Energy's Role in Reducing Poverty, United Nations Development Program, 20. Available at: http://www.undp.org/energy/docs2/ENRG-MDG_Guide_all.pdf.

WHO (2002). World Health Report: Reducing Risks, Promoting Healthy Life, World Health Organization.

World Bank, *Personal Communication*, 2008.

World Health Organization. (2007). "Choosing Interventions that are Cost Effective (WHO-CHOICE)." The World Health Organization, Accessed on 27 August, 2007, from http://www.who.int/indoorair/publications/evaluation/en/index.html.

World Resources Institute. (2008). "Climate Analysis Indicators Tool (CAIT) Version 5.0." World Resources Institute, Accessed on Feb 01, 2008.

Young, T. P. and Francombe, C. (1991). "Growth and yield estimates in natural stands of leleshwa (Tarconanthus camphoratus)." Forest Ecology and Management **41**(3-4): 309-321. http://www.sciencedirect.com/science/article/B6T6X-48XMMMY-185/2/1ea1e426d97537d52dd5fb9eb5449487

Zerriffi, H., Jihua, P., et al. (2008). Household Level Fuel Switching in Rural Hubei, Stanford University Program on Energy and Sustainable Development (PESD): Working Paper #79. Available at: http://pesd.stanford.edu/publications/household_level_fuel_switching_in_rural_hubei/.

Chapter 19
Introduction to the Ecological Dimensions of Climate Change and Disasters

Cristina Rumbaitis del Rio

Every year, a number of disasters occur and grab and hold global attention, at least for a short period of time. In 2010, the world witnessed an earthquake in Haiti at the start of the year, another one in Chile a month later, massive flooding in Pakistan, an intense summer heat wave and wildfires in Russia, and widespread flooding in Mexico. The images of destruction from these events are heartbreaking, and the damage estimates are staggering. Moreover, there are also hundreds of small-scale disasters that occur every year that are seldom reported on outside of local areas. In any given year, there are more than 700 natural catastrophic events, resulting in billions of dollars of damage and asset loss, and unquantifiable human suffering. In the first 9 months of 2010 alone, more than 236,000 people were killed and 256 million people were affected by disasters at a cost of 81 billion dollars (Center for Research in the Epidemiology of Disasters, 2010). Disasters, both large and small, erode away at a community or country's ability to develop, by diverting resources for development towards rescue, recovery, and reconstruction measures; by fraying social safety nets and networks; and reducing economic productivity. GDP losses due to disasters range from 2% to 15%.

Perhaps because of the semi-regular occurrence of natural disasters, we tend to think of disasters as inevitable but isolated events, as episodic misfortunes wholly beyond our control. But as Wisner et al. (2004) point out, this is dangerous thinking as disasters are not isolated occurrences caused by a physical trigger, but rather the product of a complex chain of social, physical, economic, and political processes. To assume that disasters are purely natural, physically determined events is to ignore many of the root causes and to ignore many of the leverage points that could be used to reduce their destructiveness.

C. Rumbaitis del Rio (✉)
The Rockefeller Foundation, 420 Fifth Ave., New York, NY 10018-2702, USA
e-mail: crumbaitisdelrio@rockfound.org

J.C. Ingram et al. (eds.), *Integrating Ecology and Poverty Reduction: Ecological Dimensions*, DOI 10.1007/978-1-4419-0633-5_19,

An individual's risk to a natural disasters is determined both by exposure and vulnerability to a given hazard. Vulnerability is defined by Wisner et al. (2004) as "the characteristics of a person or group and their situation that influence their capacity to anticipate, cope with, resist and recover from the impact of a natural hazard (an extreme natural event or process)." Where people live, what their homes are made of, what level of preparedness and hazard protection they have, what information they have, and what resources they can access all influence vulnerability. These, in turn, may be influenced by factors such as class, gender, ethnicity, age group, strength of social networks, and a myriad of other factors. In short, underlying social and political processes result in unequal exposure to hazards and unequal access to opportunities to reduce risk.

The state and management of ecosystems can influence both exposure and vulnerability to hazards. Two key services provided by ecosystems include both buffering and regulation of potential hazards. For instance, coastal mangroves can attenuate wave energy at local scales, reducing the exposure of inland populations to coastal storm surges. Similarly, wetlands regulate water flow, and reduce the incidence of flooding by retaining and slowly releasing water flow. Beyond regulating and buffering exposure to hazards, the state of natural resources can influence vulnerability or coping capacity to natural disasters. For instance, access to diversified natural resources, such as access to a variety of crops with different maturation periods, temperature tolerance thresholds, and water requirements, may increase the ability to cope with droughts.

The chapters in this section examine the complex interrelationships between humans, ecosystems, disasters, climate change, and development more closely, and use concepts and principles from ecology to identify how disaster and climate-change response and prevention measures could be structured more effectively to improve the resilience of communities, reduce human suffering, and help meet development goals.

The chapter by Ingram and Khazi, *Incorporating Ecology and Natural Resource Management into Coastal Disaster Risk Reduction*, examines coastal zone hazards, and the role of ecosystem-based disaster risk reduction approaches as opportunities to both conserve critical ecosystem services and reduce vulnerability to extreme events. The chapter reviews recent research on the role of coastal ecosystems in reducing the impact of rapid-onset disasters in the coastal zone. The authors draw on research conducted after the Indian Ocean tsunami of 2004, including personal observations from assessing coastal areas affected by the tsunami, as a primary case study to illustrate both the challenges and opportunities of using coastal conservation as a means to reduce hazard vulnerability and, in particular, to mitigate the impact of hazards on the lives of the rural poor. They assess the complex set of factors that affect the spatial and temporal distribution of disaster-regulation services of coastal ecosystems and use this to develop recommendations on the management and monitoring of coastal ecosystems to maximize the provision of disaster-regulation services.

The chapter by March, *Integrating Natural Resource Management into Disaster Response and Mitigation*, provides a development practitioner's perspective on how an ecosystems approach can be used to understand the impact of disasters, particularly slow onset and recurring disasters, on impoverished populations and help guide the development of humanitarian responses that better meet populations' needs and mitigate vulnerability to future shocks. The author examines how improving natural resource management can assist both pastoralists and farmers in developing long-term resilience to stresses such as drought and pestilence. An application of core ecological principles of the association between diversity and resilience provides greater farmer and herder choice and increased stability in the face of more variable climate conditions.

Global climate change will likely increase the incidence, severity, and extent of natural disasters, and may lead to novel natural disasters being experienced in localized areas. Rumbaitis del Rio, in *The Role of Ecosystems in Building Climate Change Resilience and Reducing Greenhouse Gases*, reviews the potential contribution of ecosystems and natural resource management in supporting poor communities to build their resilience to climate variability and long-term change. While evidence suggests that the roles that ecosystems can play in supporting human adaptation to climate change is highly contextual, dynamic, and limited to a range of conditions, ecosystem-based interventions offer a number of comparative advantages to engineered interventions. Natural infrastructure is often more cost-effective to maintain, offers multiple additive co-benefits, and thus may be, in the near term, an important "no-regrets" way of increasing resilience to climate change of particular relevance to poor communities. Similarly, ecosystem-based management approaches may be the most relevant opportunities for poor communities to contribute to global efforts to reduce the greenhouse gases that cause climate change, and also enable the poor to benefit tangibly from measures taken to increase carbon sequestered in forests and agricultural fields.

The chapter by Seimon, *Improving Understanding of Climatic Controls on Ecology in Development Contexts,* critically examines the effect that climate and climatic variability have on human livelihoods, economic development, and well-being. Seimon argues that long-term success in poverty alleviation cannot be achieved without greater comprehension and engagement with local-scale climate variability and change. Assuming static climates can lead to misapplied development interventions as well as the development of maladaptive practices in the long term. Case studies from the Peruvian Andes and equatorial Africa illustrate how a comprehensive understanding of climatology enables more proactive, rather than reactive management responses to threats and opportunities borne by climate change. Climatological analysis is not often a core skill of a development practitioner, and thus Seimon provides a practical guide as to how site-specific climatological assessments could be implemented to incorporate a more nuanced understanding of climate into development and conservation efforts.

Taken together, these chapters illustrate that while natural disasters and climate change pose a significant challenge to achieving the Millennium Development Goals, evidence from current practice suggests there is a large opportunity to leverage ecological and climatological knowledge to develop and implement disaster risk reduction, postdisaster recovery, climate adaptation, and mitigation measures in ways that target and benefit the poor. While these chapters suggest that more research is needed on the limits to, and interactions between, different ecosystem-based approaches to disaster management and building climate resilience, they also vividly demonstrate the importance of natural infrastructure for disaster risk reduction, climate adaptation, and mitigation, and for long-term holistic vulnerability reduction.

References

Center for Research in the Epidemiology of Disasters (2010). http://www.emdat.be [accessed October 28, 2010].

Wisner, B., Blaikie, P., T. Cannon, Davis, I. (2004). At Risk: Natural disasters, people's vulnerability and disasters. 2nd edition, London, Routledge.

Chapter 20
The Role of Ecosystems in Building Climate Change Resilience and Reducing Greenhouse Gases

Cristina Rumbaitis del Rio

Introduction

The fourth assessment report of the Nobel-prize winning Intergovernmental Panel on Climate Change (IPCC) provides a synthesis of hundreds of peer reviewed scientific studies and concludes that human-induced climatic change is already occurring (IPCC 2007). Observed impacts range from longer growing seasons in temperate to polar zones; changes in the timing of plant bud break, bird migrations, and egg-laying; to poleward and upward shifts in ranges in plant and animal species (IPCC 2007). Projected changes have varying degrees of associated uncertainty, but include warmer temperatures, changes in the amount and distribution and intensity of rainfall, potential changes in hurricane and cyclone frequency and intensity, fire frequency, changes in agricultural productivity, spread of temperature-related disease vectors and pests, and sea-level rise (IPCC 2007). Further evidence suggests that this climate disruption will continue for decades to centuries (Solomon et al. 2009).

In response to this evidence, climate change has been elevated as an international priority. Though once viewed as a purely environmental concern, many political leaders now realize that climate change, because of its systematic consequences, is simultaneously an economic, developmental, and even existential threat (particularly for some small island states). Military and security forces increasingly recognize climate change as a threat multiplier. As international negotiations continue under the auspices of the United Nations Framework Convention on Climate Change and through bilateral channels, there is greater commitment and resources available to support reductions in anthropogenic greenhouse gas emissions ("mitigation" in climate change parlance) and recognition of the need to simultaneously prepare for

C. Rumbaitis del Rio (✉)
The Rockefeller Foundation, 420 Fifth Ave., New York, NY 10018-2702, USA
e-mail: crumbaitisdelrio@rockfound.org

J.C. Ingram et al. (eds.), *Integrating Ecology and Poverty Reduction: Ecological Dimensions*, DOI 10.1007/978-1-4419-0633-5_20,

the now unavoidable impacts of climate change ("adaptation" in climate change parlance). The Copenhagen Green Climate Fund, a fund created through the Copenhagen Accords in 2009, pledges that developed countries will provide adequate funding for developing countries to enable and support enhanced action on both mitigation and adaptation starting at approximately $30 billion USD a year starting in 2010 and building to $100 billion USD in annual global climate change aid for developing countries by 2020.

Natural-resource-based management interventions figure prominently in the basket of strategies promoted by the international community to build resilience to climate change as well as in the measures sought to encourage low carbon development in the developing world. Many of the communities most vulnerable to climate change are communities that are highly climate- and resource-dependent for their livelihoods, and have the fewest resources to cope with the impacts of climate change. Similarly, in many instances the only way that poor communities can contribute to meeting greenhouse gas reduction goals is through improved resource management actions. These will only be adopted at a large scale if they also result in greater livelihood security, or tangible improvements in standard of living.

This chapter will briefly examine the impact of climate change on ecosystems and on the human communities dependent on those ecosystems. More detail and case studies on the interactions between climate change and variability and poverty alleviation are provided in Chap. 21, this volume. This chapter will then explore the role of ecosystems and natural resource management in supporting poor communities in building their resilience to climate change, and the contributions that the field of ecology has made and could further make in advancing climate change adaptation. We then examine the role of improved resource management in reducing greenhouse gases and poverty.

Impact of Climate Change on Ecosystems and Communities

Increasing evidence suggests that ecosystems across the globe are already changing as a result of climate change. The IPCC cites evidence from 75 studies and 29,000 data series that are consistent 89% of the time with predicted impacts of anthropogenic warming (IPCC 2007). For example, there is evidence of increasing ground instability in permafrost regions (IPCC 2007). This, combined with coastal erosion, has prompted the relocation of several native subsistence hunting communities in the US state of Alaska to more stable inland locations. In other parts of the world, increased runoff and earlier spring peak discharge in many glacier- and snow-fed rivers have been detected, which in turn affect the composition and hydrology of sub-glacier ecosystems (IPCC 2007). Glacial lake outburst floods have been documented in the Nepalese Himalayas, threatening communities that live downstream from glacial lakes (Yamaba and Sharma 1993).

In the marine environment, an average decrease in pH of the world's oceans of 0.1 pH units has been attributed to oceanic uptake of CO_2 emitted to the atmosphere from anthropogenic activities. Recent analysis suggests that ocean acidification may already be affecting shell-forming marine life (Yamamoto-Kawai et al. 2009), and may have cascading effects on phytoplankton, the base of the oceanic food chain, in low nutrient parts of the ocean (Shi et al. 2010). The consequences of climate change on marine and estuarine systems and the subsistence fishing communities that depend on them are complex and difficult to predict, but could be significant, as 60% of the world's people live on the coast and depend on oceans and marine life for food and livelihoods. Just over 100 million tons of fish are eaten world-wide each year, providing 2.5 billion people with at least 20% of their average per capita animal protein intake.

A comprehensive review of the impact of climate change on ecosystems and natural-resource-based communities is beyond the scope of this chapter, and is indeed a very active area of research. For a more complete discussion of the inter-relationships between climate, communities, and development, see Chap. 21, this volume and the IPCC Second Working Group report on Impacts, Adaptation and Vulnerability (IPCC 2007).

Enabling Ecosystem Adaptation

How ecosystems adapt to changing climate conditions will have significant consequences for the human communities dependent on those ecosystems, particularly the poorest, who are often directly dependent on natural resources for their livelihoods. The ability of ecosystems to adapt to changing climate conditions is constrained by the forces of degradation, fragmentation, and overexploitation, all of which are increasing concurrently with climate change. There are a number of measures, however, that can be taken today to help ecosystems confront this multiplicity of challenges. As a general framework, the options for assisting ecosystems in adapting to climate change include: (1) reducing pressures and stresses on ecosystems, (2) incorporation of climate concerns into ecosystem management; (3) proactive measures to accelerate/ensure that adaptation occurs, and (4) continued monitoring and adaptive management. Each of these will be explored in turn.

Perhaps the biggest opportunity to help ecosystems adapt to climate change involves implementing measures that reduce the pressures on ecosystems to minimize the compounding effects of multiple stresses. Example measures include: reducing pollution and toxin loads into ecosystems; limiting over-fishing, logging, and grazing in sensitive areas; limiting excessive water extractions or soil loss; and reducing fragmentation, encroachment, and exotic species invasions where possible. In some cases these measures require implementing new or more stringent environmental regulations and, in other cases, it might be possible to prevent introduction of a new stressor. Active restoration measures may be needed in some instances to limit stresses. The World Wildlife Fund and other conservation organizations have been actively experimenting with this approach by, for instance, implementing measures to reduce nutrient runoff to coral reefs to see if this improves the ability of reefs to tolerate increasing temperatures and acidity (Hansen et al. 2003). While this is a very promising approach toward building ecosystem resilience, interventions are difficult to implement, especially where conditions of poverty and human needs prevail, because of competing priorities and the lack of resources and skills necessary to successfully implement such interventions.

A second management approach involves the active incorporation of climate change concerns into conservation and natural resource management plans. The projected impact of climate change as well as potential changes to the configuration and health of ecosystems in the future should be considered when designing new protected areas, reviewing the management of current protected areas, and developing management strategies for buffer zones and matrix habitat surrounding protected areas. Priority measures could include expanding conservation areas to account for potential shifts in ecosystems and species ranges and distributions, protecting buffer zones and corridors that aid species migration, protecting potential climate refugia (areas likely to preserve species because the local microclimate is relatively constant), protecting special habitats that might be particularly threatened by climate change (e.g., high elevation habitats), preserving functionally important areas (such as potential breeding sites of sensitive species), and protecting areas of particularly high endemic biodiversity. A number of approaches and tools are available to assist with the integration of climate concerns into conservation planning, including downscaling of global and regional climate models, using dynamic vegetation models of species and ecosystem changes in response to climate parameters, and use of paleontological information to predict how species will behave in response to climate change. Less data-intensive approaches such as scenario planning and climate mainstreaming tools have also been used to integrate climate concerns into protected area management where more specific data is not available.

In addition to reducing stresses on ecosystems and integrating climate concerns into conservation planning, many managers will find it necessary to implement more proactive measures to ensure that ecosystem adaptation occurs at an acceptable rate. Interventions can include measures such as prescribed burning to reduce the impact of catastrophic fires, or pest control to reduce the impact of a pest outbreak that would otherwise be controlled by prolonged periods of cold temperatures. *Ex situ* conservation measures, such as germplasm, seed, sperm and propagule-banking, captive breeding in aquaria and zoos may also be needed to insure that the genetic potential for adaptation persists, especially given current extinction rates. For instance, the Crop Biodiversity Trust has recently constructed a large seed vault in Svalbard, Norway to, among other objectives, insure that future generations have the crop biodiversity needed to maintain food production in spite of potentially very different climate conditions than those experienced today (Global Crop Biodiversity Trust 2010). More controversial measures to actively enable ecosystem adaptation include assisted species migration and the reintroduction of species (Hoegh-Guldberg et al. 2008). Species introductions must be carefully considered, because of the potential that the introduction may lead to new pest problems in the introduction site, either from disruption of predator–prey dynamics, or from pests and diseases inadvertently introduced along with the target species. Because of these concerns, scientists recommend that species reintroductions and assisted migration only be implemented in cases where the target species is at high risk of decline due to climate change and where translocation and establishment are technically and practically feasible and likely to be successful, and the benefits of the translocation outweigh the biological and socio-economic costs (Hoegh-Guldberg et al. 2008). Otherwise a more prudent approach consists of prioritizing traditional conservation approaches mentioned previously, such as habitat conservation, expansion and restoration, reducing stressors, increasing connectivity, and *ex situ* conservation measures, such as creation of artificial habitat as a source of organisms to restore degraded areas. For instance, Counterpart International works with communities in the Dominican Republic to create artificial coral gardens, which serve as low-tech, cost-effective systems for growing and transplanting corals to restore degraded reef ecosystems and community-based fisheries (Counterpart International 2010).

The final approach to enabling ecosystem adaptation consists of adaptive management of ecosystems. Continuous monitoring of indicators of ecosystem health, ecosystem service delivery, and the state of climate and nonclimate stressors will be critical steps in ensuring that conservation measures are effectively enabling ecosystem adaptation to changing climate conditions.

These steps should be incorporated into planning mechanisms that require periodic reevaluation of management effectiveness and enable implementation of mid-course corrections as needed. Preservation of ecosystem flexibility should be an explicit management objective, as it may be most cost-effective to manage for a range of conservation goals today rather than suffer the consequences of too tightly constrained options in the future. For more information on how to build resilience to climate change in natural systems, please consult *Buying time: A user's manual for Building Resistance and Resilience to Climate Change in Natural Systems* (Hansen et al. 2003), an easy to follow guide produced from WWF's experience in implementing ecosystem adaptation projects, from which many parts of this discussion were drawn.

Role of Ecosystems in Helping Humans Adapt to Climate Change

While many researchers and practitioners have explored how human management can help ecosystems adapt to the consequences of climate change, surprisingly, less consideration has been given to the role of ecosystems in helping humans manage the consequences of climate change. Ecosystems provide several important services that contribute to human well-being. As was articulated in the Millennium Ecosystem Assessment (2005), ecosystems provide a range of provisioning, regulating, cultural and supporting services that directly and indirectly contribute to human well-being. Many of these services also support human adaptation to climate change. Provisioning services, such as food, freshwater, fuel and fiber production and supporting services, such as nutrient cycling and soil formation, are affected by climate conditions and need to be maintained in spite of the impacts of climate change because substitutes are either unavailable or cost prohibitive. Regulatory ecosystem services, such as flood and disease regulation and water purification, provide direct benefits to humans in terms of adapting to climate change. Across the globe, these regulating services of ecosystems have been replaced with engineered infrastructure or human systems (such as flood control infrastructure). However, these services are either non-existent or not available to the poor, in many parts of the world, making populations more reliant on ecosystem regulating functions for their safety and well-being. This natural infrastructure can, under certain conditions, be more flexible and adaptive than engineered infrastructure and may also be more cost-effective than engineered solutions. Thus, the regulatory functions of ecosystems should be included in consideration of a range of options evaluated to help communities adapt to climate change.

This section further elaborates the specific ways in which ecosystem services may support human resilience to changes in temperature, precipitation, and extreme weather events, as well as some of the indirect impacts of climate change.

Temperature

It is well established that vegetation cover can moderate land and stream surface temperatures and reduce the effects of temperature extremes. For instance, shade trees have been shown to reduce surface temperatures and energy use for cooling in urban centers (Akbari et al. 2001). The installation of a green roof atop Chicago City Hall in the United States reduced roof surface temperatures in mid-summer by as much as 10°C, as measured as the temperature difference between the green roof and an adjacent roof. Furthermore, this provides energy savings of $3600 annually (ICLEI 2010). Similarly, planting of shade trees in agricultural systems has been shown to reduce soil surface temperature extremes and reduce evaporation of soil moisture through both the direct effects of shading and the mulching effect of leaf litter (Scherr and McNeely 2007).

A modeling study calibrated for land cover and land use in Western Australia has shown that reforestation could reduce anthropogenic-induced warming by as much as 30% by 2030, but that this cooling effect declines to less than 10% past 2100 as warming intensifies (Pitman and Narisma 2005). Furthermore, findings suggest that the temperature-moderating effect of reforestation is linearly related to the spatial scale of reforestation, at least as a first-order approximation. While scale is an important factor, the temperature moderating effect of vegetation also depends on the structure and density of the vegetation canopy, the temperature profile experienced, and the effects of temperature on the vegetation canopy. This research indicates that more study is needed to further quantify the mechanisms and potential benefits of large-scale revegetation as a strategy for reducing the temperature effects of global climate change.

Flooding

Wetlands, floodplains, lakes, and coastal ecosystems play a strong role in absorbing precipitation and attenuating flooding. Mechanistically, vegetation in these systems affects the amount of rainfall that is evaporated and transpired and, consequently, the amount of water available for soil moisture storage, groundwater recharge, and streamflow, including peak streamflow which can result in flooding. Removal or fragmentation of wetlands tends to reduce flood water storage capacity and can contribute to the destructiveness of flood events (Millennium Ecosystem Assessment 2005). However, wetlands should not be viewed as singular units, but rather as a connected system of upland and lowland systems that within a catchment together provide flood regulation services. Mountain forests have been shown to contribute to flood regulation, however, the extent to which they contribute to flood regulation is difficult to generalize. Studies comparing the effects of intact and logged mountain forests on storm flow discharge show differing results. For instance a study in Madagascar showed that conversion from primary forest to swidden agriculture can increase downstream storm flow by a factor of four (The World Bank 2009). However, a study by the Food and Agriculture Organization and the Center for International Forestry Research (2005) concluded that forests are only likely to reduce flooding in relatively minor storms. In some cases, the presence of forest vegetation does not seem to have any influence in reducing peak stream flow, demonstrating the complex relationship between mountain vegetation and flood regulation. This complexity is due to the fact that the proportion of incoming precipitation that is transferred to streamflow varies according to vegetation type, structure, development, rooting depth, and health, as well as antecedent conditions such as soil moisture conditions as affected by previous weather events. Thus, the relationship between vegetation and flood regulation is in many cases nonlinear and difficult to predict. The role of vegetation in reducing the incidence and severity of flood-related landslides is similarly complex, with some research supporting the role of vegetation in stabilizing slopes and others suggesting a minor role of vegetation in preventing flood-related landslides (FAO and CIFOR 2005).

Complexity notwithstanding, many locations around the world are choosing to restore or preserve wetlands, lakes, flood plains, and montane ecosystems as a way of reducing flood risks. For instance, flood control projects in Ecuador and Argentina both make use of the natural storage of forests, wetlands, and riparian corridors to provide a cost-effective mechanism to cope with recurrent floods (The World Bank 2009). This strategy functions in two ways: (1) the water regulating function of ecosystems are used to reduce flood water volumes, and (2) conservation of floodplains and riparian ecosystems keeps economically valuable infrastructure out of those ecosystems, which are often the most dynamic and frequently flooded areas, thereby reducing economic losses.

Drought

There is a large body of literature examining the relationship between large-scale land degradation and drought and desertification. However, no specific relationship can be defined between human activities and land degradation or desertification at a large scale. The causes of desertification and long-term drought include natural climate variability, nonequilibrium dynamics of arid ecosystems, and the exacerbating effects of extractive land uses such as over-cultivation, over-grazing, and deforestation (Herrmann and Hutchinson 2006). Because of the confluence of macro-scale climate trends and local-scale management practices, there is little direct evidence of the effects of ecosystem conservation or management on reducing drought frequency or duration. However there is anecdotal evidence that improved environmental management can help moderate the impact of droughts, particularly short-term droughts. For instance, in the Sahelian country of Niger, farmer planting and protection of trees has led to an increase of tree cover over 7.4 million acres (Polgreen 2007). The increase in tree cover has reduced wind-based topsoil erosion, has increased the ability of soil to hold onto water and, thus, has benefitted crops because *Faidherbia alba*, the species of tree most often planted or conserved, shed their leaves during growing season and do not compete with crops for water. Furthermore, many of the trees are nitrogen-fixing trees, which improve soil nitrogen availability and soil fertility. The increase in tree cover was prompted by a change in the legal ownership status of trees, such that now farmers are granted ownership over the trees planted. These changes in management practices also coincided with a natural increase in precipitation, due to decadal climate variability in the region, but likely did not *cause* observed increases in precipitation. However, increased tree cover has enabled farmers to benefit from a more hospitable environment in ways that in the future may contribute more effectively buffering the impacts of drought.

On a smaller scale, vegetation cover has been shown to reduce evaporation, increase soil water holding capacity, and reduce aeolian topsoil erosion. When vegetation cover is incorporated into agricultural systems, either as ground cover or an agroforestry layer, it can reduce the impacts of minor to moderate drought on agricultural production (Scherr and McNeely 2007).

Arid ecosystems are important sources of drought resistant varieties of crops and livestock, or their wild relatives, which will be a reservoir for the diversity that will be needed over the long term to cope with increasing drought. Furthermore, when managed sustainably, these systems may provide life-sustaining sources of food (wild plants and bushmeat) when severe droughts occur and other sources of food are not available.

Coastal Storms and Erosion

Coastal ecosystems and natural structures such as near shore coral reefs, sand dunes, sea grasses, salt marshes, wetlands, lagoons, and mangroves have been shown to have a beneficial impact on reducing the impacts of coastal storms in many instances. These systems reduce impacts of coastal storms through a variety of mechanisms. Near shore coral reefs, sea grasses, mangroves, sand dunes, and coastal wetlands can reduce wave energy and, thus, the destructive power of coastal storms (Koch et al. 2009). Coastal vegetation can reduce soil erosion and inland transport of mud and debris which increases the destructive power of coastal surges. Wetlands, lagoons, and floodplains can divert and contain floodwaters. All of these coastal systems, when kept intact, mean that high-value assets (including people's homes) are located further inland and further out of harm's way. These systems, of course, have limits in their ability to buffer the impact of coastal storms and storm surges, and can be overridden and damaged by high intensity storms and storm surges. Chronic degradation of these systems can also reduce their buffering capacity. Furthermore, these systems vary in their buffering capacity spatially and temporally (Koch et al. 2009). For instance, a study of coastal mangroves in Thailand has shown that the wave attenuation function of mangroves increases nonlinearly with mangrove area (Barbier et al. 2008). This has implications for management of coastal ecosystems to maximize the regulating benefits while still accommodating sustainable coastal development (Barbier et al. 2008).

The cost effectiveness and co-benefits of ecosystem-based coastal defense are increasingly being quantified and used as justification for increased coastal ecosystem protection. For instance, mangroves in Malaysia are estimated to provide a value of $300,000 USD per km as coastal defenses relative engineered alternatives (The World Bank 2009). A Vietnamese program to plant and protect coastal mangroves as a way of buffering storms has saved an estimated $7.3 million per year in sea dyke maintenance since its inception in 1994 (The World Bank 2009). Co-benefits of preserving these ecosystems include preserving habitat for marine and estuarine fauna, improving coastal water quality, and especially in the case of mangrove preservation, conservation of carbon rich soils. A more detailed discussion of the role of coastal systems in buffering hazards and storm effects is presented in Chap. 22, this volume.

Fire

The occurrence and character of wildfires are a function of a combination of factors such as fuel load, flammability, ignition source, and fire-spreading conditions, each of which are related to land use and land cover (Millennium Ecosystem Assessment 2005). For instance, the amount of vegetation effects the fuel load as does the past disturbance history and management history of a site. Fire suppression practices can lead to the buildup of fuel loads, while selective logging or prescribed burns can reduce fuel loads. Vegetation composition can affect flammability, as some plants have highly flammable resinous saps, or alternatively high moisture contents which reduce flammability. Similarly, soil moisture conditions can increase or decrease flammability. Likelihood of ignition and fire intensity are affected by the buildup of fuels, as well as by past and current management practices. Certain management practices such as swidden agriculture or fire-based pasture management can escape management and lead to increased incidence of wildfires. Landscape structure, vegetation structure, and management interventions such as firebreaks all influence the spread of fire.

Fire regimes are likely to be modified as a result of climate change, largely a result of hotter and drier conditions. The relationship between fire incidence and drier than normal conditions associated with El Nino–Southern Oscillation (ENSO) suggests that fire may become more frequent in areas where it is rare, such as in tropical moist forests, as drying trends take effect (Dudley and Stolton 2003). Fortunately, this effect may be at least partially countered by proactive fire management techniques. Measures to reduce fuel loads, manage fire spread, or reduce the incidence of ignition may be needed to reduce the likelihood of catastrophic fires in areas where the human and ecological consequences of fire may be detrimental. However, it should be noted that fire is an important factor in regenerating ecosystem health and functioning in fire-adapted ecosystems, and efforts to control fires such as selective harvesting, prescribed burns, and other forms of management, if misapplied, can negatively affect ecosystem health and service provision (Dale et al. 2001). Thus, ecosystem-based fire management strategies must continue to be a part of forest management concerns, especially in light of climate change and the potential consequences of climate change on fire regimes.

Salinization

Increasing salinization of estuaries, surface water, and shallow coastal aquifers are expected consequences of climate-change-related sea level rise and coastal storms (IPCC 2007). This saltwater intrusion may affect drinking water supplies as many coastal communities derive drinking water from rivers upstream from where salt fronts currently occur. As salt fronts push further inland, especially during times of drought, the likelihood of drawing in salty water will increase. This will have implications on water treatment processes, industrial processes that depend on a certain water quality, and potentially human health. Shallow coastal aquifers may also become salinized as a consequence of sea level rise, decreasing the amount of groundwater available for drinking water and productive use, including irrigation. Furthermore, salinity increases in estuaries will harm aquatic plants and animals that are not tolerant of high salinity, or are not able to adapt to higher salinity conditions.

Management options to cope with increasing coastal salinization include increased releases of freshwater from reservoirs, especially during drought periods, to push back surface water salt fronts. This will require greater storage of freshwater during periods of water abundance. Groundwater injection of freshwater may also help reverse or delay groundwater salinization, though this practice must be evaluated and regulated carefully to avoid contamination and unintended side-effects. Increased use and introduction of salt-tolerant species in coastal areas may also become necessary. Already, farmers in coastal Sri Lanka are experimenting with traditional and modern rice varieties which are more saline-tolerant (Practical Action 2010).

Salinization of soils and groundwater may be exacerbated in inland areas due to climate change-related increases in aridity. Salinization in arid areas occurs where salt occurs naturally in the soil or groundwater, and becomes concentrated at the surface when trees and other deep-rooted vegetation are cleared or replaced shallow-rooted annual crops, which do not take up as much water as deeper rooted vegetation. Over time, this causes the water table to rise, passing through salt deposits and dissolving and re-depositing them at a shallower depth. Gradually, the salt raises to the topsoil layer and becomes deposited and concentrated at the soil surface. In a similar manner, irrigation in arid saline environments can result in a rising water table that brings deep salt deposits upwards to surface soils, eventually concentrating at the surface. Salinization is a slow process that can render agricultural lands and groundwater stores unusable. Salt can also damage infrastructure such as roads, pipelines, and water treatment systems. Reversal of soil salinity can be a similarly lengthy process. Options include planting salt-tolerant species, use of drip irrigation systems and drainage of irrigation water through collection canals to prevent rising of the water table. Prevention of salinization is a more cost-effective approach than repair, and can be achieved through vegetation management including preservation of trees, particularly deep-rooted native vegetation in arid areas, and careful regulation of groundwater levels through irrigation management combined with continuous monitoring of soil and groundwater salinity levels.

Food Security

Climate change impacts on agricultural production and food security are difficult to predict. In the near term, many studies conclude that limited changes in temperature and precipitation, as well as, CO_2 fertilization effects will result in moderate changes in global agricultural production (Cline 2007). However, Lobell et al. (2008) found that production impacts by 2030 could be significant for staple crops in some of the most food insecure regions of the world. For instance, in southern Africa, maize production may decline by 10–40% and wheat may decline by 5–30% relative to 1998–2002 average yields (Lobell et al. 2008). Over the longer term, impacts may be more dramatic, as models suggest that by the end of the century, average summer temperatures will exceed the hottest summer temperatures recorded to date for most of the tropics and subtropics (Battisti and Naylor 2010). This poses an increased impetus to develop crop varieties that are tolerant to heat and water stress, and accelerate development of irrigation and water management systems in some of the most impoverished regions of the world. Moreover, as agricultural development proceeds, vulnerability to climate may *increase* as recent research suggests that areas with higher yields are more susceptible to temperature-related yield decreases than areas with marginal yields as climate becomes more of a growth limiting factor than other factors such as soil nutrient depletion (Schlenker and Lobell 2010). This underscores the need to consider adaptation measures in tandem with agricultural development measures.

The role of ecosystems in assisting humans in adapting to long-term climate-related food security challenges is multifold. Development and use of new crop varieties suited to new temperature, precipitation, and salinity regimes will underpin our ability to adapt agricultural systems to climate change. New varieties, developed by either traditional breeding techniques or biotechnological methods, will to some extent depend on the availability of diverse crop genetic resources that contain desired traits—for instance, landraces or crop wild relatives with resistance to drought, heat, salinity, particular pests or diseases, and varieties with different maturation periods (Burke et al. 2009). Conservation of areas of high diversity of crop wild relatives and crop biodiversity will be indispensible to preserving future adaptive capacity. Measures include both *in situ* conservation of wild plant populations in the habitats where they naturally occur, as well as *ex situ* conservation in zoos, botanical gardens, and gene-banks. The field of conservation genetics has many tools and approaches that can be applied to improve the conservation of crop genetic diversity.

In the near term, ecosystem-oriented interventions can help make production systems more resilient to climate variability. Use of cover crops, incorporation of crop residues, and improved fallow rotation can reduce vulnerability to drought by increasing the water-holding capacity of soil and litter layers (FAO 2009; Scherr and McNeely 2007). Use of legumes in crop rotation, greater use of perennials, and crop diversification in general can increase soil fertility and decrease vulnerability to pests and diseases, the incidence and spread of which may be influenced by changing climate conditions (FAO 2009). Restoration and revegetation of pasture lands can reduce vulnerability to climate impacts by improving soil structure, reducing erosion, and maintaining soil water holding capacity (FAO 2009). For more information on ecosystem-oriented production practices, and their role in improving the stability and productivity of food production systems, please see Chaps. 2–5 in this volume.

Finally, in times of acute food scarcity, wild areas often become important sources of food and livestock forage for survival – though this can have negative health and ecological consequences, such as the spread of zoonitic diseases associated with bushmeat consumption. Additionally, if these food sources are extracted unsustainably and significantly reduce species that provide important ecological functions, the ability of social and ecological systems to adapt to changing conditions may be further undermined.

Water Security

Climate change, combined with increasing demand for water due to population growth and economic growth, is likely to exacerbate many existing water crises and create new areas of water stress (2030 Water Resources Group 2009; Chap. 9, this volume). Changes in the frequency and abundance of precipitation, though uncertain and difficult to project in global and regional climate models, are likely to increase variability in water availability and quality for human consumption and use. The role of ecosystems, particularly forests, in increasing stream water availability or stabilizing water supply is highlighted as a potential adaptation strategy, but its effectiveness is highly contextual. There is no consistent relationship between natural vegetation cover and water availability or stability, though there is some evidence the presence of cloud forests tends to increase streamflow by intercepting water from clouds (Dudley and Stolton 2003).

While no clear relationship exists between vegetation cover and water availability, there is significant evidence to suggest that natural vegetation cover improves water quality. Forests and wetland ecosystems increase infiltration, store runoff, reduce sedimentation and siltation, recharge aquifers, and contribute to streamflow. Healthy soils can, through biogeochemical cycles, remove excess nutrients and adsorb pollutants, keeping them from entering streamflow. These properties of ecosystems have been shown to significantly reduce water filtration and purification costs, as evidenced by the fact that now more than one-third of the world's top 100 cities now rely on protected areas for all or a portion of their drinking water (Dudley and Stolton 2003). For instance, the Aberdare Mountains and Mount Kenya National Parks in Kenya provide critical water to the growing city of Nairobi, and the Gunung Gede-Pangrango protected area in Indonesia provides drinking water to the Indonesian cities of Jakarta, Bogor, and Sukabumi (World Bank 2009). Conservation of natural habitats in the Upper Tuul Basin has been found to be the most economical approach to preserving future water quality for Ulaanbaatar, Mongolia's largest city (World Bank 2009). Protected areas are increasingly managed as water reservoirs to meet current and future water needs.

The cost effectiveness of conservation-based approaches for water quality maintenance is increasingly recognized by private sector interests. For instance, payment for ecosystem service schemes are used in Costa Rica to reduce siltation and ensure sufficient water supplies for hydro-electric power generation (Chap. 14, Vo. 2). In Guatemala, Pepsi Cola has paid for conservation efforts to maintain water quality for commercial use (Dudley and Stolton 2003).

Wetland ecosystems can be critical infrastructure toward maintaining water quality and even providing wastewater treatment benefits. The city of Riverside, CA, USA uses 28 ha of restored wetland vegetation to denitrify wastewater (World Bank 2009). This approach cost the city 90% less than constructing a conventional treatment facility, and provides co-benefits such as recreation, environmental education, and serving as wildlife habitat to 94 bird species (World Bank 2009).

The importance of ecosystems in providing cost-effective means to preserve or improve water quality, and in some cases water availability, will make ecosystem conservation and restoration important tools in moderating water stresses exacerbated by climate change, especially in developing country situations where resources for built water infrastructure are very limited. For more information on the role of ecosystems in managing water and poverty crises, please see Chaps. 6–9 in this volume.

Health

As described in Chap. 12, this volume, climate change will present a number of direct and indirect challenges to human health, such as the direct effects of heat waves on vulnerable populations (children, the elderly, and the infirm) and indirect impacts such as malnutrition, and spread of vectorborne and zoonitic diseases. The spread of agricultural pests should also be included in this regard, as many agricultural pests are partially regulated by climate factors such as temperature or precipitation conditions that contribute to pest population increases or declines. In many cases, the best response to these climate–health-related challenges will be early detection and response by health system actors. However, in some cases ecosystem-oriented interventions can contribute to reducing the consequence of the health threat. For instance, increasing vegetation cover in cities has been shown to reduce the heat island effect and may play a role in reducing the impact of heat waves on urban environments. Similarly, maintenance of predator–prey relationships can contribute to the regulation of certain pests and diseases. The role of ecosystems in helping human communities' adaptation to climate-related health challenges is a relatively new area of inquiry that will require much more research in the coming years.

Summary: Role of Ecosystems in Building Human Climate Change Resilience

The previous section highlights some of the key ways in which ecosystems contribute to regulating climate-change-related impacts on temperature, flooding, drought, fire, coastal storms, salinization, food security, water security, and health. In many cases the roles that ecosystems can play is highly contextual, dynamic, and limited to a range of conditions. For instance, the presence of an agroforestry layer may buffer the impacts of short-term reductions in precipitation; but it will not be able to prevent the negative impacts on crops resulting from extremely prolonged periods of drought. Our knowledge of the roles of ecosystems in reducing climate change impacts is still in its infancy and, in many cases, based on anecdotal evidence rather than experimentally derived and quantified evidence. Clearly, more research is needed to quantify and qualify the conditions under which ecosystems support the climate resilience of human communities. This will likely be a very active area of research for the ecological and development community in coming years.

Taking into account the above caveats, however, ecosystems-oriented interventions are an important component of an effective strategy to build resilience to climate change. Well functioning ecosystems contribute to overall vulnerability reduction through a number of different mechanisms, and often maintaining ecosystems and the functions they provide are the most direct way to stop maladaptive practices and contribute to general resilience-building. Specific ecosystem interventions, such as restoration of degraded systems, are often more cost-effective than implementing large-scale engineered interventions, and provide similar protective benefits plus additional environmental and social co-benefits, which are critical in developing world situations where many are directly dependent on ecosystem services for daily livelihoods. Furthermore, ecosystem interventions are often additive measures that further strengthen other kinds of interventions intended to build resilience, thereby contributing to the redundancy of the system. Consider for instance, the role of wetlands in reducing the impact of coastal storms implemented in tandem with a well-functioning disaster alert and evacuation system. While both ecosystem-oriented interventions and built infrastructure interventions have limits to their buffering capacity ecosystem interventions are often more flexible and can be incrementally adapted to changing conditions as needed. This incremental approach will be a key to building resilience to climate change as our understanding of the impacts of climate change continue to evolve. Ecosystem-oriented measures are perhaps more relevant than other kinds of engineered measures to simultaneously meet multiple poverty alleviation goals because these measures can be implemented in low resource environments *by local community groups* and will provide near term poverty alleviation benefits beyond contributing to building resilience to climate change. Thus, ecological interventions should continue to play a significant role in helping communities adapt to climate change, and will figure prominently in national and local adaptation strategies and plans alongside other kinds of measures.

Finally, it should be noted that one of the largest contributions that ecologists have made to the field of climate change adaptation is through the introduction of the concept of resilience, a term that has been diffusing through the climate adaptation community in recent years. Resilience is defined as the ability of a system to absorb shocks, to avoid crossing a threshold into an alternate and possibly irreversible new state, and to regenerate after disturbance (Resilience Alliance 2007; Chap. 22, this volume). In contrast, adaptation is defined by the IPCC as "Adjustment in natural or human systems in response to actual or expected climatic stimuli or their effects, which moderates harm or exploits beneficial opportunities" (IPCC 2007). Adaptation refers to measures taken in response to *specific* current or expected impacts. Resilience, on the other hand, is a property of a system that enables adjustment to changing conditions, including surprise conditions that cannot be anticipated. Because climate change will encompass impacts that are unpredictable, a general resilience approach that explicitly acknowledges the need to prepare for anticipated and unanticipated changes or shocks may be more effective than an adaptation approach, which focuses more narrowly on specific responses to specific impacts. Building and managing for resilience requires consideration of robustness or ability of measures to reduce the effects of a variety of impacts; the need to promote management of systems so that they reinforce each other and are redundant in productive ways; support flexible management that allows for learning and change amidst dynamic conditions; diversification and decentralization to build needed redundancy; and foresight and preparedness. Building this kind of resilience requires multi-sectoral approaches and multiple skill sets integrated within a systems analysis. Systems analysis is a central concept in the discipline of ecology that has been extended to socio-ecological systems. As such, ecological frameworks and tools to facilitate understanding of complex systems (for instance systems modeling tools) are directly applicable to understanding system vulnerabilities and identifying interlinked, pro-active approaches to build resilience to climate change. This contribution of the field of ecology to managing the climate change challenge should be developed further as a foundation for more comprehensive and extensive action to build resilience to climate change.

Role of Ecosystems in Helping Humans Reduce Greenhouse Gases

An estimated 30% of all greenhouse gas emissions are derived from the land use sector, and, thus, land-based solutions to curb greenhouse gas emission could contribute significantly to global efforts to prevent catastrophic climate change. More importantly, the land use sector provides the greatest set of opportunities within the range of mitigation actions to directly involve poor communities in reducing greenhouse gases and in ways that tangibly benefit them, too. It should be noted that certain sectors of the international community has been slow to support land-based climate change mitigation solutions because, as opposed to more technologically oriented solutions, land-based solutions are viewed as impermanent and volatile, too complex, too diffuse and too slow to yield benefits on the scale needed to tackle the climate problem. Coordinated efforts among NGOs, academics, and donors to address these issues over the past decade have resulted in important policy inroads, particularly with respect to Reducing Emissions from Deforestation and Degradation (REDD).

Land-based greenhouse gas mitigation activities can be generally divided into two categories of activities: (1) activities that sequester or remove carbon from the atmosphere, and (2) activities that reduce the amount of greenhouse gases emitted to the atmosphere. Each of these approaches consists of specific measures to be undertaken in both the forestry and agriculture land use sectors, and will be examined in turn.

Land-based activities that sequester carbon from the atmosphere include fixing carbon in terrestrial vegetation and in enriching the amount of carbon stored in soil. This can be accomplished through a range of activities including: restoring vegetation cover on degraded land, afforestation or other means of increasing terrestrial biomass, low till soil management, and other soil management techniques that increase soil organic matter. Restoring degraded lands has multiple ecological benefits in addition to increasing the amount of carbon stored, including improved watershed functioning such as groundwater recharge, reducing soil erosion, and providing habitat for wildlife. Benefits to humans can include improved availability of fuel wood, improved pasture areas, improved hunting grounds, as well as the productive and health benefits of improved watershed functioning and soil conservation. Afforestation of previously unforested areas can have the same effects, depending on how projects are implemented and how communities are allowed to utilize afforested areas. However, it should be noted that afforestation of previously unforested areas, such as conversion of grasslands to forestlands, can have negative biodiversity impacts and degradation of associated ecosystem services, depending on the scale of conversion, species used to afforest, and other factors. The Clean Development Mechanism (CDM) of the Kyoto Protocol provides incentives for increased carbon sequestration through both reforestation and afforestation, though only a limited number of these kinds of projects have actually received accreditation under the CDM.

Climate concerns have recently led to an emerging movement to incent carbon rich farming practices. Carbon rich farming practices include both on-farm practices to improve carbon storage on farm, such as low tillage soil management to increase soil carbon, greater use of perennial crops, use of fodder and rotational grazing, and also off-farm practices such as conservation or restoration of forests and grasslands within the broader agricultural landscape (Scherr and Sthapit 2009). Practices that improve soil carbon include use of green manures, compost, crop residues, or livestock waste as fertilizer, use of mulch and crop residues to maintain soil moisture, no or low tillage (FAO 2009). Use of cover crops, and water management structures such as terraces and contour farming can reduce soil erosion, and thus improve the amount of soil carbon retained on farm. Use of cover crops, extended fallow rotations, and incorporation of legumes in crop rotation can also increase the amount of carbon stored in soils and in aboveground biomass. Greater use of perennial crops, shrubs, and trees can also increase the amount of carbon-stored aboveground and belowground, as at a minimum these species keep their root biomass from year to year, whereas annually tilled crops turn over every year (Scherr and Sthapit 2009). Perennials can be incorporated through agroforestry approaches, where tree species are incorporated into the farming system as shelterbelts, shade trees, fertilizer trees, or to produce a harvestable product (e.g. fruits, nuts, wood, medicines, fuel, and fodder for animals). These practices also have potential positive ecosystem benefits such as providing habitat for wildlife, including wild pollinators, or improving water infiltration and groundwater recharge, and can help reduce the amount of inorganic fertilizer used where fertilizer use is excessive and polluting. Over time, many of these practices can also increase farm productivity and stability of production in spite of increasing climate variability, though it is not uncommon for productivity to decline initially upon implementing some of these measures. For instance, introduction of tree species may out-compete crop species for light, even while enhancing soil fertility and soil water availability. Furthermore, many of these practices are more labor intensive for the farmer which, combined with potential initial yield reductions, present a significant barrier to widespread adoption of these practices. Climate finance, such as carbon credit schemes, could provide an important mechanism to incent greater adoption of these environmentally friendly production practices by helping farmers overcome high opportunity costs. Payments to farmers for increasing the amount of carbon stored in farming landscapes could take various forms. These could include climate-financed subsidies for agricultural inputs, crop insurance, extension services, or, alternatively, could take the form of price premiums for carbon-rich agricultural products that provide farmers with the resources needed to rationalize adoption of these more labor-intensive farming practices.

A number of different land use practices can be applied to reduce emissions from the land use sector. The largest opportunity, and the category of activities gaining the most political traction, is the use of payments for Reduced Emission from Deforestation and Degradation (REDD). Deforestation accounts for 18% of total anthropogenic greenhouse gas emissions – a larger volume than from the entire transportation sector combined (Meridian Institute 2009). The international climate policy community has by and large recognized that without providing financial incentives to decrease deforestation, it will be difficult to reduce global emissions to the amount needed to stabilize the global climate system at no more than 2°C of warming. Sufficient and effective financing of large-scale forest protection to reduce greenhouse gas emissions could have benefits for wildlife, and poor communities who could receive payments or other incentives necessary to curb deforestation, which can result in the maintenance of a whole suite of other ecosystem services such as watershed regulation, disaster protection, food, fuel, wood, and medicine. Chapter 11, Vol. 2 discusses payments for greenhouse gas emissions reductions in the forestry sector, and their potential poverty alleviation implications.

Reducing greenhouse gas emissions from agricultural land uses could take several forms. The largest source of greenhouse gas emissions from agricultural systems, apart from deforestation for agriculture, is due to the over-application of nitrogenous fertilizers and manure, which through microbial processes leads to nitrous oxide emissions. Nitrous oxide has a global warming potential nearly 300 times greater than CO_2. Globally, it is estimated that this accounts for 2,100 million tons of CO_2 equivalents of greenhouse gas emissions. However, there is a great range of variability in rates of fertilizer application globally. For instance, fertilizer application rates in Northern China and the Midwestern US are two and one orders of magnitude, respectively, greater than in Western Kenya, where harvest removals of nitrogen and phosphorous exceed inputs from fertilizer and biological nitrogen fixation (Vitousek et al. 2009). Over time, negative nutrient imbalances such as those in Western Kenya lead to soil depletion and declining yields. Positive nutrient imbalances, such as those described for northern China lead to high rates of nitrogen oxide emissions and nitrogen pollution of aquatic systems. Better-targeted timing and placement of nutrient inputs, modifications to livestock diets to reduce the nitrogen content of manures, and preservation or restoration of riparian vegetation buffers can all be used to reduce agricultural nitrous oxide emissions and nitrogen pollution of aquatic systems (Vitousek et al. 2009).

The next largest source of agricultural greenhouse gas emissions is from methane produced by livestock through the fermentation process occurring in the rumens of animals. An estimated 1,800 million tons of CO_2 equivalents are produced annually through this process. Livestock also contribute to greenhouse emission through the land cleared for animal pasture, the nitrous oxide and methane produced from manure, and from the consequences of over grazing (resulting potentially in reduced carbon sequestration in vegetation and soils).

Options for improving the greenhouse gas footprint of livestock rearing include: use of nutrient supplements and feed mixes with a higher starch content which leads the animal to produce less methane, improved rotational grazing of livestock which allows pastures to recover before reintroducing livestock offsetting a portion of the emissions produced by livestock, and better storage and management of animal manure, including conversion of manure to energy for household or community use through use of biogas digestors (Scherr and Sthapit 2009).

The last large source of greenhouse gas emissions from the agricultural sector is from paddy rice production. Anaerobic decomposition in flooded paddy rice systems emits approximately 600 million tons of CO_2 equivalents in the form of methane. Research is underway through the Rice and Climate Change Consortium of the International Rice Research Institute and collaborating institutions to develop rice production systems that enhance rice production while reducing the global warming impact of rice systems. Research will include measuring and modeling greenhouse gas emissions from paddy systems and experimenting with rice soil management measures to reduce greenhouse gas emissions (International Rice Research Institute 2010).

As the previous sections have illustrated, there are a variety of mechanisms through which carbon sequestration can be enhanced and greenhouse gas emissions can be reduced from forestry and agriculture sectors, some with added benefits for poverty reduction. Ecologists have made a number of contributions to the identification, articulation, and implementation of these approaches. For instance, knowledge of species traits and rates of biomass accumulation could be used to design ecologically sound afforestation and reforestation systems that meet local needs, and do not over-exploit local water resources (Jackson et al. 2005). Ecologists have a further role in helping to identify where reforestation or afforestation schemes are not ecologically appropriate (Farley et al. 2008). Landscape ecological knowledge and spatial analysis have led to the development of the "landscape carbon" concept, wherein the different components of an agricultural landscape are used to maximize carbon storage and other critical ecosystem processes such as watershed services (Sarah Scherr, personal communication). This approach is gaining favor as a rational way to aggregate and manage carbon components in heterogeneous landscapes. Biogeochemical research on agricultural production systems is critical to identifying practices that could be implemented more widely to reduce nitrous oxide and methane emissions from production systems. Field-based and remote sensing approaches to carbon measurement and forest carbon modeling are key contributions to the development of international carbon accounting protocols, and form the basis for the verification of land-based emissions reductions in future international agreements (D. James Baker, personal communication). These are just a few examples of the modes in which ecological knowledge has influenced the development of policy and practice around land-based climate change mitigation measures in recent years.

Greater use of land-based mitigation options is an indispensible component of global efforts to prevent dangerous climate change. These mechanisms enable developing world countries to make and meet greenhouse gas emission targets. Any future global climate agreement will necessitate participation of the developing world as well as the developed world, and thus will likely require improved land use management approaches at a much larger scale. The discipline of ecology is well placed to help identify ecosystem-based practices and approaches to meet mitigation targets, in a locally environmentally sound manner that ideally also delivers poverty reduction benefits.

Conclusions

This chapter has reviewed the many ways in which ecosystem-oriented approaches contribute to building climate change resilience and to reducing greenhouse gas emissions to the atmosphere. In some cases ecosystems provide direct benefits, while in others, ecosystems either reduce pressure on human systems or play a contributing role to supporting climate change adaption or mitigation efforts. Of course, efforts to both mitigate and adapt are often complimentary and can be pursued in parallel in the same landscape. Ecosystems can be thought of as natural infrastructure to be managed to flexibly and cost-effectively reduce human vulnerability to climate impacts. However, there are limits to this natural infrastructure, which need to be better understood and documented. Similarly, ecosystems may be preserved and managed differently to make significant contributions to global efforts to curb climate change. Importantly, this chapter has highlighted that there are number of measures that can be implemented *today* to meet climate change adaptation and mitigation needs while also contributing to poverty reduction. Ecological contributions to developing these range of options include everything from providing empirical data, measurement approaches, and modeling frameworks to assess and design ecologically oriented interventions to the development conceptual basis for approaches to building resilience. As the impacts of climate change continue to unfold and the impetus to slow or stop global warming gains strength, the role of ecological knowledge in further developing a set of effective and equitable adaptation and mitigation measures will only increase in importance, as is the need to educate decision makers and the public about the role of ecosystems in addressing society's needs with respect to climate change.

References

2030 Water Resources Group. 2009. Charting our Water Future: Economic Frameworks to Inform Decision-making. The Barilla Group, The Coca-Cola Company, The International Finance Corporation, McKinsey & Company, Nestlé S.A., New Holland Agriculture, SAB Miller plc, Standard Chartered Bank, and Syngenta AG. Washington, DC. USA. http://www.mckinsey.com/clientservice/water/charting_our_water_future.aspx.

Akbari, H., M. Pomerantz, and H. Taha. 2001. Cool surfaces and shade trees to reduce energy use and improve air quality in urban areas. Solar Energy **70** (3):295–310.

Battisti, D.S. and R.L. Naylor. 2010. Historical Warnings of Future Food Insecurity with Unprecedented Seasonal Heat. 2009. Science **5911**: 240–244.

Barbier, E.B., E.W. Koch, B.R. Silliman, S.D. Hacker, E. Wolanski, J. Primavera, E.F. Granek, S. Polasky, S. Aswani, L.A. Cramer, D.M. Stoms, C.J. Kennedy, D. Bael, C.V. Kappel, G.M.E. Perillo and D.J. Reed. 2008. Coastal ecosystem-based management with nonlinear ecological functions and values. Science **319**: 321–323.

Burke, M.B., D.B. Lobell and L. Guarino. 2009. Shifts in African crop climates by 2050, and the implications for crop improvement and genetic resources conservation. Global Environmental Change **19**: 317–25. doi:10.1016/j.gloenvcha.2009.04.003.

Center for International Forestry Research and Food and Agriculture Organization of the United Nations. 2005. Forests and Floods: Drowning in Fiction or Thriving on Facts? RAP Publication 2005/03. Bogor Barat, Indonesia.

Cline, W.R. 2007. Global Warming and Agriculture: Impact Estimates by Country. Center for Global Development and Peterson Institute for International Economics. Washington, DC, USA.

"City of Chicago, USA - City Hall Rooftop Garden Pilot Project (Urban Heat Island Initiative)" ICLEI – Local Governments for Sustainability Official Website. http://www.iclei.org/index.php?id=2743 [accessed 24 January 2010].

"Climate Adaptation Program" World Wildlife Fund for Nature Official Website. http://www.worldwildlife.org/climate/adaptationprograms.html [accessed 15 January 2010].

Dale,V.H., L.A. Joyce, S. Mcnulty, R.P. Neilson, M.P. Ayres, M.D. Flannigan, P. J. Hanson, L.C. Irland, A.E. Lugo, C.J. Peterson, D. Simberloff, F.J. Swanson, B.J. Stocks, and B.M. Wotton. 2001. Climate Change and Forest Disturbances. Bioscience **51**(9): 723–734. doi:10.1641/0006-3568(2001)051[0723:CCAFD]2.0.CO;2.

Dudley, N. and S. Stolton. 2003. "Ecological and Socio-economic Benefits of Protected Areas in Dealing with Climate Change." Pages 215–231 *in* L.J. Hansen, J. L. Biringer, and J. R. Hoffman, editors. Buying time: A user's manual for building resistance and resilience to climate change in natural system. World Wildlife Fund, Gland, Switzerland.

Farley K.A., G. Piñeiro, S.M. Palmer, E.G. Jobbágy and R. B. Jackson. 2008. Stream acidification and base cation losses with grassland afforestation. Water Resources Research **44** (12): Cite ID W00A03. doi:10.1029/2007WR006659.

"Fighting salinity through traditional practices" Practical Action Official Website. http://practicalaction.org/disaster-reduction/climatechange_ricebiodiversity [Accessed 14 February 2010].

Food and Agriculture Organization, 2009. Food Security and Agricultural Mitigation in Developing Countries: Options for Capturing Synergies. Rome, Italy. ftp://ftp.fao.org/docrep/fao/012/ak596e/ak596e00.pdf.

Hansen, J.L., J.L. Biringer and J.R. Hoffman. 2003. Buying time: A user's manual for building resistance and resilience to climate change in natural system. World Wildlife Fund, Gland, Switzerland.

Herrmann, S. M., and C.F. Hutchinson. 2006. The Scientific basis: links between land degradation, drought and desertification. Pages 11–27*in* P.M. Johnson, K. Mayrand, and M.Paquin, editors. Governing Global desertification: Linking Environmental Degradation, Poverty and Participation. Ashgate Publishing, Hamphsire, England.

Hoegh-Guldberg, O., L. Hughes, S. McIntyre, D. B. Lindenmayer, C. Parmesan, H. P. Possingham and C.P. Thomas. 2008. Assisted Colonization and Rapid Climate Change. Science **321** (5887): 345–346.

IPCC, 2007: Summary for Policymakers. Pages 7–22 *in* M.L. Parry, O.F. Canziani, J.P. Palutikof, P.J. van der Linden and C.E. Hanson, Eds., Climate Change 2007: Impacts, Adaptation and Vulnerability. Contribution of Working Group II to the Fourth Assessment Report of the Intergovernmental Panel on Climate Change. Cambridge University Press, Cambridge, UK and New York, USA.

Jackson, R.B., E.G. Jobbágy, R. Avissar, S. Baidya Roy, D. Barrett, C.W. Cook, K.A. Farley, D.C. leMaitre, B.A. McCarl, B. Murray. 2005. Trading water for carbon with biological carbon sequestration, Science **310**: 1944–1947.

Koch, E.W., E.B. Barbier, B.R. Silliman, D.J. Reed, G.M.E. Perillo, S.D. Hacker, E.F. Granek, J.H. Primavera, N. Muthiga, S. Polansky, B.S. Halpern, C.J. Kennedy, C.V. Kappel, E. Wolanski. 2009. Non-linearity in ecosystem services: temporal and spatial variability in coastal protection. Frontiers in Ecology and the Environment **7**(1): 29–37.

Lobell, D.B., M.B. Burke, C.Tebaldi, M.D. Mastrandrea, W. P. Falcon and R.L. Naylor. 2008. Prioritizing Climate Change Adaptation Needs for Food Security in 2030. Science **319**: 607–610.

Meridian Institute. 2009. Reducing Emissions from Deforestation and Forest Degradation (REDD): An Options Assessment Report. Prepared for the Government of Norway, by Arild Angelsen, Sandra Brown, Cyril Loisel, Leo Peskett, Charlotte Streck, and Daniel Zarin. Washington, DC. USA. http://www.REDD-OAR.org.

Pitman, A.J., and G.T. Narisma. 2005. The role of land surface processes in regional climate change: a case study of future land cover change over south western Australia. Meteorology and Atmospheric Physics **89**: 235–249. DOI - 10.1007/s00703-005-0131-1.

Polgreen, L. 2007. In Niger, Trees and Crops Turn Back the Desert. New York Times. February 11, 2007.

Resilience Alliance. 2007. Resilience Alliance web site:http://www.resalliance.org/576.php. Accessed 28 October 2007.

"Rice and Climate Change Consortium" International Rice Research Institute Official Website: http://beta.irri.org/rccc/index.php?option=com_wrapper&Itemid=26 [Accessed 23 March, 2010].

Scherr, S, and J. McNeely. 2007. Farming with Nature: The Science and Practice of Ecoagriculture. Island Press, Washington, DC, USA.

Scherr, S. and S. Sthapit. 2009. Mitigating Climate change through Food and Land Use. Worldwatch Report 179, L. Mastny, editor. World Watch Institute. Washington, DC. USA.

Schlenker, W., and D.B. Lobell. 2010. Robust negative impacts of climate change on African agriculture. Environmental Research Letters **5**(1): 014010–014018. doi:10.1088/1748-9326/5/1/014010.

Shi, D., Y. Xu, B.M. Hopkinson, and F.M.M. Morel. 2010. Effect of Ocean Acidification on Iron Availability to Marine Phytoplankton. Science **327**(5966): 676–679.

Solomon, S., G.K. Plattner, R. Knutti and P. Friedlingste. 2009. Irreversible climate change due to carbon dioxide emissions. Proceedings of the national Academies of Sciences. **106** (6):1704–1709.

"Svalbard Global Seed Vault" Global Crop Biodiversity Trust Official Website. http://www.croptrust.org/main/arctic.php?itemid=211 [Accessed 17 January 2010] Millennium Ecosystem Assessment, 2005. *Ecosystems and Human Well-being: Synthesis*. Island Press, Washington, DC. USA.

"Tourism and Conservation: Coral Gardens" Counterpart International Official Website:http://www.counterpart.org/ourWork/tourismConservation/coralGardens/index.html [Accessed 22 March, 2010].

Vitousek, P.M., R.L. Naylor, T. Crews, M.B. David, L.E. Drinkwater, E. Holland, P.J. Johnes, J. Katzenberger, L.A. Martinelli, P.A. Matson, G. Nziguheba, D. Ojima, C.A. Palm, G.P. Robertson, P.A. Sanchez, A.R. Townsend, and F.S. Zhang. 2009. Nutrient Imbalances in Agricultural Development. Science **324**:1519–1520.

World Bank. 2009. Convenient Solutions to an Inconvenient Truth: Ecosystem-based Approaches to Climate Change. The World Bank, Washington DC. USA.

Yamaba, T and C.K. Sharma. 1992. Glacier Lakes and Outburst Floods In the Nepal Himalaya. Snow and Glacier Hydrology (proceedings of the Kathmandu Symposium, November 1992). **218**: 319–330.

Yamamoto-Kawai, M., F.A. McLaughlin, E.C. Carmack, S. Nishino, and K. Shimada. 2009. Aragonite Undersaturation in the Arctic Ocean: Effects of Ocean Acidification and Sea Ice Melt. Science **326**(5956): 1098–1100.

Chapter 21
Improving Understanding of Climatic Controls on Ecology in Development Contexts

Anton Seimon

Introduction

In contexts of development, climate is one of the several inherited background states, a geographic endowment which, along with factors such as soil types, water availability, and geographic location, exerts a strongly coercive influence on ecological system types and character, and consequentially, on human settlement patterns, health characteristics, livelihoods and economies. Climatic determinism has long been offered as an explanation (and with considerable resultant controversy) to explain the impoverishment of tropical regions relative to the economic vibrancy and affluence of societies in the mid-latitudes (Sachs 2000). However, it is indisputable that climate and climatic variability play a highly influential role in human livelihoods, economic development and health outcomes. For the global poor, the role of climate is magnified due to omnipresent vulnerability related to lower levels of the ability to cope with climatic stress. Consequentially, long-term success in poverty alleviation cannot be achieved without comprehension and engagement with issues of climatic variability and climate change.

Climate Change in Development Contexts

By the close of the first decade of the twenty-first century, climate change has gained wide acceptance as a complex and rapidly mounting threat to ecosystems and humanity. Climate change impacts have become fully integrated into development discourse and are increasingly considered in development planning.

A. Seimon (✉)
Wildlife Conservation Society, 2300 Southern Boulevard, Bronx, New York, 10460-1099, USA
e-mail: aseimon@wcs.org

J.C. Ingram et al. (eds.), *Integrating Ecology and Poverty Reduction: Ecological Dimensions*, DOI 10.1007/978-1-4419-0633-5_21,

The term "climate change" as used in traditional climatology is the shift from baseline means established in recent multi-decadal experience, as well as, changes in climatic variability characteristics. In development studies and many other contexts, the term is almost universally used in reference to anthropogenically influenced climate change and its perturbation of "natural" climate and climatic variability. The United Nations Framework Convention on Climate Change (UNFCCC) definition states that: "Climate change … means a change of climate, which is attributed directly or indirectly to human activity that alters the composition of the global atmosphere and which is in addition to natural climate variability observed over comparable time periods." By either definition, climate change represents growing disequilibrium over time between climate and the ecosystems and human livelihoods that co-existed in some assumed quasi-stable balance (though neither the climate nor ecological science communities currently maintain that these systems were ever stationary). Climatic variability and climate change bring stresses that affect that quantity, quality, and reliability of ecosystem services which can have deleterious effects on vulnerable populations, and none more so than the global poor (IRI 2005).

Critical to anticipating climate change impacts upon ecosystems, human livelihoods, and socioeconomic development is holistic understanding of how of climatic means, variability, and extremes exert control over ecology. In the absence of such understanding, climate model change projections for the future cannot be properly evaluated in relation to environmental conditions. Despite this, understanding of climatological baselines is frequently taken as a given in development projects without due diligence assessment of how well the climatological context is incorporated. This can lead to unpleasant surprises: dams that fail to fill; levees that fail; hydropower generation plants with insufficient river flows; agriculture schemes that fall short on yield; supposedly moist forests that suffer catastrophic fires; proliferations of pests and pathogens; and so on.

Through processes of industrialization and untrammeled natural resource extraction, humans have changed the atmosphere's composition, and in doing so, have committed the planet to an inexorable climate change regime that is beginning to register profound ecological consequences. The ecological contexts of development into the future will inevitably be characterized by increasing disruption among the myriad complex interactions that are the evolved organizational structures of ecosystems and the species contained within them. Terms such as *range shifts*, *migrations*, *disequilibria*, and *asynchronies* are increasingly applicable to characterize the disruption of ecological systems in years to come relative to past experience that has shaped our understanding to date.

In terms of contemporary development efforts intended to provide socioeconomic uplift to the poor in an ecologically sustainable manner, this means that past experience may not provide representative analogs for future outcomes. For example, natural geographic endowments of favorable climate and soils that might have described a given landscape as suitable for cultivation of maize, which, in turn, thus shaped traditional livelihoods and nutrition and subsequently influenced development planning for the region are in a process of alteration that might render such cultivation unfeasible within several decades. Or, from a sustainability perspective, continued utilization of this region for similar land use practice would necessarily require increasing bio- and geo-engineering interventions to sustain maize cultivation. Furthermore, modeling results project that by the century's end, local combinations of climate parameters are likely to develop that are not known to exist anywhere in our current experience, giving rise to new descriptors as *novel climates*, along with their ecological counterparts, *no-analog communities* (Williams et al. 2007; Battisti and Naylor 2009).

While it is difficult to anticipate such changes, the means to comprehend the possibilities and build scenarios around them can be greatly abetted by rigorous assessments of the relationship of present day climates with ecology in any given setting. A common shortcoming in present day development planning is incomplete or flawed understanding of contemporary climatology and attendant climatic disturbance regimes for a given context. Ecosystems and human livelihoods are influenced by climate and conform to climatic norms; climate anomalies are therefore, significant as stressors, and have impacts often proportional to their magnitude (e.g. Lyon and Barnston 2005). Infrequent outlier events of high-magnitude – which nonetheless are part of the envelope of "normal" climatic variability – yield short-term shocks that severely stress ecological and human livelihood systems (i.e. disasters), and may for stronger events surpass natural or anthropogenically fortified resilience and coping mechanisms.

Thus, the propensity for climate change to extend the envelope of variability to encompass extremes beyond the realm of recent experience is of mounting concern. Recent high impact disasters such as the 2003 summer heat wave in Western Europe (more than 70,000 fatalities; Robine et al. 2008), Hurricane Katrina in the southern United States in 2005 (more than 1,800 fatalities, $81 billion in losses; Knabb et al. 2006), and Australian wildfire outbreaks in 2009 are often presented in the media and in some scientific circles as likely manifestations of climate change. A case could be made for the Australian case, somewhat concordant with the perspective of Williams et al. (2007), in that meteorological conditions that spawned the fires near Melbourne, in terms of concurrent extremes of temperatures (48°C), low humidity (7%), and wind (gusts above 20 m/s), were of an extreme that has no parallel in more than a century of climatological records from the region.

Climatology as Represented in Development Initiatives

A multitude of efforts are underway to map the world – and the developing world, in particular – to identify regions of enhanced risk to livelihoods and ecosystems to climate-enhanced disasters, perturbed disturbance regimes, and secondary impacts such as altered disease and pests patterns (e.g. IPCC 2007; Lenton et al. 2008; McGrananhan et al. 2006). Increasingly, the development community is applying tools to reevaluate existing developments strategies and planning through the lens of climate change (Agrawala 2005; Noble 2005). Such screening methods are generally combined with cost–benefit analysis and scenario building according to climate model output, and yield outcomes that commonly stress the necessity of building resilience into livelihoods and safeguarding ecosystems, while avoiding specifics on how change will proceed due to the inherent uncertainties of climate, and environmental modeling. The challenge, therefore, is seen to rest in constraining vulnerability to accommodate an increasingly climatically stressed future.

The urgency to address climate change in development planning is now well recognized and frequently viewed as fundamental to many ongoing initiatives. Almost invariably, such initiatives emphasize climate modeling as the principal means of understanding future climate and environmental conditions. Frequently absent, however, is retrospective analysis and interpretation of current climatic baselines that will form the reference frame for assessing changing conditions. The role of the climatologist in development projects related to climate change, where found at all, is often mostly a technical role, principally involving model manipulations and conforming outputs to user specifications.

What is nonetheless still required to make the ecological significance of climate model projections meaningful is detailed examination of the reference base for assessing change: the associations between climate and environment in the present day and in the recent past. The emphasis on modeling presupposes that a comprehensive baseline understanding of the role of climate in influencing ecosystems and human livelihood is already well developed. Unfortunately, this is often not the case, and is furthermore hindered by poor data resources and difficulties in gaining access to them in Least Developed Countries. Furthermore, while it is a straightforward exercise to compare observed climate parameters of the present with modeled projections of the future, it is a far more complex endeavor to derive the ecological and functional significance of such changes and to understand where threshold values exist that represent tipping points to abruptly altered states (Lenton et al. 2008; Adger et al. 2009). Current climatic trends at a given locale might be strongly at odds with model projections, making assertions about the future particularly problematic without comprehensive baseline understanding of multi-decadal variability. Such is the case in East Africa, where an overall drying trend culminating in a catastrophic drought in 2009 is apparently contrary to projections of a moistening climate through the course of the twenty-first century in the consensus of general circulation models utilized in the IPCC 4th Assessment (IPCC 2007). However, the modeled increases are projected to become significant only after several decades, so, in this case, the seemingly opposed patterns should not be used to reject the modeled projections.

Assessments performed at large scales (regions to continents) are inevitably of relatively limited utility in more restricted geographic contexts where adaptation actions are ultimately applied, especially in regions of complex topography, which are improperly represented in models. Methodologies used to refine model output, such as statistical downscaling and more localized simulations utilizing regional climate models, are designed to incorporate local influences, but their efficacy is largely predetermined by the availability of representative climatological data within a given area of interest. This becomes a particular hindrance for addressing climate changes in the areas afflicted with high levels of poverty: Least Developed Countries generally have sparse climate observing networks and data archives available to serve as inputs for focused climate modeling studies and the reference base for assessing model outputs. New remote sensing and data analysis techniques are being used to fill in these gaps, though they vary in efficacy as substitutes for actual *in situ* observations.

Ecological Impacts upon Biodiversity

A multitude of climate change impacts upon biodiversity and ecosystems have now been identified as likely for the future. Many have already been demonstrated to already be measurable and significant, and are comprehensively reviewed in Parmesan (2006). Profound rearrangement of global biogeography forced by climate change will inevitably impact humanity in a multitude of ways, with the global poor being those most likely to suffer the most adverse consequence due to their persistently high state of vulnerability and high reliance on ecosystem service provision for survival and livelihoods. One of the greatest challenges of incorporating a changed ecological future into development planning is anticipating the synergies that will result from and within climatically perturbed natural systems, many of which are likely unforeseeable (Lawler et al. 2009). In particular, in many geographic contexts the associations and feedbacks among disturbance regimes of factors such as fire, pathogens, invasive species and how they will influence ecological systems and functioning individually is probably beyond predictive capacity at the present time.

Spatial and Temporal Scale Considerations

At global scales poverty is largely a low-latitude phenomenon, and tropical climatology is clearly among a number of causative factors underlying this pattern (Sachs 2000). Unlike the climates of the more temperate latitudes, tropical climates are not thought of as being highly variable: it is merely the base state and occasional extremes and attendant natural disasters that drive environmental and socioeconomic outcomes. In contrast, close examination of low-latitude climatological observations sites evaluated by the author across a broad range of geographic domains reveals a largely unrecognized level of detail suggesting more significant influences of climate on both tropical ecosystems and human activity than generally recognized.

A shortcoming constraining how tropical climates are understood is a product of the conventional analysis approach: the near universal utilization of climatological statistics at monthly or annual resolution only. While observations are recorded at daily or hourly intervals, aggregated monthly sums and averages are almost invariably used as climatological data by environmental researchers, development workers, and others concerned with poverty alleviation to characterize climate conditions. This practice might have origins in long-held perceptions that tropical climates are largely invariant, and the only climate phenomena of significance are the extreme hazards that occasionally occur. This is understandable: in contrast to mid-latitude weather patterns characterized by seasonal change, tropical climates seem dull in comparison, with long sequences of days featuring little apparent variation. Case studies offered here from Peru and Uganda show how more highly resolved analysis might be of benefit to development studies and practice by revealing dynamic linkages between climatic behavior and ecological responses.

Peruvian Andes

The tropical Andes are renowned as Earth's foremost biodiversity "hotspot" (Myers 2000), yet are also characterized by widespread rural poverty and low overall socio-economic development. The region has a rich legacy of ecological research, particularly with regard to human land use in the extremely diverse environments created by the complex topographic mosaic and abundant biodiversity (Halloy et al. 2005). Yet as recently as 1986, a seminal compendium of studies on tropical biogeography (Vuilleumier and Monasterio 1986) with emphasis in the Andean region lacked any mention of the El Niño-Southern Oscillation (ENSO), which has since been demonstrated to be the leading driver of inter-annual tropical climatic variability. After a sequence of regional crises related to ENSO events, climatic variability and its corresponding need for risk management, ENSO is fully appreciated to be a dynamic driver of human response and ecosystem functions throughout the tropical Andes (Glantz 1996). In this light, climate change represents a potential exacerbating influence upon a region already beset with endemic climatic stress from factors such as ENSO (Sperling et al. 2008). At high elevations, the impacts of strong multi-decadal warming are already amply evident in terms of rapid deglaciation and related ecological response (Seimon et al. 2007; Vuille et al. 2008; Hole et al. 2010).

Long-term climate predictions presented as changes in mean parameters are of limited value without a comprehensive understanding of how present-day climate and its variability relate to ecosystems and the services they provide. This requires evaluation of measured climatic variability and change trends to ecological and social impacts and responses.

In Peru, until very recently daily climate statistics from national agencies have been effectively unavailable without high purchasing cost, so the utilization of monthly data for analysis is less one of choice than availability. Climatological knowledge is therefore drawn largely from time series of monthly sums of accumulated precipitation and monthly mean temperature, yet this analysis approach misses much valuable detail. For example, analysis of a 42-year data set of daily records from a high elevation climate station in Cusco, Peru (3,365 m ASL) compiled by the author offers insight into the complexity of the regional climatology beyond what can be ascertained with conventional monthly statistics. When examined at high temporal resolution, it is evident that maximum, mean, and minimum temperatures have markedly different seasonal behavior and multi-decadal trends (Fig. 21.1). Furthermore, warm and cold ENSO events (i.e., El Niño and La Niña) shown are

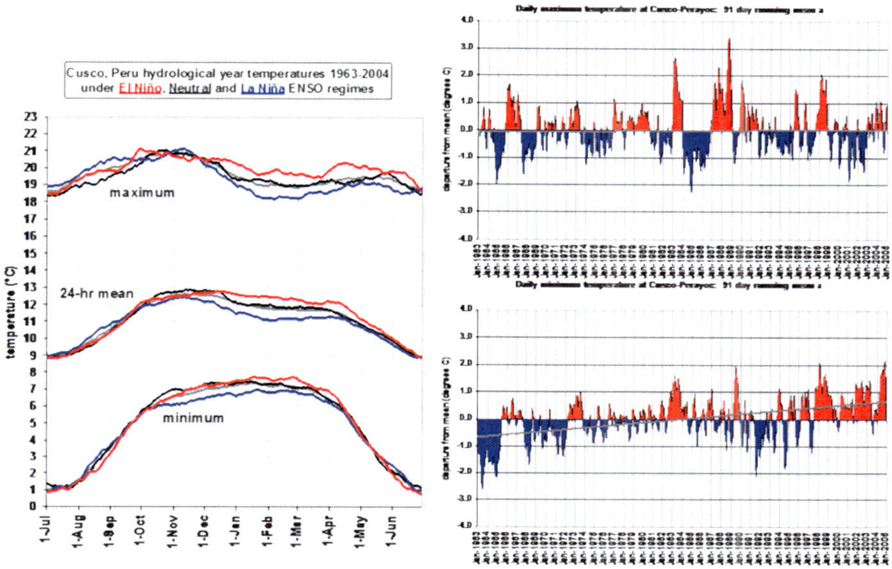

Fig. 21.1 (*left*) The annual cycles of daily average maximum, 24-h mean and minimum temperature at Cusco, Peru (3,365 m ASL) recorded between 1963 and 2004 according to the El Niño-Southern Oscillation (ENSO) phases of El Niño (*red*), Neutral (*black*), and La Niña (*blue*), while the means of each for all years are shown in gray. The distinct and contrasting patterns demonstrate the complexity of thermal climatological variability at this low latitude, high elevation site that is representative of a broad domain of the tropical Andean region. (*right*) Seasonal-scale (91-day) running mean anomalies at Cusco for daily maximum (*upper*) and minimum (*lower*) temperature from 1963 to 2004. Positive anomalies are shown in red, negative in blue. The overall trend (*gray line*) is neutral for maximum yet rapidly increasing for minimum (+1.4°) over the 42-year period. The differences between the two series are distinct and complex, while the magnitude of variation is far larger than generally expected for a location deep in the tropics (latitude 12.5°S)

associated with thermal anomalies of opposite signs that are most apparent during the austral summer wet season. In addition, the preponderance of El Niño events relative to La Niña since the mid-1970s has contributed to the upward trend in maximum temperature, whereas the minimum temperature shows a neutral trend.

The pattern from Cusco is representative of a large expanse of the high Andean region where seasonal temperature anomalies tend to be regionally synchronous due to the dominance of ENSO, whereas the pattern for precipitation varies spatially (Vuille et al. 2003). At Cusco, the inter-annual range of variation in daily maximum temperature over the summer (December–March) growing season months has been as much as 3.8°C in recent years (from +2.5°C accompanying a strong El Niño event in 1983, to −1.3°C under La Niña conditions in 1984). The large magnitude of such variability pre-dating expected anthropogenic influences identifies the likelihood of considerable resilience inherent in species and ecosystems to sequential climatic extremes. In this example, the range in seasonal maximum temperature means over 2 years exceeds the expected net thermal increase projected for the end of the twenty-first century under all but the most extreme emissions scenarios depicted by climate models for the Andes.

This ENSO-related variability exerts strong control over a range of ecological processes related to temperature as a function of elevation. Under La Niña conditions, nocturnal minima in the growing season average 1°C colder than under El Niño. While this difference may seem minor, in cultivated crops, and especially tuberiferous plants sown widely in the high Andes, the lower nocturnal temperatures favor storage over respiration in plants, thereby benefitting crop yields. This suggests marked inter-annual variability in the partition of daily energy flows for respiration and storage, and a trend under the warming climate regime toward less favorable conditions for storage. This would translate to a reduction in potato crop yield, which could serve as a diagnostic indicator of a climate change impact for similarly evolved plant taxa. However, a correlation between thermal conditions and potato yields in the high Andes has yet to be demonstrated, so this is merely speculation.

While the thermal conditions associated with El Niño events should, therefore, be associated with reduced yields, the same may not hold true near the upper limits to cultivation. The seasonal mean freezing level altitude is likewise controlled by ENSO phases, creating significant year-to-year variation in cultivation success at high elevations (>4,000 m), where despite the omnipresent risk of losses from frost, potatoes and other tubers are nonetheless extensively cultivated. At even greater elevations, the thermal and solar radiation variability related to ENSO exert considerable control over the degree of snow and ice melt and resultant runoff from glacial margins (Sicart et al. 2005) with downstream impacts on streamflow, hydropower generation potential and irrigation.

In these contexts, climatic warming represents introduced pressure upon a highly variable system with hypersensitive ecological responses that strongly affect highland dwellers. Adaptive responses are already evident in national-level actions, such as concerted efforts to buffer highland water supplies through a network of reservoirs in order to artificially enhance dry season flows, and also shifting from electricity generation from glacier-fed hydroelectric plants, a strategy in place for more than half a century, to other sources. More local adaptive responses include cultivation being practiced at ever increasing elevations as the warming climate brings a growing season to highland zones hundreds of meters above the apparent limit of the recent past (Hole et al. 2010).

Uganda–Rwanda Case

In the Greater Virunga Landscape of the African equatorial tropics of southwestern Uganda, neighboring northwest Rwanda and the eastern Democratic Republic of Congo, is a complex landscape of densely settled agricultural regions abutting highly biodiverse protected mountain forests. These forests house the totality of remaining populations of Earth's largest primate, the critically endangered Mountain Gorilla (*Gorilla gorilla beringei*). Five decades of ecological research on the gorilla and its environment have produced comprehensive understanding of the species and its native habitat, and also of the intense human pressure and habitat degradation occurring across the mountain gorilla's range related to pervasive poverty of the region's human populations. Conservation efforts to save this species that currently number around 750 individuals are closely tied to poverty alleviation initiatives in the densely settled landscape surrounding the national parks of the tri-national border, where the remnant gorilla groups are clustered.

In northwestern Rwanda and adjacent southwestern Uganda, the understanding gained now underpins what is widely perceived as a successful development response where high-priced tourism provides benefits and employment opportunities in the local economies. Immediately outside the protected areas designated for conservation of wildlife, however, human population densities are among the highest in Africa, averaging 300–600 km^2 in Rwanda. The dense human population imparts great pressure upon natural resources: available land for cultivation, animal protein, and timber. Within this domain as elsewhere, climatic variability is a significant influence on outcomes to wildlife and human interests alike. The dire conservation predicament of the gorillas, constraining them to sharply delineated protected forests abutting intensely settled farmlands, yields a high environmental sensitivity to climatic stress. Any reduction in farm output increases pressure for illegal exploitation of protein (i.e. bushmeat), both for subsistence and for income, from protected forest resources. Similarly, climate anomalies such as droughts can yield rapid environmental responses such as fire outbreaks in other moist forests (Fig. 21.2), highlighting the susceptibility of this region to climatic variability in the present day, and especially, the adverse impacts of climate change in years to come (Seimon and Picton-Phillipps 2010).

Given the high degree of development interest and conservation planning focused on obtaining successful outcomes for both wildlife and people in this local region of tropical Africa, it is surprising to find that understanding of the area's climatology is very poorly developed. In common with much of equatorial Africa, the annual climate cycle here is defined by twin wet seasons separated by drier periods. At regional to local scales, this characteristic rainfall pattern is clearly evident on monthly pluviograms; such data typically comprise the climatological reference utilized by researchers, agricultural interests, and others concerned with environment and development. A classic representation of this is demonstrated in a conventional pluviogram showing monthly means developed from daily rainfall records from Bwindi National Park in Uganda (Fig. 21.3, upper panel) (Seimon and Picton Phillipps 2010).

However, pluviograms of the same data at higher temporal (daily and weekly) resolution reveal a remarkably different rainfall climatology characterized by robust intra-seasonal variability to precipitation, whereby each rainy season is revealed to be interrupted by intense maxima flanked by temporary minima (Fig. 21.3, lower panel). The exceptionally large-magnitude fluctuations in rainfall rate centred in early May and September are strongly evident in the daily and 7-day smoothed data,

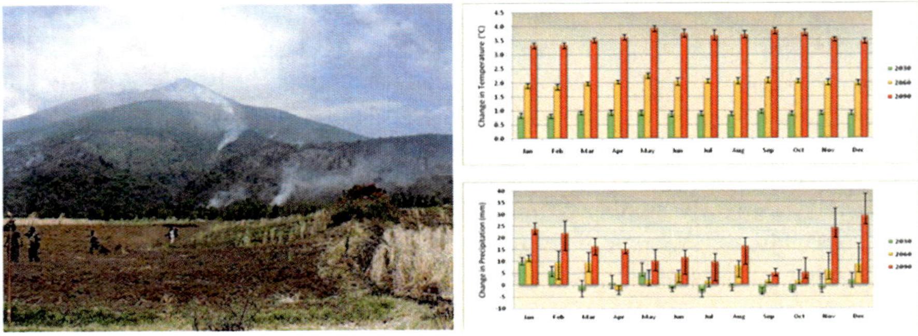

Fig. 21.2 (*left*) Fires within the boundary of Volcanoes National Park, Rwanda viewed from agricultural lands below during exceptionally dry conditions associated with a major East African drought in July 2009 (Photo courtesy of James Kemsey, International Gorilla Conservation Program). (*right*) Projected monthly mean temperature (*upper*) and precipitation (*lower*; in mm per month) changes across a geographic domain encompassing Volcanoes National Park under the IPCC "A2" emissions scenario, relative to baseline conditions set at the year 1990 for 2030 (*green*), 2060 (*yellow*), and 2090 (*red*). The range of values in the domain for each bar is depicted by the thin lines. In these projections strong temperature increases under anthropogenic greenhouse gas-induced climatic warming will greatly increase evaporative losses from this moist tropical forest. In the latter of the twenty-first century these losses will be largely offset by comparatively strong increase in precipitation; however, closer to the present, in 2030 temperature gains in the June–August dry season coincide with precipitation reductions, indicative of the building potential for increasing stress and propensity for fire in this most tropical forest (Source for modeling output: Picton Phillips and Seimon 2010)

but entirely masked by averaging in the monthly means. Such signals are likely of considerable significance to local ecology, yet would remain invisible and undetected using conventional climatological analysis. Agricultural concerns might look to the arrival of pollinators that have grown accustomed to the climatic triggering of flowering phenologies associated with the time-specific patterns during these periods. Public health concerns could monitor for short-term positive or negative trends in climatically influenced diseases outbreaks such as malaria and Rift Valley fever, along with other pests and pathogens.

Fig. 21.3 (continued) a geographic domain encompassing Volcanoes National Park under the IPCC "A2" emissions scenario, relative to baseline conditions set at the year 1990 (*blue*), in 2030 (*green*), 2060 (*yellow*), and 2090 (*red*). The range of values in the domain for each bar is depicted by the thin lines. The data correspond with rainfall projections shown in Fig. 21.2. At this temporal scale the model output appears to represent the Bwindi pluvial data very well. (*bottom*) Pluviogram showing rainfall climatology for Bwindi at monthly and daily resolution for the period 1991–2006. The monthly data are the same as shown in the top panel. The higher resolution data show highly pronounced climatological behavior at sub-monthly scales that is no longer apparent when aggregated into monthly means according to convention. The sub-monthly information is of potentially high significance to ecological systems and human interests, yet is effectively unrecognized. Bwindi data courtesy of the International Tropical Forestry Center, reproduced from Seimon and Picton Phillipps (2010)

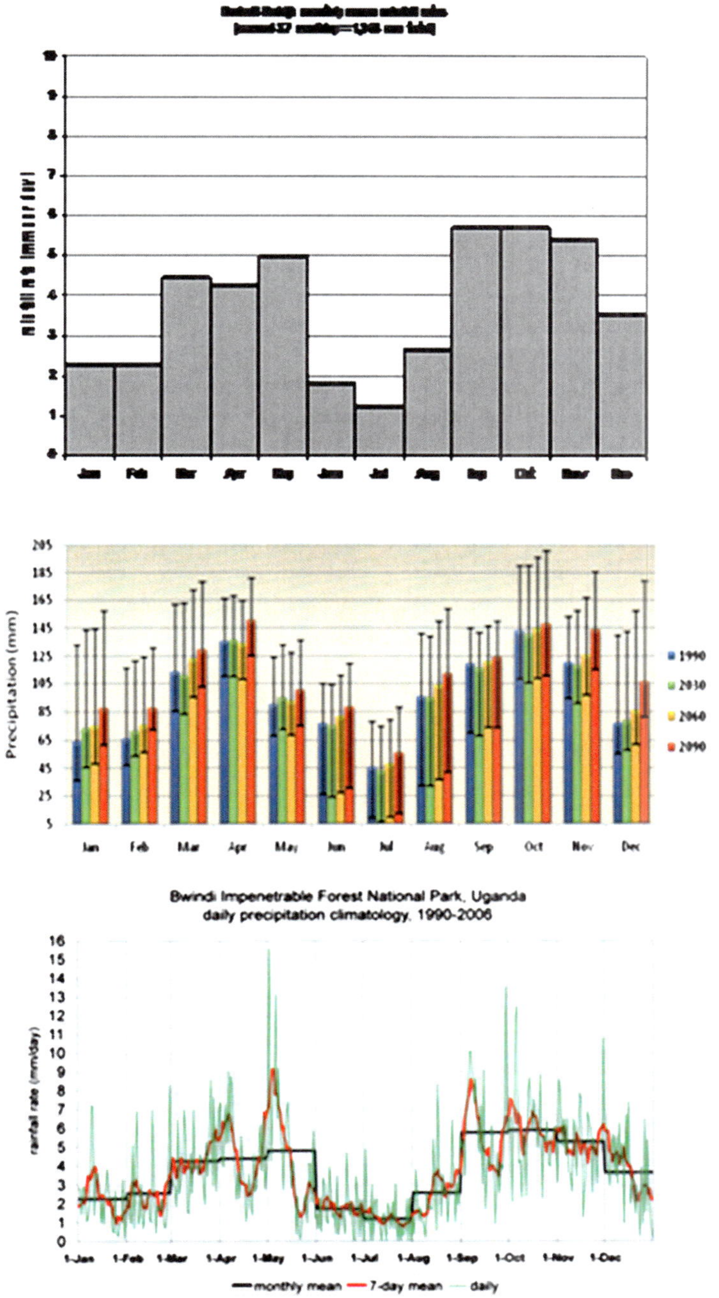

Fig. 21.3 (*top*) Conventional pluviogram showing monthly mean rainfall rates (mm/day) based on data from 1990 to 2006 at Ruhija in the Bwindi Impenetrable Forest National Park, Uganda. (*middle*) Projected monthly mean precipitation (mm) across the Greater Virunga Landscape,

Much as shown from the Andean case, the full ecological significance of the distinct climatic patterns that characterize present day climatology in the Greater Virunga Landscape has yet to be determined. However, their identification is also an invitation to reconsider climate change in contexts of shorter term phenomena than generally considered. For example, in the Cusco case, can climate models offer meaningful guidance on durational changes in the annual wet and dry seasons; will the pluvial peak late in January shift earlier or later; will the amplitude of peaks change, and what is the future of the rapid rise period of late December when daily precipitation rate effectively doubles over a 5-day period? (see Jiménez et al. 2010). Regarding thermal variability, will ENSO-related temperature anomalies follow similar patterns to those shown in Fig. 21.1 but from higher baselines yielded by greenhouse gas warming, or will the nature of ENSO's local imprint change dynamically in the radically perturbed atmosphere of the future? If the latter, what aspects of the temperature parameters will undergo shifts, and to what consequence? These are not merely minor details: such questions address climatic factors that strongly influence a wide variety of environmental and socioeconomic outcomes and efforts to adapt to climate change.

It is, therefore, desirable that climate modeling continues to be improved, and refined to a point whereby such sub-monthly climatological detail can be reproduced in output generated for contemporary climate. If so, it would be of great interest to examine the temporal evolution of distinct characteristics such as those exhibited at Bwindi into the model's depictions of the future. For example, in a rain-fed agricultural landscape, a shift by several weeks in the start and/or end date of the rainy season, and the timing of its pluvial peak, might prompt a shift in the assemblage of cultivars planted by farmers. Having foresight of such climatological changes could in time become integral in development strategies for that region. In contrast, climate model projections are currently presented mostly as changes in aggregated quantities; e.g. across region × climate models depict reductions in rainfall of 20% by year Y (see, e.g. regional assessments in IPCC 2007). This "area under the curve" summary statistic is prevalent, though it becomes a challenge to translate a single parameter value into ecological outcomes. But what an Andean farmer and Ugandan conservation manager and a myriad of other interests are probably much more concerned with is the *shape* of the curve and its *location* along the calendar timeline, as well as the net accumulation and cascading consequences, for which we have only just begun to build predictive skill. The IPCC model projections for the greater Virunga landscape, shown in Fig. 21.2, are informative in this regard, yet still very crude compared to what might be possible if climatology at daily resolution could be meaningfully projected for the future.

Conclusions

A comprehensive understanding of climatology increases the potential for proactive rather than reactive management responses to threats and opportunities borne by climate change. Modeling approaches and capabilities are powerful tools, and as such are having the inevitable consequence of shaping how we view and understand climate change impacts and threats. In these depictions of the future, climate is largely conflated into temperature and hydrological changes, and other parameters are only considered when specific objectives or concerns are tied to them. In contrast, improved understanding of present day climatology can improve comprehension of climate changes ongoing and those projected for the future. Fundamental to this is appreciation that important unknown and unrecognized elements of contemporary climate exist for almost any geographic context, and that such factors might prove to be of major significance and most prone to perturbation as a consequence of anthropogenically forced climate change.

Therefore, there exists considerable potential for site-specific climatological assessments to rectify incomplete or simplistic understanding of climate within development, and conservation contexts. This would, in turn, increase the capacity to anticipate and plan for climate change in truly meaningful ways. Interests concerned with poverty alleviation and development planning more generally could examine the basis for current knowledge of a locale's climatology by addressing the following:

1. Determine if the climatological context has been developed from site-specific assessments or from generalizations based upon broader scale studies. If assessments are available, are they perfunctory studies or conducted with diligence by persons with appropriate experience?
2. Evaluate the quantity and quality of the data used in these determinations. Site-specific observations may not always be available, in which case gridded analysis fields are often used as an alternative, but should be evaluated for their representation of actual climates at specific sites of interest.
3. When possible, inquire of those with local knowledge of the particularities of the climate from their perspectives, and consider how well this information matches that discussed in published reports and analyses.

To improve the level of understanding one might consider the following actions:

1. Data mining to obtain site-specific climate records. These are often available from both conventional climatological data archives (e.g. the Global Hydro Climatological Network (GHCN: http://www.ncdc.noaa.gov/oa/climate/ghcn-monthly/index.php); the Climate Research Unit at the Hadley Center (CRU http://www.cru.uea.ac.uk/cru/data/); and the International Research Institute for Climate and Society (IRI: http://iridl.ldeo.columbia.edu/), as well as from local observational sites and networks, which often contain additional records not found in the global archives.
2. Prioritize analysis of daily observations, rather than conventional summary statistics at monthly or greater intervals.
3. Include a climatologist or other specialist with requisite experience in working with "raw" climatological data to help develop comprehensive climatologies as inputs into development planning.
4. Tailor the products of climatological analysis as possible to address ecological and/or socioeconomic questions tied to the projects objectives. The same applies to climate model predictive outputs to make them more meaningful for development applications.

References

Adger, W.N., S. Dessai, M. Goulden, M. Hulme, I. Lorenzoni, D.R. Nelson, L.O. Naess, J. Wolf and A. Wreford. 2009. Are there social limits to adaptation to climate change? Climatic Change. **93**: 335–354.

Agrawala, S., editor. 2005. Bridge over Troubled Waters: Linking climate change and development. Organization for Economic Cooperation and Development. Paris, France.

David S. Battisti and Rosamond L. Naylor. 2009. Historical Warnings of Future Food Insecurity with Unprecedented Seasonal Heat. Science. **323**: 240–244.

Glantz, M.H. 1996. Currents of change: El Niño's impact on climate and society. Cambridge University Press, Cambridge, UK.

Halloy, S.R.P., A. Seimon, K. Yager and A. Tupayachi. 2005. Multi-dimensional (climatic, biodiversity, socioeconomic, and agricultural) context of changes in land use in the Vilcanota watershed, Peru. *In* E.M. Spehn, M. Liberman and C. Körner, editors. Land Use Changes and Mountain Biodiversity. CRC Press, Boca Raton, Florida, USA.

Hole, D.G., K.R. Young, A. Seimon, C. Gomez Wichtendahl, D. Hoffmann, K. Schutze Paez, S. Sanchez, D. Muchoney, H. R. Grau and E. Ramirez. 2010 *in press*. Adaptive management for biodiversity conservation under climate change – a tropical Andean perspective. *In* S.K. Herzog, R. Martínez, P.M. Jørgensen and H. Tiessen, editors. Climate change effects on the biodiversity of the tropical Andes: an assessment of the status of scientific knowledge. Inter-American Institute of Global Change Research (IAI) and Scientific Committee on Problems of the Environment (SCOPE), São José dos Campos, Brazil and Paris, France.

IPCC, 2007. Climate Change 2007. The Physical Science Basis. Contribution of Working Group I to the Fourth Assessment Report of the Intergovernmental Panel on Climate Change. S. Solomon, D. Qin, M. Manning, Z. Chen, M. Marquis, K.B. Avery, M. Tignor and H.L. Miller, editors. Cambridge University Press, Cambridge, UK.

IRI. 2005: Sustainable development in Africa: is the climate right? International Research Institute for Climate and Society, IRI technical report no. IRI-TR/05/1.

Jiménez, I. L. F. Aguirre, S. K. Herzog, G. H. Kattan, E. P. Anderson, G. Brehm, P. M. Jørgensen, M. Maldonado, J. L. Mena, R. Martínez, J. D. Pabon, A. Seimon and C. Toledo. 2010 *in press*. Phenology and interspecific ecological interactions of Andean biota in the face of climate change. *In* S.K. Herzog, R. Martínez, P.M. Jørgensen and H. Tiessen, editors. Climate change effects on the biodiversity of the tropical Andes: an assessment of the status of scientific knowledge. Inter-American Institute of Global Change Research (IAI) and Scientific Committee on Problems of the Environment (SCOPE), São José dos Campos, Brazil and Paris, France.

Knabb. R.D., J.R. Rhome and D.P. Brown. 2006. Tropical Cyclone Report, Hurricane Katrina, 23–30 August 2005. National Hurricane Center, Coral Gables, Florida, USA.

Lawler, J. J., S. L. Shafer, D. White, P. Kareiva, E. P. Maurer, A. R. Blaustein, and P. J. Bartlein. 2009. Projected climate-induced faunal change in the western hemisphere. Ecology. **90**: 588–597.

Lenton, T., H. Held, E. Kriegler, J. Hall, W. Lucht, S. Rahmstorf and H.J. Schellnhuber. 2008. Tipping elements in the Earth's climate system. Proceedings of the National Academy of Sciences. **105**: 1786–1793.

Lyon, B., and A.G. Barnston. 2005. ENSO and the Spatial Extent of Interannual Precipitation Extremes in Tropical Land Areas. Journal of Climate. **18**: 5095–5109.

McGranahan, G., D. Balk and B. Anderson. 2006. Low coastal zone settlements. Tiempo **59** (see also: http://sedac.ciesin.columbia.edu/gpw/lecz.jsp).

Myers, N. 2000. Biodiversity hotspots for conservation priorities. Nature. **403**: 853–858.

Noble, I. 2005. "Screening development projects for risks of climate change." Presented at the US Climate Change Science Program Workshop "Climate Science in Support of Decision Making." Arlington, 14–16 November 2005, www.climatescience.gov/workshop2005/presentations/EC1.6_Noble.ppt.

Parmesan, C. 2006. Ecological and evolutionary responses to recent climate change. Annual Review of Ecology Evolution and Systematics. **37**: 637–669.

Robine, J.-M., S. L. K. Cheung, S. Le Roy, H. Van Oyen, C. Griffiths, J.-P. Michel, and F. R. Herrmann. 2008. Death Toll Exceeded 70,000 in Europe during Summer of 2003. Comptes Rendus Biologies **331**: 171–78.

Sachs. J.D. 2000. Globalization and patterns of economic development. Review of World Economics. **136**: 579–600.

Seimon, T. A., A. Seimon, P. Daszak, S. Halloy, P. Sowell, B. Konecky, L.M. Schloegel, C.A. Aguilar, J. Simmons, and A. Hyatt. 2007. Upward range extension of Andean anurans to extreme elevations in response to tropical deglaciation. Global Change Biology. **13**: 288–299.

Seimon, A. and G. Picton Phillipps. 2010. Climatological Assessment of the Albertine Rift for Conservation Applications. WCS Albertine Rift Climate Assessment. Whitepaper report no. 1.

Sicart, J. E., P. Wagnon and P. Ribstein. 2005. Atmospheric controls of the heat balance of Zongo Glacier (16°S, Bolivia). Journal of Geophysical Research **110**: D12106.

Sperling, F., C. Validivia, R. Quiroz, R.Valdivia, L. Angulo, A. Seimon, and I. Noble. 2008. Transitioning to Climate Resilient Development: Perspectives from Communities in Peru. Environment Department Papers. Climate Change Series paper number 115. The World Bank, Washington DC, USA.

Vuille, M., R.S. Bradley, M. Werner and F. Keimig. 2003. 20th century climate change in the tropical Andes: observations and model results. Climatic Change. **59**: 75–99.

Vuille, M., B. Francou, P. Wagnon, I. Juen, G. Kaser, B.G. Mark, and R.S. Bradley. 2008. Climate change and tropical Andean glaciers – Past, present and future. Earth Science Reviews. **89**: 79–96.

Vuilleumier, F. and M. Monasterio. 1986. High Altitude Tropical Biogeography. Oxford University Press. New York, USA.

Williams, J.W., S.T. Jackson and J.E. Kutzbach. 2007. Projected distributions of novel and disappearing climates by 2100 AD. Proceedings of the National Academy of Science. **104**: 5738–5742.

Chapter 22
Incorporating Ecology and Natural Resource Management into Coastal Disaster Risk Reduction

Jane Carter Ingram and Bijan Khazai

Introduction

In the wake of major recent disasters such as the Indian Ocean tsunami in 2004 and Hurricane Katrina in 2005, international awareness and concern about hazards and the potential role of natural resources in reducing their impacts has grown considerably. Although natural *hazards* may not be stopped, the impacts of *disasters* often can be reduced and, in some cases, prevented. This chapter will address these issues by exploring if and how species and ecosystems may contribute to disaster risk reduction.

Economically poor or marginalized communities are often the most vulnerable to natural hazards through exposure, sensitivity to the event and the ability to recover (Kaplan et al. 2009). Impoverished communities often live at the interface of nature and society in weak, poorly designed structures, which may increase their vulnerability. In many cases, low income communities have no option but to develop settlements in hazard prone environments that are not ideal for human habitation (Maskrey 1989; Degg and Chester 2005). Additionally, many subsistence-based communities depend directly on ecosystem services for their livelihoods and, thus, any shock that alters the availability of or disrupts the accessibility to natural resources critical for survival may have devastating impacts. Furthermore, low income populations typically have few resources or assets available for recovery or rebuilding and, thus, may be driven deeper into poverty by a disaster. For these and many other reasons, disasters disproportionately impact the poor (Mutter 2005).

Ecological degradation is a key factor contributing to poverty traps (Sachs 2005) and has been observed to have amplified the severity of many recent disasters.

J.C. Ingram (✉)
Wildlife Conservation Society, 2300 Southern Boulevard, Bronx, New York 10460, USA
e-mail: cingram@wcs.org

J.C. Ingram et al. (eds.), *Integrating Ecology and Poverty Reduction: Ecological Dimensions*, DOI 10.1007/978-1-4419-0633-5_22,

For example, in the aftermath of the Indian Ocean tsunami in 2004, scientists and researchers observed that intact mangroves, coral reefs and sand dunes appeared to buffer the impacts of the waves throughout many of the affected countries. In a similar manner, following Hurricane Katrina in 2005, heightened attention was placed on the role wetlands might have played in buffering the event had they not been degraded (Fischetti 2005). The impacts of other recent disasters, such as Hurricane Mitch in Central America, Tropical Storm Jeanne in Haiti and flash flooding in the Philippines, to name but a few examples, may have been moderated had deforestation on hillsides been less severe (Hammill et al. 2005). However, it has been difficult to translate observations regarding the role of ecosystems in mitigating hazards into tangible disaster risk reduction and recovery policies, because many reports have been anecdotal. In cases where observations have been quantitative, they have often been met with skepticism and debate in the literature. Findings have also been contextually specific, making it difficult to transfer relationships between ecosystem condition and disaster mitigation services at one site to other sites.

Despite these challenges, the World Conference on Disaster Reduction, held in Kobe, Hyogo, Japan during January 2005, adopted the Hyogo Framework of Action (HFA) for the years 2005–2015, which underscored the need for, and identified ways of, building the resilience of nations and communities to disasters. Specific actions outlined in the HFA focused on the role of ecosystems in disaster risk reduction (UNISDR 2010). Priority areas for action included "Environmental and natural resource management" and "Land-use planning and other technical measures." One of the areas where such actions are especially needed is the coastal zone, where both populations and disasters are likely to increase in future years.

People are attracted to coastal zones due to the convergence of multiple, highly productive ecosystems – coral reefs, lagoons, seagrass beds, sand dunes, mangrove forests and other types of coastal vegetation – that are rich in biological resources and of high economic importance. Yet, the world's coastal zones are especially at risk of disasters due to increasing population densities living in a highly dynamic environment. Recent estimates suggest that approximately 41% of the world's population lives within 100 km of the coast, which represents only 20% of the planet's land surface (Martinez et al. 2007). It is estimated that that the number of people living in and around coastlines will increase by another 34% to 3.1 billion people by 2025 (Duxbury and Dickinson 2007). As population and development grow in coastal areas, natural hazards common to this dynamic zone may claim more lives and cause more damage. Of particular concern are poor small island states and coastal zones in developing countries where people are highly and directly dependent on coastal resources for their livelihoods, are physically vulnerable to hazards due to their proximity to the coasts, and may be less able to recover from shocks. Due to the high dependency of poor coastal populations on natural resources and the high likelihood of extreme events in this area, an ecosystem-based approach to disaster risk reduction may offer important, cost-effective opportunities to both conserve critical ecosystem services and reduce vulnerability to extreme events. However, coastal zones can be challenging to define (Duxbury and Dickinson 2007), meaning that implementing ecosystem-based management across various ministries and user-groups associated with coastal systems may be difficult. All of these challenges have served as the inspiration for this chapter, which will address if and how coastal hazards may be mitigated by ecological communities and through natural resource management.

This chapter discusses recent evidence for ecosystem-based disaster regulation services in the coastal zone; explores the challenges of integrating natural resource management into disaster risk reduction planning; and addresses a few ways in which ecological science may be leveraged more effectively to reduce the impacts of coastal hazards. To begin, this chapter will review recent research on the role of ecological systems, namely coastal vegetation, in mitigating against sudden onset disasters in the coastal zone, drawing heavily from research conducted after the Indian Ocean tsunami of 2004 as a primary case study, which revealed opportunities for and challenges of using natural resource management to reduce the impacts of natural hazards on the rural poor. In light of research findings, the second section of this chapter will address how the application of ecological science and tools might be leveraged towards coastal disaster risk reduction.

The Indian Ocean Tsunami of 2004

The tsunami that hit Indian Ocean countries on December 26, 2004 was the third largest, sudden onset disaster in recorded human history and the biggest to be caused by sea waves (Levy and Gopalakrishnan 2005). Approximately, 225,000 people were killed in the event (AusAID 2007). Following the tragedy, unprecedented research and debate ensued about the role of coastal ecosystems in mitigating the impacts of the tsunami. The results of studies conducted in the months and years following the Indian Ocean tsunami have suggested complex relationships between the condition of coastal ecosystems, such as mangroves, and protection from the wave(s). These relationships varied throughout the tsunami-affected zone since the capacity of an obstacle to withstand and buffer any impact depends heavily on the magnitude of the impact, the profile of the waves in this case, which were not uniform throughout the region. For example, in Aceh, where the tsunami force was overwhelming, the buffering role of coastal ecosystems was probably fairly negligible, while, in areas farther from the epicenter, the protective role of ecosystems may have been more important (Cochard et al. 2008). This section will review and explore the evidence generated from recent studies on the role of coastal ecosystems, particularly vegetation, in protecting people from the Indian Ocean tsunami of 2004.

What Evidence Exists on the Role of Ecosystems as a Natural Defense Against Tsunamis?

Coastal Vegetation

A variety of experimental studies have suggested that coastal ecosystems could play an important role in reducing the effects of coastal hazards by demonstrating the ability of littoral vegetation, namely mangroves, to reduce wave action. For example, Quartel et al. (2007) used field instrumentation to test current velocity and water level at an open tidal flat, at the beginning of mangrove vegetation and inside the mangrove stand. Their results showed that mangroves reduced wave height 5–7.5 times more effectively than non-vegetated beach plains, which they say clearly indicates the effectiveness of mangrove forests for buffering wave action. The dense network of trunks, branches and above ground roots of the mangrove vegetation create a high drag force, although, the degree of this force depends on mangrove species composition and the density of the stems. Massel (1999) also found that the structural features of a mangrove stand influence its ability to attenuate waves. They used numerical modeling and field observations in Australia and Japan to show that the rate of wave energy attenuation by mangroves was a function of the density of stems in the stand, the diameter of mangrove roots and trunks, and the spectral characteristics of the incident waves. Research also has shown that the wave buffering potential of mangroves can be comparable to that provided by man-made structures. For example, Harada et al. (2002) conducted a hydraulic experiment to study the tsunami reduction effect of coastal permeable structures using models of mangroves, a coastal forest, a wave dissipating block, a rock breakwater and houses. This work concluded that mangroves can be as effective as concrete seawall structures for reduction of tsunami effects on house damage. Similarly, more recently, Teh et al. (2009) used numerical and analytical modeling to show that mangroves can be effective at reducing wave heights and velocities. However, they state the degree of reduction depends on wave period, wavelength, and mangrove characteristics including forest width and density, as other researchers have reported.

While the results from these studies are compelling, field observations on the role of coastal vegetation, such as mangroves, in decreasing the impacts of the Indian Ocean tsunami of 2004 have been varied, with a paucity of quantitative studies on the protective services provided by coastal vegetation during the event (Khazai et al. 2007; Cochard et al. 2008). Even some of the more quantitative field studies on the role of mangroves and/or coastal vegetation in mitigating the tsunami have been questioned due to the statistics and analytical techniques used (see Dahdouh-Guebas et al. 2005; Kerr et al. 2006; Baird and Kerr 2008; Iverson and Prasad 2008 for critiques of approaches used and responses to critiques). Nevertheless, results from these studies have helped elucidate some of the possibilities and limitations of using ecological systems for disaster protection.

One of the first studies on these issues after the tsunami was conducted by Dahdouh-Guebas et al. (2005) who used a semi-quantitative assessment technique to assess the protective capacity of mangroves in relation to the tsunami. They conducted surveys in 24 mangrove lagoons and estuaries along the South-West, South and South-East coasts of Sri Lanka in January 2005. They found that mangroves did indeed afford protection in sites where they occurred, but the degree of ecological degradation of the mangroves was a critical factor influencing a mangrove stand's ability to protect human communities from the tsunami waves: mangrove species associated with degraded stands were found to offer less protective capacity than species found in more ecologically intact stands. This suggests that conservation of mangrove composition, as well as extent, is critical for retaining their protective capacity. In a similar study, Danielsen et al. (2005) assessed the role of vegetation in mitigating the impact of the tsunami along the affected coastline in Tamil Nadu, India. The results of their analysis of damage in areas with different vegetation densities suggested that mangroves helped decrease damage from the tsunami because the villages located behind the mangroves were much less damaged compared to villages not located behind mangroves. However, most of the villages located behind the mangroves were also a farther distance from the shore than villages that were not located behind mangroves, which the authors do not account for in their analysis. Danielsen et al. (2005) also showed that five villages located within *Casuarina* plantations experienced only partial damage, which the authors interpret to mean that *Casuarina* plantations provided considerable protection from the waves. The authors note a major caveat to their results (in the supplementary material and methods section): they do not know what the structure of the wave was like in this area or if it was completely uniform across the coastline, although they infer this by the relatively homogenous continental shelf proximate to the affected coastline. In a study on the protective role of vegetation in Sri Lanka, Kaplan et al. (2009) note the important of context and location on the landscape when considering wave reduction potential of coastal ecosystems. For example, they noted that a village located 1.7 km from the sea on an inlet and surrounded by *Rhizophora* species was severely damaged because the inlet acted to amplify the wave's forces allowing the wave to penetrate farther inland than otherwise would have been expected at that location.

Tanaka et al. (2007) conducted field surveys across 29 sites in Thailand and Sri Lanka on the effectiveness of different species of trees/coastal vegetation in mitigating against the tsunami. The field surveys included assessments of vegetation types, direction of the tsunami, and height of waves. They also calculated the drag co-efficients of each species recorded, which included: (a) *Casuarina equisetifolia*, a representative tree that grows in beach sand; (b) *Anacardium occidentale*, a plantation species found in the coastal zone; (c) *Cocos nucifera*, commonly known as the coconut tree and a plantation species found widely throughout the coastal zone; (d) *Avicennia alba* and *Avicennia marina*, hereafter *Avicennia* spp., a representative mangrove species found in small tidal zones; (e) *Pandanus odoratissimus*, a species that grows in beach sand; and (f) *Rhizophora apiculata*, a mangrove species that produces a dense network of aerial roots and *Rhizophora mucronata*, a mangrove species representative of large tidal zones. Their results of field observations combined with modeling show that the ability of vegetation to attenuate the tsunami

Fig. 22.1 Stand of *Causarina* trees approximately 20 m from the high tide mark on the coast of Arugam Bay, Sri Lanka in May 2005, 4 months after the tsunami hit the coastline. These young trees withstood the tsunami and there is no other vegetation community between this stand and the ocean, but its protective features are questionable. This structure was under construction at the time this picture was taken by the author, JCI

waves was a trade-off between stem diameter and spacing of trees, which is why mangroves can be effective defences (because of a high stem density in mangrove forest stands). As stem diameter gets larger, the individual tree becomes more resistant to wave forces, but with an increased diameter there is less room for a higher number of stems and, so, stem density decreases. This problem was particularly evident in *C. equisetifolia* stands: when the stand is young, there is a high density of small stems which makes it effective at attenuating wave forces, but the individual stems can break easily at this age because they are small; when the stems get bigger they are more resistant to the waves, but there is more space between each tree meaning that the stand's overall ability to attenuate waves is lower (*see* Fig. 22.1). Similarly, despite the abundance of *Cocos nucifera* throughout the tsunami affected zones and many coastal areas, too much space exists between each stem to reduce wave force significantly. The authors suggest that landscape planning that accounts for proper spacing of vegetation throughout the coastal zone, from seaward to inland areas, can maximize the potential benefits that different coastal tree species may offer with respect to tsunami protection, even if, individually, they cannot effectively attenuate wave forces. For example, a landscape structured with small- and large-diameter trees may buffer the impact of large waves because the densely populated small-diameter trees ($d > 0.1$ m) collectively could reduce the velocity of the tsunami current, while the large-diameter trees ($d > 0.3$ m) could trap the broken branches and man-made debris. The vertical structure of tree stands may also provide an effective soft landing for people carried away by large waves or surges or for climbing when waves are approaching. Thus, each species may play different roles in providing disaster protection services and their effectiveness at providing these services will be determined by abundance, density, demography, and ecology of the species in addition to their location on the landscape. An understanding of species' traits that are important for certain ecological functions, such as wave attenuation, could be used to inform landscape planning aimed at disaster risk reduction. Some species and communities may only exist in certain conditions, meaning many of these functions will be context specific. Thus, the challenges and opportunities of leveraging natural ecosystems and coastal landscapes to provide disaster protection services must be considered on a case by case basis and, of course, in relation to the prevailing hazard and its characteristics.

Due to the many social, physical, and ecological variables that influence the protective capacity of natural systems against disasters, it can be difficult to collect data and select analytical techniques that generate robust results on the relationships between vegetation and mortality for informing policy and/or land use planning. For example, in a fairly controversial study of 25-km of tsunami affected coastline in Tamil Nadu, India, Kathiresan and Rajendran (2005) collected information on distance from shore, elevation, area of mangroves/coastal vegetation, number of deaths and per capita loss of life in 18 hamlets. Their results showed a significant negative correlation between the human death toll and the distance of human habitation from the sea ($r^2=0.61$, $p<0.01$), the elevation from mean sea level ($r^2=0.63$, $p<0.01$) and the area of mangrove and other coastal vegetation ($r^2=0.58$, $p<0.01$). However, their results were challenged by Kerr et al. (2006) who reanalyzed their data and found that distance from sea and elevation together explained 87% of the variation of mortality in the area with vegetation contributing less than a 1% increase in the explanatory power. Similarly, they found that distance from sea accounted for 61% of the variation in wealth lost, with elevation and vegetation combined accounting for only 6.5% of the total variation. Vermaat and Thampanya (2006) then challenged the approach taken by Kerr et al. (2006) and reanalyzed the data to derive results that suggested that fewer lives and less property were lost from hamlets that were in the shelter of mangrove stands, even when they corrected for distance from the sea and elevation. Similarly, Iverson and Prasad (2007) built empirical vulnerability models of damage/no damage based on elevation, distance from shore, vegetation, and exposure and determined that forested zones could help reduce damage to an event like the tsunami. These findings were criticized by Baird and Kerr (2008) due to their failure to account for many variables important in determining the impact of the tsunami (also see reply by Iverson and Prasad 2008). Collectively, these and other studies have led to conflicting results on the role of coastal vegetation in mitigating against the tsunami, which can be related to the complex relationships among social, ecological, and physical factors at each site; unique characteristics of the wave(s) in a specific location; and the application of different methods and analytical techniques for testing relationships.

While these studies explore ways in which ecological systems may influence wave forces associated with a tsunami or other coastal extreme event, few of the available studies on this topic have made use of the vast amount of data that is now available for reconstructing, modeling, and interpreting the wave energy at each site and its impact on land. Future analyses using these data will be important for furthering understanding of important processes and the protective role of ecosystems (Cochard et al. 2008).

The Role of Other Coastal Ecosystems in Protecting Against the Tsunami

Although most of the focus following the Indian Ocean tsunami of 2004 was on the role of mangroves and other coastal vegetation types in protecting inland communities, other coastal features were also observed to have contributed to protecting coastal human populations from the wave(s), namely sand dunes, coral reefs and seagrass beds.

Sand dunes were thought to have played an important role in buffering against the wave(s) (Liu et al. 2005; Ingram et al. 2006; Mascarenhas and Jayakumar 2008), especially in places where sand dunes were tall in height and densely vegetated. In Yala National Park, for example, sea incursion by the tsunami occurred where dunes were deficient, as in lagoons or river outlets (Fernando et al. 2006) or in places where dunes had been removed. This finding was supported by observations of the authors Ingram and Khazai in Yala National Park, where one hotel, which had removed sand dunes for an unobstructed beach view was completely destroyed by the tsunami with an almost complete loss of life (at high occupancy) while another lodge, a few hundred meters away, was virtually undamaged due to the protective barriers of sand dunes that had been conserved (Fig. 22.2). Dunes were also observed to be inefficient at reducing the tsunami's force when located at the center point of an arc-shaped bay (authors' observations).

Following the Indian Ocean tsunami, it was recorded that communities located inland of coral reefs degraded from years of coral mining suffered higher damage and loss of life than communities located a short distance away on the same coastline, but sited inland of intact coral reefs (Fernando and McCulley 2005). Simulation experiments have also demonstrated the role of coral reefs in buffering tsunami wave action and have shown that their effectiveness is determined by the amplitude and wavelength of the incident tsunami; the geometry and health of the reef; and the offshore distance of the reef (Kunkel et al. 2006). However, in contrast, a review by the United Nations Environment Program (UNEP) World Conservation Monitoring Center (WCMC) review found little evidence that the presence of reefs reduced tsunami damage on shore. Some studies have even suggested that inundation was greater on coastlines with reefs than on those without, due to bathymetric factors and the way in which the tsunami can gain force as it approaches certain types of shorelines (Wells and Kapos 2006).

Chatenoux and Peduzzi (2007) used a Geographic Information System (GIS) modeling technique and multiple regressions to assess the effects of seagrass, coral reefs and mangroves for mitigating the impact of the tsunami in sites across

Fig. 22.2 Tsunami protection provided by sand dunes and the importance of landscape context in determining their functional performance. (**a**) Large, vegetated sand dunes surrounding a hotel, which was almost unaffected by the tsunami due to the protective dune system. (**b**) The site of a hotel located a few hundred meters from the hotel pictured in Fig. 22.2a, where dunes had been removed to create an unobstructed view of the ocean. Occupancy in the hotel was high and there was an almost complete loss of life. (**c**) Large, vegetated dunes that were breached and located at the center point of a bay ringed by sand dunes

Fig. 22.2 (continued)

Indonesia, Thailand, continental India, Sri Lanka and Maldives. Their results indicate that the width of the flooded plain (the proxy for damage in this study) was strongly correlated with distance from the subduction fault line; near-shore geomorphology and length of proximal slope; percentage of coral; and percentage of seagrass beds. These factors accounted for 65.5% of the variance in the width of the flood plain at 56 sites with a significance of $p<0.05$. The authors hypothesize that seagrass may have helped reduce the tsunami impact by mechanical influences that attenuate the wave; however, they state that the results could also be an artifact of the distribution of seagrass beds that may only occur in areas where wave energy is naturally lower. The researchers also found that damage was higher behind coral reefs because areas where coral was growing were in shallow waters with small slopes, two conditions leading to higher waves. Although few doubts exist regarding the positive role coral reefs play in providing coastal protection from typical waves, caution should be kept, the authors warn, when rebuilding facilities inland of coral reefs if future tsunami risks are high.

The Role of Ecological Systems in Protecting Against Other Coastal Hazards

Coastal zones are naturally dynamic and prone to a variety of hazards, thus, the role and management of coastal ecosystems with respect to one type of hazard must be considered alongside the many other hazards characteristic of this eco-tone. In fact, the role of mangroves in attenuating waves associated with cyclones and hurricanes is thought to be as, if not more, important than their role in attenuating waves from tsunamis due to the higher frequency of weather related coastal hazards. For example, Blasco et al. (1992) used satellite imagery to assess the impacts of flooding resulting from cyclonic activity in Bangladesh in the deltaic complex of the Ganges. In mangrove forests, there was no sign of destructive effects by floods or winds, which the authors interpret to mean that these tidal forests are adapted for annual cyclones. The authors also note that a coastal afforestation programme carried out in the intertidal zone 15 years before the event, created a safety belt of trees, sufficiently thick to protect embankments facing open sea, which were undamaged by the storm. Similarly, Das and Vincent (2009) used data on mortality from hundreds of villages following a massive cyclone that hit the Orissa coast in 1999 to assess the protective functions of mangroves. They concluded that the width of mangrove stands was a powerful predictor of mortality, with fewer deaths recorded in villages located behind wider stands, but mangroves were no substitute for the lives saved from early warning systems. In India and the Philippines, fishermen have also recognized the importance of intact mangroves in protecting against coastal hazards such as cyclones and flooding. In a study on the value of mangroves to a fishing community in the Philippines, more than 90% of fishermen surveyed, regardless of where they fished, thought that mangroves provided protection from storms and typhoons, acted as a nursery site and should be protected. Those people fishing only in the mangrove forests perceived more benefits from these habitats and were prepared to pay more to protect them than those fishing outside of them (Walton et al. 2006). For these reasons, in some cases, mangroves may be more beneficial than built coastal defence systems because they also offer provisioning services, such as fish and fuel wood, as well as regulating services, such as storm protection, whereas built systems only provide the latter (Tri et al. 1998). In addition, they may be more capable of adapting to dynamic conditions, such as sea level rise associated with climate change, than static, engineered defences. Several studies have estimated that the economic value of the coast-line protection services provided by mangroves and coral reefs can be quite high. In an analysis of the economic value of storm protection services provided by coastal ecosystems in Belize, the World Resources Institute estimated that mangroves contribute approximately US$111–167 million/year in avoided damages while coral reefs contribute approximately US$120–180 million in avoided damages (Cooper et al. 2009). In a similar study in Tobago and St. Lucia, Burke et al. (2008) estimated that the annual value of shoreline protection services in avoided damages provided by coral reefs is approximately US$18–33 million for Tobago and US$28–50 million for St. Lucia. They estimate that coral reefs provide 20–40% of shoreline stability in the places where they occur, although they state the degree of protective services offered by coral reefs are a function of coastal context and determined by factors such as elevation and slope of the shore, the geologic origin of the area (and resistance to erosion) and the wave energy along the coast (Burke et al. 2008).

Wetlands also play important roles in buffering the effects of coastal hazards such as hurricanes. Much attention was given to the protective services provided by wetlands following Hurricane Katrina in 2005 (Fischetti 2005). Results from an analysis of data on hurricane damage since 1980, demonstrated that coastal wetlands in the United States provided an estimated US$23.2 billon/year in storm protection services (Costanza et al. 2008).

Intact natural forests located inland of the coastal zone may also play important roles in protecting coast-lines from storms by moderating soil filtration rates and, thus, preventing runoff, erosion and flooding associated with cyclones and hurricanes (Hammill et al. 2005). For example, after Hurricane Jeane hit the island of Hispaniola in 2004, the island's two countries experienced significantly different death tolls: in Haiti approximately 2,700 people were killed and in the Dominican Republic less than 20 people were killed. These stark contrasts have been attributed to Haiti's extensive deforestation, which has left less than 2% of the island's land area under forest cover. In comparison, the Dominican Republic retains forest on approximately 28% of its land area (Peduzzi 2005).

Opportunities and Challenges for Incorporating Ecology into Coastal Disaster Reduction

Recent disasters such as the Indian Ocean tsunami of 2004 have provided tragic, yet important, opportunities to explore the disaster regulation services provided by coastal ecosystems. While multiple analyses have been conducted on the role that ecosystems played in reducing the impacts of the Indian Ocean tsunami, many reports have been qualitative or semi-quantitative and, when quantitative, have often been based upon experimental simulations or cannot be easily extrapolated to other sites. The difficulty in producing robust analyses of disaster regulating ecosystem services is largely due to the complexity of biotic and abiotic factors that comprise the coastal zone and interact to determine an ecosystem's ability to resist the impact of an extreme event.

From previous studies, it is clear that implementing natural resource management activities to help reduce disasters and achieve the HFA can be far from straightforward: relationships between natural features and disaster reduction are not universal, as demonstrated from research after the tsunami. Nevertheless, research frameworks proposed for studying ecosystem services and many existent ecological tools could be helpful for achieving the goals proposed by the HFA framework and furthering our knowledge on how ecosystems can be managed to protect people against prevailing hazards, while providing other critical ecosystem services.

Developing a Common Language Between Ecologists and Disaster Specialists

For ecologists to effectively inform hazard management and planning, it will be important to establish common approaches for looking at the various dimensions of complex systems. In this regard, it may be helpful to examine disaster regulation functions of ecosystems through the lens of well established hazard risk assessment nomenclature and methods. Accordingly, the functional performance of ecosystems in regulating disaster risk could be defined as a function of the hazard, exposure and resilience (or vulnerability), where each part has to be considered by itself, as well as, related to the other. This could be expressed as:

$$\text{Ecosystem functional performance} = f\left(\text{hazard, exposure, resilience}\right)$$

In reviewing the role of ecological systems in regulating disasters, it is important that their performance be related to the characteristics of the prevailing hazard. Specifically, it is useful to look at hazards in terms of three components: the spatial, dimensional and temporal components. The spatial characteristics of the hazard determine the geographic area which is potentially threatened. The dimensional attributes of the hazard provide information on the intensity and magnitude of the hazard. The spatial and dimensional characteristics of a hazard at a particular locality are heavily controlled by local site conditions such as backshore and offshore topography, geometry, sediment supply, relative sea-level and ecological characteristics, all of which vary between coasts and even along adjacent sections of one coastline. Furthermore, the spatial and dimensional attributes must be considered for *each* hazard type as each hazard will have different characteristics that will be relevant for disaster risk reduction. For example, the base physical processes governing tsunamis, tropical cyclones and storm surges are very different: ordinary storm waves or swells break and dissipate most of their energy in a surf zone, while tsunamis break at shore. Finally, the temporal description of the hazard expresses when and how often future hazards are to be expected. This characteristic of the hazard can be expressed as the recurrence period, which are intervals (or cycles) of recurrence between hazard events that demonstrate the frequency and the probability of occurrence of the hazard. When evaluating the performance of any coastal protection system, engineered or natural, it is critical that the spatial and dimensional hazard parameters are related to the recurrence period of the hazard. Exposure refers to the degree to which an entity is in contact with the hazard (i.e. distance from hazard). Exposure should not be viewed as a static property as it can change throughout the day, seasonally and over longer periods of time.

Resilience has been defined as the ability of a system to absorb shocks, to avoid crossing a threshold into an alternate and possibly irreversible new state, and to regenerate after disturbance (Resilience Alliance 2010). Walker (2009) has identified two kinds of resilience that may be helpful for understanding how to apply these concepts to ecosystem-based hazard management: "specified" resilience deals with the resilience "of what, to what" (e.g., the resilience of crops to a drought); "general" resilience considers resilience of the system as a whole, but does not consider any particular kind of shock, or any particular aspect of the system that might be affected. It is important that both be considered in parallel for natural resource management aimed at reducing poverty, so that enhancement of one ecosystem service (i.e. storm protection) does not compromise the provisioning of other key ecosystem services (i.e. food production). However, the components that comprise resilience for a particular system must be better understood in relation to the intensity of the prevailing hazard and the level of exposure of the system. As discussed in the previous sections, this information is not always readily available and, in many cases, the way in which post-disaster data are collected and analyzed may not be consistent. Consistent historic post-damage databases are invaluable for constructing the needed empirical relationships that provide information on if and how ecological systems can contribute to disaster risk reduction. The establishment of internationally recognized standard procedures and guidelines for collecting post-disaster ecological data in the field could be very helpful in this regard.

Moving Forward: Relevant Ecological Tools, Principles, and Theories for Informing Disaster Risk Reduction

Generally, ecological knowledge and approaches have been under utilized for explaining the role that natural features and components may play in mitigating against single and multiple hazards, yet, multiple researchers have outlined priorities for deepening understanding of other ecosystem services and how to manage them (Kremen 2005; Kremen and Ostfeld 2005; Bennett et al. 2009). Much of this work has focused on ecosystem services such as pollination, regulation of water quality, and climate regulation through carbon sequestration and storage. This section will draw upon these frameworks and research and will explore how they can be adapted to disaster regulation services.

Understanding the Ecological Components That Contribute to Hazard Mitigation

Even for some of the most studied ecosystem services, a paucity of information exists on the role of biodiversity in supporting their provisioning in real world contexts or at scales of relevance to ecosystem management (Kremen 2005; Kremen and Ostfeld 2005; Balvanera et al. 2005). Thus, it is not surprising that comparatively little is known about the role of biodiversity in providing disaster regulation services, although, recent events have drawn more attention to the need for this information and have resulted in increased research on this issue, as described in the previous sections.

Research on the relationship between biodiversity and ecosystem functioning (BEF) has identified ways in which biodiversity influences ecosystem functions (Naeem 2002; Naeem and Wright 2003; Loreau 2010). Studies on BEF have helped explain how ecological functions are influenced by species richness, biomass, abundance, and density of species (Kremen 2005). This type of research has already been useful for illustrating the ecological attributes that are important for providing disaster regulation functions. For example, species diversity acting in combination with other ecological attributes, such as density and stem size, has been shown to be important for determining the effectiveness of mangrove forests at attenuating waves (Massel 1999).

Conceptual frameworks for demonstrating how biodiversity influences ecological functions and the provisioning of ecosystem services have been proposed by Luck et al. (2003), Kremen (2005), Kremen and Ostfeld (2005) and Luck et al. (2009). This body of work outlines ways to identify how ecosystem services are provided and the conditions under which they are most efficient to better inform ecosystem-based management. One approach is to identify ecosystem service providers (ESPs) (Kremen 2005) or service providing units (SPUs) (Luck et al. 2003), which have more recently been united into the Service Provider concept (Luck et al. 2009). The appropriate ecological level for defining SPs is service-dependent and might be at the level of species, populations, habitats, or ecosystems. In general, the functional importance of each SP will depend both on its effectiveness at providing the service (efficiency) and its abundance (Balvanera et al. 2005; Kremen 2005). Functional contributions of SPs have been measured or estimated for pollination, bioturbation, dung burial, water flow regulation, carbon sequestration, leaf decomposition, and disease dilution, but this analysis has not been done widely for disaster regulation services. Drawing from the classification of well-researched examples, Table 22.1 presents a preliminary classification of SPs that might be helpful for identifying how biodiversity is related to ecological functions (wave attenuation in this case) that contribute to disaster regulation services.

Theories and tools emerging from BEF and Ecosystem Service research will be helpful for answering critical questions pertaining to ecosystem-based disaster reduction strategies such as: which species or landscape components provide functions that protect people from prevailing hazards? In what abundances, densities and/or what degrees of biomass are needed for these components to provide these functions and how do those vary throughout space and time? Are the SPs similar across hazards with different spatial, dimensional and temporal attributes? If not, what composition of species and/or landscape features should be conserved to optimize resilience to a range of potential hazards?

Assessing the Spatial and Temporal Distribution of Disaster Regulation Services

Increasingly, ecologists are modeling the spatial and temporal distribution of ecosystem services across landscapes, including disaster regulation services, such as flood mitigation (for example, see the Natural Capital Project 2010). Berger and Rey (2004) map the disaster protection services provided by mountain forests in France and suggest these services should be better integrated into zoning laws and policies because the protective role of forests is rarely considered in hazard risk mapping.

Table 22.1 Coastal disaster regulation ecosystem service providers (ESPs) of importance for Tsunami protection (Adapted from frameworks developed by Kremen (2005) and Luck et al. (2009))

Ecosystem service (s)	Function(s)	Ecosystem service provider	Functional trait (s)	Functional unit	Spatial scale	Service provider characteristics	Supporting element	References (not exhaustive)
Tsunami protection; Storm protection	Wave attenuation	Mangrove trees	Branching, aerial roots; Strong stems	Habitat	Local	Spatial extent of mangrove stand, stem density and species composition/degradation status of stand	Topography of coastal shelf and shoreline; Location in landscape	Massel (1999), Harada et al. (2002), Dahdouh-Guebas et al. (2005), Quartel et al. (2007), Kaplan et al. (2009), and Das and Vincent (2009)
Tsunami protection; Storm protection	Wave attenuation	Seagrass	Leaf area index	Habitat	Local	Species composition and density	May vary temporally-seasonally and daily; Depth of seafloor	Koch et al. (2009) and Chatenoux and Peduzzi (2007)
Tsunami protection (sometimes); Storm protection	Wave attenuation	Coral reefs	Geometry: Flat (no branches)	Habitat	Local	Health of reef/condition and location	Depth of coastal floor and distance from shore	Fernando and McCulley (2005), Kunkel et al. (2006), but see Chatenoux and Peduzzi (2007), Wells and Kapos (2006), Cooper et al. (2009), and Burke et al. (2008)
Tsunami protection	Wave attenuation	Casuarina trees	Resistant to breaking; drag co-efficient	Habitat	Local	Age of stand and stem density	Location in landscape	Tanaka et al. (2007) and Kaplan et al. (2009)
Tsunami protection	Wave attenuation	Sand dune	N/A	Habitat-community	Local	Height, width, and location (not at the center point the bay)	Vegetation to stabilize sands	Liu et al. (2005) and Mascarenhas and Jayakumar (2008)

Note: "sand dunes" have been included for illustrative purposes even though they are not a species

Conducting this kind of work for a variety of hazards will require landscape level assessments that identify the mechanisms by which ecosystems in the coastal zone provide disaster regulation functions (as outlined in the previous section); how these functions are related to the spatial, dimensional and temporal attributes of the hazard; and how geometry and spatial configuration of land and sea scape components influence the disaster regulating functions provided by ecosystems. The latter is critical because context may strongly influence the strength of relationships between ecological components and disaster regulation services that may seem robust in one location, but may not be in another. For example, Kaplan et al. (2009) found that bays and estuaries acted to amplify the strength of the tsunami waves such that houses located far from the coast and protected by forests, which under normal conditions may have offered protection to the waves, were still damaged if they were close to the edge of an inland estuary or bay. Thus, it is important to acknowledge that the provisioning of disaster regulation services provided by an ecological community is dependent upon multiple factors including location in the landscape. It is also critical to understand the temporal provisioning of disaster regulation services. Since ecosystems are inherently dynamic, the provisioning of ecosystem services may be variable throughout time (Koch et al. 2009). Thus, it is important to predict which periods of the day, month and/or year when disaster regulation services provided by ecosystems are lowest and, thus, when disaster risks are highest. Temporal assessments of ecosystem service provisioning combined with spatial mapping of coastal disaster regulation services could be helpful for creating a dynamic model of coastal zone vulnerability that could be analyzed in relation to the spatial, dimensionsal and temporal attributes of a hazard.

An understanding of SPs that comprise different ecosystem services and models of how multiple ecosystem services are distributed temporally and spatially across a landscape may help guide management aimed at fostering general resilience of the system by revealing ecosystem service synergies or trade-offs associated with different natural resource management practices. Tools that are being developed such as InVEST (Natural Capital Project 2010), which can model and map the delivery, distribution and economic value of ecosystem services, provide useful resources for doing this. Such tools may be helpful when managing a coastal-scape to provide multiple ecosystem services, such as biodiversity, food, and disaster risk reduction. Since some land uses, such as biodiversity for conservation and disaster risk reduction, may require large, ecologically intact areas, management efforts and resources for those two goals could be combined to be more cost-effective. In contrast, land use for shrimp farming requires considerably less space (Barbier et al. 2008), but often results in habitat destruction, indicating it should be placed in a part of the landscape that does not compromise biodiversity or disaster risk reduction services. Information on the spatial and temporal distribution of disaster risk reduction services may help inform more strategic placement of competing land uses, such as fish farms, so that high hazard risk spots in the landscape are not made more risky by loss of important habitat.

Managing and Monitoring Dynamic Disaster Regulation Services

Once species have been identified with respect to their functional traits (Naeem and Wright 2003; Balvanera et al. 2005; Petchey and Gaston 2007), new opportunities may arise for monitoring and management. This might include classifying species by traits that are relevant for wave attenuation such as high leaf area index (which can create drag forces) or soil stabilization potential (such as root structure that helps stabilize soil). Trait based classification of species in combination with data on abundance, biomass or density, for example, can be used to calculate functional diversity metrics that permit assessments of how functional differences among species are distributed throughout a community (Symstad et al. 2003; Diaz et al. 2007; Walker et al. 2008; Griffin et al. 2009). A corresponding feature of communities that is also useful in this context is the measure of "functional redundancy," which indicates the degree of overlap in species that provide a certain function within a community (Petchey and Gaston 2007). Low redundancy of species with traits that provide important functions related to disaster regulation at a specific site, for example, may indicate that this area could be more vulnerable, if the species that provide those functions decrease in abundance, biomass or density. Values of functional diversity and redundancy could be mapped to assess the aggregate disaster regulation functions provided by a coastal land/sea-scape, to identify how vulnerable certain parts of the land/sea-scape may be to losing important disaster regulation functions (as indicated by redundancy), and, thus, where conservation and development activities should be focused across a land/sea scape.

In general, both efficiencies and abundances of SPs may vary as a result of changes in resources, predators, competitors and mutualists, as well as, responding to changing physical or biophysical parameters (Kremen 2005); they are by no means static. This is especially true in the dynamic coastal zone. Thus, managing ecosystem-based disaster regulation services will require monitoring the resilience of a system, to ensure key ecosystem functions are not lost, rather than management aimed at maintaining a static state, as would be the goal of maintaining built infrastructure aimed at disaster reduction.

Monitoring and management of disaster regulation services will require an understanding of not only the natural dynamism of coastal ecosystems, but also of how they respond to external pressures. Yet, different observations have emerged with respect to how biodiversity and ecosystem functions respond to change making it difficult to preemptively manage declines in ecosystem services resulting from external stressors. In some cases, biodiversity may offer stability to disturbance through the portfolio (Tilman et al. 1998) or insurance effect (Naeem 1998), but in other cases this pattern may not hold (Balvanera et al. 2005; Loreau 2010). While these patterns may differ across contexts, ecosystem services are unlikely to change in a linear fashion in response to an increasing pressure or a stressor. Barbier et al. (2008) have demonstrated non-linear shifts in the provisioning of several coastal ecosystem services, including wave attenuation as a result of habitat loss.

Furthermore, different ecosystem services respond differently to change, which has important implications for management charged with providing multiple ecosystem services. Thus, more research is needed on how individual ecosystem services respond to change and how these changes impact and are impacted by changes in other ecosystem services.

Designing management and monitoring plans for ecosystem-based disaster regulation functions will require baseline information on coastal ecosystem services. This information can be used to develop targets for management that indicate when critical disaster regulating ecosystem services are declining. Setting targets and monitoring for management that will help optimize the resilience of a system will require a sound understanding of the ecological components that comprise disaster regulation services and an understanding of the baseline spatial and temporal distribution of these components in relation to hazard risk. Perhaps, because of gaps in knowledge on SPs critical for disaster regulation services and the complexities of how they work with respect to abiotic factors (such as topography and landscape context), few monitoring programs have focused on measuring ecologically provided disaster services; rather, monitoring and management have typically focused on provisioning or cultural services (Carpenter and Folke 2006). In the midst of uncertainty about the nature of a range of ecosystem services and their relationships with one another, maintaining regulating services, such as disaster regulation services, may be a key precautionary way to build "general resilience" in a system: research suggests that declines in regulating ecosystem services can result in declines in overall ecosystem resilience, even when there are not substantial or apparent reductions in other ecosystem services (Bennett et al. 2005). However, since very little is known about the relationship among multiple ecosystem services or interactions between them (Bennett et al. 2009), in parallel to efforts focused on better understanding and conserving disaster regulation services, general research on the relationships among various coastal ecosystem services is also of critical importance for informing natural resource management aimed at reducing overall vulnerability of coastal communities.

Conclusions

The Framework for Action that came out of Hyogo called for natural resource management and land-use planning to play an important role in building resilience to disasters, which represents clear opportunities for ecologists. Hazard prone coastal zones represent ideal places to adapt established hazard risk assessment methods to account for the protective capacity offered by coastal ecosystems. Much work on this topic proliferated after the Indian Ocean tsunami in 2004, but due to conflicting results and a lack of quantitative studies across different sites, a paucity of information exists for guiding decision makers on the role and management of ecosystems for disaster risk reduction. We must not wait for more major disasters to refocus attention and catalyze further work on this topic in hazard prone areas, many of which have already been mapped extensively (Dilley et al. 2005).

To contribute towards these challenges, ecologists will need to deliver relevant scientific knowledge on how ecosystems function and change, how they are linked to human well-being and how humankind can use them in a sustainable way (Loreau 2010). Well-developed tools such as Geographic Information Systems and satellite remote sensing can be applied in new ways to map and model the distribution and economic value of ecosystem services. Additionally, research on BEF and ecosystem service dynamics could help foster a better understanding of the ecology of the disaster regulation functions provided by nature and inform ecosystem-based management that aims to maintain or restore the resilience of coastal ecosystems. However, it is important to not focus on maximizing the specific resilience of one ecosystem service (such as wave attenuation) at the expense of general resilience of the ecosystem, which may require a wide range of ecological functions. This is especially true in the coastal zone, where people depend directly on coastal habitats for many ecosystem services that support their livelihoods and daily needs such as food, shelter, construction materials, recreation, and cultural purposes.

Thus, minimizing overall vulnerability of poor communities in hazard prone coastal areas will require not only providing physical protection against hazardous events, but also ensuring communities have access to a diversity of ecosystem services. Seawalls and built infrastructure may offer physical protection against extreme events, but may not ensure the provisioning of other ecosystem services and, in some cases, coastal infrastructure can negatively affect coastal ecosystems that provide services important for human livelihoods (Ingram and Dawson 2001). Ecological tools and principles can help conservationists and development practitioners maximize the provisioning of multiple ecosystem services important for poverty reduction in the coastal zone. Ensuring that this happens will require ecologists to participate in disaster relief missions, recovery, and rebuilding, so, that natural resource management lessons can be learned and applied towards recovery in disaster affected areas and towards risk reduction in hazard prone areas. If we think about ecosystems as natural infrastructure, then, it will be as important to include ecologists on these missions as it is to include engineers and others who assess damage to built infrastructure.

Because coastal ecosystems are dynamic by nature and often used by people for a variety of purposes, it is unlikely that they will be able to provide complete physical protection from all hazards, all of the time. In some cases where ecosystems have been so degraded that their protective capacity has been lost and restoration is not possible, engineered defenses may be the only option, although even these do not provide perfect protection, as demonstrated in the case of Hurricane Katrina. No disaster risk reduction technique is sufficient on its own, but must be part of a holistic strategy that integrates structurally sound buildings, early warning systems, education, and evacuation plans. Including ecosystem-based management in such strategies could provide an effective way to reduce human vulnerability to coastal hazards, while also providing other critical ecosystem services important to coastal populations in the long-term.

References

AusAID, 2007. Indian Ocean Disaster. http://www.ausaid.gov.au, accessed January 2010.

Baird, A. H. and A. M. Kerr. 2008. Landscape analysis and tsunami damage in Aceh: comment on Iverson and Prasad (2007). Landscape Ecology 23:3–5.

Balvanera, P., C. Kremen, and M. Martinez-Ramos. 2005. Applying community structure analysis to ecosystem function: Examples from pollination and carbon storage. Ecological Applications 15:360–375.

Barbier E.B., Koch E.W., Silliman B.R., Hacker S.D., Wolanski E., Primavera J., Granek E.F., Polasky S., Aswani S., Cramer L.A., Stoms D.M., Kennedy C.J., Bael D., Kappel C.V., Perillo G.M.E.and D.J. Reed. 2008. Coastal ecosystem-based management with nonlinear ecological functions and values. Science 319: 321–323.

Bennett, E. M., G. D. Peterson, and L. J. Gordon. 2009. Understanding relationships among multiple ecosystem services. Ecology Letters 12:1394–1404.

Bennett, E. M., G. D. Peterson, and E. A. Levitt. 2005. Looking to the future of ecosystem services. Ecosystems 8:125–132.

Berger, F. and F. Rey. 2004. Mountain protection forests against natural disasters and risks: New French developments by integrating forests in risk zoning. Natural Disasters 33:395–404.

Blasco, F., M. F. Bellan, and M. U. Chaudhury. 1992. Estimating the Extent of Floods in Bangladesh Using Spot Data. Remote Sensing of Environment 39:167–178.

Burke, S., Greenhalgh, D.P., Cooper, E., 2008. Coastal Capital – Economic Valuation of Coral Reefs in Tobago and St. Lucia. WRI, Final Report – June, 2008. World Resources Institute, Washington, D.C.66 pp. Available online at http://www.wri.org/publications

Carpenter, S. R. and C. Folke. 2006. Ecology for transformation. Trends in Ecology & Evolution 21:309–315.

Chatenoux, B., Peduzzi, P. 2007. Impacts from the 2004 Indian Ocean Tsunami: analysing the potential protecting role of environmental features. Natural disasters 40:289–304.

Cochard, R., S. L. Ranamukhaarachchi, G. P. Shivakoti, O. V. Shipin, P. J. Edwards, and K. T. Seeland. 2008. The 2004 tsunami in Aceh and Southern Thailand: A review on coastal ecosystems, wave hazards and vulnerability. Perspectives in Plant Ecology Evolution and Systematics 10:3–40.

Cooper, E., L. Burke, and N. Bood, 2009. "Coastal Capital: Belize. The Economic Contribution of Belize's Coral Reefs and Mangroves." WRI Working Paper. World Resources Institute, Washington, DC. 53 pp. Available online at http://www.wri.org/publications.

Costanza, R., O. Perez-Maqueo, M. L. Martinez, P. Sutton, S. J. Anderson, and K. Mulder. 2008. The value of coastal wetlands for hurricane protection. Ambio 37:241–248.

Dahdouh-Guebas, F., L. P. Jayatissa, D. Di Nitto, J. O. Bosire, D. Lo Seen, and N. Koedam. 2005. How effective were mangroves as a defence against the recent tsunami? Current Biology 15:R443–R447.

Danielsen, F., M. K. Sorensen, M. F. Olwig, V. Selvam, F. Parish, N. D. Burgess, T. Hiraishi, V. M. Karunagaran, M. S. Rasmussen, L. B. Hansen, A. Quarto, and N. Suryadiputra. 2005. The Asian tsunami: A protective role for coastal vegetation. Science 310:643–643.

Das, S. and J. R. Vincent. 2009. Mangroves protected villages and reduced death toll during Indian super cyclone. Proceedings of the National Academy of Sciences of the United States of America 106:7357–7360.

Degg, M. R. and D. K. Chester. 2005. Seismic and volcanic hazards in Peru: changing attitudes to disaster mitigation. Geographical Journal 171:125–145.

Diaz S. L.S., de Bello F.,Quetier F., Grigulis K. and M.Robson. 2007. Incorporating plant functional diversity effects in ecosystem service assessments. Proceedings of the National Academy of Sciences of the United States of America 104: 20684–20689.

Dilley, M., Chen, R.S., Deichmann, U., Lerner-Lam, A.,and Arnold, M. 2005. Natural Disaster Hotspots: A Global Risk Analysis. World Bank, 145 pages.

Duxbury, J. and S. Dickinson. 2007. Principles for sustainable governance of the coastal zone: In the context of coastal disasters. Ecological Economics 63:319–330.

Fernando, H. J. S., McCulley, J.L. 2005. Coral Poaching Worsens Tsunami Destruction in Sri Lanka. Eos 86:301, 304.

Fernando, P., E. D. Wikramanayake, and J. Pastorini. 2006. Impact of tsunami on terrestrial ecosystems of Yala National Park, Sri Lanka. Current Science 90:1531–1534.

Fischetti, M. 2005. Protecting against the next Katrina – Wetlands mitigate flooding, but are they too damaged in the gulf? Scientific American 293:18–+.

Griffin J.N., Mendez, V., Johnson, A.F., Jenkins, S.R. and A. Foggo. 2009. Functional diversity predicts overyielding effect of species combination on primary productivity. Oikos. 118: 37–44.

Hammill, A., Brown, O. and A. Crawford. 2005. Forests, Natural Disasters and Human Security arborvitæ 27:8–9.

Harada, K., Imamura, F., Hiraishi, T. 2002. Experimental study on the effect in reducing tsunami by the coastal permeable structures. Final Proceedings of the International Offshore and Polar Engineering Conference:pp. 652e658.

Ingram, J. C. and T. P. Dawson. 2001. The impacts of a river effluent on the coastal seagrass habitats of Mahe, Seychelles. South African Journal of Botany 67:483–487.

Ingram, J. C., G. Franco, C. Rumbaitis-del Rio, and B. Khazai. 2006. Post-disaster recovery dilemmas: challenges in balancing short-term and long-term needs for vulnerability reduction. Environmental Science & Policy 9:607–613.

Iverson, L. R. and A. M. Prasad. 2007. Using landscape analysis to assess and model tsunami damage in Aceh province, Sumatra. Landscape Ecology 22:323–331.

Iverson, L. R. and A. M. Prasad. 2008. Modeling tsunami damage in Aceh: a reply. Landscape Ecology 23:7–10.

Kaplan, M., F. G. Renaud, and G. Luchters. 2009. Vulnerability assessment and protective effects of coastal vegetation during the 2004 Tsunami in Sri Lanka. Natural Disasters and Earth System Sciences 9:1479–1494.

Kathiresan, K. and N. Rajendran. 2005. Coastal mangrove forests mitigated tsunami. Estuarine Coastal and Shelf Science 65:601–606.

Kerr, A. M., A. H. Baird, and S. J. Campbell. 2006. Comments on "Coastal mangrove forests mitigated tsunami" by K. Kathiresan and N. Rajendran [Estuar. Coast. Shelf Sci. 65 (2005) 601–606]. Estuarine Coastal and Shelf Scienc 67:539–541.

Khazai, B., Ingram, J.C., Saah, D. 2007. The Protective Role of Natural and Engineered Defense Systems in Coastal Hazards. Report Prepared for the State of Hawaii and the Kaulunani Urban and Community Forestry Program of the Department of Land and Natural Resources by the Spatial Informatics Group

Koch, E. W., E. B. Barbier, B. R. Silliman, D. J. Reed, G. M. E. Perillo, S. D. Hacker, E. F. Granek, J. H. Primavera, N. Muthiga, S. Polasky, B. S. Halpern, C. J. Kennedy, C. V. Kappel, and E. Wolanski. 2009. Non-linearity in ecosystem services: temporal and spatial variability in coastal protection. Frontiers in Ecology and the Environment 7:29–37.

Kremen, C. 2005. Managing ecosystem services: what do we need to know about their ecology? Ecology Letters 8:468–479.

Kremen, C. and R. S. Ostfeld. 2005. A call to ecologists: measuring, analyzing, and managing ecosystem services. Frontiers in Ecology and the Environment 3:540–548.

Kunkel, C. M., R. W. Hallberg, and M. Oppenheimer. 2006. Coral reefs reduce tsunami impact in model simulations. Geophysical Research Letters 33.

Levy, J. K., Gopalakrishnan, C. 2005. Promoting Disaster-Resilient Communities: The Great Sumatra-Andaman Earthquake of 26 December 2004 and the Resulting Indian-Ocean Tsunami. Internationa Journal of Water Resources Development 21:543–559.

Liu, P. L. F., P. Lynett, H. Fernando, B. E. Jaffe, H. Fritz, B. Higman, R. Morton, J. Goff, and C. Synolakis. 2005. Observations by the International Tsunami Survey Team in Sri Lanka. Science 308:1595–1595.

Loreau, M. 2010. Linking biodiversity and ecosystems: towards a unifying ecological theory. Philosophical Transactions of the Royal Society B-Biological Sciences 365:49–60.

Luck, G. W., R. Harrington, P. A. Harrison, C. Kremen, P. M. Berry, R. Bugter, T. P. Dawson, F. de Bello, S. Diaz, C. K. Feld, J. R. Haslett, D. Hering, A. Kontogianni, S. Lavorel, M. Rounsevell, M. J. Samways, L. Sandin, J. Settele, M. T. Sykes, S. van den Hove, M. Vandewalle, and M. Zobel. 2009. Quantifying the Contribution of Organisms to the Provision of Ecosystem Services. Bioscience 59:223–235.

Luck, G. W., G. C. Daily, and P. R. Ehrlich. 2003. Population diversity and ecosystem services. Trends in Ecology & Evolution 18:331–336.

Martinez, M. L., A. Intralawan, G. Vazquez, O. Perez-Maqueo, P. Sutton, and R. Landgrave. 2007. The coasts of our world: Ecological, economic and social importance. Ecological Economics 63:254–272.

Mascarenhas, A. and S. Jayakumar. 2008. An environmental perspective of the post-tsunami scenario along the coast of Tamil Nadu, India: Role of sand dunes and forests. Journal of Environmental Management 89:24–34

Maskrey, A. 1989. Disaster mitigation: a community based approach. In: Development Guidelines, No. 3, Oxfam, Oxford, UK

Massel, S. R. 1999. Tides and Waves in Mangrove Forests: In Fluid Mechanics for Marine Ecologists. Springer-Verlag.

Mutter, J. 2005. The Earth Sciences, Human Well-Being, and the Reduction of Global Poverty. Eos 86:157,164–165.

Naeem, S. 2002. Ecosystem consequences of biodiversity loss: The evolution of a paradigm. Ecology 83:1537–1552.

Naeem, S. and J. P. Wright. 2003. Disentangling biodiversity effects on ecosystem functioning: deriving solutions to a seemingly insurmountable problem. Ecology Letters 6:567–579.

Naeem, S. 1998. Species redundancy and ecosystem reliability. Conservation Biology 12:39–45.

Peduzzi, P. 2005. Tropical cyclones: paying a high price for environmental destruction. Environment & Poverty Times 3:3.

Petchey O.L. and K.J. Gaston. 2007. Dendrograms and measuring functional diversity. Oikos. 116: 1422–1426.

Quartel, S., A. Kroon, P. Augustinus, P. Van Santen, and N. H. Tri. 2007. Wave attenuation in coastal mangroves in the Red River Delta, Vietnam. Journal of Asian Earth Sciences 29:576–584.

Resilience Alliance, 2010. Key Concepts. http://www.resalliance.org/.

Sachs, J. D. 2005. The End of Poverty: Economic Possibilities for Our Time. Penguin, New York.

Symstad A.J. C.F.S., Wall D.H., Gross K.L., Huenneke L.F., Mittelbach G.G., Peters D.P.C. and Tilman D. 2003. Long-term and large-scale perspectives on the relationship between biodiversity and ecosystem functioning. Bioscience 53:89–98.

Tanaka, N., Sasaki, Y., Mowjood, M.L.M., Jinadasa, K. B. S. N.,Homchuen, S. 2007. Coastal vegetation structures and their functions in tsunami protection: experience of the recent Indian Ocean tsunami. Landscape and Ecological Engineering 3:33–45.

Teh, S. Y., H. L. Koh, P. L. F. Liu, A. I. M. Ismail, and H. L. Lee. 2009. Analytical and numerical simulation of tsunami mitigation by mangroves in Penang, Malaysia. Journal of Asian Earth Sciences 36:38–46.

Tilman, D., C. L. Lehman, and C. E. Bristow. 1998. Diversity-stability relationships: Statistical inevitability or ecological consequence? American Naturalist 151:277–282.

The Natural Capital Project, 2010. http://www.naturalcapitalproject.org/home04.html. Accessed January 2010.

Tri, N. H., W. N. Adger, and P. M. Kelly. 1998. Natural resource management in mitigating climate impacts: the example of mangrove restoration in Vietnam. Global Environmental Change-Human and Policy Dimensions 8:49–61.

UNISDR (United Nations International Strategy for Disaster Reduction), http://www.unisdr.org/, Accessed January 2010.

Vermaat, J. E. and U. Thampanya. 2006. Mangroves mitigate tsunami damage: A further response. Estuarine Coastal and Shelf Science 69:1–3.

Walker S.C., Poos M.S. and Jackson D.A. 2008. Functional rarefaction: estimating functional diversity from field data. Oikos 117: 286–296.

Walker, B. 2009. Specifed and General Resilience, Module 1.5. Assessing and managing resilience in social-ecological systems: A practitioners workbook. Resilience Alliance.
Walton, M. E. M., G. P. B. Samonte-Tan, J. H. Primavera, G. Edwards-Jones, and L. Le Vay. 2006. Are mangroves worth replanting? The direct economic benefits of a community-based reforestation project. Environmental Conservation 33:335–343.
Wells, S. and V. Kapos. 2006. Coral reefs and mangroves: implications from the tsunami one year on. Oryx 40:123–124.

Chapter 23
Integrating Natural Resource Management into Disaster Response and Mitigation*

Julie A. March

Introduction

No population on this planet is immune to the threat of disaster. Whether it takes the form of a rapid onset disaster like the 2004 Indian Ocean tsunami which devastated the coastal zone of Aceh Indonesia, or the slow onset drought of 2006 which crept across the Horn of Africa, leaving poor harvests and weakened herds throughout the region, nearly all communities are at risk. The full impact of a disaster depends on the scale of the disaster and the ability of the affected population to both withstand the shock and to recover. When populations cannot recover on their own, national governments and at times, the international humanitarian community in the form of donors and implementing agencies must provide support to the people in need.

The United States and other nations regularly respond to disasters and provide relief assistance to affected communities. These responses address rapid onset disasters such as cyclones, tsunamis, and earthquakes in addition to slower onset events such as drought and complex emergencies characterized by conflict and population displacement. The immediate goal of a humanitarian response to a disaster is to save lives and reduce human suffering. Quite often, the conditions which precipitate the disaster are not targeted in the initial life saving response. This can create potential for future shocks because the economic, social, environmental or ecological conditions which contributed to the initial shock persist. For some populations, recovery becomes increasingly difficult as their resiliency erodes with

---

*The views expressed in this article/chapter are those of the author and do not necessarily reflect the views or opinions of the US Agency for International Development or the US Government.

J.A. March (✉)
Office of U.S. Foreign Disaster Assistance, United States Agency for International Development, Washington, DC, USA
e-mail: jmarch@usaid.gov

J.C. Ingram et al. (eds.), *Integrating Ecology and Poverty Reduction: Ecological Dimensions*, DOI 10.1007/978-1-4419-0633-5_23,

successive shocks. This is especially true for vulnerable populations who rely on the natural environment for their survival. For this group, the disruption of the access to or availability of environmental goods, such as seeds, crops, pasture, livestock, farm land or fish following a disaster is especially challenging as they have few other resource options. Prevalence of recurrent disasters highlights the importance of saving lives while at the same time building resiliency of the people and the environment to prevent and withstand future shocks.

Ecological principles provide a starting point for creating such a response. Echoing the systems concepts put forth by Eugene Odum in which he suggested that the "whole is greater than the sum of its parts" (Odum 1964), disaster response must consider the system within which the beneficiaries operate. The reductionist view which considers only the immediate needs of the individual at a given point in time must expand to consider the individual as one part of a much larger system. Understanding how a disaster impacts different facets of this system can guide a response to better meet the populations' needs, avoid exacerbating conditions on the ground, and mitigate vulnerability to future shocks.

Environmental Resources and Disasters

Resource Dependency and Vulnerability

Poverty and a high dependency on natural resources are closely linked throughout much of the developing world (Prakash 1997). Degradation of the natural environment can most negatively affect those who have few livelihood options beyond the use of natural resources. The relationship between poverty, environmental degradation and vulnerability to shocks/disasters is closely integrated and, although there is often debate about how each component influences the other, their connection is not disputed (Duraiappah 1996).

Those who are most vulnerable to economic stress and who have greatest dependency on natural resources often face significant challenges to restoring their livelihoods post disaster. This group, which includes subsistence farmers and pastoralists, frequently has limited access to formalized safety nets such as insurance, large amounts of mainstream credit or the ability to seek alternative and lucrative employment elsewhere. The more their asset base is eroded, the more difficult recovery can be.

Access to resources can be restricted following a disaster for a few days to many years. For nearly 20 years, protracted conflict in Northern Uganda between the Lord's Resistance Army (LRA) and the Government of Uganda led to the displacement of almost 1.5 million people, a great number of them subsistence farmers. When insecurity restricted access to land, the ability to exist as a subsistence farmer was greatly challenged because their remaining resources were not sufficient to sustain more costly options, such as renting land in safer areas or moving out of the region permanently, which could have enabled them to continue farming for the duration of the conflict. Upon leaving their land, the majority of displaced farmers did not immediately have access to formal safety nets and informal safety nets (such as relying on family and friends or host communities for food, shelter or resources) were stressed with the number of people affected in the region. Ultimately, many of the displaced resided in Internally Displaced Persons (IDP) camps and depended upon food assistance for survival until security improved. As the security situation improves and farmers begin to return to their fields, the loss of assets over time, such as stored seed resources, tools and animals, combined with limited available resources to replace assets has contributed to a slow recovery process.

Coping Capacity

In response to a disaster or a shock, people utilize a variety of strategies and mechanisms to enable them to survive until conditions improve. At the onset of the first shock, those affected are more likely to have a diverse suite of coping mechanisms available to them. These could include moving to an unaffected area, seeking additional work, and/or relying on family or friends for food and shelter. As the disaster wears on (as in the Northern Uganda example) or additional shocks ensue (such as renewed conflict, a delayed rainy season, or repeated flooding), some coping mechanisms may become fully exhausted, leaving only the less desirable strategies which may negatively affect their long term livelihood sustainability. Among these less desirable coping strategies are the sales of assets, which are essential to their well being and to their recovery. For example, to satisfy the need to buy food prior to planting, the farmer may be forced to sell other vital implements like axes, hoes or animals or take out a loan for a percentage of his harvest to enable his family to access the inputs they need. The challenge between meeting short term needs while making decisions which benefit longer term recovery are evident when considering the risks involved in this endeavor. The immediate need is for food, followed by farming inputs to plant in the upcoming season. The cost of purchasing seed (if there are not saved seed resources), may not be offset by gains from the harvest. In the event that the harvest is successful, resources might be recuperated. In the event that it is not, the farmer then finds himself in a cycle of debt and depleted assets which can potentially hamper future production.

In other cases, the coping mechanisms, such as removing children from school when school fees cannot be paid, reducing the number of meals consumed per day, or spending long hours in search of "famine foods" – wild foods only harvested in times of extreme food insecurity – can have long term impacts on the next generation with outcomes such as stunting or illiteracy.

Other coping strategies fulfill short-term needs for cash and food, yet ultimately contribute to an overall reduction of long term environmental health (Scott 2006). For instance, when people are displaced to temporary camps or settlements normal livelihood activities are not possible, resulting in a notable rise in sales of firewood and charcoal from trees in the immediate area. The increased pressure on resources opens the area to soil compaction, erosion and, in some areas at severe risk of degradation, desertification. An effective response to mitigate degradation around camps would be designed to meet immediate needs while preventing complete depletion of environmental assets or adoption of negative coping strategies. Humanitarian programs supporting alternative income generating means such as "cash for work" are one way to reduce pressure on resources in displacement situations.

A Systems Approach

One of the most promising ways to slow the downward spiral of disaster, vulnerability, and degradation is to respond to disasters by addressing not only the symptoms but also the components of the larger complex system which eroded under the strain of the disaster. In developing countries where survival often involves a reliance on natural resources, looking at the system from an ecological perspective can provide clues as to how to respond effectively in the short-term and fortify vulnerable populations against future vulnerability at the same time.

Persistent vulnerability after a disaster response suggests that restoring an asset base to its predisaster condition without examining the context in which the disaster occurs or improving the underlying conditions which contributed to the disaster may not produce lasting positive change (Helland 1980). Additionally, adding inputs without careful analysis can actually weaken the degraded system (Sperling 2008) and does not necessarily lessen the likelihood that the population will be vulnerable to a similar disaster in the future. For example, adding livestock immediately after animals are lost to drought without considering the potential impact the new or additional animals will have on available water and fodder resources can hamper the recovery of the remaining animals. Because of the high volume of repeat emergencies and the desire to do no harm in a response situation, donors and implementing agencies like non-governmental organizations (NGOS) are increasingly seeking responses which address not only the immediate needs in the aftermath of the disaster but which also facilitate progress toward greater sustainability (UNISDR 2005).

The general nature of many emergency humanitarian responses is immediate and short term programming. Even so, short term responses should consider the long term implications of response, especially in the case of input provision (Reaymaekers 2008). In the "developmental relief" ideology promoted since the 1990s, relief should facilitate transition from an emergency state toward a more stable recovery phase (Campanaro et al. 2002). When relief actions do not enable this, they do more to encourage dependency and stagnation than they do to promote real progress away from vulnerability.

Looking at the effects of a disaster on an individual within a larger system rather than considering only their immediate needs post disaster can help better target the underlying issues which are crucial to recovery. If a vulnerable farmer in Southern Africa loses a maize crop to drought, simply replacing hybrid maize seed is a flawed response if it does not consider and address underlying factors that may have caused the loss. There are many factors to consider – maize may not be the optimal crop choice for low rainfall, poor soil zones, and further, promotion of hybrid crops which require purchased seed season after season, may put cash strapped farmers at still higher levels of vulnerability. On the other hand, in many parts of Africa, there are local variations of the saying that "if you have not eaten maize, then you haven't yet eaten for the day," stressing the importance maize symbolizes in many African communities. Thus, the loss must be analyzed in the context of the climate, the soil, the productive capacity of the land, the access to land and inputs and markets, cultural preferences and the other assets the farmer has to rely upon for survival. In this example, the farmer and their interests, abilities and limitations are equally important to respond successfully. A system-based perspective also implies that multiple options which promote resilience, rather than single linear solutions need to be considered. Support that allows farmers to strategize in relation to a variety of risks might help the farmer emerge stronger compared to support which is very prescriptive in the options it provides to the farmer to mitigate a variety of stresses. After programming assistance ends, the farmer remains in the midst of a dynamic situation and therefore is better served by options than by solutions which offer no flexibility as the environment changes.

Understanding Linkages for Better Response

A robust ecological system is often more able to support the populations depending upon them through disasters, just as degradation of the environment can increase the negative effects of a disaster. For example, the heavy rains and wind which accompanied Hurricane Gustav in August of 2008 in Haiti were a significant force on their own, but became devastating for the local populations when followed by the mudslides which destroyed or damaged agricultural land and nearly 10,000 homes. Deforestation throughout an estimated 98% of Haiti has been blamed for the scale and force of the mudslides (Boutrous 2008). The high deforestation rates have been linked to the pressure on forests to supply approximately 80% of the Haitian population with wood for cooking and building. In the case of Haiti, it is not entirely possible to pinpoint whether poverty intensifies the increased environmental degradation or vice versa; the relationship is difficult to assess if we are trying to identify causation. What is clear, however, is that they are connected, and by better understanding the linkages, we can begin to better understand how to respond in a meaningful and sustainable manner.

Following are two examples where relief responses included methodologies, which supported the goals of sustainability and resiliency from initial short term responses to longer term recovery. The first example discusses the emergency conditions brought about by the 2006 drought in the Horn of Africa and provides an example of the ability to respond to emergency needs while considering the broader system and its limitations. The second example discusses methodological changes in input (seed) delivery to better support seed systems.

Pastoralism Under Pressure

Background

Each year pastoralists migrate within the Horn of Africa from areas of rainy season pasture to dry season pasture and back again with their herds. As the rainfall moves, so do the herders and their animals. Even in a year of normal rainfall, stress on the system is evident as traditional practices and grazing patterns are increasingly disrupted by changing human-environment conditions, altering the balance herders were able to achieve for many many years. Although there have always been struggles to maintain the delicate balance between animals and resources, new challenges make it ever more difficult to achieve. A growth in population and competition for land increases conflict between herders and farmers. Land rights issues disrupt traditional grazing patterns and at the same time, conflict in border areas of a number of countries restricts traditional routes. There is also the issue of long fought battles over access to water and who can access this scarce resource, with government policies on land use and expansion of agricultural production frequently playing a role in restricting grazing (McCarthy et al. 2004). Pastures are further degraded by deforestation for charcoal production, overstocking of pastures, encroachment by invasive species, and stocking of animals based on economic value rather than suitability to the environment. Access to animal health care is also often hard to locate or impossible to afford for many herders (Trench et al. 2007). These factors combined with misdirected aid (Oba and Lusigi 1987) have shifted the balanced system herders have evolved over many generations toward one which is difficult to sustain.

Enduring challenges and weaknesses in the system are intensified under the added strain of low rainfall. This was exemplified during the drought in 2006 when Kenya, Somalia, and Ethiopia experienced an extended period of less than normal precipitation, eventually resulting in emergency needs for an estimated 2.5 million people in Kenya in 2006 alone (World Food Program 2006).

Assessing the Damage and Planning a Response

Field-based assessments by international NGO's, community-based organizations, government ministries, and international donors often provide the bulk of information used by the humanitarian community to evaluate the potential for disaster and the number of people likely to be affected. Early warning and monitoring systems also play a large role in identifying and reporting on indicators which signal an oncoming emergency. The Famine Early Warning System Network (FEWSNET) is one such early warning system operational in the region. By January 2006 the reports from the field in international forums were indicating that the drought was likely to have serious consequences for the pastoralist populations. Reduced access to sufficient pasture, increased animal mortality and a deviation of "normal" migration patterns were all cited as evidence that conditions were deteriorating in the region (FEWSNET 2006). As the period of reduced rainfall continued, the animal mortality increased along with the "distress sale" of animals. The increase in sales generally occurs at a point in time when the animals are in poor physical condition, the markets are glutted with people trying to sell emaciated cows or goats or their hides, and the terms of trade for the animal and animal products continue to decline. Opportunistic buyers can buy and fatten the animals for resale, taking advantage of the low market prices at the time of sale. In some areas such as northern Kenya, there were reports of 10–25% of camels and goats and up to 40% of cattle and sheep being lost due to lack of feed and water by March 2006. In a region without a multitude of income generating options, the loss of animals can have severe impacts on the livelihoods of those who depend on them. Pastoralists rely on their herds as their source of ongoing food security (for milk, blood, meat, and fat) and as a form of capital. Livestock act as a walking account which can be drawn upon to meet household needs for additional food, medicines, school fees, and other living expenses.

The answer to this problem appears to be simple – provide fodder and water or wait for the rains to fail and then replace the animals – yet neither fodder provision nor restocking were viable options in light of the limitations of the system. Quite often the first instinct in the face of human suffering is to desperately attempt to retain status quo. Doing so in this case would only serve to exacerbate the poor conditions. Examination of the system prior to the drought revealed an ecosystem out of balance and under strain. This area of northern Kenya had been experiencing more frequent droughts – accompanied by livestock herd decimations. There have been three large scale droughts in the past decade compared to gaps of 9–12 years between devastating droughts in the decades preceding it (Beaumont 2009). The limiting factor in the environment was not only the lack of rainfall, but also the multiple issues related to the condition of the pastures and the animals, such as lack of sufficient rangeland area, insufficient animal healthcare, underdeveloped markets, and undue reliance on a single source of livelihood (Ndikumana et al. 2000). Deneve (1995) simply, but truly states "When grazing, animals reduce the source of their survival gradually and cannot survive once vegetation has disappeared." Providing fodder to sustain all remaining animals would eliminate the natural mechanism in place which controls grazing pressure, further throwing natural control mechanisms out of alignment.

Supporting this type of intervention would not have addressed the underlying ecological and economic issues which contributed to the negative impacts of the drought. So the challenge for the humanitarian community was and still is to find ways to support the sustainability of the systems as a whole. For example, rather than adding only cattle back to the pastures, concentrating on more optimal use of pasture resources by promoting diversification of herds could have a greater positive impact. Increased herd diversity has several benefits. First, it spreads risk for the pastoralists – if one species is wiped out by a disease or if one does not have access to their required vegetation, the others may still survive. Secondly, more diverse herds make better use of available forage. Each animal has different requirements and mixed herds of cattle, camels and goats, for example, allows better usage of the grasses, scrub and tree vegetation. Third, diversification gives pastoralists access to many more food security products, available at different intervals, such as, goat milk, cow milk, and camel milk. For example, restocking with cattle only loads the environment with animals that are less adapted to lack of water and poor fodder quality conditions (compared to goats and camels). Instead, supporting a transition toward a more diverse herd has a great potential to spread risk while reducing competition for the same feed resources. Some pastoral populations, in response to recurring stress, have dramatically changed the composition of their herds, having to adapt their management styles accordingly. In the case of the northern Kenya Samburu cattle herders, the drought susceptible cattle herds have been mixed with camels.

Programs which ensure that surviving animals are able to utilize the forage they consume are essential. Promoting programs which increase access and availability of animal health services through training community animal health workers and promotion of small scale veterinary drug stockists presented a greater opportunity for sustainability than those programs which increased competition for scarce resources by increasing stocking rates. Ensuring that animals are healthy and free of parasites enables them to better utilize what food they do consume. Animal health interventions are most effective prior to the peak of a crisis like drought. Being proactive rather than reactive increases the chance that an animal will survive the crisis, whereas once the animals are emaciated or sick, routine vaccinations could have limited benefit.

Because the pastoralists rely fully on animals for their livelihoods, a response must balance the vulnerability of the human population with the fragility of the natural system. While it was clear that a full restocking of animals, even several months after the drought, would increase grazing pressure too significantly, providing support for the pastoralists to maintain their breeding stock would not. By retaining breeding stock, the pastoralists would have the potential to select which animals to keep while at the same time maintaining their capacity to regenerate their herds once pasture conditions improve. This would provide a genetic pool for breeding when conditions improve. Thus, supplementary feeding of breeding stock was supported in light of the benefits to the herders. Saving breeding stock potentially eliminated the need to sell assets or borrow funds to reestablish herds. This also allows some buffer against dropout from pastoralism. Because of the time and cost related to recovery, there is a high degree of pastoralist dropout following a large loss of animals and the pastoralists find themselves in urban centers locking for work (Oxfam 2008). While this is not always a worse option financially, more often than not, the pastoralist skill set does not guarantee a successful transition from herding animals to either working in agriculture or in an urban center.

Finally, interventions which can increase the efficiency of pastoralism will also help build resiliency of the pastoralists by lessening pressure on the resource base. Animal health interventions have already been mentioned as one means to achieve this. Others include better control of invasive pasture species which are unpalatable to livestock or which out-compete preferred varieties. Also crucial is addressing the tendency many pastoralists have of holding animals as an investment rather than regularly selling the animals or selling in response to predicted shortfalls in fodder before the situation becomes dire. Weak market linkages exacerbate this trend and are increasingly being addressed in a number of emergency and development programs. These programs might include improved market facilities, improved roads to market, price reporting and monitoring, and formation of cooperatives for sales.

On another level, supporting access to grazing land would also contribute to reducing pressure on pastoral areas. The pastoralists frequently find themselves marginalized when it comes to negotiating land rights and are being channelized to make way for a variety of other land uses, including game parks, infrastructure development, and farming. Ensuring that pastoralists have the ability to maintain their livelihoods and grazing routes will contribute to a closer approximation of balance.

By the end of 2006, the US Agency for International Development's Office of US Foreign Disaster Assistance had provided more than $4,813,521 in response to the Horn of Africa drought in Kenya alone and supported many of the interventions mentioned in this section.

Supporting Agricultural Rehabilitation Post Disaster (or During)

A similar response choice can be seen in the post disaster support options offered to subsistence farmers throughout the world. Farmers' livelihoods are susceptible to the full range of disasters, from volcanic eruptions, tsunamis and floods to drought, conflict, and displacement. Because farmers are closely tied to the agricultural calendar and many of the most vulnerable populations depend on rainfed agriculture, any disruption of the planting or harvesting cycle can have negative consequences for livelihoods and food security.

Response Strategy

For many years, disruptions of the agricultural cycle due to disasters were treated with the distribution of seed to farmers through a direct seed distribution methodology. These distributions often provided packets of the same seed to all affected farmers regardless of farmer preference, varietal suitability or individual need. Direct seed distributions assume that there is no seed available. In some cases, especially when formal and informal markets have been disrupted, this is accurate. However, when this is not the case and seed is available, by supplying seed rather than supporting the local formal and informal seed supply channels, a short term response can undermine long term stability of the seed systems.

With food insecurity on the rise, especially in Africa where the number of malnourished populations is projected to continually increase, the seed delivery methodology and its efficacy have been receiving a timely reconsideration. Pioneers in relief seed distribution methodology and the science of seed system assessment have been working since the 1990s to take seed response to a systems level (Sperling 2000; Remington et al. 2001).

Correct diagnosis of the constraints to farmer's food security is a seemingly logical yet often missing step. Seed need is often assumed when in reality, the need for food does not equate to seed need. For effective response, there needs to be a differentiation between food insecurity and seed insecurity. Within seed insecurity, differentiation between seed access and seed availability is also critical to ensure an appropriate response (Sperling 2002). There are many reasons why populations do not have seed. The seed may be available but farmers cannot access it due to lack of money, breakdown of barter relationships, or insecurity which prevents movement. Alternatively, for whatever reason, the seed may not be available locally or in local markets – this could be due to consecutive years of poor harvest, disease or pest infestation. Even during one of the worst conflicts of recent times, the Rwandan genocide, which peaked in the middle of the harvest season, farmers were able to maintain significant seed supplies, including access to nearly all of the varieties of beans which they preferred to plant (Sperling 1997). Even in the most difficult of times, farmers may be able to save their seeds for planting by bringing them with them or accessing them from other farmers and seed sellers through formal and informal seed systems. Varieties which were not brought into displacement were later sought from other farmers.

With a more nuanced understanding of farmer's seed need, the shortfalls of a one-size-fits-all distribution have become more apparent and the need for a more tailored response has been appreciated by development and relief agencies (Eberdt 2003). Since 2000, there has been a shift away from the direct seed distribution methodology in which all farmers receive a standard package of seeds with the same amount and seed type regardless of need or preference, when seed supply and security conditions allow. Instead, the seed fair and voucher program initially piloted by Catholic Relief Services (CRS) (Remington et al. 2002) or some form of it, is widely used. In this method, vulnerable farmers with assessed seed need are provided vouchers to spend at the seed fair which is populated by local and regional merchants of seed. The goal is to have a wide range of seed varieties available. These include local landraces and sometimes improved seed from government extension agencies, international agricultural research centers, or private sector seed companies. The role of farmer choice is significant in this model and varieties which are well suited to an individual's plot are likely to be chosen (Longley et al. 2002).

Responding to disaster with an intervention that allows for farmer choice and a more tailored package of support strengthens the resiliency of the system in many ways. Rather than having every farmer in an affected area planting the same packet of beans distributed in a direct seed distribution, agrobiodiversity at the plot and village level can be enhanced. Having more seeds of more varieties on offer can allow for different crop suites to be grown with the selected seeds. This allows farmers to have more options and to strategize. The approach spreads risk for the farmer and provides more chance of optimal utilization of soil nutrients and moisture. It also presents greater resistance against the spread of crop pests and diseases. A system of greater biodiversity is more able to withstand both of these. It also provides economic advantages in the event of surplus production. Greater variety in the event of marketing will prevent glutting the market with a harvest of one crop, allowing a better price at time of sale.

The seed fair also offers the opportunity to maintain genetic diversity by supporting local varieties and local seed markets, or new varieties that have been locally confirmed (Sperling et al. 2006). Because relief seed in the direct distribution model is brought in from elsewhere, it has the potential to be less well adapted to local conditions or preferences for taste and cooking properties. Supporting the work of local seed producers allows for local well adapted varieties to return to the market once the crisis has passed.

Finally, the quest for sustainability in disaster response has led many to move away from the outright provision of high cost inputs to farmers. Even a short term disaster response program has the capacity to provide training, where appropriate, on low cost sustainable methodology such as integrated pest management and use of organic fertilizers. This could have a more lasting effect on supporting agricultural system health and function than a one time delivery of production enhancing, yet high cost inputs.

Conclusion

Both examples illustrate how ecological principles can support the design of disaster responses that not only support immediate livelihood needs, but also enhance livelihood resiliency over the long term. Without addressing the root causes of vulnerability through disaster response, the likelihood increases that demand for assistance will expand as livelihoods degrade and groups who do not recover from an initial disaster slip from a phase of acute need to a more chronic state of need. With a finite amount of funding resources available to support disaster response, it is becoming increasingly important that even short term relief programming incorporates long term planning (Maxwell et al. 2008).

Ideally, the most effective response is going to be one which allows the affected populations to recover from the emergency and to better cope with future stress, either through strengthened livelihoods or improved environmental conditions. The best responses seek a balance between promoting survival today and planning for survival in the future. Response options must be examined to determine what is ecologically sound as well as what will survive in the face of recurring climatic, political, and economic shocks?

By nature, this is a moving target. The challenges vulnerable populations face are similar, yet each situation deserves its own careful consideration. Ultimately, the power to decide upon the best course of action should rest with the target beneficiaries. Equipping them with options and skills will do far more to promote long term resiliency than supplying inputs that may last only until the next shock. This is the model that has been in place for many years and that thankfully, due to flexibility on the part of donors and innovation on the part of populations and implementing partners, is beginning to change.

Resiliency is built when coping mechanisms are strengthened against future shocks. In Mali, a farmers group began drying their okra and other vegetables to provide a food source during the long dry season.

Rather than benefitting from relief by receiving seed inputs to sow, this woman was supported through a program that focused on training for successful multiplication of locally adapted seed in Northern Uganda. This season, she will be able to sell the seed because of its high quality and because the surrounding farmers witnessed its success.

Scarce resources can become the source of tension between herders and other users when rainfall is especially low in pastoral areas.

References

Beaumont, P., 2009. " The last nomads: drought drives Kenya's herders to the brink". The Observer, September 13, 2009.

Boutrous J., UN Press Conference by UN Coordinator in Haiti –October 15, 2008.

Campanaro, G., Hepburn, G., Kowalska, M., Wang, A.,2002. Developmental Relief : the European Perspective. Prepared for the Transition Working Group. InterAction, Washington D.C.

Deneve, R. 1995. Sahel Studies IUCN: Sahel-Sahel: A controversial Vision. The World Conservation Union, Island Press.

Duraiappah, A. 1996. Poverty and Environmental Degradation: a Literature Review and Analysis. CREED Working Paper 8.

Eberdt, C. 2003, Questioning seeds and tools: emerging strategies in post-disaster seed relief and rehabilitation. The Journal of Humanitarian Assistance. February, 1–16.

FEWSNET Kenya Monthly Food Security Update : January 2006. FEWSnet, GOK, World Food Program.

Helland, J. 1980. Five essays on the study of pastoralists and the development of pastoralism. African Savanna Studies Occasional Paper No. 20, Bergen, Norway.

Kirkbride, M., Grahn, R., August, 2008. Survival of the fittest, Pastoralism and climate change in East Africa, Oxfam Briefing Paper 116, Oxfam International.

Longley, C., Dominguez, C., Saide, M.A., Leonardo, W.J., (2002) Do Farmers Need Relief Seed? A Methodology for Assessing Seed Systems. *Disasters* **26**(6): 343–355.

Maxwell, D., Sadler, K., Sim, A., Mutonyi, M. Egan, R., Webster, M. 2008. Emergency Food Security Interventions: Good Practice Review Number 10. Overseas Development Institute, London, U.K.

McCarthy, N., Dutilly-Diane, C., Drabo, B., Kamara, A., Vanderlinden, J., Managing Resources in Erratic Environments, An Analysis of Pastoralist Systems in Ethiopia, Niger and Burkina Faso, 2004. International Food Policy Research Institute Research Report 135. Washington D.C.

Ndikumana J., Stuth J., Kamidi R., Ossiya S., Marambii R. and Hamlett P. 2000. Coping mechanisms and their efficacy in disaster-prone pastoral systems of the Greater Horn of Africa. Effects of the 1995–97 drought and the 1997–98 El Niño rains and the responses of pastoralists and livestock. ILRI Project Report.

Oba, G. and Lusigi, W. 1987. 'An Overview of Drought Strategies and Land Use in African Pastoral Systems'. *ODI Pastoral Development Network Paper* 23a, 1986–7, Agricultural Administration Unit. 33p. London: Overseas Development Institute.

Odum, E.P. 1964. The New Ecology. BioScience **14**:14–16.

Prakash, S. 1997.Poverty and Environment Linkages in Mountains and Uplands: Reflections on the 'Poverty Trap' Thesis. CREED Working Paper 12.

Reaymaekers, T., 2008. Conflict and food security in Beni – Lubero: Back to the future? 169–196 *in* Alinovi, L., Hemrich, G., Russo, L.,. Beyond Relief, Food Security in Protracted Crises. Practical Action Publishing, U.K.

Remington, T., Maroko, J., Walsh, S., Omanga, P. and Charles, E. 2001. Getting off the seed and tools treadmill with CRS seed vouchers and fairs. *Disasters* **26**(4): 302–315.

Scott, L. 2006. Chronic Poverty and the Environment: a Vulnerability Perspective.Overseas Development Institute Chronic Poverty research Centre Working Paper 62.

Sperling, L. 1997. The effects of the Rwandan war on crop production and varietal diversity: a comparison of two crops. in, L. Sperling, ed. *War and Crop Diversity*. AGREN Network Paper no.75. London: Overseas Development Institute, pp. 19–30.

Sperling, L. 2000. Emergency Seed Aid in Kenya: a Case Study of Lessons Learned completed under a United States Agency for International Development contract (#LAG-4111-00-3042-00).

Sperling, L., Remington, T., and Haugen, J. 2006. Seed Aid For Seed Security: Advice for Practitioners, Practice Briefs. International Center for Tropical Agriculture and Catholic Relief Services 1–10. Rome, Italy.

Sperling, L. 2008. When disaster strikes: a guide for assessing seed security. Cali: International Center for Tropical Agriculture.

Trench, P., Rowley, J., Diarra, M., Sano, F., Keita, B., 2007. Beyond Any Drought, International Institute for Environment and Development London, England.

UNISDR 2005 World Conference on Disaster Reduction, January 2005, Lobe Hyogo Japan: Hyogo Framework for Action 2005–2015: Building the Resilience of Nations and Communities to Disasters.

World Food Program, Jan 2006, "World food program warns of catastrophe in Horn of Africa". http://www.wfp.org/node/630.

Conclusion: Integrating Ecology and Poverty Reduction

Jane Carter Ingram, Fabrice DeClerck, and Cristina Rumbaitis del Rio

Conclusion

As discussed throughout the chapters of the two volumes comprising this series on *Integrating Ecology and Poverty Reduction*, in recent years an increasing amount of global attention has focused on the role of the natural environment in contributing to poverty reduction (McNeely and Scherr 2003; Ash and Jenkins 2007; World Bank 2007; Tekelenburg et al. 2009; Chivian and Bernstein 2008; Galizzi and Herklotz 2008). These volumes complement and build upon this growing body of work, but look specifically at the ecological dimensions of multiple development challenges related to rural poverty and the ways in which ecological science can be applied to address some of these challenges. The majority of the chapters comprising the two volumes have focused on these issues in poor, rural areas, where approximately 70% of the developing world's 1.4 billion extremely poor people live (IFAD 2011). In these places, direct dependence on nature for subsistence is often high, and access to social services, markets, and employment opportunities is often limited. However, several chapters in these volumes, such as the chapter on water supply planning by Fitzhugh et al. (Chap. 8, Vol. 1) and population by Marcotullio et al. (Chap. 8, Vol. 2) suggest that experiences gained from the successes and failures of natural resource management in developed countries and nuanced understandings of how urbanization influences poverty and the environment may be useful for informing decisions and polices in rural areas of developing countries.

Several recurrent messages have surfaced from this body of work that illustrate the importance of ecological science for understanding challenges related to poverty reduction and the enabling conditions that influence the effective application of ecological science and tools for addressing those challenges. Broadly, these messages can

J.C. Ingram (✉)
Wildlife Conservation Society, 2300 Southern Boulevard, Bronx, NY, 10460, USA
e-mail: cingram@wcs.org

J.C. Ingram et al. (eds.), *Integrating Ecology and Poverty Reduction: Ecological Dimensions*, DOI 10.1007/978-1-4419-0633-5, be encapsulated by three overarching themes: the challenges of preventing and managing complex trade-offs; the importance of social and economic contexts for determining the application and utility of ecological science; and the paradigm shifts that will be required to effectively integrate ecology into development practice and planning. These themes are certainly not new ones within development or environmental fields; however, the collective chapters in these volumes focus on these themes through the lens of ecology, as it relates to multiple development challenges and potential solutions.

Understanding and Managing Complex Trade-Offs

These chapters encourage a careful consideration of potential ecological and social trade-offs that may result from projects aimed at poverty reduction and conservation. The importance of addressing trade-offs lies in the risks of losing biodiversity and critical ecological functions; crossing thresholds, beyond which it may be difficult to restore ecosystem services; and causing unintended consequences on poor communities, who may be left more vulnerable as a result of even the most well-meaning actions. Others have also recognized the importance of these issues and, consequently, efforts to identify and model trade-offs to inform decision making have grown in recent years (Tallis and Polasky 2009). Despite this increasing attention on trade-offs and the proliferation of new tools to address them, as these chapters reveal, it remains a challenge to identify and manage them, especially, when they occur across interconnected social and ecological systems and across a range of spatial and temporal scales. These chapters demonstrate these challenges in the context of several types of trade-offs: trade-offs between and within development goals; spatial and temporal tradeoffs; economic trade-offs; trade-offs among natural resource user groups; and trade-offs between technology and nature, as a provider of ecosystem services that contribute to poverty reduction.

Many of the chapters in Volume 1, *Integrating Ecology and Poverty Reduction: Ecological Dimensions*, explore the difficulty of navigating the temporal and spatial trade-offs associated with managing farms, watersheds, landscapes, and coastal zones for the multiple needs of the poor such as food, nutrition, water, fuel, health, and physical security. Historically, development efforts have often focused on maximizing one ecosystem service, such as food production, at the expense of other services that are critical for the livelihoods of rural populations in the long term, as discussed by Smuckler et al. with respect to food production, Ganz et al. with respect to energy production, and Randhir and Hawes with respect to watershed management (Chaps. 3, 7, and 17, Vol. 1). Even within single development sectors, trade-offs have occurred: for example, in some cases, total food production has been enhanced at the expense of overall nutrition (Chap. 4, Vol. 1). Collectively, these chapters emphasize that projects designed to reduce rural poverty should consider the many ecosystem services that are important to rural households and how interventions aimed at improving one service may impact other key components of people's livelihoods and well-being. Such a holistic way of thinking may be facilitated by a social-ecological systems approach that includes an appreciation of the connectedness of natural, social, and economic systems across multiple spatial and temporal scales. Ecologists are equipped to think this way and can contribute to such approaches by fostering understanding on the ecological relationships among ecosystem services at multiple spatial and temporal scales (Carpenter et al. 2009; Bennett et al. 2009). Tools, models, and frameworks that can elucidate the social, economic, and ecological trade-offs occurring across different spatial and temporal scales as a result of natural resource management practices and policies are critical for informing decision making. Much work in this area is currently under way by groups such as the Natural Capital Project (NatCap 2007).

The chapters in Volume 2, *Integrating into Ecology and Poverty Reduction: The Application of Ecology in Development Solutions*, address how trade-offs related to balancing conservation and poverty reduction can be managed, negotiated, and, in some cases, avoided through education, gender equality, demography, innovative financing, and ecosystem governance. In some cases, mechanisms such as payments for ecosystem services (PES) can address potential economic trade-offs associated with implementing environmentally sustainable practices by paying people, in cash or in kind, to protect or enhance ecosystem services, as described by Jenkins (Chap. 10, Vol. 2) and as demonstrated by Sachedina and Nelson (Chap. 12, Vol. 2) in Tanzania. While payments through PES (or PES-like mechanisms) primarily aim to incentivize conservation and enhancement of ecosystem services, they may also provide an important source of income or other resources to poor, rural communities, where few other markets or income-generating opportunities exist. However, as Fisher (Chap. 13, Vol. 2) discusses, it can be difficult to achieve multiple policy goals, such as conservation and poverty reduction, through a single instrument such as PES, without compromising success in one or both of the goals. For example, the distribution, stocks, and flows of ecosystem services may vary across a landscape, as Estrada and DeClerck describe (Chap. 14, Vol. 2), and do not coincide necessarily with areas on a landscape that are the highest priorities for poverty reduction. Thus, some trade-offs in the level of ecosystem services conserved or rural income generated, for example, may be inevitable if a tool like PES is being applied to achieve conservation and contribute to development goals. While tools like PES can be useful for addressing economic trade-offs associated with conservation, some ecosystem services currently do not have value in traditional markets (deGroot et al. 2010) such as the health benefits provided by ecosystems as described in the chapters by Myers; Keesing and Ostfeld; and Levi et al. (Chaps. 11, 13 and 14, Vol. 1) and protection from extreme events, as explored in chapters on climate change and disasters (Chaps. 20–23, Vol. 1). Thus, it is important to quantify and articulate potential trade-offs involving a range of ecosystem services, some of which may be very localized and non-monetary in value, in ways that are clear and meaningful to affected stakeholders. To do this, ecologists will need to work in combination with professionals from other disciplines, policy makers, and local communities to identify the suite of ecosystem services that are important to different user groups; to generate knowledge on how services are provided ecologically; and to develop guidance on how to manage and monitor the condition of key services important at local, national and international scales.

Trade-offs may be inevitable in landscapes or seascapes where multiple stakeholders have competing and conflicting uses for the ecosystem services available. In many cases, it will be necessary to negotiate compromises that require some or all stakeholders to yield their ideal patterns of ecosystem service "consumption" to ensure that a wide range of a landscape or seascape's biodiversity and ecological functions are conserved in the long-term. These negotiations must be grounded in the reality that the diverse benefits provided by nature may accrue at very different spatial or temporal scales to different groups of people. In cases where nature cannot provide services in the quantity or at the scale desired, production or consumption of one service may be increased through management practices and/or technology. However, this could result in a change in the provisioning of other ecosystem services, which must be explored carefully and communicated clearly with all affected people. As the chapters on the governance of ecosystems (Chaps. 16–18, Vol. 2) discuss, balancing competing claims for natural resources can be aided significantly by credible science to support decision makers in developing natural resource management plans and identifying science-based targets to monitor progress toward those plans.

One common decision that often involves trade-offs is the choice between using engineered or technological providers of services rather than relying on ecosystems to supply those services. For example, a seawall may be able to provide highly predictable, measurable protection against coastal storms, but may degrade coastal ecosystems that provide a range of critical services for local communities. Thus, in some cases holistic vulnerability of poor, coastal communities may be reduced more effectively in the long-term through conserving coastal ecosystems, such as mangroves, that provide storm protection services, in addition to other important services, such as food production from fisheries, fuelwood, construction materials, climate regulation through carbon storage, and biodiversity for tourism and cultural purposes (see Chap. 22, Vol. 1). Shifting toward more holistic approaches and away from static engineered approaches may be feasible and preferable in certain contexts, but will require research to quantify the social, economic, and ecological advantages of nature-based versus traditional, built approaches (Chaps. 9 and 22, Vol. 1). In cases where ecological thresholds have already been crossed and the ecosystem may no longer be able to provide key services, engineered substitutes for services may be the only option. As Smuckler et al. discuss (Chap. 3, Vol. 1), inorganic fertilizers may be needed in some cases to "jump-start" extremely degraded soils and rebuild the productivity of the system again, even if they are not desirable as a singular approach in the long term. However, the authors note that financial and ecological trade-offs may still occur in such dire situations, even when few other alternatives are available. While ecological science and management can help avoid trade-offs and their negative consequences in some cases, in other cases, trade-offs may be inevitable and ecological science can merely contribute to identifying what they might be so that people can plan accordingly.

The Importance of Social and Economic Context for Applying Ecological Science and Tools

Many of the chapters in these volumes emphasize that prevailing social and economic conditions must be considered in order to understand how humans utilize and value ecosystems, so that ecological science and tools can be more effectively applied. For example, multiple, interacting scales of governance influence the use of natural resources, such as local rules, national policies, and international treaties. To be effective, political processes and institutions of influence should be compatible with the scale of the ecological processes and services they are intending to conserve, although this does not always happen, as McClennen points out with respect to fisheries management and as Holland demonstrates with respect to protected areas (Chaps. 16 and 18, Vol. 2). Furthermore, relationships that may influence decisions about natural resource management change across spatial and temporal scales. For example, Bremner et al. (Chap. 6, Vol. 2) state that relationships between population, poverty, and environmental degradation that exist at a national scale may be difficult to identify at local scales. Thus, an ongoing challenge for ecologists is to translate observations and findings related to important ecological processes and patterns into information that is meaningful to stakeholders and resonates at scales of relevance for decision making. To do this, information must be obtained, collected, and applied within the context of how prevailing environmental, social, cultural, and economic conditions influence choices about natural resource use or the results may be irrelevant for addressing problems, as discussed in Chaps. 16, 17 and 18 with respect to energy challenges in Volume 1 and Chap. 11 with respect to carbon projects in Volume 2. A failure to do this could lead to perverse policies that exacerbate poverty and environmental degradation in the long-term rather than improving them. Collectively, these chapters emphasize that for ecological science and tools to be useful, it is crucial for ecologists to engage with local people, natural resource users, and groups who are addressing social, governance, and economic challenges in the places where they work to better understand the broader framework within which environmental problems are situated, and the spatial and temporal scales at which information will be useful to support decision making over natural resources.

Fostering a New Paradigm

Throughout these volumes, authors have called for a new way of thinking about the relationships between people and nature if ecological science and tools are to be more regularly and seamlessly integrated into development practice and policies. In Naeem's language (Chap. 19, Vol. 2), paradigms such as "nature is our friend" or "nature is our foe" have been critical for shaping economic and environmental debates throughout history and continue to dominant many discussions regarding conservation, sustainable development, and poverty reduction. However, such simplistic views of the ways in which people interact with their environment have contributed little to our ability to conserve ecosystems and reduce poverty of the world's poorest people. In contrast, Naeem (Chap.19, Vol. 2) uses several ecological concepts to illustrate the nature of poverty and, based on this understanding, demonstrates how human poverty consists of the interactions between ecological poverty and social poverty. Other chapters in these volumes encourage new thinking about the impact of human population on natural resources (Chaps. 6 and 7, Vol. 2), the environmental benefits of urbanization (Chap. 8, Vol. 2), the role of women in natural resource management (Chap. 4, Vol. 2) and the importance of rural education for addressing poverty and ecosystem degradation (Chap. 3, Vol. 2). Many of these chapters have promoted holistic, interdisciplinary frameworks for illustrating the relationships between poverty and the natural environment that require a nuanced understanding of social and ecological interactions (for example, Chaps. 7 and 17, Vol. 1; and Chap. 6, Vol. 2). Naeem presents an example of this with the Ecologically Sustainable Development framework (Chap. 19, Vol. 2). Such guiding frameworks and principles that promote consilience across disciplines will be helpful for revisiting deeply held ideas regarding the ways in which humans use and are embedded in ecological systems and will be required to design and implement novel, innovative, and lasting solutions to environmental degradation and poverty reduction in the coming decades of the twenty-first century. As Naeem states in Volume 2, *ecologically* sustainable development should be our goal and, as such, must be founded on a better understanding of how humans interact with and affect an ecosystem like any other species, by contributing to ecosystem functions that influence the flow of nutrients, energy, and water, for example.

Summary

Ecological systems are a source of the many diverse services that make life on earth possible. In rural areas of many developing countries, the major geographical focus of these volumes, it can be difficult to access affordable and/or appropriate substitutes for many of the services provided by nature such as food production; pest control; soil fertility; water for drinking, bathing, and cooking; energy for cooking, heating, and electricity; shelter; disaster protection; and medicine. If multiple development challenges are to be solved simultaneously and sustainably, solutions for poverty reduction must not undermine the persistence of species and ecological functions that generate the many ecosystem services that rural communities currently depend upon for their livelihoods and well-being and that will be critical for future generations. These volumes have attempted to address the ecological nature of some of the major challenges related to poverty reduction and the ways in which ecological science can be more effectively leveraged within political, economic, and cultural processes mechanisms, and institutions to address some of these problems.

We hope that the chapters included within these volumes have catalyzed discussions and ideas that will ultimately foster a world in which extreme poverty is a concept of the past and ecological sustainability is a guiding principle of the future.

References

Ash, N. and M. Jenkins. 2007. Biodiversity and Poverty Reduction: The importance of biodiversity for ecosystem services. United Nations Environment Programme World Conservation Monitoring Centre (UNEP-WCMC).

Bennett, E. M., G. D. Peterson, and L. J. Gordon. 2009. Understanding relationships among multiple ecosystem services. Ecology Letters 12:1394–1404.

Carpenter, S. R., H. A. Mooney, J. Agard, D. Capistrano, R. S. DeFries, S. Diaz, T. Dietz, A. K. Duraiappah, A. Oteng-Yeboah, H. M. Pereira, C. Perrings, W. V. Reid, J. Sarukhan, R. J. Scholes, and A. Whyte. 2009. Science for managing ecosystem services: Beyond the Millennium Ecosystem Assessment. Proceedings of the National Academy of Sciences of the United States of America 106:1305–1312.

Chivian, E. and Bernstein, A. eds. 2008. Sustaining life: How human health depends on biodiversity. New York, NY: Oxford University Press.

de Groot, R. S., R. Alkemade, L. Braat, L. Hein, L. Willemen. 2010. Challenges in integrating the concept of ecosystem services and values in landscape planning, management and decision making. Ecological Complexity 7:260–272.

Galizzi, P. and Herklotz, A. 2008. The Role of the Environment in Poverty Alleviation (People and the Environment). Fordham University Press, New York, New York.

IFAD, 2011. International Fund for Agricultural Development. The Rural Poverty Report, 2011. (http://www.ifad.org/rpr2011/report/e/rpr2011.pdf, accessed on January 12, 2011).

McNeely, J. A. and Scherr, S. J. 2003. Ecoagriculture: Strategies To Feed The World And Save Wild Biodiversity. Island Press, Washington, D.C.

NatCap 2007. The Natural Capital Project, Stanford University. http://www.naturalcapitalproject.org/InVEST.html accessed on 27 December 2010.

Tallis, H. and Polasky, S. 2009. Mapping and valuing ecosystem services as an approach for conservation and natural resource management. The Year in Ecology and Conservation Biology. Annals of the New York Academy of Science 1162:265–283.

Tekelenburg, A., B.J.E. ten Brink, M.C.H.Witmer. 2009. How do biodiversity and poverty relate? – An explorative study. Netherlands Environmental Assessment Agency (PBL), Bilthoven.

World Bank, 2007. Poverty and the Environment: Understanding Linkages at the Household Level. World Bank.

Index

E

Printed by Publishers' Graphics LLC
MO20120518